SCOPE 27

Climate Impact Assessment

SCOPE 27

Climate Impact Assessment

Studies of the Interaction of Climate and Society

Edited by
Robert W. Kates
Center for Technology, Environment, and Development, Clark University, Worcester, Massachusetts

with

Jesse H. Ausubel
National Academy of Engineering, Washington DC

and

Mimi Berberian
Center for Technology, Environment, and Development, Clark University, Worcester, Massachusetts

Published on behalf of the
Scientific Committee on Problems of the Environment (SCOPE)
of the
International Council of Scientific Unions (ICSU)
by
JOHN WILEY & SONS
Chichester · New York · Brisbane · Toronto · Singapore

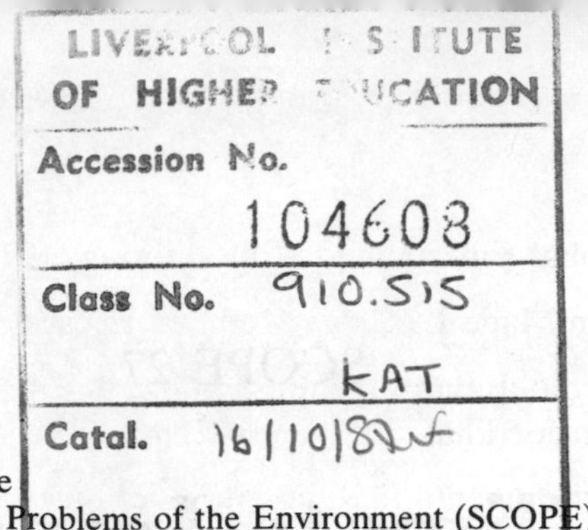

Reprinted August 1986

Library of Congress Cataloging in Publication Data:

Main entry under title:

Climate impact assessment.
(SCOPE 27)
Includes index.
1. Climatology—Social aspects. 2. Environmental
impact analysis—Social aspects. I. Kates, Robert
William. II. Ausubel, Jesse. III. Berberian, Mimi.
IV. International Council of Scientific Unions.
Scientific Committee on Problems of the Environment.
V. Series: SCOPE (Series: Chichester, West Sussex) ; 27.
QC981.C45 1985 304.2′5 84–20922

ISBN 0 471 90634 4

British Library Cataloging in Publication Data:

Climate impact assessment : studies of the
interaction of climate and society.—(SCOPE ; 27)
1. Climatic changes—Social aspects
2. Climatic changes—Economic aspects
I. Kates, Robert W. II. Ausubel, Jesse
III. Berberian, Mimi IV. International Council
of Scientific Unions, *Scientific Committee on
Problems of the Environment* V. Series 304.2′5
QC981.8.C5

ISBN 0 471 90634 4

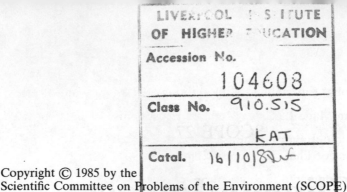

Printed and Bound in Great Britain

SCOPE 1: Global Environmental Monitoring, 1971, 68pp (out of print)

SCOPE 2: Man-Made Lakes as Modified Ecosystems, 1972, 76pp

SCOPE 3: Global Environmental Monitoring System (GEMS): Action Plan for Phase 1, 1973, 132pp

SCOPE 4: Environmental Sciences in Developing Countries, 1974, 72pp

Environment and Development, proceedings of SCOPE/UNEP Symposium on Environmental Sciences in Developing Countries, Nairobi, February 11–23, 1974, 418pp

SCOPE 5: Environmental Impact Assessment: Principles and Procedures, Second Edition, 1979, 208pp

SCOPE 6: Environmental Pollutants: Selected Analytical Methods, 1975, 277pp (available from Butterworth & Co. (Publishers) Ltd., Sevenoaks, Kent, England)

SCOPE 7: Nitrogen, Phosphorus, and Sulphur: Global Cycles, 1975, 192pp (available from Dr Thomas Rosswall, Swedish Natural Science Research Council, Stockholm, Sweden)

SCOPE 8: Risk Assessment of Environmental Hazard, 1978, 132pp

SCOPE 9: Simulation Modelling of Environmental Problems, 1978, 128pp

SCOPE 10: Environmental Issues, 1977, 242pp

SCOPE 11: Shelter Provision in Developing Countries, 1978, 112pp

SCOPE 12: Principles of Ecotoxicology, 1978, 372pp

SCOPE 13: The Global Carbon Cycle, 1979, 491pp

SCOPE 14: Saharan Dust: Mobilization, Transport, Deposition, 1979, 320pp

SCOPE 15: Environmental Risk Assessment, 1980, 176pp

SCOPE 16: Carbon Cycle Modelling, 1981, 404pp

SCOPE 17: Some Perspectives of the Major Biogeochemical Cycles, 1981, 175pp

SCOPE 18: The Role of Fire in Northern Circumpolar Ecosystems, 1983, 344pp

SCOPE 19: The Global Biogeochemical Sulphur Cycle, 1983, 495pp

SCOPE 20: Methods for Assessing the Effects of Chemicals on Reproductive Functions, 1983, 568pp

SCOPE 21: The Major Biogeochemical Cycles and Their Interactions, 1983, 554pp

SCOPE 22: Effects of Pollutants at the Ecosystem Level, 1984, 443pp

SCOPE 23: The Role of Terrestrial Vegetation in the Global Carbon Cycle: Measurement by Remote Sensing, 1984, 272pp

SCOPE 24: Noise Control, 1984

SCOPE 25: Appraisal of Tests to Predict the Environmental Behaviour of Chemicals, 1985, 400pp

SCOPE 26: Methods for Estimating Risks of Chemical Injury: Human and Non-human Biota, 1985, 712pp

SCOPE 27: Climate Impact Assessment: Studies of the Interaction of Climate and Society, 1985

Funds to meet SCOPE expenses are provided by contributions from SCOPE National Committees, an annual subvention from ICSU (and through ICSU, from UNESCO), an annual subvention from the French Ministère de l'Environment et du Cadre de Vie, contracts with UN Bodies, particularly UNEP, and grants from Foundations and industrial enterprises.

International Council of Scientific Unions (ICSU)
Scientific Committee on Problems of the Environment (SCOPE)

SCOPE is one of a number of committees established by a non-governmental group of scientific organizations, the International Council of Scientific Unions (ICSU). The membership of ICSU includes representatives from 68 National Academies of Science, 18 International Unions and 12 other bodies called Scientific Associates. To cover multidisciplinary activities which include the interests of several unions, ICSU has established 10 scientific committees, of which SCOPE is one. Currently, representatives of 34 member countries and 15 Unions and Scientific Committees participate in the work of SCOPE, which directs particular attention to the needs of developing countries. SCOPE was established in 1969 in response to the environmental concerns emerging at that time; ICSU recognized that many of these concerns required scientific inputs spanning several disciplines and ICSU Unions. SCOPE's first task was to prepare a report on Global Environmental Monitoring (SCOPE 1, 1971) for the UN Stokholm Conference on the Human Environment.

The mandate of SCOPE is to assemble, review, and assess the information available on man-made environmental changes and the effects of these changes on man; to assess and evaluate the methodologies of measurement of environmental parameters; to provide an intelligence service on current research; and by the recruitment of the best available scientific information and constructive thinking to establish itself as a corpus of informed advice for the benefit of centres of fundamental research and of organizations and agencies operationally engaged in studies of the environment.

SCOPE is governed by a General Assembly, which meets every three years. Between such meetings its activities are directed by the Executive Comittee.

R.E. Munn
Editor-in-Chief
SCOPE Publications

ICSU PRESS

Executive Secretary: V. Plocq

Secretariat: 51 Bld de Montmorency
75016 PARIS

Contents

Preface

The decade of the 1970s was marked by a growing climate consciousness, both popular and scientific. The new interest was sparked by a series of extreme climate events and related disruptions, and by scientific speculation as to increased climate variability and possible climate change. Two sets of events during this period attracted both scientific and public interest. The first, in 1972, was the apparent simultaneous occurrence of unfavorable weather in many parts of the globe and its speculative relationship to a wide variety of socioeconomic events, including the quadrupling of various commodity prices around the world, food shortages in the Sahel of West Africa and in South Asia, a drastic fall in the anchovy fishery of the Pacific, and even changes in government in Ethiopia and Niger. The second was an emerging scientific consensus that human-induced alterations in the chemical constituents of the atmosphere could lead to large regional, and even global, changes of the atmosphere in the form of more acidic rain, greater ultraviolet radiation, and altered temperatures.

The early eighties again found persistent drought in northeast Brazil and in many countries of Africa; a warming of sea-surface Pacific temperatures leading to the most remarkable 'El Niño' event recorded to date; and the warmest years in a century of northern hemisphere temperatures. A scenario for a new and most serious set of climatic consequences following a major international exchange of nuclear weapons, the 'nuclear winter', was postulated. The science of human-induced alterations of the atmosphere became more complex with the slowing of the rate of fossil fuel use and with improved understanding of the way the many 'greenhouse' gases contribute to global warming, while at the same time reducing the net destruction of the ozone shield against ultraviolet radiation.

In sum, the diversity of novel climatic experience continues unabated, the recognition of potential sources of human-induced alteration has increased, and the pace and degree of change are questioned and debated. Nonetheless, it is widely agreed that one such change, a long-term global warming derived from the enrichment of the atmospheric content of the 'greenhouse' gases, is underway. Within the time period of the projected global average warming, measured in tens of hundreds of years, sustained variation of climate will occur in many places, and lesser periods of favorable or unfavorable climate will occur in most places—a function of normal variability. Where these changes are large—the extremes greater than what is customary—where people and places are vulnerable, or where human activity meshes poorly with natural opportunity, significant climate impacts to people, ecosystems and societies are likely to occur. How to respond to such impacts—adjusting to changing climate, coping with extremes, matching human needs to climate endowment—are issues of considerable import. The scientific study of climate and society will inform societal response. Concepts of climate impact assessment are new, the methods are still under development. This volume is an authoritative review of these methods and concepts, a contribution of the Scientific Committee on Problems of the Environment to the World Climate Program.

The World Climate Program (WCP) was initiated in February of 1979 at a meeting of 350 scientific experts held under the aegis of the World Meteorological Organization. The Program is directed at four goals:

1. improving our understanding of the physical climate system;
2. improving the accuracy and availability of climate data;
3. expanding the application of current climate knowledge to human betterment; and
4. advancing our understanding of the relation between climate and human activities.

Organized to address the fourth goal was the World Climate Impact Program (WCIP), of which this study is an initial effort. WCIP's purpose was well-stated in its founding document:

> ... the ultimate objective of the Impact Study Programme within the World Climate Programme will be to insert climatic considerations into the formulation of rational policy alternatives. In areas of the world characterized by different natural environmental conditions, social structures or economic systems, and differing levels of development, there can be different interactions and responses to climatic variability. The basic studies should aim at an integration of climatic, ecological and socio-economic factors entering into complex problems of vital importance for society, such as availability of water, food, and energy. Specifically, the Programme should strive for:
>
> (a) Improvement of our knowledge of the impact of climatic variability and change in terms of the specific *primary responses* of natural and human systems (such as agriculture, water resources, energy, ocean resources and fisheries, transportation, human health, land use, ecology and environment, etc.);
> (b) Development of our knowledge and awareness of the *interactive* relations between climatic variability and change and human socio-economic activities;
> (c) Improvement of the *methodology employed* so as to deepen the understanding and improve the simulation of the interactions among climatic, environmental and socio-economic factors;
> (d) Determining the characteristics of human societies at different levels of development and in different natural environments which make them especially

resilient to climatic variability and change and which also permit them to take advantage of the opportunities posed by such changes;
(e) Application of this new knowledge of techniques to practical problems of concern to developing countries or which are related to a common need for all mankind.
(World Meteorological Organization [1980], *Outline Plan and Basis for the World Climate Programme, 1980–1983*, 32–34)

At the time of the World Climate Conference in 1979, many of the major impacts of climate variation were already well known and others were under study. Participants reported on this sampling of impacts that were then known:

- Natural disasters claim an average of $40 billion in global resources and at least 250,000 deaths a year. Of this dollar amount, $30 billion accrue from three major events with a significant atmospheric component: floods (40 percent), tropical cyclones (20 percent), and drought (15 percent). The distribution of deaths and damages is widely skewed, with 95 percent of deaths occurring in poorer nations and 75 percent of economic damages in wealthy nations. (Kates [1979], in World Meteorological Organization's *Proceedings of the World Climate Conference*, 683)
- The current grain-producing systems of the world are still highly sensitive to the occurrence of large climatic anomalies. The trend toward higher grain yields has slowed down, or even leveled out, since the early 1970s. Some agriculturalists attribute this to worsening weather. Evidence also exists to suggest that grain yields have been subject to more weather-related variability during the 1970s than during the previous two decades, when dramatic increases in yields suggested that technological inputs were overcoming the variability of weather. (McQuigg [1979], in *Proceedings of the World Climate Conference*, 420–421)
- The increasing use of air conditioning and electric heating in homes has increased the sensitivity of energy demand to temperature changes. Results of studies in the United States showed that in one year out of 100 years one should expect the total demand for heating fuel to exceed the long-term average demand (for constant economy) by as much as 10 percent and at least 3.6 percent of an average of one heating system in five. The probable extreme deviations are larger when regions are considered; for example, for the South Atlantic states of the United States, in one year out of 100 years one should expect a total demand for heating fuel to exceed its long-term average demand by 20.4 percent. (Williams, Häfele, and Sassin [1979], in *Proceedings of the World Climate Conference*, 281–282)
- A substantial amount of the production of any economy is directly or indirectly used to offset or negate the economic effects of climatic variation. Considering only the purchases by consumers in the northern hemisphere above 40° latitude, the amount spent may be as high as 10

percent of per capita income. (d'Arge [1979], in *Proceedings of the World Climate Conference*, 656)

Along with impacts, the World Climate Conference was informed about practical actions that are taken to anticipate, prevent, reduce or mitigate undesired impacts or to take advantage of desirable ones. These include:

- For agriculture, crops are planted late or harvested early, and are partially stored for use during exceptionally severe periods of drought or cold. Through genetic selection, hardier or heat resistant varieties of crops are obtained and applied. Farm operators plant a mixture of crops to protect against climate extremes and thereby avoid the possible loss of a single weather-sensitive crop. Energy-intensive machinery is utilized to reduce time for seeding or harvesting. (d'Arge [1979], in *Proceedings of the World Climate Conference*, 656)
- Industries stockpile raw materials to avoid shortages due to reduced deliveries during inclement weather. Employers hire additional workers and adjust working hours to reduce production stoppages due to employee illness or inability to travel to work during periods of extreme climate. Special snow removal equipment is purchased and stored in case of severe storms. (d'Arge [1979], in *Proceedings of the World Climate Conference*, 656)

Additional actions were proposed. For example, it was reported that in the opinion of the authors the:

- Use of the best practice currently available in developing countries could reduce the world death toll from drought, flood, and tropical cyclones by 85 percent and similar use of best practice in industrial countries could reduce property damage by 50 percent. (Burton *et al.*, 1978; quoted in Kates [1979], in *Proceedings of the World Climate Conference*, 687)

Thus the World Climate Impact Program addressed practical necessity: preventing and mitigating the disasters of extreme events; tuning climate-sensitive sectors of the economy (such as energy, food, fiber, water) to accommodate climate variation better; and anticipating, preventing and adapting to natural and human-induced climate change. The work at that date was at best suggestive. Systematic efforts at climate impact assessment were recent; methods, however, are evolving rapidly.

The Scientific Committee on Problems of the Environment (SCOPE) of the International Council of Scientific Unions undertook to prepare the authoritative review of the methodology of climate impact assessment called for in the World Climate Impact Program. The objectives of the review were:

1. to examine existing methodology;

2. to foster the development of new methodological approaches; and
3. to inform a broad range of disciplines as to the available concepts, tools and methods beyond their own specialty.

This volume is a major product of that review. But of equal importance is the network of scientific interest that has been created in the course of the project. Over three hundred individual researchers and administrators from thirty-six countries have participated, exchanged publications, or expressed continuing interest in the review and the general field. Contact with them has been maintained by two SCOPE newsletters devoted to climate impact assessment and will be continued through informational letters coordinated by the US National Climate Program Office.

Included in this network are the thirty-eight authors of papers (from thirteen countries), the one hundred invited reviewers of papers (from twenty-three countries), the eight members of the SCOPE-appointed Scientific Advisory Committee, and the fourteen national correspondents from SCOPE National Committees. The network itself and the production of this volume were coordinated by the Clark University Climate and Society Research Group, using funds from the US National Science Foundation, the United Nations Environment Program (UNEP), the Scientific Committee on Problems of the Environment (SCOPE), the US National Academy of Sciences, and the Oak Ridge Associated Universities. In addition, through SCOPE, the Andrew W. Mellon Foundation provided funds for William Riebsame to serve as postdoctoral research fellow.

The SCOPE effort began with a Scientific Advisory Committee, chaired by F. K. Hare, whose members were J. de Vries, J. Escudero, H. Flohn, A. Mascarenhas, W. J. Maunder, J. L. Monteith, and R. Slatyer. At the same time, the national correspondents (V. V. Alexandrov, USSR; J. Aragón, Spain; A. P. M. Baede, The Netherlands; C. Capel-Boute, Belgium; P. K. Das, India; M. Glantz, USA; W. Kuhnelt, Austria; S. C. Lu, Taiwan; K. Meyer-Abich, West Germany; J. L. Monteith, United Kingdom; J. Neumann, Israel; Y. Omoto, Japan; A. B. Pittock, Australia; and D. Rosell, The Philippines) appointed by National Committees were solicited for information and suggestions. These were reviewed at a meeting of the Scientific Advisory Committee at St Hilda's College, Oxford University, in September 1980. There the Committee adopted a framework to select topics, authors, and a common set of instructions for contributors. Another meeting, hosted by the Atmospheric Environment Service of Canada a year later, brought together review authors to discuss either preliminary drafts or their plans for papers.

Papers were submitted over the following two years and each received an international, interdisciplinary review from a group consisting of J. L. Anderson, W. Bach, C. L. Bastian, E. Bernus, A. K. Biswas, E. Boulding, R. Bryson,

M. I. Budyko, J. J. Burgos, J. C. Caldwell, L. J. Castro, C. Caviedes, R. Chen, L. S. Chia, W. C. Clark, D. Cushing, G. Dahl, K. Devonald, J. Dooge, M. El-Kassas, N. J. Ericksen, H. Flohn, H. D. Foster, W. J. Gibbs, E. S. Gondwe, G. Gunnarsson, J. C. Hock, C. S. Holling, M. A. Islam, W. W. Kellogg, A. Khosla, A. V. Kneese, V. A. Kovda, M. Lechat, T. A. Malone, A. Mani, B. Martin, G. A. McKay, D. H. Meadows, D. Mileti, J. K. Mitchell, S. H. Murdock, T. Murray, W. Nordhaus, J. S. Oguntoyinbo, P. O'Keefe, T. O'Riordan, J. P. Palutikof, C. Pfister, J. D. Post, T. Potter, T. K. Rabb, C. Sakamoto, S. H. Schneider, W. R. D. Sewell, M. M. Shah, M. S. Swaminathan, J. A. Taylor, T. Vasko, R. A. Warrick, G. F. White, G. D. V. Williams, J. S. Winston, B. Wisner and C. P. Wolf. These reviews were followed by several rounds of revision.

At Clark University, Mimi Berberian, Thomas Downing and William Riebsame organized the initial stages of the review, and Maggie Grisdale of Trinity College, Toronto directed the ensuing meeting that brought together review authors. The many drafts of papers were patiently typed by Jane Kjems, Joan McGrath, and Lu Ann Renzoni. I served as volume editor, aided by Jesse Ausubel and Mimi Berberian. Jeanne Kasperson provided invaluable bibliographic assistance.

If one adds to the group of participants the many scientists and publishers that shared material and illustrations with us, literally hundreds of people have generously tendered assistance, motivated by the common bonds of science that transcend discipline and nationality and by shared concerns for climate and society. I hope they can take much satisfaction from our collective activity, as I relieve them of responsibility for any fault or error. I am particularly grateful for the endless patience of the authors, the universal helpfulness of the reviewers, and the generous understanding of funding agency program officers. Relieved also of fault, but tendered special gratitude arising from five years of close collaboration, I thank Ken Hare for his counsel and support; Bill Riebsame for his energy and insight; Jesse Ausubel for his knowledge and versatility; and for all of us, Mimi Berberian, for her skill and sense.

A project, a network, a volume: this is also a set of papers. The individual papers transcend professional boundaries and examine climate impact assessment in a non-disciplinary fashion as a set of linked analytic components, as techniques of case study and modeling, and as reviews of past experience. Each author has sought to review the state of his or her art, not for peers, but for scientific colleagues who are interested in climate impact assessment but schooled in a different discipline or lacking experience in a particular technique. The achievements, weaknesses, and capabilities of the various methodologies are set forth with candor, tempered by empathy. It is our hope that workers new to climate impact assessment will be realistic in their expectation of the various methods, sympathetic with the common scientific problems faced, and challenged both by their practical necessity and intellectual adventure.

ROBERT W. KATES

Worcester, Massachusetts (USA)
June, 1984

List of Contributors

J. H. AUSUBEL — National Academy of Engineering
2101 Constitution Avenue, N.W.
Washington DC 20418, USA

JAN DE VRIES — Department of History & Economics
University of California at Berkeley
Berkeley, California 94720, USA

DENNIS EPPLE — Graduate School of Industrial Administration
Carnegie-Mellon University
Pittsburgh, Pennsylvania 15213, USA

JOSÉ CARLOS ESCUDERO — Department of Social Medicine
Universidad Autónoma Metropolitana
Xochimilco-Mexico, POB 23-181
Mexico 23, DF

BARBARA FARHAR-PILGRIM — 4600 Greenbriar Court
Boulder, Colorado 80303, USA

G. FARMER — Climatic Research Unit
University of East Anglia
Norwich NR4 7TJ, England

MICHAEL H. GLANTZ — Environmental & Societal Impacts Group
National Center for Atmospheric Research
Boulder, Colorado 80307, USA

F. KENNETH HARE — Trinity College in the University of Toronto
6 Hoskin Avenue
Toronto, Ontario M5S 1H8, Canada

R. L. HEATHCOTE — The Flinders University of South Australia
Bedford Park, South Australia 5042

N. J. HUCKSTEP — Climatic Research Unit
University of East Anglia
Norwich NR4 7TJ, England

M. J. INGRAM — Department of Modern History
The Queen's University of Belfast
Belfast, NT7 1NN, Northern Ireland

JILL JÄGER — Fridtjof-Nansen-Strasse 1
D-7500 Karlsruhe 41
Federal Republic of Germany

N. S. JODHA — International Crop Research Institute for the Semi-Arid Tropics
Andhra Pradesh 502 324 Patanchnu P.O., India

ROBERT W. KATES — Center for Technology, Environment, and Development
Clark University
Worcester, Massachusetts 01610, USA

TSUYOSHI KAWASAKI — Faculty of Agriculture
Tohoku University
Sendai 980, Japan

V. F. KRAPIVIN — Computing Center of the USSR Academy of Sciences
40 Vavilova Street
Moscow 117333, USSR

MARIA E. KRENZ — Environmental & Societal Impacts Group
National Center for Atmospheric Research
Boulder, Colorado 80307, USA

LESTER B. LAVE — Graduate School of Industrial Administration
Carnegie-Mellon University
Pittsburgh, Pennsylvania 15213, USA

HENRI NOEL LE HOUÉROU — CNRS, CEPE
Louis Emberger
B.P. 5051
Montpellier Cedex 34033, France

C. A. KNOX LOVELL — Department of Economics
University of North Carolina
Chapel Hill, North Carolina 27514, USA

A. C. MASCARENHAS — Institute of Resource Assessment
University of Dar es Salaam
Dar es Salaam, Tanzania

W. J. MAUNDER — New Zealand Meteorological Service
P O Box 722
Wellington 1, New Zealand

DAVID MILLER — Department of Geological Sciences
The University of Wisconsin-Milwaukee
Milwaukee, Wisconsin 53201, USA

N. N. MOISSEIEV — Computing Center of the USSR
Academy of Sciences
40 Vavilova Street
Moscow 117333, USSR

R. MORTIMER — Climatic Research Unit
University of East Anglia
Norwich NR4 7TJ, England

HENRY A. NIX — Commonwealth Scientific & Industrial
Research Organization
Division of Land & Water Resources
Canberra City
A.C.T. 2601, Australia

BÉLA NOVÁKY — Institute for Water Management
Alkotmány u. 29
Budapest 1054, Hungary

A. E. J. OGILVIE — Climatic Research Unit
University of East Anglia
Norwich NR4 7TJ, England

CSABA PACHNER — West Transdanubian Water Authority
Vörösmarty u. 2
Szombathely 9700, Hungary

MARTIN L. PARRY — Department of Geography
University of Birmingham
P O Box 363
Birmingham B15 2TT, England

WILLIAM E. RIEBSAME — Department of Geography
University of Colorado
Boulder, Colorado 80309, USA

JENNIFER ROBINSON — Department of Geography
University of California at Santa Barbara
Santa Barbara, California 93106, USA

V. Kerry Smith — Department of Economics
Vanderbilt University
Nashville, Tennessee 37235, USA

Yu. M. Svirezhev — Computing Center of the USSR
Academy of Sciences
40 Vavilova Street
Moscow 117333, USSR

Károly Szesztay — Institute for Water Management
Alkotmány u. 29
Budapest 1054, Hungary

A. M. Tarko — Computing Center of the USSR
Academy of Sciences
40 Vavilova Street
Moscow 117333, USSR

Anne V. T. Whyte — Institute for Environmental Studies
University of Toronto
Toronto, Ontario M5S 1A1, Canada

T. M. L. Wigley — Climatic Research Unit
University of East Anglia
Norwich NR4 7TJ, England

Part I
Overviews

Running as a thread through the entire volume, linking together the sectoral studies, the analytic methods and the case examples, are conceptual models of the interaction of climate and society, definitions of climate variability and change, and assumptions as to the state of knowledge concerning climate processes. These concepts and definitions, presented by Kates in Chapter 1 and Hare in Chapter 2, provide to all authors a common vocabulary for describing climatic events, consequences and human responses, a common framework for linking climate and societal impacts, and a common interest in both industrialized and developing countries.

Within this framework, climate variability and change provide three types of events of interest: extreme weather events, persistent periods, and little ages. These events impact on exposed social, areal, or activity units of human or ecological organization, leading to ordered biophysical, social, or ecological consequences. In turn these impacts are modified by cultural adaptation and adjustment responses that may amplify or dampen the consequences of climate events. In the simplest of frameworks, the links between events, units and consequences of climate impact models are linear. In the more realistic and complex interaction model framework, causality is jointly determined by climate and society. As with all such frameworks, relationships are linked in ordered flows that belie the reality and simultaneity of the real world.

The degree to which the authors employ these common concepts, concerns and vocabulary differs, as considerable translation of disciplinary or sectoral practice or tradition is often required. Nonetheless, all have tried, and brief editorial introductions to the major sections guide the reader to the connections between a particular chapter and the overarching schema of climate impact assessment.

The overview on research by Riebsame in Chapter 3 serves a different function, providing a common conceptual and historical review of climate–society research organized under four key concepts: climate as setting, as determinant, as hazard, and as resource. Riebsame's view that research, both past and future, flows directly from these different, but not exclusive, concepts of climate–society interaction serves not only to organize the diverse literature of this interdisciplinary field, but to analyze its structure as well.

The final overview, Chapter 4 by Maunder and Ausubel, links directly to the rest of the volume by posing the question of how one begins to undertake specific climate impact studies. They suggest that one major way to begin is by assessing the overall climate sensitivity of activities, places, or groups of interest. Past experience and current methods for determining overall sensitivity are presented. It emerges from many studies that agriculture and water resources are activities and sectors that are clearly sensitive to climate. Methods appropriate to the study of these and other sensitive sectors follow in Part II of this volume.

Climate Impact Assessment
Edited by R. W. Kates, J. H. Ausubel and M. Berberian

CHAPTER 1
The Interaction of Climate and Society

ROBERT W. KATES

Center for Technology, Environment, and Development
Clark University
Worcester, Massachusetts 01610 USA

1.1 INTRODUCTION

Climate impact assessment is one of a family of interdisciplinary studies that focus on the interaction between nature and society or the study of the human environment. As such it must draw upon theory, methods and research findings from all the great domains of science: physical, biological and social–behavioral. While hyphenated science (such as bio-physical, socio-biological) is common along the research frontier, integration of the three domains is rare. Although integration is necessary for nature–society problems, it poses a special challenge to science and scientific workers: to integrate their theory and methods with those of neighboring domains in ways that maintain the common ethos of scientific method.

Achieving that integration is a major objective of this volume. This introductory overview explores the underlying assumptions as to the relationship between climate and society common to most disciplinary or methodological approaches. The individual authors have addressed their colleagues in neighboring scientific domains, attempting to share with them the opportunities, limits, and problems of specific disciplines or methods. Nonetheless, there are at least two inherent problems with such scientific integration.

1.1.1 'Hard' and 'Soft' Science

The first and larger question relates to the apparent differences between the great domains. Most scientific workers would agree that as one moves from the physical to the biological to the behavioral–social sciences, one moves from older to younger sciences with less experience, consensual theory, and ability to experiment and replicate results. Thus there may be less predictability, more speculation and greater uncertainty.

These differences may be amplified further. First, it can be argued that as one progresses across the domains, complexity increases, reaching its height in the human sciences. If this is so, then it is not clear whether it is the youth and experimental limits of the human sciences or the complexity of the phenomena that they study that leads to the lack of consensual theory and predictability.

Second, a common sequence in climate impact assessment is for the ordering of impacts to parallel the ordering of scientific domains. A range of possibilities, outcomes and human choices is attached to each link in a chain of impacts, and uncertainty accumulates along the chain. Thus as one moves, for example, from an increase in global CO_2, to a shift in global heat balances, to regionally differentiated changes in climate, to impacts on primary production in agriculture or fisheries, to impacts on nutrition, trade and economic growth and development, the second- and third-order impacts become more diffuse, the possible outcomes multiply, and the level of precision diminishes. This ordering of impacts—from the physical (CO_2, heat balance, regional climate) to the biological (biomass, agriculture, fisheries yield) to the human (nutrition, trade, economic development)—parallels the major domains of science. Thus larger uncertainties are inevitably associated with the human sciences, although specific relationships (such as seasonal consumption *vs* seasonal weather) may actually be better understood.

These broad differences between the domains of science—whether attributable to the science, to the complexity of the phenomena studied, or to the ordering of impacts—all contribute to the pejorative distinction between 'hard' and 'soft' sciences. This distinction, however, should not dominate the practice of climate impact assessment. Volcanic eruptions and revolutions, jet stream tracks and business trends, may be equally difficult to predict. The coupling of the ocean of air to the ocean of water may be as poorly understood as the coupling of the biosphere to human society. Indeed, in each of the domains necessary for integrated impact assessment, a mixture of well- and poorly-understood phenomena will be found.

1.1.2 Linking Methodologies

A second problem of integration is the problem of linkage between very differing methodologies. For example, consider one important causal chain of impacts, the effect of some chlorine compounds on the stratosphere. For this chain, production data for chlorofluorocarbons are linked to models of chemical transformation, stratospheric diffusion and ozone depletion, informed by laboratory studies of chemical rate constants, and checked by rocket and satellite observations. These, in turn, are linked to varied epidemiological correlations with both melanoma and non-melanoma cancers, laboratory experiments with plants and animals, and economic extrapolation of the losses in productivity and costs of treatment and mitigation. Each of these methods of study and analysis has its own scientific style and language and differing standards of performance, replicability, uncertainty, and significance. Linking these diverse methodologies and estimating their cumulative error is a major task for the emerging quasi-discipline of risk assessment. As yet, there has been no comprehensive study of the problems of integrating such scientific apples and oranges.

1.1.3 Models of Relationships

This paper serves as an overview of impact methodology and it is organized by the assumed relationship between climate and society. All assessments of climatic impact assume, explicitly or implicitly, certain underlying relationships among climatic events and impacted people and places, often grouped together as populations of humans, plants or animals; social or economic activities; or regions defined by size or national boundaries. These relation-

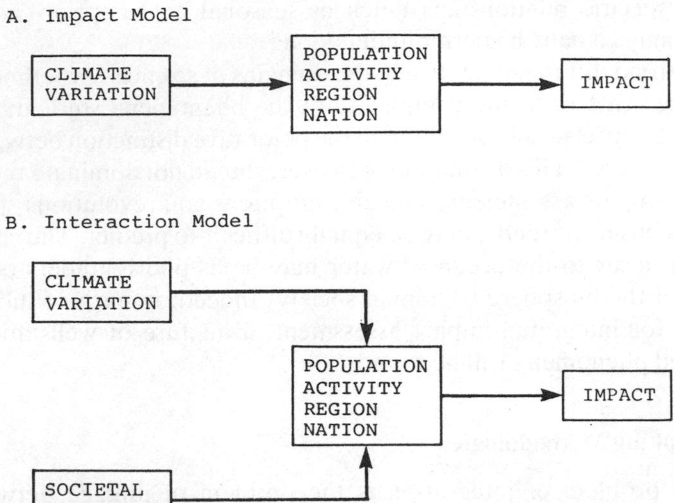

Figure 1.1 Schematics of impact and interactive models are highly simplified graphic depictions of types of study methodologies. They illustrate differences in assumed relationships and are not an attempt to 'model' or necessarily depict the 'real world' relationship. The boxes describe possible study elements and the arrows the direction of assumed relationships. Both models attempt to identify 'cause and effect' relationships, with climate the 'cause' in the impact model 1.1A and climate and society the joint 'causes' in the interaction model 1.1B

ships fall into two sets of nested models: impact models and interaction models. In their basic form they are shown in Figure 1.1.

In the simplest of assumed relationships, the impact model (Figure 1.1A), variation in one or more aspects of climate affects a defined population, activity, sector, region or nation and 'causes' impacts—changes in state that would not have occurred in the absence of the variation in climate state. Methodologies utilizing this model (and its elaborations) are overviewed in Section 1.2. Following that is Section 1.3, on studying interactions. The interaction model (Figure 1.1B) recognizes that impacts are joint products of the interaction between climate and society and that similar climatic variations will yield different impacts under different sets of social conditions. These social conditions determine whether a society is more or less vulnerable to an undesirable variation in climate or more or less able to utilize the opportunity provided by a favorable variant.

1.2 STUDYING IMPACTS

1.2.1 Impact Models

The basic impact model is part of a family of relationships (Figure 1.2) based on the assumption of direct cause and effect. Climate events impinge upon populations, activities, regions or nations and 'cause' impacts. An example from the literature can help describe that sequence.

Warrick (1980) and his colleagues set out to study the impacts of repeated droughts (five times since 1890) on the Great Plains of the midcontinental United States. His five-step study method is shown in Figure 1.3. In this model the sequence is as follows: drought (defined by specific space–time measures)

IMPACT MODELS

A. Basic Model

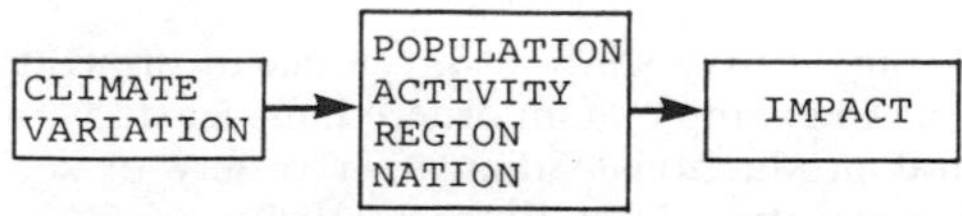

B. Ordered Impacts

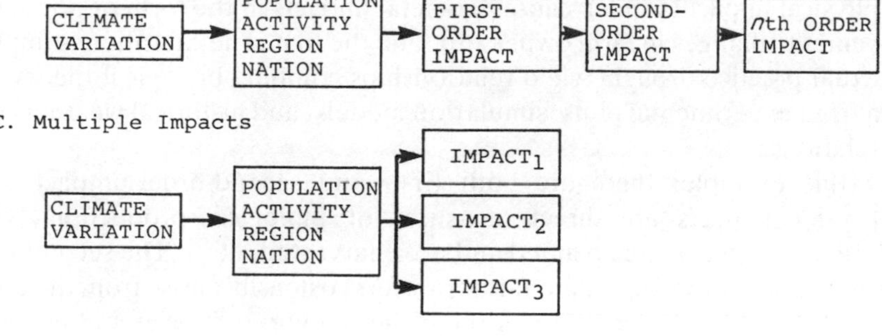

D. Anthropogenic Climate Impact

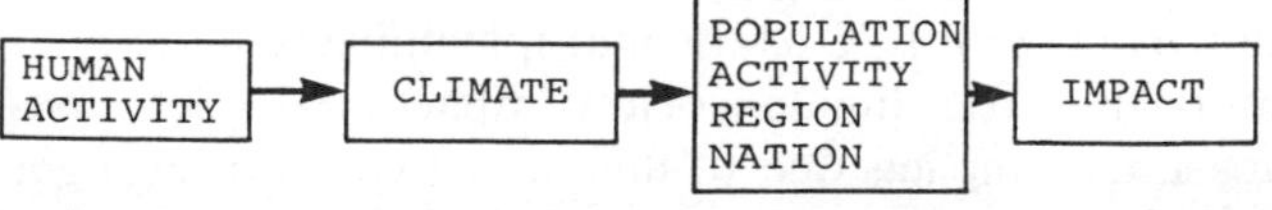

Figure 1.2 Four types of impact models: (A) the basic simplified model; (B) ordered impacts assessed as they cascade through physical and social systems; (C) multiple impacts in different sectors or within a given area; and (D) the impacts of human-induced climatic change

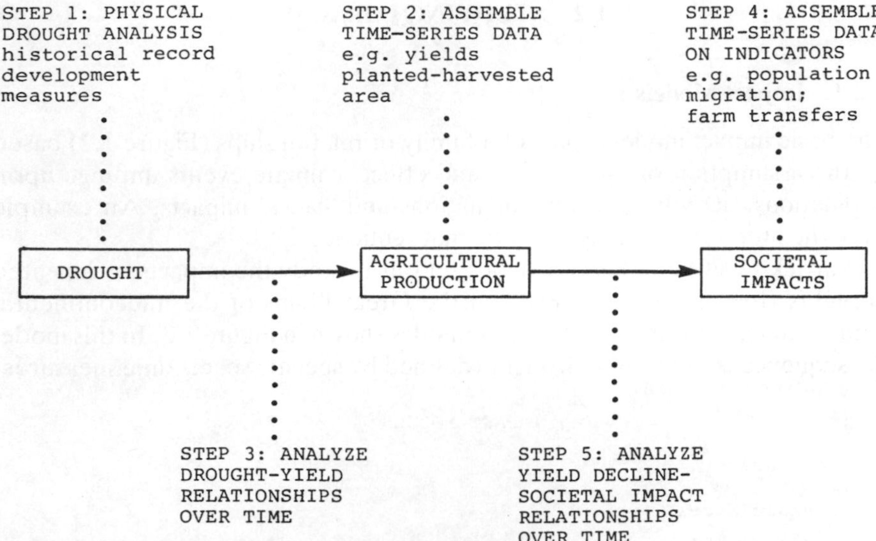

Figure 1.3 Methods used in a comparative study of a century of drought on the Great Plains of the United States. An example of an ordered impact study, with the first-order impacts of agricultural production leading to a variety of social and economic impacts impinging upon society. (After Warrick, 1980)

'causes' decreases in yield (production per unit area). In turn, this primary biophysical impact (yields) 'causes' societal impacts in the form of population movement, changes in farm ownership, and the like. The causation is implied; in actual practice drought–yield relationships combine biological theory with data from experimental plots, simulation models, and historical weather–yield correlations.

In this example, there are both first- and second-order impacts. The first-order impacts are direct measures of biological production (yield, differences between area planted and area harvested, etc.). The second-order impacts (population migration, farm transfers) ostensibly arise from the effects of yield decreases propagating through the network of social and economic relationships. This, then, is the expanded variant of the impact model with ordered impacts (Figure 1.2B). Other first-order impacts, such as yields of other crops, of grass, of insects and the like, could also be studied. This type of assessment is shown in Figure 1.2C as the multiple impacts model.

A final variant of the basic climate–society impact model describes climate changes and subsequent impacts due to human activity. For changes such as those due to fossil fuel consumption, chlorofluorocarbon production, or the use of the supersonic transport, there is an added step (Figure 1.2D) in the basic model. This is the model that underlies the Climate Impact Assessment Program (CIAP) study of the use of supersonic transport aircraft (Grobecker *et al.*, 1974) described in Chapter 22, and many studies on assessing fossil fuel burning (National Research Council, 1983).

1.2.2 Study Elements

Even with only three elements, as in the basic impact model (Figure 1.2A), the analyst must make many choices. In the Great Plains study, for example, drought had to be defined, the Great Plains areally bounded, agricultural production expressed as specific crops, impacts categorized, and indicators of these chosen. These study elements can be generalized as

1. climate events,
2. exposure units, and
3. impacts or consequences.

1.2.2.1 Climate Events

The nomenclature of climate is in dispute, depending in part on two somewhat different conceptualizations focusing on state and process (see Chapter 2). In one, climate is a statistical distribution of a set of atmospheric states, some of which are called weather. Climate variability is a measure of the dispersion of the distribution of such states. Climate change is a change in the parameters of the distribution. In the other concept, climate is a set of interacting geophysical processes that generate atmospheric states. Variation in atmospheric states arises from different interactions within the boundaries of those processes; climate change involves new and different boundary conditions.

As a statistical distribution, climate is continuous, yet it is helpful in describing impact methodologies to think of such distributions as being composed of three different scales of events: between-year weather extremes; persistent periods or decade-long episodes; and century- or multi-century-long climate trends.

For the *extreme weather events*—flood-producing rainfall, frost, fog, seasonal drought, storm, snowfall, and tropical cyclone—climate impact studies focus on the effects of their recurrent variation. When these between-year events recur consecutively, they appear as a *persistent period*. Intense precipitation events that cause floods, for example, are conveniently studied as annually-occurring events generally assumed not to be persistent, whereas droughts are studied primarily as persistent and cumulative multi-year events. Flohn (1979) and Kukla *et al.* (1977) have argued for a recent period of multi-year persistence of cooler weather. When century-long periods of above-or below-normal temperature and precipitation tend to recur in the period between glaciations, paleo- and historical climatologists delineate 'little ages', or epochs. The most notable of such recent ages are the 'little ice age' of the fifteenth to nineteenth centuries and the medieval warm epoch, or 'little climatic optimum', of the tenth to thirteenth centuries (Lamb, 1978).

In Chapter 2, our current understanding of each type of event and its controlling geophysical processes is reviewed. It is likely that each is generated by a different set of geophysical processes and boundary conditions—a different climate, in effect. They overlap, however; a 'little ice age' is composed of several persistent periods which in turn will be marked by cool and wet years, perhaps with early frosts or major floods. Understanding these relationships is very much a part of the research effort of the World Climate Programme. For the purpose of most impact study, however, this threefold distinction may be sufficient. The analyst needs to choose the appropriate scale of event to be considered and then describe its expected (normal or changed) variation.

1.2.2.2 Exposure Units

Whatever the choice of events for which impacts are to be studied, an impacted group, activity or area exposed to those events must be selected. In general the focus is on individuals, populations (human or non-human) or species; activities in the form of livelihoods; specific sectors (in more differentiated economies); or on both the groups and activities found within a specific society, region, or nation-state. Some impact studies use a nested approach, building, for example, on models of a regional economy based on economic sectors, which in turn are based on the activity of individual people, households, or economic units participating in that sector. As with the choice of climate events, it is not easy to decide how to select the exposed groups, which activities to include or exclude, or how extensive the area of impact should be.

One strategy for selecting units is to focus on the climate event and to enclose the group, area or activity within a climatic boundary (such as semi-arid regions, hurricane-prone coasts, coastal upwelling zones, or floodplains). Conversely, one may use society as a starting point (as described in Section 1.3.1) and encompass especially sensitive societal groups or activities (see Chapters 10, 14 and 17) or areas engaged in certain activities (Chapters 5, 6, 7, 8 and 9). A third strategy is to select a unit relevant to some exogenous purpose, for example a national boundary, which might be neither climatically nor societally homogenous. Finally there is selection by opportunity, such as determining the unit to be studied by an existing model (Chapter 18) or data source, or by a particular extreme event (Chapter 15). The latter approach is particularly prominent in historic studies, where the availability of data seems to dictate the unit to be studied (Chapters 11, 21).

1.2.2.3 Impacts and Consequences

The most difficult choices of study elements are the choices of impacts and consequences. Here it is helpful to assign an order of propagation (first, second, ... *n*th order) to events, although these may be arbitrary in the sense that the real time process actually takes place simultaneously or that the sequence, if any, is unknown. Nonetheless, it is useful to distinguish first-order impacts, usually of a biophysical nature, from higher-order impacts consisting of socioeconomic valuation, adjustment responses, and long-term 'change'. It is also important to recognize the dual nature of impacts: gains as well as losses are experienced, and growth as well as decline takes place. One version of such ordering to describe historic analyses is given in Figure 1.4 from Ingram *et al.* (1981).

First-order or primary impact studies are the most common form of impact study and have the best developed methodology. In several activity sectors—food and fiber, water, heating and cooling energy—quantitative functional relationships in the form of weather–yield models (Chapters 5, 8), rainfall–runnoff models (Chapter 8) and energy demand models (Chapter 9) have been constructed for many parts of the world. Midlatitude tree growth is carefully calibrated with climate indices because it serves as proxy data for inferring climates, and can be used for forest growth studies. In other primary production areas, such as fisheries (Chapter 6) and pastoralism (Chapter 7), relationships are often qualitative, both species- and locale-specific, or in need of study.

The most common secondary impacts studied are hunger and malnutrition (Chapter 10), economic disruption (Chapter 12) and other social impacts (Chapter 13). Detailed historic examples of the depopulation of marginal lands during the fourteenth to seventeenth centuries in Scotland (Parry, 1978), widespread hunger in Europe in 1816–17 (Post, 1977) and other historical case studies are given in Chapter 21. Studies of contemporary problems (Chapter 22) include hunger, starvation and death, and economic loss in the Sahel (Seifert and Kamrany, 1974; García, 1981); or the socioeconomic effect of diminished crop yields (National Defense University, 1980).

The chain of impacts can be extended to the wider economic and social impacts, as shown in Figure 1.4. One set of such impacts involves adjustment responses, activities designed to alter potential impacts by modifying climatic processes or altering biospheric vulnerability. For convenience these impacts are discussed within the framework of interaction studies (see Section 1.3), but they should also be viewed as direct consequences of climatic variability and change. Indeed, in many areas (such as urban water supply) the major impacts of climate variation are the very large efforts and expenditures made to prevent adverse effects.

The use of the basic impact model is criticized in this volume and elsewhere for its literal determinist implications, and for its use of climate as a major determinant of human events (Chapters 3, 11, 14 and 21). While there are many examples of such usage, most analysts use it not for simplistic thinking but as a simplifying assumption. Through a variety of research designs and

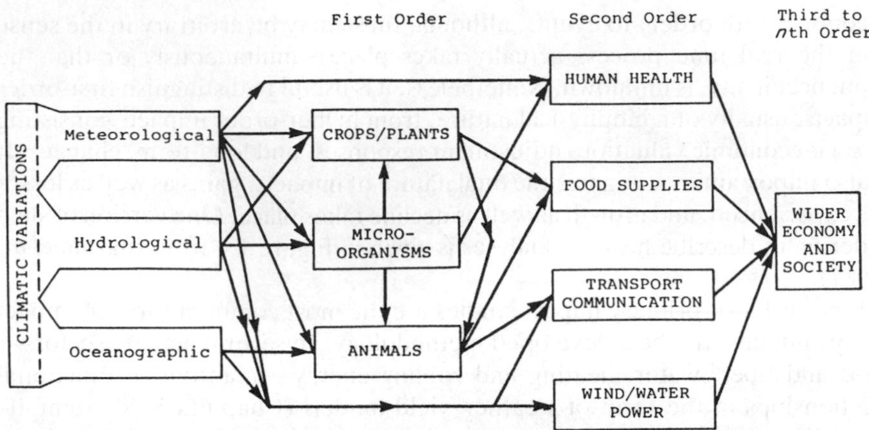

Figure 1.4 First- to *n*th-order effects of climate on the environment and society. As the specific impact becomes more removed from the 'cause', progressively more interactions intervene to disguise and modify the link. (Reproduced by permission of Cambridge University Press from Ingram *et al.*, 1981)

statistical devices they seek not to ignore societal interaction, but to hold it constant in order to examine the climatic contribution to impacts in isolation from societal influence. Alternatively, other analysts seek to include societal variation as a specific element in interactive studies.

1.3 STUDYING INTERACTIONS

The specific inclusion of some aspects of human activity and social organization along with climatic variation is the essence of interactive models. Such an approach is explained in the observation made by the SCOPE *Workshop on Climate/Society Interface* in 1978:

> ... that a change in *either* climate *or* society affects impact. Since both climate and society are constantly changing, the magnitude and character of impacts is also not constant. Impact studies must, therefore, involve investigations of climatic variability *and* of social change. The question to be addressed to any society at any time is: 'Is the society becoming more or less vulnerable to climatic variability?' Or to put the matter differently, 'Will any specific proposed social change or development have the effect of increasing or decreasing vulnerability?' (SCOPE, 1978, 17).

1.3.1 Interaction Models

The basic interactive model is sketched in Figure 1.5A. Such an interactive model forms the basis of the historical work of Rosenof (1973) on Ellis County, Kansas (USA) and to a lesser degree that of Parry (1978) on Scotland, as well as a number of case studies in the International Federation of Institutes for

INTERACTION MODELS

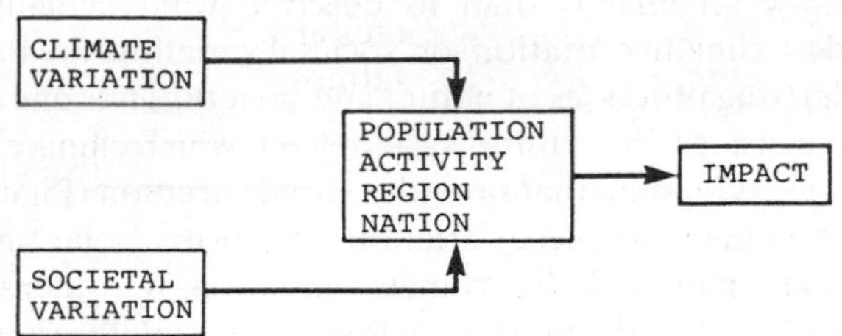

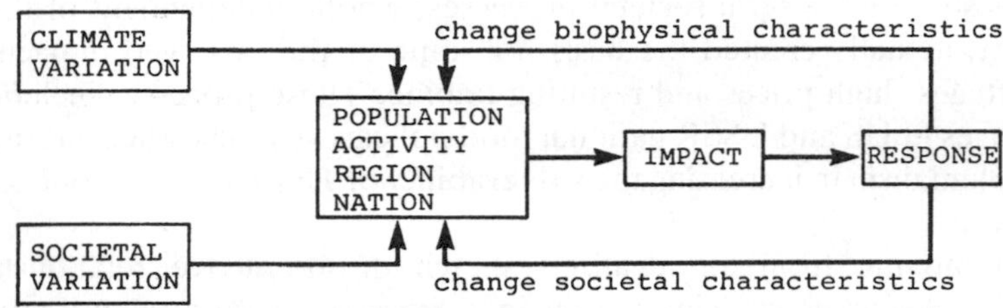

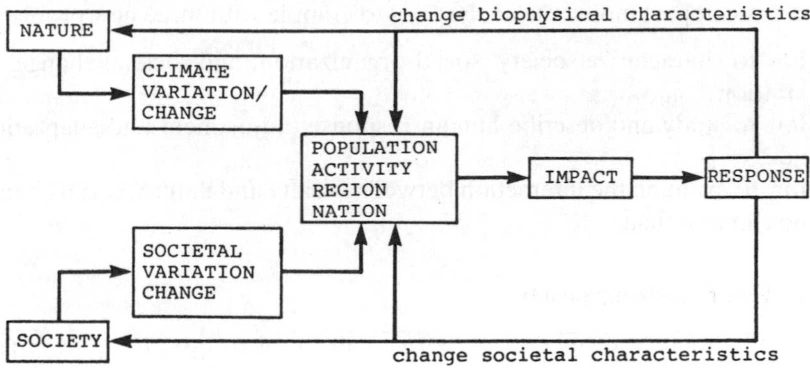

Figure 1.5 Three types of interaction models stressing: (A) simple interaction, (B) feedback to the social activity, and (C) feedback to underlying physical and social processes and structures

Advanced Study (IFIAS) project, *Drought and Man, 1972* (García, 1981). But studies that limit themselves to considering only a single sequence of climate–society interactions are actually few in number. Combinations of human activity or social organization invariably lead to the consideration of human responses. Such responses, in the form of adaptations or adjustments, act functionally to change either the biophysical characteristics or the societal characteristics of the interaction (Figure 1.5B).

It is easier to draw schematics than to describe what actually occurs. Variables described as 'climatic variation' or 'societal variation' are themselves products of the underlying processes of nature and society. Thus one can move further upstream to consider the 'nature' system from which climate variation is derived and the 'society' system that provides social variation (Figure 1.5C).

In the natural system there are basic variations in process (solar fluctuation, orbital variation) that may underlie climate variation. Or there may be anthropogenic activity, such as the burning of fossil fuels or deforestation, with a resultant increase in the CO_2 content in the atmosphere. In the recent IFIAS study of the impact of drought worldwide in 1972 (García, 1981), it is suggested that several societal processes, wholly independent of the climate event, actually created the basis for some of the observed impacts of food shortages, high prices and resulting famine. These processes included major changes in US and USSR national food policies as well as the historical effects of colonialism in increasing the vulnerability of Third World people to external forces.

In moving from an idealized sketch of interactive relationships with feedback—acknowledging the distinct processes of society and nature—to a concrete design for an assessment, a process of oversimplification inevitably seems to take place. When societal interaction and human response are added to the basic impact model the analyst has to grapple with three new problems:

1. How to characterize society, social organization, and societal change and variation?
2. How to study and describe human response, adjustment and adaptation? and
3. How to examine the interaction between society and nature, social change, and climatic change?

1.3.2 Characterizing Society

The term 'society' is commonly used for small social groups, as a generic term for human organization, and as a reference to particular ethnic, livelihood, geographic or political entities that range from groups greater than families to archetypes of socioeconomic organizations which include many nations. Thus while conveniently ambiguous, it requires specific characterization in assessment studies.

1.3.2.1 Between Societies

On a global basis, the division of societies focuses on issues of development. This is done in two ways, by dividing societies into 'developed' and 'developing' nations, either as individual nations or regional clusters, or by generalized archetypes of development that do not necessarily coincide with political boundaries. Both developed and developing nations are frequently differentiated by whether they are market- or centrally-planned economies. Developing nations may be classified by income (per capita, gross national product) and/or other measures (such as percentage of literacy, percentage of employment in manufacturing), yielding such classes of nations as 'least developed' or 'most seriously affected'. The World Bank (1980) groups nations into a sixfold hybrid classification (low income, middle income, industrialized, capital surplus, oil exporters, and centrally planned) and the Organization for Economic Cooperation and Development employs a similar sevenfold classification. Beyond the simplicities of North–South and East–West regional groupings, Mesarovic and Pestel (1974) employed 10 regions in their World Model and Leontiev (1977) employed 15 regions in his.

In lieu of dividing the world, matched samples have been a convenient device. For example, studies of the national impact of three major atmospheric hazards have used matched country samples: drought (Australia, Tanzania); flood (Sri Lanka, United States); and tropical cyclone (Bangladesh, United States) (Burton *et al.*, 1978). Although these studies are in many ways an odd assortment (arranged for the most part by the availability of cooperative investigators), summary results as shown in Table 1.1 reveal a general and consistent set of differences. Deaths are found primarily in the three developing countries studied, and although the absolute amounts of economic costs and losses are greater in the cases of Australia and the United States, the relative impact of the events is still felt most heavily in the developing countries. In terms of the percent of per capita GNP, these three atmospheric hazards extort an economic impact ten times greater in the developing countries than in the two industrialized countries.

1.3.2.2 Within Societies

Within a particular nation or region several different approaches have been used to characterize society, including

1. modeling the overall society, economy or particular activity;
2. identifying perturbations—extreme or unusual social events that are the equivalent of extreme natural events or climate anomalies;
3. differentiating groups, livelihoods, subregions or activities by their potential vulnerability to climate change and variability; and
4. examining social factors, mechanisms, and trends that lead to greater or lesser vulnerability.

1.3.2.3 Society, Economy, Activity

For most societies, economies, or livelihood activities, there already exists a

Table 1.1 Selected estimates of natural hazard losses

Hazard	Country	Total population (millions)	Population at risk (millions)	Annual death rate/million at risk	Losses and costs *per capita* at risk: Damages losses ($)	Costs of loss reduction ($)	Total costs ($)	Total costs as percent of GNP
Drought	Tanzania	13	12	40	0.70	0.80	1.50	1.84
	Australia	13	1	0	24.00	19.00	43.00	0.10
Floods	Sri Lanka	13	3	5	13.40	1.60	15.00	2.13
	United States	207	25	2	40.00	8.00	48.00	0.11
Tropical cyclones	Bangladesh	72	10	3000	3.00	0.40	3.40	0.73
	United States	207	30	2	13.30	1.20	14.50	0.04

(After Burton *et al.* 1978)

body of scientific description of structure and function. In some cases, primarily economic activities, quantitative models have been built. Some type of national economic model exists for most countries. Thus, one starting point for an interactive study is to draw on existing societal theories, descriptions or models and attempt to relate these to climate variations. In practice, there is only limited experience with the use of existing economic models (Chapter 12), biophysical models (Chapter 18), and social theory (Chapter 13).

The *Global 2000 Report* by the US Council on Environmental Quality (1980) attempted to link climate data to available government models with little success. The Computing Center of the USSR Academy of Sciences is currently constructing a global biosphere model that would link the global atmosphere and oceans, economy and ecology (see Chapter 19). Some success in using existing models has been obtained with agricultural and food models, such as the Model of International Relations in Agriculture (MOIRA) (Linneman *et al.*, 1979); the US Department of Agriculture grain, oilseed and livestock (GOL) model (USDA, 1978); the University of Southern California interactive model (Enzer *et al.*, 1978); and the International Futures Simulation (IFS) model (Liverman, 1983).

Johnson and Gould (1984) have used the extensive archeological/anthropological theory and description of Mesopotamian floodplain irrigated agriculture to construct a model of the society along the Tigris and Euphrates Rivers during the last 5000 years. This working model (abstracted in Figure 1.6) was then used to examine the impact of climate variability on the size of population. In their simulation, a variable climate as opposed to a constant climate was found to limit the 'carrying' capacity of the system, but actually extreme periods of drought lead to an increase in population carrying capacity by stimulating the extension of the irrigation system.

1.3.2.4 Vulnerability Concepts

An alternative to trying to describe or model an entire society, economy or region is to focus on particular aspects of social structure that the analyst thinks are related to the issue of increased vulnerability to or utility of climate. Several related methods have been used to approach the issue; underlying each is a common conceptual basis somewhat akin to the engineering concepts of stress and strain.

In this view, societies experience *perturbations* that stress the system(s) under consideration. How the society responds to the perturbation is determined by its *vulnerability*, literally meaning its capacity to be wounded. In the sense of societies impacted by climatic or social perturbation, vulnerability is the capacity to suffer harm or to react adversely (Timmerman, 1981).

Societies have as well a capacity to resist such harm in quite different ways, which may be suggested by the contrast in western and eastern metaphors of

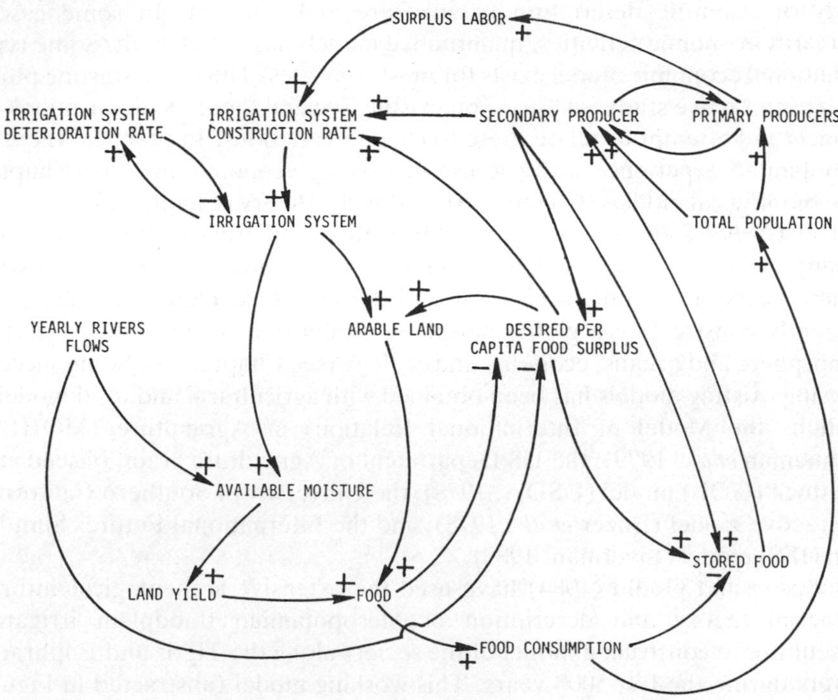

Figure 1.6 Major feedback loops of a model construction of an irrigation-based village or tribal society on the floodplain of the Tigris–Euphrates rivers *c.* 5000 years ago. The model is used to test the impact of climate variability and change on human population numbers. (Source: Johnson and Gould, 1984)

martial arts—the strength of the oak tree versus the resilience of the bamboo. Thus some societies respond to perturbation with properties that resist displacement from an equilibrium variously characterized as *stability* (Holling, 1973); *absorptive capacity* (Burton *et al.*, 1978); and *endurance* (García, 1981); or by such terms as *inelasticity* or *resistance*. Other societies respond by being *elastic* or highly *resilient* (Holling, 1973; García, 1981); they may be displaced easily from an equilibrium but they eventually rebound to that equilibrium position without *structural change* (García, 1981); *extinction* (Holling, 1973); or *societal collapse* (Timmerman, 1981).

An extension of these basic concepts finds differing emphases by various authors. García (1981) notes that these modes of response are not mutually exclusive and societies might possess properties of both high or low resilience and endurance. Alternatively, Holling (1973) contrasts resilience and stability, suggesting an implied tradeoff. This tradeoff is carried a step further by Bowden *et al.* (1981), who hypothesize that various technological and social responses that lessen the impacts of minor climatic stresses may serve to increase the catastrophic vulnerability of the society to a major perturbation.

The value of this conceptual framework beyond the heuristic level of metaphor or illustration is not yet clear and is not helped by the profusion of somewhat different definitions accorded to the same concept (resilience, for example, is variously interpreted by Holling [1973], García [1981], and Timmerman [1981]). But as a guide towards characterizing society, these concepts lead naturally to three methodological emphases: on perturbation; on vulnerable people, regions and activities; and on societal actions that increase or decrease vulnerability.

1.3.2.5 Perturbations and Extreme Events

The first approach attempts to deal with a major criticism of the basic impact model; namely, that it assumes that the observable impact results from a climatic variation or change, whereas in reality it may have been produced by another cause or by the joint action of climate with another event.

For example, in the Great Plains case study (Warrick, 1980), the social impacts in the form of population migration and farm transfer were surely affected by the great worldwide economic depression during the 1930s that preceded and coexisted with the drought conditions. The 'year without summer' in North America and Europe, 1816–17, which was linked to the eruption of Mt Tambora in Indonesia, similarly saw widespread agricultural failure coupled with a post-Napoleonic-war economic slump (Post, 1977).

Identifying such events, then, is one form of societal description. The analyst, particularly in historical studies, abstracts from the massive social experience those events that would seem to reinforce or buffer the impacts of climatic variation or that serve as alternative explanations for observed impacts. Parry (1981) has labeled such events 'proximate factors'. Such events can be thought of as perturbations in the system's environment.

In this view of climate–society interaction, adverse perturbations may arise from either climate or society, and in rare cases they may be functionally the same even if their origin differs. The clay tablets of the Kingdom of Larsa, which existed 4000 years ago on the floodplain of the Euphrates River, tell of the dependence of the city–state on its major irrigation canals and the efforts of invading forces to throw dikes across the canal in order to sever the water network (Walters, 1970). In terms of perturbations, such warlike activity is functionally indistinguishable from a major flood that severed the network by washing out the canal. Similarly, the loss of water from a sustained drought could be functionally equivalent to the loss of water from failure to organize the cleansing of the irrigation channels of silt and vegetation.

1.3.2.6 Vulnerable People, Regions and Activities

A second method of characterizing society's relationship to vulnerability is to focus on vulnerable people, groups, places or livelihood systems. Climate variation may be expected to impact especially those groups, activities or regions that under 'normal' climate conditions are already stressed (Chapter 14). Thus peoples whose social or economic position is already marginal are especially vulnerable. Similarly, places that during 'normal' climate conditions are already on the margin of cultivation or pastoralism are most susceptible to adverse impacts from unfavorable climate anomalies or changes. And within economies, activities that at best are hardly remunerative cannot withstand further stress from climatic variation.

To give but one example, two great droughts have occurred in the Sahelian-Sudanic region of West Africa in this century (Kates, 1981). In brief, there are two major livelihood groups in this area whose activities overlap—nomadic pastoralists and sedentary agriculturalists. Between 1910–15 when the first drought occurred, and 1968–74, the period of the most recent drought studied, a major change in the social status of the two groups took place. During the first drought, the nomadic pastoralists, especially the camel-herding Tuareg, were still reasonably independent of colonial rule and had a complex set of vassal relationships with agriculturalists, who tilled in their behalf. By the time of the most recent drought, the numbers of nomads had diminished, former slaves and vassals had been freed, raiding for food was no longer an option, and the Tuareg role in trans-desert transport had almost disappeared. Independent governments administered by agriculturalists existed in each Sahelian country. This shifting fortune was strongly mirrored in the impacts of hunger, starvation, malnutrition and economic loss suffered during the recent great drought. And in the region as a whole the same pattern held. The Sahelian countries—arid, poor, mostly landlocked—suffered more than their neighbours in Nigeria or Senegal.

1.3.2.7 Cumulative Changes

The shifts in vulnerability among Sahelian peoples exemplify a third analytic approach, one which focuses on trends that may consist of small, cumulative changes in either lessening or increasing vulnerability. For example, let us return to the Great Plains case study. Commercial agriculture on the Plains is about 125 years old, and during this period five major droughts have been experienced. Warrick (1980) and colleagues hypothesized that if one could equalize or normalize the physical extent of these droughts, their negative impacts would appear diminished over time because of long-term societal trends that tend to lessen the impact.

Analysts frequently assume these trends in an unspecific way. For example, technology in agricultural studies is often represented simply by a linear time trend (see Chapter 5). The assumption is that technological changes account for yield increases and can be simulated by the passage of time. Comprising 'technology' are many small changes in plants, cultivation techniques, farming machinery, and farm organization that increase yields over time. Thus yields during a drought occurring late in a time-series undergoing such a trend will surely be larger than yields during an earlier drought. To that extent, the impact of drought is lessened.

Conversely, the analysis might argue for increased vulnerability. The shifts in vulnerability observed in the Sahel over time, for example, were part of a larger trend in which colonialism increased the commercialization of agriculture in many parts of the world. The result was decreasing proportions of crops grown for local consumption and increasing amounts of export crops, thereby reducing the stores of food crops and of labor available to buffer drought (Kates, 1981).

Thus societal trends related to vulnerability can be described in such unspecific terms (technological change, commercial development, or even 'colonialism'), or in very specific terms. In a major US review of natural hazards (including nine atmospheric hazards), a number of major and minor societal trends were examined for their impact on hazard vulnerability and included: population size and distribution; the establishment of second homes; the use of mobile homes; corporate organization; health, safety, consumer protection and environmental impacts leglislation; the legal system; tax policy; and communication patterns (White and Haas, 1975).

1.3.3 Describing Human Response

Purposive action is the distinctive characteristic of human behavior. Almost all climate events, if they recur or persist, evoke some individual and collective human activity intended to prevent, reduce or mitigate undesired impacts or to enhanced desired outcomes. The World Climate Programme and this volume are part of that process, but so are nomadic movements, granaries, futures markets, and collecting firewood.

The ways in which such activities are described are quite numerous. For example Ingram *et al.* (1981) describe four classes of adjustments: the adjustments of climate, of the biosphere, of society, and inadvertent, unplanned or uncontrolled effects (see Figure 1.7). In dealing with anthropogenic change, Kellogg and Schware (1981) distinguish between measures designed to avert change and those designed to mitigate the effects of change. Meyer-Abich (1980) distinguished between prevention of, compensation for, and adaptation to CO_2-induced climate change. Schelling (1983) suggests four categories of response to CO_2-induced climate change—options involving

1. the production of CO_2,
2. its removal from the atmosphere,
3. the modification of climate and weather, and
4. adaptation to anticipated and experienced change.

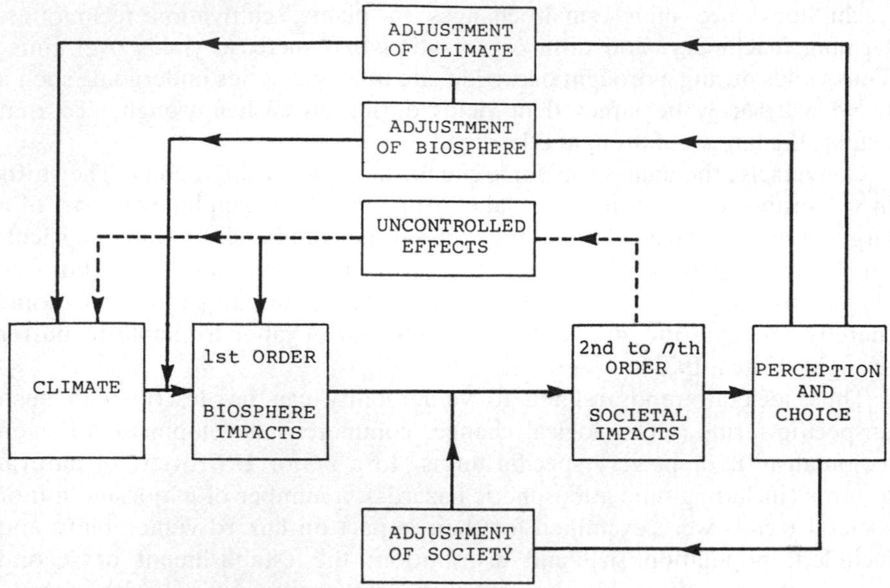

Figure 1.7 Conceptual model of the impact of climate on man and society, showing possible feedbacks via adaptive strategies. Uncontrolled effects refer to inadvertent or unplanned modifications of the biosphere and/or the climate as a result of climatic stress or otherwise. (Reproduced by permission of Cambridge University Press from Ingram *et al.*, 1981)

These logical distinctions have been elaborated on in a typology of human responses developed by geographers in their collaborative work on natural hazards (Burton *et al.*, 1978).

1.3.3.1 Adaptation and Adjustment

This typology distinguishes between cumulative long-term responses to natural hazards, called *adaptations*, and short-term responses, described as *adjustments*. Adaptation can be biological or cultural. There are a few well understood biological adaptations to climate (for example, body heat and evaporation control). But it is primarily through culture that human societies adapt to the many different climates that humans inhabit.

The distinction between adaptation and adjustment is not very clear. Burton *et al*. (1978) give the following examples:

> Building a dam to store additional water for irrigation during a drought period would be classed an adjustment. A system of cut-and-slash farming in the Laotian highlands, with all its requirements for appropriate social organization in timing the cutting, burning, cultivation, and revegetation of forest lands, would be counted an adaptation. Designing a house to resist a storm surge would be an

adjustment; locating and organizing a community over a long period of time so that its houses are beyond the reach of storm surge would be an adaptation. An individual or group may, however, choose to apply as an adjustment a practice that long has been an adaptation elsewhere, as when a home owner builds a house flood-proofed with a design imported from a distant place. (p. 37)

Adjustments are both incidental and purposeful. *Incidental adjustments* are those actions that functionally serve to reduce vulnerability, although their origin is for a non-hazard-related purpose. Thus a community that restricts mobile home usage for taxation, social class or amenity purposes also lowers its vulnerability to tornadoes and windstorms. *Purposeful adjustments* take three general forms: accepting losses and distributing their impacts in various ways; reducing losses by trying to modify events or prevent their effects; and basic changes in location or livelihood systems. The range of theoretical adjustments for an individual or small group is shown in Figure 1.8. In actual practice, this range may not be available, especially to marginal social groups or areas.

In all these classifications, their authors acknowledge the somewhat arbitrary typology and the overlapping categories involved, but in practice there seem to be few problems of interdisciplinary understanding. The problems of studying adaptation and adjustment are not those arising from

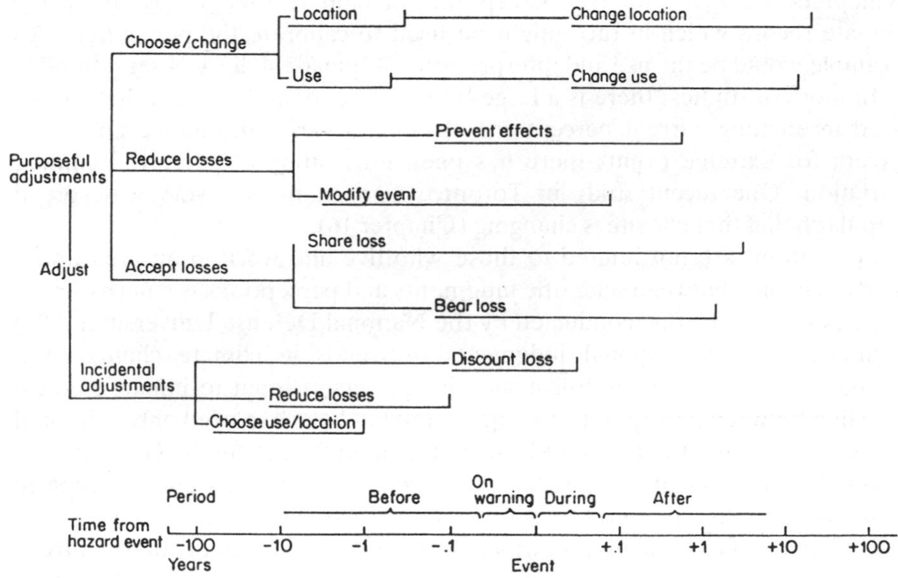

Figure 1.8 A Choice Tree of Adjustment. Adjustment begins with an initial choice of a resource use, livelihood system, and location. For that choice various incidental and purposeful adjustments are available, at somewhat different time scales for initiation. The most radical choice is to change the original use or location. (Reproduced by permission of Oxford University Press from Burton *et al.*, 1978)

different terminology or categories. Rather they involve basic problems of perception, function analysis, and social theory.

1.3.3.2 Understanding Perception

To understand retrospectively the utilization of adjustments, one needs to reconstruct the perceived climate, its variability and changes (Chapter 16). Such historical reconstruction (Wigley *et al.*, 1981) requires documentation, which limits such work to literate societies and the literate classes within societies. Ideally, such perception data (from reports, letters and journals, for example) should be calibrated with a climatological record. When such calibration can be done, surprising results may follow. For example, Lawson and Stockton (1981) demonstrate that the widely denigrated fictional 'myth of the Great American Desert' was actually supported by the experience of the Long expedition, whose report identified an area of the Great Plains of the United States midcontinent as desert. By happenstance the expedition took place in the core region of a severe drought, providing in a relative sense a drought greater than that of the 1930s. In general, however, there is a considerable circular involvement in the reconstruction of past climates (Chapters 11, 21), with the perceptions of climate used to reconstruct a climate record which in turn might be used to calibrate the perception. An example would be the use and interpretation of personal diary observations.

In modern studies, there is a large body of theory and method that can be used in eliciting current perceptions of climate variability and change, but except for extreme events there has been little study of perceived climate variation. One recent study in Toronto, Canada finds a very widespread popular belief that climate is changing (Chapter 16).

Perceptions are not limited to those who live and work in an area under study. The line between scientific judgments and perceptions is a narrow one. Studies such as the one conducted by the National Defense University (1980) formally elicit professional judgments of trends in climate change in a probabilistic format, then utilize such judgments as input to impact studies. The link between perception and adjustment has been explored only in limited ways. Warrick and Bowden (1981) have shown, in the case of the Great Plains, that recurrent droughts seem to have created a scientific and public perception that leads to relief measures in anticipation of actual drought, to the point of providing relief to local areas in which drought does not actually materialize.

1.3.3.3 Attributing Function

A second major problem in adjustment–response analysis is the attribution of function to discrete actions which are part of a cultural whole. Porter (1976) has posed this question in his description of Kamba farming practice in Kilungu,

Kenya. To the external observer the fields of the Kamba are jumbled and disordered—maize, beans, cow peas, sorghum, groundnuts, bullrush millet, red millet, cassava, pumpkins, callabashes and pidgeon-peas planted together in a single field. Porter describes the functional adjustments:

> The crops come up all together in a riotous profusion of vines, leaves and stems. Weeds have no chance. The phosphorus flush in the soil that comes with the first rains is taken advantage of by these crops. Crops, such as maize, which might be indolent about putting down roots and strong tillers in a soil well supplied with moisture, have to compete with the other plants, and thus they put down roots to a depth of several feet. These deeper roots come into play later on in the season. Although the rate of moisture use of all these crops planted together is high, it occurs at the one time of the season when there is likely to be enough moisture available. Further, the interplanting provides a good continuous canopy within which photosynthesis can proceed. Since tropical areas have a relatively uniform radiation income all year around, a continuous leaf canopy, one with a moderate to high leaf-area index, is the most efficient user of radiant energy for photosynthesis....
>
> After about seven or eight weeks some of the crops are harvested. Beans may be harvested as green vegetables. As the season progresses the millet is harvested; beans harvested for seed are taken and the vines pulled up. The number of plants per square meter of soil begins to decline and only crops requiring a longer time to mature are left. The bare weedless soil between the plants is dry, which forms a barrier to the movement of moisture to the surface. Evapotranspiration thus is reduced. The sparse plant population remaining is able to tap moisture from a larger volume of soil without competition from adjacent plants. This moisture, combined with the lesser amounts of rain which come at the end of the 'grass rains', many times is sufficient to bring the maize and the longer growing millets to harvest. The plants also provide some shade for crops set out to get a start on the main rains. The thick mat of plant cover in the first weeks of the 'grass rains' and the second rains also serve to hold the soil. There is also the adaptive fact that if the rains do give out, some crops will have been harvested and the agricultural effort will not have been a total loss. Another fact is that interplanting reduces the amount of work considerably by eliminating the need to weed, which often is the most serious impediment to agriculture and the management of larger acreages. (Porter, 1976, 134–135)

This agricultural system has considerable built-in capacity to absorb the effects of drought. The complex of practices may be described as a cultural adaptation or a mix of incidental adjustments. In either case they are largely unconscious on the part of the farmer; in a sense he farms better than he knows. This presents two problems to the analyst: how to characterize the complex of practices made up of so many distinctive and interwoven plants, activities, cultural practices and timing, and how to attribute individual or aggregate purpose to the cumulative successful function. These problems are exacerbated in historical case studies (Chapters 11, 21), where the complex is preserved only sketchily by archeological and historical reports and the farmers are beyond inquiry.

Such perplexities are not limited to societies where formal agricultural science is absent. In industrialized countries similar analytic problems exist. Riebsame (1981), as part of the Great Plains Case Study, sought to examine the mechanisms of adaptation and adjustment that have diminished the adverse impacts of drought in spring wheat production in the state of North Dakota, USA. From 32 technical and 30 social adjustments employed to some extent in North Dakota between 1890 and 1980, five had a major effect on reducing vulnerability: summer fallowing, stubble mulching, crop insurance, disaster payments, and farm diversification; for several other widely promulgated adjustments, like marginal land retirement, shelterbelts and maintaining reserves, it was difficult to demonstrate whether their employment had any effect at all.

1.3.3.4 Using Theory

Finally, there are competing theoretical models of the major social processes in adjustment and adaptation. For example, neoclassical economics emphasizes market processes in which the 'invisible hand' will encourage or discourage various adjustments (Lave, 1981). Neo-Marxist analysis emphasizes the differential access to adjustments by social class or marginal group and the interrelationship between decreased vulnerability for one group and increased vulnerability for others (Spitz, 1980). Social–psychological or policy-oriented decision-making studies emphasize processes of conscious choice by rational or 'boundedly' rational decision-making individuals or groups (Slovic *et al.*, 1974). In practice, all of these activities—markets, class conflict, rational choice and many more—can be found in societal structures. But the selection of emphasis has strong implications for what might be learned from past experience or which public policies to suggest for future action.

1.3.4 Examining Interaction

The characterization of climate and society and the identification of feedback mechanisms is but a first step towards exploring the interaction between climate and society. At least three methods of investigating such interaction are in common use:

1. causal and correlational explanations,
2. experiments, and
3. modeling and simulation.

1.3.4.1 Causal and Correlational Explanations

In causal explanation, the chain of causality with the varying contributions of climate and society would be carefully detailed and verified. There are no ready examples of interactions for which the physical, biological and social principles are so well understood that a causal analysis is explicable. Perhaps closest to such an ideal are experimental agricultural studies of the interaction of moisture (climate) and fertilizer (society), but even these principles are derived more from experiments and correlational data than from causal theory. Correlational explanation lacks the assurance of causal explanation; its convincing quality must rest on associations so strong as to dominate chance or alternative explanations. As with all correlational analysis the description of relations can take various forms: descriptive, mathematical, or graphic.

A good example of descriptive correlational analyses is Post's (1977) study of the widespread food shortages that developed in Europe following the eruption of Mt Tambora in 1816 and the associated poor weather. Within the general theme of the impact of this event, Post tries in qualitative fashion to estimate the impacts on the European population of the extreme climate event, the post-Napoleonic-war crisis, and the imperialist attitudes of the British and Austro–Hungarian empires.

In mathematical form, correlational analysis centers around multiple regression, in which impacts are functions of various climate indicators and societal indicators. For example, Johnson and McQuigg (1974) try to examine the joint impact of temperature, precipitation, soils, population density, property taxes and government subsidy on US Great Plains land prices. They use principal components analysis to characterize the climate and combine these climate components with the societal variables in multiple regression equations that account for 60 percent of the variation in prices.

In graphic form, Parry (1981) sketches in Figure 1.9 the relationship between cultivation limit, a measure of farm abandonment (his impact variable), the areal level of resource availability (the amount of cultivable land determined by climate variation), and proximate factors (disruptive societal events).

1.3.4.2 Quasi-experiments

Correlational studies could be enhanced if the factors to be compared were placed within an experimental design. Although it is possible to study experimentally the response of plants to the doubling of CO_2 in the greenhouse (Lemon, 1983), it is not possible to study the response of society. However, it is possible to use matched 'before and after' or 'with and without' situations as controls in natural quasi-experiments. Such methodologies have been used in studies of long-term local impact following natural hazards (Friesema *et al.*, 1979; Wright *et al.*, 1979; see also Chapter 15). In another example, as discussed in Section 1.3.2.6, the two great droughts in the Sahel were compared for impacts, although population size, societal organization and activity differed considerably. With drought held constant, the variability in impact was ascribed to changing social organization and trends.

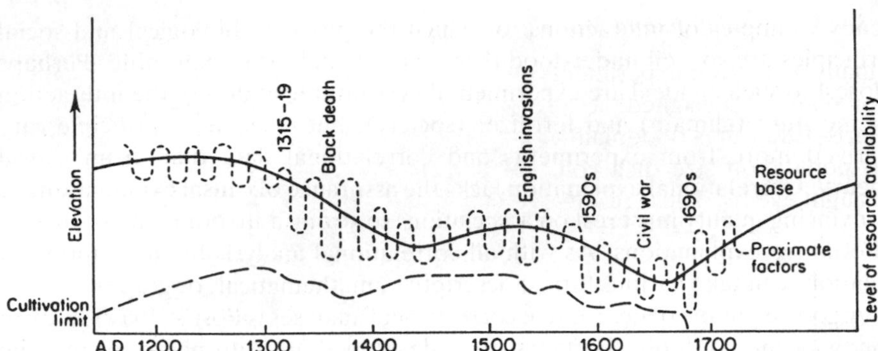

Figure 1.9 Proximate and indirect factors behind farmland abandonment in upland Scotland. Between 1200 and 1700 AD there was a long-term downward trend in the level of potential resource availability, i.e. the available amount of cultivable land determined by adverse climate experienced at lower and lower elevations in the upland regions. Concurrent with this trend are various societal fluctuations in fortunes called 'proximate factors'. These perturbations combine to depress agricultural activity as shown in the dips in the cultivation limit, which measures the level of actual cultivation use and its inverse, farm abandonment. The recurrent shocks of the seventeenth century combined with the adverse climate lead to total abandonment of the area under study. (Reproduced by permission of IIASA from Parry, 1981)

Inadvertent climate modification has provided yet another quasi-experimental format. Urbanization changed the climate downwind of St Louis, Missouri in the United States, providing an opportunity to trace the 'before and after' primary impacts on soil, streamflow, agriculture and buildings, and secondary impacts on the economy, institutions, and public health and welfare (Changnon *et al.*, 1977).

Recently it has also been shown that sharp discontinuities in regional climate, sustained for one or two decades, take place (Karl and Riebsame, 1984). Such changes arise from normal variation in the climate system and do not require fundamental climate change as explanation. Yet the populace affected experiences a real change in its climate, on the order of 1 °C in mean temperature and 25 percent in mean precipitation between periods. Such changes provide an opportunity to investigate impacts perception and adjustments, and quasi-experimental designs are being developed for this in several countries (Kates *et al.*, 1984).

Finally, the World Climate Applications Programme provides still another experimental format. This program, designed to provide new or newly disseminated climate data tailored to specific applications, can change people's understanding of their climate, with subsequent changes in perception and adjustment. Baseline studies of perceptions and adjustment can be made prior to introducing the new data and the effect of the data and their presentation in varying formats can be carefully monitored.

1.3.4.3 Simulation and Modeling

Finally, computer experiments employing mathematical modeling and simulation can be designed. Using the model of irrigated agriculture in the Tigris–Euphrates floodplain, Johnson and Gould (1984) examined the impact on population size of a constant climate, a variable climate, and extreme drought events. These results are shown in Figure 1.10.

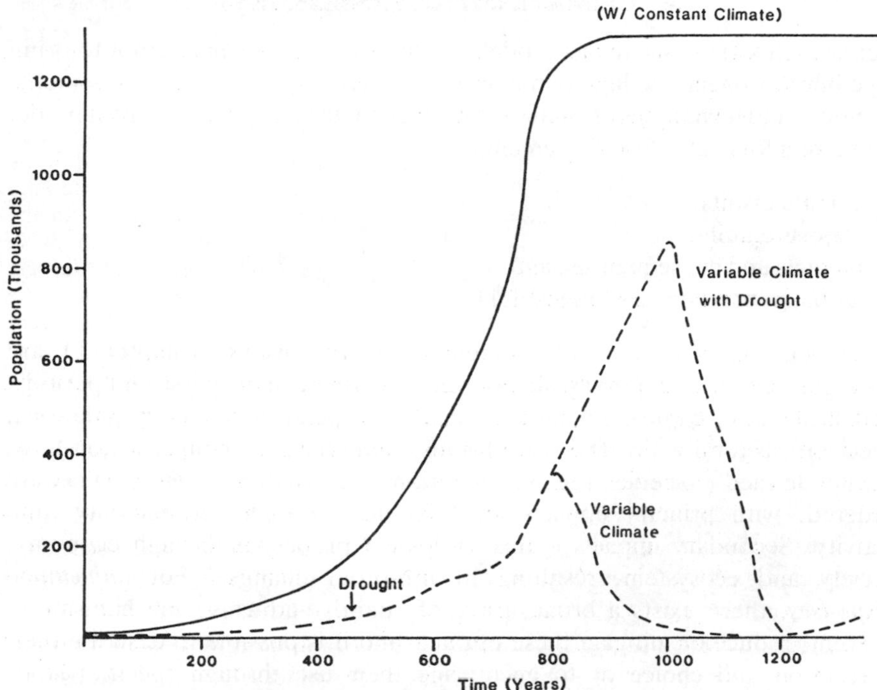

Figure 1.10 Three simulation traces of an irrigation-based village/tribal society model of the Tigris–Euphrates floodplain. Under *constant climate* (mean annual streamflow available each year in irrigation system), population grows, then stabilizes in classic form of the demographic transition. Under *variable climate* (by normal distribution of streamflow) population rises to 25% of constant climate and collapses (as in historical record). If a major persistent drought period is inserted into the variable climate (*variable climate with drought*), this leads to an adaptive response (extension of irrigation system) and an expanding population which, nonetheless, eventually collapses. (Source: Johnson and Gould, 1984)

A recent unpublished effort by Warrick, simulating a recurrence of 1930s drought with current technology in the United States, utilized the University of Southern California food model (Enzer *et al.*, 1978). Warrick found only minor global impacts if the recurrence were limited to the wheat-growing regions, but major impacts if the drought simultaneously affected both corn-and wheat-growing areas. Liverman (1983) has carefully investigated the use of a basic global model, the International Futures Simulation version of the Mesarovic and Pestel (1974) world model for climate impact experiments. She found that models of this type have little predictive value, limited policy application, but considerable promise as research tools.

1.4 CLIMATE IMPACT ASSESSMENT

Returning to the basic impact model (Section 1.2.1), we can attempt to bring together this overview, linking the model to a typology of impact assessments or modes and techniques of analysis. We begin with an expanded impact model containing four sets of study elements:

1. climate events,
2. exposure units,
3. impacts and consequences, and
4. adjustment responses (Figure 1.11).

Each set has a number of elements. *Climate events* (Chapter 2) are distinguished on a temporal scale: extreme weather events, persistent periods, and 'little ages'. *Exposure units* are socially grouped (individual, populations, species), sectorally divided (livelihoods, activities, economic sectors), or areally defined (societies, regions or nations). *Impacts and consequences* are ordered, with primary impacts on biological systems, productivity, and activity. Secondary impacts (gains or losses) propagate through economy, society and ecosystem, resulting in *n*th-order changes. For *adjustment responses*, there exists a broad array of adaptive-adjustive mechanisms to prevent, reduce or mitigate these impacts and it is possible to describe their perception and choice or to encourage their use through specific policy prescriptions.

Each set of elements is linked by an analytic mode, a way of studying the connections between sets. These modes serve as the framework for this volume.

Sensitivity studies (Chapter 4) attempt to identify climate-sensitive groups, activities and areas, linking them to the varied levels of climate events (Chapter 2). The direct impacts experienced by such exposed groups, activities or areas are identified through *biophysical impact studies* and include studies of climate impact on biological or physical productivity: in agriculture (Chapter 5), fisheries (Chapter 6), pastoralism (Chapter 7), water resources (Chapter 8) and energy resources (Chapter 9). Examining how biophysical impacts are propagated into human socioeconomic and political systems is the task of *social impact assessment*. The focus for such studies can be human populations (Chapter 10), past societies (Chapters 11 and 21), the economy (Chapters 12 and 20), current societies (Chapter 13), marginal locations or groups (Chapter

14), or the area impacted by an extreme event (Chapter 15). *Adjustment response studies* link impacts with responsive behavior, and analytic methods may focus on the perception and choice of adjustments (Chapter 16) or on their availability and efficiency (Chapter 17).

Integrated assessments include at least three links—sensitivity studies, biophysical impact studies, and social impact studies—and may correspond to the simple input–output model. Because of the scale and linkages of integrated assessments, some form of modeling is frequently used, including global modeling and simulation (Chapter 18), biosphere modeling (Chapter 19) and scenarios (Chapter 20). When such studies include feedback in the form of adjustment response, they are designated as *assessment-adjustment* studies, coresponding to the interactive model with feedback. Examples of both types of integrated studies are found in the reviews of historical and recent integrated studies, (Chapters 21 and 22, respectively). Missing, both from the diagram (Figure 1.11) and from experience, is a mode of analysis labeled *comprehensive impact assessment*, envisioned as corresponding to the interactive model with feedback and studies of underlying historical natural and social process.

In practice many attempts at assessment do not follow this carefully linked causal system; rather, they attempt to 'jump' study elements, going directly from climate events to inferences of higher-order consequences. These are probably less reliable than the more carefully specified and linked analyses (see further Chapters 11, 14, 20 and 21). In general, reliability seems greatest at the beginning of the causal chain—more is known about biophysical impacts, less about long-term social or ecological change.

This volume, the product of over 100 authors and reviewers, is an authoritative up-to-date statement of what-we-know about climate impact assessment. Viewing the 22 chapters as a whole leads to four major conclusions:

- More and better methods exist to assess the impacts of climate variability and change than are currently being employed. This is partly due to disciplinary isolation and partly to the limited effort expended to date on the study of the interaction of climate and society as compared to the study of the dynamics of climate itself. Recent, ongoing or planned second-generation studies, however, evidence levels of conceptual sophistication, scientific clarity and methodological integration not previously found (see, for example, Nordhaus and Yohe, 1983; Parry and Carter, 1983; Waggoner, 1983; Kates *et al.*, 1984; Santer, 1984). Moreover, there is evidence in these studies of an emerging consensus as to the use of key concepts and terminology.
- While it is convenient to consider climate impact assessment methods as ordered by classes of impact or interaction models, even the simplest of impact models use some underlying concept of climate–society interaction. The best-defined climate impact relationships are relative, rather than

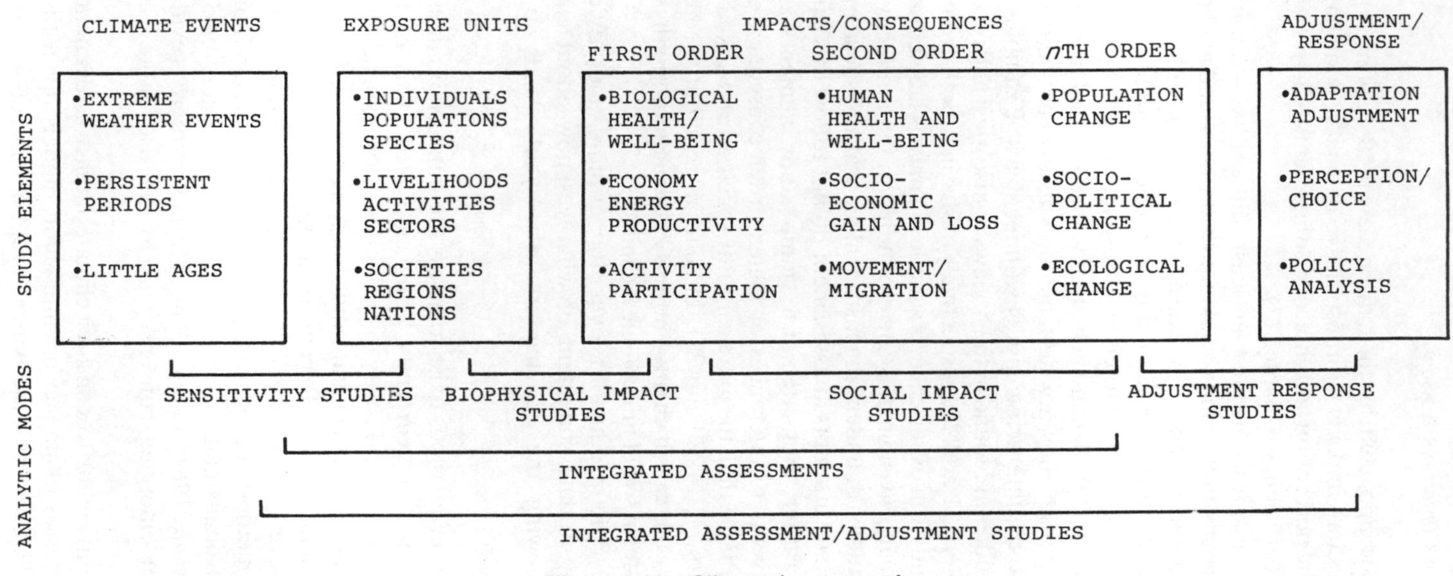

Figure 1.11 Climate impact study

absolute, appropriate to a given technology, social organization, group vulnerability, and the like. Invoking climate as a sole determinant of human events is rarely if ever justified. Scientific studies of climate impacts always need to consider alternative, joint and multiple hypotheses for postulated impacts.

- The largest single impact of climate is the human effort expended to adjust or adapt to the seasonality, variability and change of climate. Most case studies are negatively biased, excessively preoccupied with the residual damages and losses experienced by society, and ignore the significant costs of human adjustment. In practice, low residual losses or damages may be a sign of relative insensitivity to climate, or conversely, of very great sensitivity accompanied by effective, successful, albeit costly, human adaptation. Concepts and methods to elicit loss or damage are better developed than concepts and methods that assess the social cost of adaptation. Least developed are concepts to assess and utilize the opportunities presented by climate as a resource.
- There is a great disparity in both methods and effort of climate impact assessment between the industrialized and the developing nations. Developing nations are clearly more vulnerable to climatic perturbations and suffer greater damages and extraordinarily greater losses of life, yet the scientific effort in identifying and mitigating climate impacts is not centered where it is needed most. Methods do exist, however, to identify the major climate sensitivities of all societies, and current knowledge of climate variability and potential human adjustment can serve to diminish the vulnerability of developing nations.

If there is a single conclusion to this volume, it is that more is known about the interaction of climate and society than is utilized and much of what remains to be known is discernible. Scientists in all parts of the world who recognize the practical necessity of informed adjustment to climate are invited to take up the challenge of integrated assessment, and to share in the intellectual adventure of developing theories and methods on the borders of current thought.

REFERENCES

Bowden, M. J., Kates, R. W., Kay, P. A., Riebsame, W. E., Warrick, R. A., Johnson, D. L., Gould, H. A., and Weiner, D. (1981). The effect of climate fluctuations on human populations: Two hypotheses. In Wigley, T. M. L., Ingram, M. J., and Farmer, G. (Eds.) *Climate and History*, pp. 497–513. Cambridge University Press, Cambridge, UK.

Burton, I., Kates, R. W., and White, G. F. (1978). *The Environment as Hazard*. Oxford University Press, New York.

Changnon, S. A., Jr., Huff, F. A., Schickedanz, P. T., and Vogel, J. L. (1977). *Summary of Metromex. Vol. 1: Anomalies and Impacts*. Illinois State Water Survey Bulletin 62, Urbana, Illinois.

Enzer, S., Drobnick, R., and Alter, S. (1978). *Neither Feast Nor Famine*. Lexington Books, Lexington, Massachusetts.

Flohn, H. (1979). Short-term climate fluctuations and their impact. Introduction at *International Conference on Climate and History, 8–14 July, 1979, University of East Anglia, Norwich, UK*.

Friesema, H. P., Caporaso, J., Goldstein, J., Lineberry, R., and McCleary, R. (1979). *Aftermath: Communities after Natural Disaster*. Sage Publications, Beverly Hills, California.

García, R. (1981). *Drought and Man: The 1972 Case History. Vol. 1: Nature Pleads Not Guilty*. Pergamon, New York.

García, R., and Escudero, J. C. (1982). *Drought and Man: The 1972 Case History. Vol. 2: The Constant Catastrophe: Malnutrition, Famines, and Drought*. Pergamon, New York.

Grobecker, A. V., Coroniti, S. C., and Cannon, R. H., Jr. (1974). *The Effects of Stratospheric Pollution by Aircraft*. CIAP Report of Findings, DOT-TST-75-50. US Department of Transportation, Washington, DC.

Holling, C. S. (1973). Resilience and stability in ecological systems. *Annual Review of Ecology and Systematics*, **4**, 1–22.

Ingram, M. J., Farmer, G., and Wigley, T. M. L. (1981). Past climates and their impact on man: A review. In Wigley, T. M. L., Ingram, M. J., and Farmer, G. (Eds.) *Climate and History*, 3–50. Cambridge University Press, Cambridge, UK.

International Meteorological Institute (1984). Progress report for Problem Area IV: Impacts on managed and unmanaged terrestrial ecosystems. *Project on International Assessment of the Impact of an Increased Atmospheric Concentration of Carbon Dioxide on the Environment*. Draft, 11 April 1984. IMI, Stockholm.

Johnson, D., and Gould, H. (1984). Effects of climate fluctuation on human populations: Study of Mesopotamian society. In Biswas, A. K. (Ed.) *Climate and Development*. Tycooly International Publishing, Dublin.

Johnson, S. R., and McQuigg, S. (1974). Some useful approaches to the measurement of economic relationships which include climatic variables. In Taylor, A. S. (Ed.) *Climatic Resources and Economic Activity*. John Wiley & Sons, New York.

Karl, T. R., and Riebsame, W. E. (1984). The identification of 10- to 20-year temperature and precipitation fluctuations in the contiguous United States. *Journal of Climate and Applied Meteorology*, 23 (6 June), 950–966.

Kates, R. W. (1981). Drought in the Sahel: Competing views as to what really happened in 1910–14 and 1968–74. *Mazingira*, **5**, 72–83.

Kates, R. W., Changnon, S. A., Jr., Karl, T. R., Riebsame, W., and Easterling, W. E. (1984). *The Climate Impact, Perception, and Adjustment Experiment (CLIMPAX): A Proposal for Collaborative Research*. Climate and Society Research Group, Center for Technology, Environment, and Development, Clark University, Worcester, Massachusetts.

Kellogg, W. W., and Schware, R. (1981). *Climate Change and Society: Consequences of Increasing Atmospheric Carbon Dioxide*. Westview Press, Boulder, Colorado.

Kukla, G. L., Angell, J. K., Korshover, J., Dronia, H., Hoshiai, M., Namias, J., Rodewald, M., Yamamoto, R., and Iwashima, T. (1977). New data on climatic trends. *Nature*, **270**, 573–580.

Lamb, H. H. (1978). *Climate: Present, Past, and Future. Vol. 2: Climatic History and the Future*. Methuen, London.

Lave, L. B. (1981). *Mitigating Strategies for CO_2 Problems*. CP-81-14. International Institute for Applied Systems Analysis, Laxenburg, Austria.

Lawson, M. P., and Stockton, C. (1981) The desert myth evaluated in the context of climatic change. In Smith, C. D., and Parry, M. L. (Eds.) *Consequences of Climatic Change*, pp. 106–118. Department of Geography, University of Nottingham, UK.

Lemon, E. R. (Ed.) (1983). *CO_2 and Plants: The Response of Plants to Changing Levels of Carbon Dioxide*. AAAS Selected Symposium No. 84. Westview Press, Boulder, Colorado.

Leontiev, W. (1977). *The Future of the World Economy: A United Nations Study*. Oxford University Press, New York.

Linnemann, H., Deltooge, J., Kayser, M., and Van Heemst, H. (1979). *MOIRA: Model of International Relations in Agriculture*. North Holland Publishing Co., Amsterdam.

Liverman, Diana M. (1983). *The Use of a Simulation Model in Assessing the Impacts of Climate on the World Food System*. University of California, Los Angeles-National Center for Atmospheric Research Cooperative Thesis No. 77. NCAR, Boulder, Colorado.

Mesarovic, M. D., and Pestel, E. (1974). *Mankind at the Turning Point*. E. P. Dutton, New York.

Meyer-Abich, Klaus M. (1980). Chalk on the white wall? On the transformation of climatological facts into political facts. In Ausubel, J., and Biswas, A. K. (Eds.) *Climatic Constraints and Human Activities*, pp. 61–92. Pergamon, New York.

National Defense University (1980). *Crop Yields and Climate Change to the Year 2000*. NDU, Fort Lesley J. McNair, Washington, DC.

National Defense University (1983). *World Grain Economy and Climate Change to the Year 2000: Implications for Policy*. NDU, Fort Lesley J. McNair, Washington, DC.

National Research Council (1983). Board on Atmospheric Sciences and Climate. *Changing Climate* (Report of the Carbon Dioxide Assessment Committee). National Academy Press, Washington, DC.

Nordhaus, W. D., and Yohe, G. W. (1983). Future paths of energy and carbon dioxide transmissions. In National Research Council, *Changing Climate* (Report of the Carbon Dioxide Assessment Committee), pp. 87–153. National Academy Press, Washington, DC.

Parry, M. L. (1978). *Climatic Change, Agriculture and Settlement*. Archon Books, Hamden, Connecticut.

Parry, M. L. (1981). Evaluating the impact of climatic change. In Smith, C. D., and Parry, M. L. (Eds.) *Consequences of Climatic Change*, pp. 3–16. Department of Geography, University of Nottingham, UK.

Parry, M. L., and Carter, T. (1983). *Assessing Impacts of Climatic Change in Marginal Areas: The Search for Appropriate Methodology*. IIASA Working Paper WP-83-77. International Institute for Applied Systems Analysis, Laxenburg, Austria.

Porter, P. W. (1976). Climate and agriculture in East Africa. In Knight, C. G., and Newman, J. L. (Eds.) *Contemporary Africa: Geography and Change*, pp. 112–139. Prentice-Hall, Englewood Cliffs, New Jersey.

Post, John D. (1977). *The Last Great Subsistence Crisis in the Western World*. The Johns Hopkins University Press, Baltimore, Maryland.

Riebsame, W. E. (1981). *Adjustments to Drought in the Spring Wheat Area of North Dakota: A Case Study of Climate Impacts on Agriculture*. PhD Dissertation, Department of Geography, Clark University, Worcester, Massachusetts.

Rosenof, T. (1973). *Cultural Sensitivity to Environmental Change: The Case of Ellis County, Kansas, 1870s–1900*. Institute for Environmental Studies, Institute for Humanities (Report 5), University of Wisconsin, Madison, Wisconsin.

Santer, B. (1984). Impacts on the agricultural sector. In Meinl, H., and Bach, W., Socioeconomic Impacts of Climatic Changes Due to a Doubling of Atmospheric CO_2

Content. Contract No. CL1-063-D, Commission of the European Communities, Brussels.
Schelling, T. C. (1983). Climatic change: Implications for welfare and policy. In National Research Council, *Changing Climate* (Report of the Carbon Dioxide Assessment Committee), pp. 449–482. National Academy Press, Washington, DC.
SCOPE (1978). *Report of the Workshop on Climate/Society Interface*, held 10–14 December, 1978 in Toronto. SCOPE Secretariat, Paris.
Seifert, W. W., and Kamrany, N. M. (1974). *A Framework for Evaluating Long-term Strategies for the Development of the Sahel-Sudan Region. Vol. 1: Summary Report: Project Objectives, Methodologies, and Major Findings*. MIT Center for Policy Alternatives, Cambridge, Massachusetts.
Slovic, P., Kunreuther, H., and White, G. F. (1974). Decision processes, rationality, and adjustment to natural hazards. In White, G. F. (Ed.) *Natural Hazards: Local, National and Global*, pp. 187–205. Oxford University Press, New York.
Spitz, P. (1980) Drought and self-provisioning. In Ausubel, J., and Biswas, A. K. (Eds.) *Climatic Constraints and Human Activities*, pp. 125–147. Pergamon, New York.
Timmerman, P. (1981). *Vulnerability, Resilience and the Collapse of Society*. Environmental Monograph No. 1. Institute for Environmental Studies, University of Toronto, Canada.
US Council on Environmental Quality (1980). *The Global 2000 Report to the President*. US Government Printing Office, Washington, DC.
US Department of Agriculture (1978). *World GOL Model Analytic Report*. Foreign Agriculture Economic Report No. 146. USDA, Washington, DC.
Waggoner, P. E. (1983). Agriculture and a climate changed by more carbon dioxide. In National Research Council, *Changing Climate* (Report of the Carbon Dioxide Assessment Committee), pp. 383–418. National Academy Press, Washington, DC.
Walters, S. D. (1970). *Water for Larsa: An Old Babylonian Archive Dealing with Irrigation*. Yale University Press, New Haven, Connecticut.
Warrick, R. A. (1980). Drought in the Great Plains: A case study of research on climate and society in the USA. In Ausubel, J., and Biswas, A. K. (Eds.) *Climatic Constraints and Human Activities*, pp. 93–123. Pergamon, New York.
Warrick, R. A., and Bowden, M. J. (1981). The changing impacts of drought in the Great Plains. In Lawson, M. P., and Baker, M. E. (Eds.) *The Great Plains: Perspectives and Prospects*. Center for Great Plains Studies, University of Nebraska, Lincoln, Nebraska.
White, G. F., and Haas, J. E. (1975). *Assessment of Research on Natural Hazards*. MIT Press, Cambridge, Massachusetts.
Wigley, T. M. L., Ingram, M. J., and Farmer, G. (Eds.) (1981). *Climate and History*. Cambridge University Press, Cambridge, UK.
Wright, J. D., Rossi, P. H., Wright, S. R., and Weber-Burslin, E. (1979). *After the Clean-up: Long Range Effects of Natural Disasters*. Sage Publications, Beverly Hills, California.

Climate Impact Assessment
Edited by R. W. Kates, J. H. Ausubel and M. Berberian

CHAPTER 2
Climatic Variability and Change

F. KENNETH HARE

Trinity College in the University of Toronto
6 Hoskin Avenue, Toronto, Ontario M5S 1H8
Canada

2.1 INTRODUCTION

Climate has been a neglected factor in world affairs (Fedorov, 1979; White, 1979; Hare, 1981a; Hare and Sewell, 1984). In most people's thinking climate is a vague idea, without much power to evoke reaction. As long as it performs reliably it can be ignored. Yet the 1970s showed that national economies and world trade can be shaken by extreme climates, and that our defences against such rude episodes are flimsy. The problem is to articulate the nature of climate's impact on society. And that implies an understanding of the nature of climate itself.

Climate relates to many sectors of the human economy, and to many aspects of the world's natural ecosystems. A broad concept of climate is needed for such an exercise as the World Climate Programme. This chapter seeks that broad concept, and tries to set out the facts about climate itself as an environmental factor. To assist the reader, words with a defined technical meaning have been italicized the first time they appear in the text.

Human societies are highly adapted to the ordinary *annual* rhythms of climate, which we call the seasons. Housing, clothing, health measures and recreational habits are adjusted to make good use of the seasonal changes, or to cope with their rigors. So also are most forms of technology—above all agriculture, forestry and transportation. The annual variation of the weather elements is familiar to all; it is taken for granted, and is in some ways a resource to be tapped.

Failure of the expected seasonal changes, however, often leads to dislocation, discomfort and economic loss. The bulk of what is loosely called climatic impact probably comes from disturbances of the expected annual rhythms. Strictly speaking these impacts belong to the *interannual* scale; they arise from differences between the actual and the expected, or between one year and the next. Almost all climates display these differences, which have hitherto appeared unpredictable. Wise economic strategy takes them into account essentially by built-in safety margins. But few economies do so as effectively as possible. Much of this volume has to do with this imperfect adjustment, and asks the question: can it and should it be improved?

The *decadal* scale of climate presents much the same paradox. Socioeconomic adaptation to climate is high, but implicit. On this scale there is time for significant technological change—new varieties of crop, new systems of power transmission, and many others—though here again it is the *interdecadal* changes that matter most. Climatic anomalies on such scales are usually not large, but there are noteworthy exceptions. Some decades, for example, may be marked by high variability; others may remain monotonously reliable. Climatic fluctuations lasting several decades are not uncommon, and are often strongly regional in their effects. They, too, are as yet unpredictable, and may remain so indefinitely.

At the scale of the *century* one is in the realm of sweeping technological change. Human resources with which to exploit or withstand climatic behavior can be expected to change significantly, especially in advanced industrial societies. Century-long fluctuations of climate may occur. Their impacts are long-enough enduring for societies to become partially adapted to them—and for the anomaly to be thought a permanent change.

This chapter is concerned not with the adjustments, but with the nature and behavior of climate itself. One cannot expect to learn much about the impact of climate if climate itself remains an unexplored and undervalued idea. But what is relevant to society alters as society innovates. One must seek to understand climate in the light of what mankind can perceive, understand and adapt to.

2.2 DEFINITIONS OF CLIMATE

The word climate is often seen as ambiguous, because of confusion with related words like weather, or vague climatic conditions. In fact it is capable of fairly exact definition. In recent years, moreover, public usage seems to have moved closer to that of the professional.

In lay usage, climate usually stands for the expectation of weather on time-scales comparable with a human lifetime; it is the layman's sense of the sequence of weather he or she may expect at a given locality (see Whyte, Chapter 16). As such it governs countless daily decisions, from choice of personal clothing to the work calendar of the farmer. Habitually this expectation is taken for granted. Only when unexpected weather occurs does the ordinary citizen become acutely aware of the stresses that the atmosphere's behavior can bring—sometimes to the point where the question is asked (as it is below): 'is the climate changing?' This question arises from the suspicion that recent weather lies outside normal expectation, that is, outside the present climate.

Professional usage is still erratic, but usually starts with the same idea of expectation. It is assumed

1. that a climate exists at any given moment; and
2. that the integrated experience of the recent past specifies this climate.

We tend to assume that the near future will resemble this recent past. For lack of any better guide we accept a dictum attributed to Whitehead: how the past perishes is how the future becomes. We think it highly probable that past experience will repeat itself. In effect, we assume that the climate of today will endure for an undefined period.

In Figure 2.1 are sketched idealized time-series of a representative physical parameter, such as air temperature, or atmospheric humidity. Curves A to D illustrate modes of *variation* (i.e., change with time) typical of such series:

1. *Curve A* displays strictly periodic variation about a mean value, or *central tendency*, with a well-defined period of recurrence, and finite amplitude. Variation of this sort is rare in atmospheric time-series, except in association with daily and annual solar rhythms (themselves almost exact). Simple statistical analysis can, however, identify these periodic signals, if they exist, in the midst of large non-cyclical variation (i.e., noise).
2. *Curve B* shows the short-term variation typical of atmospheric time-series. Temperature, for example, varies in reality on all time-scales from fractions of a second to millennia or more. The high-frequency variation in curve B is distributed about a central tendency that changes almost impulsively in a short period to a new régime, within which *quasi-periodic variation* occurs. All such variation can be assigned, by means of spectral

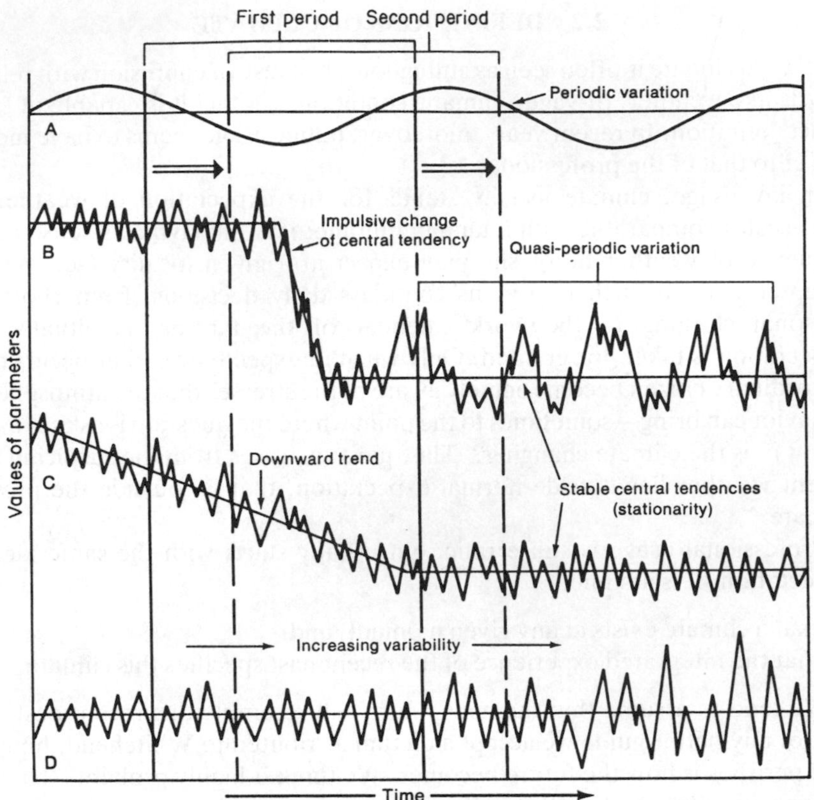

Figure 2.1 Idealized time-series (curves A to D) of a representative parameter of a climatic element that is continuous in time (such as temperature or pressure). Vertical bars indicate arbitrary averaging or integrating periods (usually 30 years) which are recalculated each decade (see dashed bars)

analysis, to exact periodic frequencies, but to do this is often artificial and misleading; hence the use of the adjective quasi-periodic.

3. *Curve C* shows a series in which a downward *trend* occurs up to the midpoint; afterwards the series is *stationary*. The short-period variations appear unchanged throughout.
4. *Curve D* displays a constant central tendency, but short-term variations appear to increase in amplitude as time progresses.

Figure 2.1 implies that the parameter chosen is continuous in time, as are temperature and pressure (which are also continuous in space). In fact many climatic parameters are discrete in both realms. Rainfall is an example. Others include cloud amount and type and some aspects of atmospheric electricity. To handle such parameters climatologically we usually sum or average them over time (for example, monthly or annual rainfall) and sometimes over space as well (as with runoff); the discreteness then usually vanishes.

Formally to specify a climate, the climatologist assembles time-series of the necessary parameters (see below), and then seeks to generalize them statistically over a suitable period—the choice of which is largely arbitrary, since nature offers no clue as to correct length. Recent practice, as standardized by the World Meteorological Organization, is to use 30-year periods, which are updated every decade: 1951–80 *normals* (the misleading term in use) have recently replaced the 1941–70 set. In Figure 2.1 the vertical bars through the diagram, labelled first and second periods, indicate this procedure. In practice (though not in the hypothetical cases in Figure 2.1) the use of 30-year normals gives fairly stable values of central tendency. Differences between successive averaging periods are small.

2.3 CLIMATIC NOISE, VARIABILITY AND CHANGE

This arbitrary but necessary procedure makes possible a formal definition of the terms *climatic noise*, *variability* and *change*:

1. *Climatic noise* may be defined as that part of the variance of climate attributable to short-term weather changes. The chance location of the start and end of the averaging periods with respect to short-term variations may produce small statistical differences between successive periods. Leith (1975) confines the term climatic noise to this usage. Others use it more loosely to include, for example, anomalous weather extremes, or all short-term variation.
2. *Climatic variability* is best thought of as the manner of variation of the climatic parameters *within* the typical averaging period. Suitable measures of variability include such parameters as the standard deviation of continuous elements like temperature and pressure, and the frequency spectrum of the observed variations (giving such useful measures as return-period, or the probability of successive anomalous months or years). A proper statement of the climate of an averaging period should hence include, in addition to measures of central tendency (means, medians or totals), suitable estimates of the variability.
3. *Climatic change* is said to occur when the differences between successive averaging periods exceed what noise can account for, i.e., when a distinct signal exists that is visible above the noise. A common usage is to refer to short-term changes lasting only a few decades as *climatic fluctuations*, especially if conditions then return to the earlier state. Many authorities reserve the term change for longer-term variation extending over centuries (such as the Little Ice Age, or the mid-Holocene desiccation of much of the

subtropical world). All such usage depends on the time-scale employed as standard.

Figure 2.2 shows estimated annual rainfall over England and Wales from 1766 until 1970 (after Wigley *et al.*, 1984). Clearly annual rainfall is a variable element. Within any 30-year period there are likely to be yearly totals as high as 1000 mm, or as low as 800 mm. In the 205 years of the record, rainfall exceeded 1150 mm five times and fell below 700 mm five times. The time-series shows no significant trend over the period of record, and little autocorrelation (that is, rainfall in a given year is unrelated to that of the years before and after). There is obviously high variability, and the mean value obtained for any arbitrary 30-year period is likely to differ from that derived from another choice. In the long perspective, however, the record justifies the generalization that annual mean rainfall over England and Wales, though very variable from year to year, has not changed greatly over the period.

Figure 2.3 shows a similar estimate (after Parthasarathy and Mooley, 1978) for spatially averaged rainfall over monsoon India. Again there is no significant trend over the entire period 1865–1970. If, however, one inspects the curve starting in 1900 and ending in 1960, a distinct upward trend can be seen. This was a vital climatic fluctuation that had a real bearing on India's capacity to increase food production in pace with her population explosion. The abrupt downward sag in the mid-1960s was an unwelcome surprise. The record shows a tendency for wet or dry years to follow one another, and also a small but statistically significant 2½- to 3-year periodicity. Here again, however, the most central fact is that no lasting change in mean annual rainfall can be demonstrated. It is obviously dangerous to base long-term conclusions on short-term records—or to assume that trends will continue indefinitely.

These two records illustrate another general principle. *Any trends that may be present in a climatic time-series, and especially any persistent climatic changes, are usually very small by comparison with short-term variations*. The latter obscure the signal corresponding to the change, just as radio noise may obscure the desired radio signal. It is not yet easy, for example, to detect the signal corresponding to the expected effects of carbon dioxide increase on the trend of air temperature (Madden and Ramanathan, 1980; Wigley and Jones, 1981), because short-term variations dominate the record.

Public recognition of these facts is very partial. A sequence of wet or dry years, as displayed in Figure 2.3, often raises false hopes or fears, and leads to speculation about lasting change. Institutional memory of severe extremes tends to fade quickly, but the public at large retains a recollection of events like cold winters or dry summers.

Complete uniformity of usage of the terms defined in this section is not to be expected. They are not dictated by the properties of nature, but respond to our need for a language to describe the atmosphere's steady state and modes of variation. Terms like event, episode and anomaly are used qualitatively by

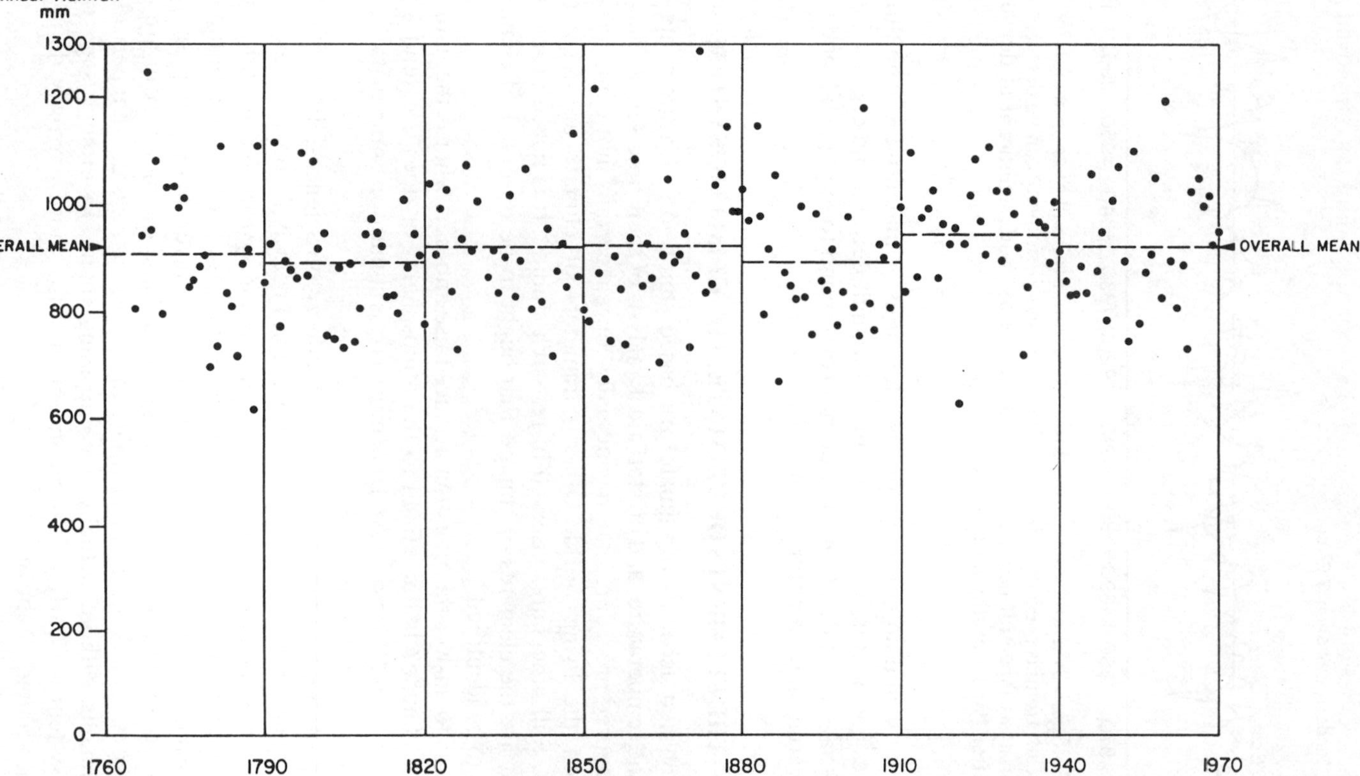

Figure 2.2 Spatially averaged and standardized annual rainfall for 1766–1970 over England and Wales, showing a typical, highly variable climatic element. No significant trend or periodicity is present, but very large interannual differences are common. Differences in variability between periods are also visible. (After Wigley *et al.*, 1984)

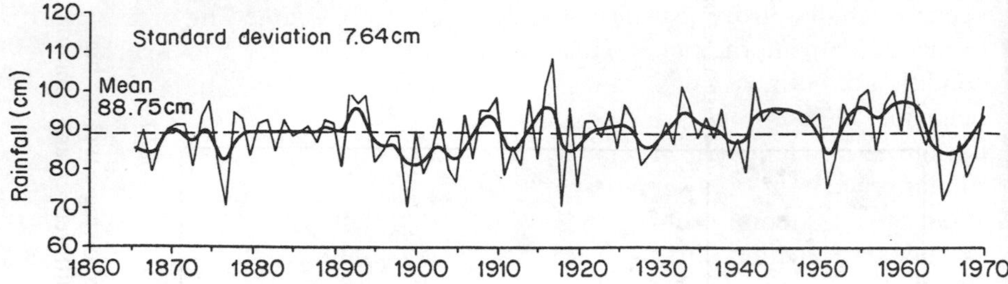

Figure 2.3 Spatially averaged and standardized summer rainfall over monsoon India. No significant trend is present, but a weak 2½- to 3-year periodicity can be detected by spectral analysis (spectral peaks at 2.72 and 2.85 years, both significant at the 90% level). (After Parthasarathy and Mooley, 1978)

most workers, and do not imply precise durations or magnitudes. The three time-scales given attention in this volume—annual, decadal and century-long periods—have exact durations, but do not correspond to any special mode of climatic variation (except for the annual seasons).

2.4 THE ELEMENTS OF CLIMATE; THE CLIMATIC SYSTEM

Most published accounts of climate are based simply on measurements of rainfall and temperature, and on statistical analysis of their time-series. This is a time-honored procedure that may actually conceal certain subtle impacts of climate on the living world. Other parameters of climate are regularly observed. Still others are observed in research studies. It is hence proper to ask: what are the elements of climate that ought to be observed, if climatic impact is to be identified?

Attempts to answer this question at once raise another: what is the natural domain of climate? Is it confined to the atmosphere, or does it extend into other environmental sectors? And one must ask: is climate an entity in itself, or is it part of a larger system?

Some professional opinion (National Academy of Sciences, 1975; World Meteorological Organization, 1975, for example) heavily favors the view that the atmospheric climate is part of a larger *climatic system* that includes the domain within which there are significant exchanges of mass, momentum or energy during the typical year. In this view the upper layers of the ocean, sea ice (and to lesser extent land ice, or glaciers), the soil and the biotic cover are all parts of the system. Alternatively climate can be seen as a system that pervades these domains without actually encompassing them. In either usage the climatic system is clearly a larger concept than the more narrowly defined climate of Section 2.2 above.

In practice it is still usual to confine the idea of climate to the atmosphere and ocean (including frozen surfaces), which are closely coupled by processes of exchange and interaction. Similar parameters can be used to express the condition of both domains, and similar bodies of theory exist to explain their behavior. Meteorologists and physical oceanographers are drawn from similar backgrounds. Hence they readily understand one another, and tend to adopt similar procedures of observation and compilation.

On environmental grounds, however, the concept of climate needs to be extended to include surface hydrology, soil processes and the systems of exchange between the atmosphere and the biota. Hydrology presents few problems, since its practices are fairly close to those of the atmospheric scientists; elements such as precipitation and evapotranspiration are actually common to both scientific areas. Soil moisture, carbon and nitrogen storage, and their exchanges with the atmosphere are also vital, but are not often included in the idea of climate. The rôle of the biota presents the most difficulty; bioclimatology is a specialized field that has not yet yielded measures that can be systematically built into a continuous monitoring of the climatic system. Nevertheless it is highly desirable, but not operationally usual, to include suitable bioclimatic parameters in the list of measured and analyzed climatic elements. Examples might include the phenology of crops and natural vegetation, transfers of heat, moisture and carbon dioxide within crop layers, and data concerning biological productivity (which is related to water and energy incomes).

It is possible, however, to get much closer to the reality of climate–society interaction by analysis of the actual processes of linkage. One can distinguish between human physiological linkages, the field of medical climatology (Weihe, 1979; Flach, 1981) and indirect linkages involving natural vegetation, crops, surface waters, the seas and their fish populations. In all cases the linkages involve actual transfers that can, in principle, be measured. The main linkages can be simply classified:

1. *The natural energy system*, dominated by the flux of solar energy. This system includes the fluxes of solar, terrestrial and atmospheric radiation, the convective heat fluxes (latent and sensible) in the atmosphere, the fluxes of heat in the soil and ocean, and the complex conversions of energy between one form and another—including the internal transformations between potential, thermal and kinetic forms within the atmosphere and ocean. All these fluxes and transformations are in principle measurable, but few are actually monitored. Instead, the usual practice is to measure the temperature of the air, water or soil. Temperature is a parameter of heat storage, though its gradient (i.e., direction and rate of spatial variation) is a determinant of heat transfer. Preference for measurements of temperature, rather than of energy fluxes, arose two centuries ago because thermometers were cheap and easy to use—and because meteorologists and oceanogra-

phers set great store by the so-called parameters of state (temperature, pressure and density). Climatic impact nevertheless depends on modifications of the fluxes of energy, as well as on the levels of energy storage.

2. *The hydrologic cycle*, which describes the exchanges of water substance between sea, air, soil, rock, plants and animals. In this case it is usual to measure precipitation and river discharge, both of which are transfer mechanisms. Evapotranspiration is also measured, but often with too little attempt to simulate its actual behavior in nature. Again the choice of measured parameter has been influenced by cost: rain gauges, evaporation pans and stream gauging points are cheap and easy to install and operate. Storage parameters such as soil water storage have had less attention. Water is made available to crops and vegetation from stored water, for which we have fewer statistics. Here again the choice of measured element has obscured the study of climate–biota–economy interactions.
3. *The carbon cycle* describes the exchange of carbon between biotic, atmospheric, soil, crustal and oceanic reservoirs. Only gradually has it been realized that this cycle is of fundamental importance to climate, and is indeed part of the climatic system itself. Only one climatic element has yet been recognized—the concentration of carbon dioxide in the atmosphere, continuously monitored since 1957 at a handful of stations. Levels of carbon storage in reservoirs other than the atmosphere can only be approximated, and none of the transfer mechanisms is routinely measured. Understanding is growing, but adequate detail is still lacking (Olson, 1982; Woodwell, 1982).
4. *The other biogeochemical cycles* describe the exchange systems for nitrogen, sulfur and a few others where there are significant atmospheric storages and transfer pathways (Bolin, 1979). This includes the transfer of recently added pollutants, such as the chlorofluoromethanes and other halocarbons, which may have significant climatic and health effects. Few if any of these cycles are monitored sufficiently well to include them in analyses of climatic impact. Sufficient detail is also lacking concerning the particulate and aerosol load of the atmosphere, which affects visibility and turbidity (or ability to transmit solar radiation), and may have a long-term effect on surface temperatures.

It is obvious that the analysis of climate–society interaction must currently be based on what is actually observed, and especially on those elements for which we have long records, and hence some idea of natural background levels and variability. Only in certain specific research areas is it sometimes possible to go beyond this limitation. Gradually, as the nature of climatic impact becomes clearer, the observational habits of the climatologist will change. Meanwhile one must use what is available.

2.5 IS PRESENT-DAY CLIMATE CHANGING?

2.5.1 Public Perception

There is a widespread belief that climate is unstable, and may be undergoing change. This is not a new impression. Literary sources show that speculation about climatic change is a venerable human tradition. Yet the changes in world climate during mankind's period of literacy have been small by comparison with those of Pleistocene times (Budyko, 1977). What accounts for the persistence of what looks, in hindsight, like an exaggeration of these changes?

Public speculation about such issues has been hampered by a failure to distinguish between local and global factors in the control of climate. Major land use changes, which are highly visible, have significant impacts on local radiative balances, and to some extent on temperature. Their visibility has led to speculation that still greater changes may alter local or even global climate. Actually, however, global processes are on such a scale that they are not easily affected. It is the general circulation of the atmosphere and the planetary scale of energy exchanges that ultimately control non-seasonal temperature and humidity variations even at local sites, via the mechanism of advection. Precipitation, too, comes largely from systems controlled by global rather than local processes. Nevertheless popular apprehensions persist, perhaps with some reason, that human manipulation of the earth must lead to climatic change, perhaps of a hostile kind.

Present-day speculation seems to spring from certain recognizable sources, such as popular memory of climatic extremes, widespread media comment on the subject, gloomy statements by professional climatologists (or by other scientists), and real fluctuations of temperature during the past century. In particular, certain specific changes seem to have impressed political opinion:

1. The decrease in northern hemisphere surface air temperature between 1938 and 1965 (see Figure 2.4), which was of an order of 0.5 °C. This decrease has sometimes been announced by the media as the onset of a new glacial period. In actual fact the decrease has reversed since 1965, and is in any case strongly regional, affecting some areas much more than others (Van Loon and Williams, 1976–77; Kukla *et al.*, 1977; Jones *et al.*, 1982). A parallel downward trend of sea-surface temperature has been documented (Kukla *et al.*, 1977). Neither sea nor air temperature shows the same tendency in the southern hemisphere; in fact a reverse movement has been claimed.
2. The major west-European drought of 1975–76, which affected densely populated countries with vital agricultural areas. The drought was the most severe in more than two centuries of record in some districts, where rainfall over the crop season was as little as 40 percent of normal. The drought was

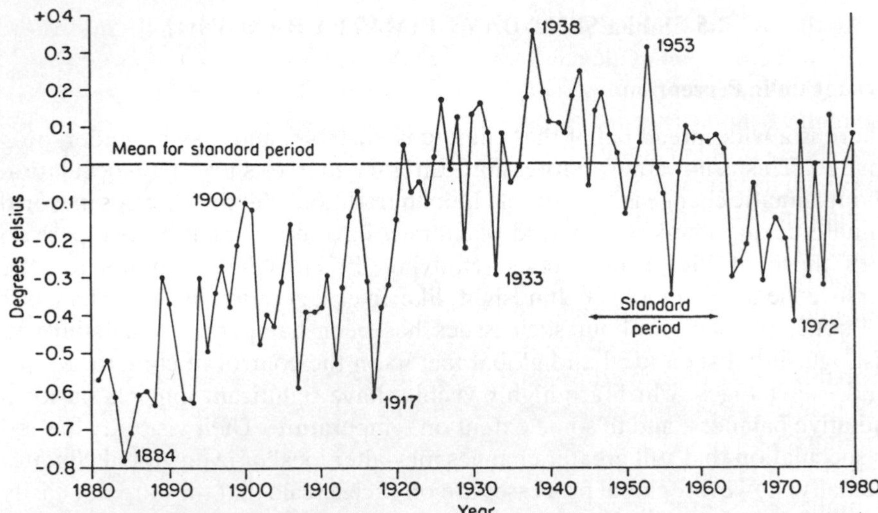

Figure 2.4 Variation of surface air temperature for northern hemisphere since 1881, to show the recent fluctuation. (After Jones *et al.*, 1982)

short by comparison with, for example, the Sahelian drought, but exposed the vulnerability of water-supply systems and crops to rainfall anomalies.

3. Severe winters in eastern North America during the middle 1970s, which badly stressed the energy delivery systems of the world's most extensively industrialized region. In some areas temperatures were more than 5 °C below normal over the entire 3-month season.
4. Disturbances during the 1970s of the world's food system due to quite small reductions of cereal supply caused by climatic anomalies occurring at a time of increasing use of cereal as livestock feed. The crop deficiencies in the Soviet Union in 1972, 1974, 1975, 1979, 1980 and 1981 were particularly significant because they led to large Soviet grain purchases on the world market, and to dramatic fluctuations of price.
5. The Sahelian drought, popularly supposed to have occurred in the years 1968–73, but actually a gradual desiccation of much of north and east sub-Saharan Africa that began in 1955–60, an effect that has continued until the present. Because it imperilled many fledgling African states this desiccation achieved high visibility, leading in 1977 to the UN Conference on Desertification, the first world political conference about a climatically related stress.

2.5.2 Professional Opinion

It is a common professional view that items 2 to 5 above lie within the variability of the present climate; they ought, that is to say, to have been expected by prudent societies that kept an eye on the climatic record.

At the World Climate Conference (World Meteorological Organization, 1979) in Geneva, no evidence was brought forward from any part of the world to suggest that present climate is in the process of change, although it displays the high variability normal to its behavior. The most serious impacts on society have been brought about by short and drastic anomalies of the sort listed above. The time-series of Figure 2.3 emphasize, for example, that in drought-prone monsoon India the rainfall anomalies are chiefly short but large departures from a basically stable régime.

The point is often made, however, that lasting climatic change, if it occurs, will mimic short fluctuations. We may not recognize the change until it has been in progress for decades. A decrease in mean rainfall, for example, will probably be accompanied by a tendency towards more frequent dry years. It will be extremely difficult to recognize climatic change in time to do anything about it.

Just as there is little evidence for significant changes of mean temperature, rainfall or most other measured elements, there is also little evidence that their variability is changing significantly. Many investigations show that variability does in fact fluctuate in magnitude and character. Thompson (1975), Haigh (1977), McQuigg (1979) and others have suggested that reduced rainfall variation in the US Midwest in the late 1950s and 1960s was responsible for part of the observed increase in corn yields in that period. Haigh's analysis does indeed demonstrate that an upward trend of mean July precipitation and a downward trend of mean August temperature (both favoring good corn yield) affected the state of Illinois between 1930 and 1976, and that variability of rainfall was markedly lower in the middle of that period. But there is no consensus that overall variability of climate has recently changed in any large region of the earth. Such variability is related in most regions to persistent anomalies of the general circulation of the atmosphere. The analysis by Van Loon and Wiliams (1976–77) cited above shows that such anomalies can have effects on temperature over 15-year averaging periods. What is lacking is any conclusive demonstration:

1. that variability of climate on a world scale has recently changed; and
2. that there is any real connection between rises or falls of mean air temperature and climatic variability, as is often claimed (Angell and Korshover, 1978; Ratcliffe *et al.*, 1978; Van Loon and Williams, 1979).

2.5.3 The Search for Predictability

Clearly the impact of climatic variability on all time-scales will depend on the ability to predict individual variations with sufficient skill to make precautionary measures possible. Unfortunately there is a professional consensus that such predictions are as yet rarely possible, even under favorable circumstances. Present day short-term numerical weather prediction models are effective only up to 4- or 5-day periods, except in areas where weather changes are slow (for example in the tropical deserts). Increases in computer capability are slowly improving this situation, but the cost of even a day's extension in reliable forecasts is very high. Moreover the nature of weather systems makes it doubtful *in principle* whether such an approach can be pushed very far. Tentative foreshadowing on the monthly or seasonal scale can be based on analogue methods, on assumed sun-spot or orbital variations, and on assumed relationships with sea-surface temperature anomalies. Some of these attempts achieve better than chance results, but none has yet reached the point where management decisions can be firmly based on the outcome.

The search for such predictability is the major objective of the research component of the World Climate Programme. It is premature to prejudge the results of the large-scale effort now being launched, but the following generalizations appear sound:

1. Attempts to model future climatic variation resulting from the build-up of carbon dioxide and other infra-red absorbers yield predictions of *trend*, on the order of 0.1 °C a^{-1} or less. Though this is not negligible, it has relevance mainly for long-range planning of such things as plant breeding, navigation in ice-infested waters, and dam construction. The models do not predict future variability with any assurance.
2. The best hope for seasonal or interannual prediction may well rest on two existing bodies of active research: sea-surface temperature anomaly to atmosphere relationships, and possible large-scale oscillations in atmosphere–ocean behavior. Among the latter the most promising is the so-called *southern oscillation*.

The southern oscillation has been known, though not understood, for 60 years, having served for many years as the basis for statistical forecasting of Indian monsoon rainfall. It consists of a large longitudinal quasi-periodic oscillation of atmospheric pressure, representing gigantic transfers of mass in east–west directions between the South Atlantic and South Pacific at one pole, and a broad area between the Red Sea, East Africa and Australasia at the other. Pressure is negatively correlated between these two large areas, but unfortunately the period of oscillation varies. The average period is 38 months, but individual oscillations have varied between 2 and 10 years (see Philander, 1983, for a recent review under the auspices of the Commission for the Atmospheric Sciences of the World Meteorological Organization).

The southern oscillation is highly correlated with sea temperature behavior over the equatorial Pacific. Shortly after the onset of the atmospheric event (a fall of pressure in the western Pacific, and a slackening of the southeast trades) the normal cold upwelling off the Peruvian coast is weakened, and much warmer water appears nearshore, which badly affects the highly productive marine ecosystem—the so-called *El Niño* event. In the typical case the warmest water shifts westwards along the equatorial belt of the Pacific, and within 18 months cold upwelling is re-established off Peru. The dramatic effect of these events on fisheries has given them high visibility in recent decades. (See Chapter 6 for further discussion of the El Niño event.)

This entire phenomenon is a vast longitudinal perturbation of atmosphere and ocean that has hitherto eluded adequate theoretical understanding. The World Climate Research Programme will give such understanding very high priority. The phenomenon seems to be meaningfully correlated with climatic variation outside the directly affected regions—for example with air pressure and temperature over North America. Unfortunately the regularities are in most cases insufficient for successful forecasting. There is first the irregular period and erratic behavior of the southern oscillation events, and second the weakness of the observed correlations. Only theoretical understanding and effective modelling can serve to reduce these uncertainties and make reliable prediction possible.

Predictability may well come in the future, for at least some of these phenomena. Meanwhile, there is abundant and conclusive evidence that variability of the main climatic elements is a normal and necessary part of nearly all climates. This variability extends to significant fluctuations over several decades. But most long climatic time-series show an underlying stability, suggesting that the variability is distributed around a fundamentally unchanging normal state. Investigations have failed to show that changes in variability tend to occur as mean temperatures rise and fall. They have also failed to confirm truly periodic anomalies. The recurrence of drought somewhere on the Great Plains of the United States in the 1890s, 1910s, 1930s, 1950s, and, to a limited degree, the 1970s, for example, is not seen as truly periodic (see, for example, Stockton and Meko, 1983). Such quasi-periodic happenings are probably misleading as to future climate, as evidenced in the unrealized expectation of widespread drought on the Great Plains in the 1970s.

2.6 IS FUTURE CLIMATIC CHANGE LIKELY?

If present and recent past climates show little sign of change, why is there so much concern about the possibility of climatic disruptions of the economy?

Two answers are clearly possible, both of which may be correct. One is that human society is becoming more vulnerable to climatic impact as population grows, as the demand for food increases, and as other pressures on the food and energy systems increase. This view is widely debated in other chapters of this volume (see Chapters 9, 10). The other answer says that processes are now at work that will upset the stability of present climate. Conceivably we may be able to predict the resulting changes, so that they will not catch the world unawares.

The second view is widely held among climatologists. Since the major disruptions of the international grain trade following the crop year 1972–73, much research has been conducted to test the stability of present climate. This has consisted of one-, two- and three-dimensional modelling exercises representing the present climate as a function of present-day controls, and examination of the models' sensitivity to hypothetical changes in these controls. The sensitivities tested include the dependence of world temperature and precipitation on:

1. variations in solar energy input (the solar constant);
2. variations in atmospheric optical properties due to altered carbon dioxide concentration, the presence of pollutants such as the chlorofluoromethanes, and higher levels of particle load due to human activity or vulcanism; and
3. changes in air–ocean interactions, or in the extent of glaciation.

On the time-scales of importance here—interannual or interdecadal—only the second kind of disturbance seems likely to be effective. In any case these turn out to be similar to those that might follow a very small change in the solar constant (Manabe and Wetherald, 1980). There have been many attempts to establish air–sea interactions by empirical methods, and some of these have been effective (for example, Namias, 1976; Hastenrath, 1978). But these appear to relate most closely to fairly brief fluctuations of climate. Specific patterns of sea temperature anomaly are found to be connected with air temperature or precipitation anomalies elsewhere (teleconnections), occasionally with some useful time lag that may make prediction possible. As such the sensitivities of type 3 above relate more to the internal variability of existing climate than to lasting climatic change. For the latter to occur it is now fairly well agreed that changes in external forcing* are necessary.

2.6.1 Carbon Dioxide

The change in forcing seen most likely to occur is that to be expected from the build-up of carbon dioxide (CO_2) in the atmosphere (see, for example, Kellogg and Schware, 1981; Hare, 1981b). Figure 2.5 shows the observed change of CO_2 concentration in the atmosphere since serious monitoring began in 1957–58. The observational network is now worldwide, and it is clear that the build-up is a planetary process. The 1983 concentration was near 342 ppmv (parts per million by volume), equivalent to 724×10^9 metric tons of elemental carbon in the atmospheric reservoir (World Climate Programme, 1981 and subsequent data

*'Forcing' is useful jargon for an influence or factor whose behavior is external to the model itself: it 'forces' the model, but is not thereby affected.

in Clark, 1982). The rate of increase varies from 0.5 to 2.2 ppmv a^{-1}, averaging recently near 1.2 ppmv a^{-1}. In all, a little under 3×10^9 metric tons of carbon are being added to the atmospheric reservoir each year. The source is probably chiefly fossil fuel consumption, currently a little below 6×10^9 metric tons a^{-1}, which implies an atmospheric retention rate of a little over 50 percent. The balance is disappearing into some other reservoir, probably the surface and intermediate waters of the ocean. Atmospheric mixing is so efficient that both additions to and subtractions from the atmospheric store can be thought of as global processes; the CO_2 build-up is not, like acid rain, a local or regional phenomenon.

Such a rapid increase of CO_2 necessarily changes the optical properties of the atmosphere. CO_2 is largely transparent to solar radiation, but it strongly absorbs the return long-wave terrestrial radiation near 15 μm, a wavelength at which water vapor is ineffective as an absorber. The effect of CO_2 build-up is thus to increase the resistance to the upward radiative transfer of heat in the atmosphere. To achieve the necessary global balance between incoming solar and outgoing terrestrial radiation the earth's surface must hence become a little warmer, and the stratosphere a little cooler. This is the *greenhouse effect* so extensively discussed in the popular press.

It seems clear that the build-up will continue, since its presumed causes—fossil fuel consumption, probably reinforced by the effects of forest clearance and soil impoverishment—are unlikely to abate. Estimates by Häfele (1978) and his associates at the International Institute for Applied Systems Analysis suggest for the year 2030 a total world power consumption at the rate of $18–36 \times 10^{12}$ Watt (terawatt = TW). Rotty and Marland (1980) give a similar estimate of 27 TW in 2025, a figure which allows for economic restraints to rapid expansion. If correct, this represents a growth from present consumption (8.2 TW) of only 2 percent per annum. Recent data show that between 1950 and 1973 fossil fuel consumption grew at 4.58 percent per annum; since then it has risen at only 2.25 percent (Rotty, 1983).

Rotty and Marland (1980) estimate that about three-quarters of the 2025 power demand will be met by fossil fuel burning—a level of CO_2 release almost three times the present value. Allowing for a 25 percent error on either side, they estimate a doubling of present CO_2 levels (i.e., from 342 to 684 ppmv) by 2060 at the earliest, and most probably after 2075. An expert World Climate Programme panel chaired by Bolin (World Climate Programme, 1981) concluded that in 2025 the atmospheric concentration will actually be in the range 410 to 490 ppmv, with 450 ppmv the most likely value. This is clearly far below earlier estimates, some of which visualized a doubling of present values by 2020 or 2030. Nevertheless it is vital to remember that the use of fossil fuels may proceed more rapidly than these estimates suggest, and the build-up of CO_2 with it.

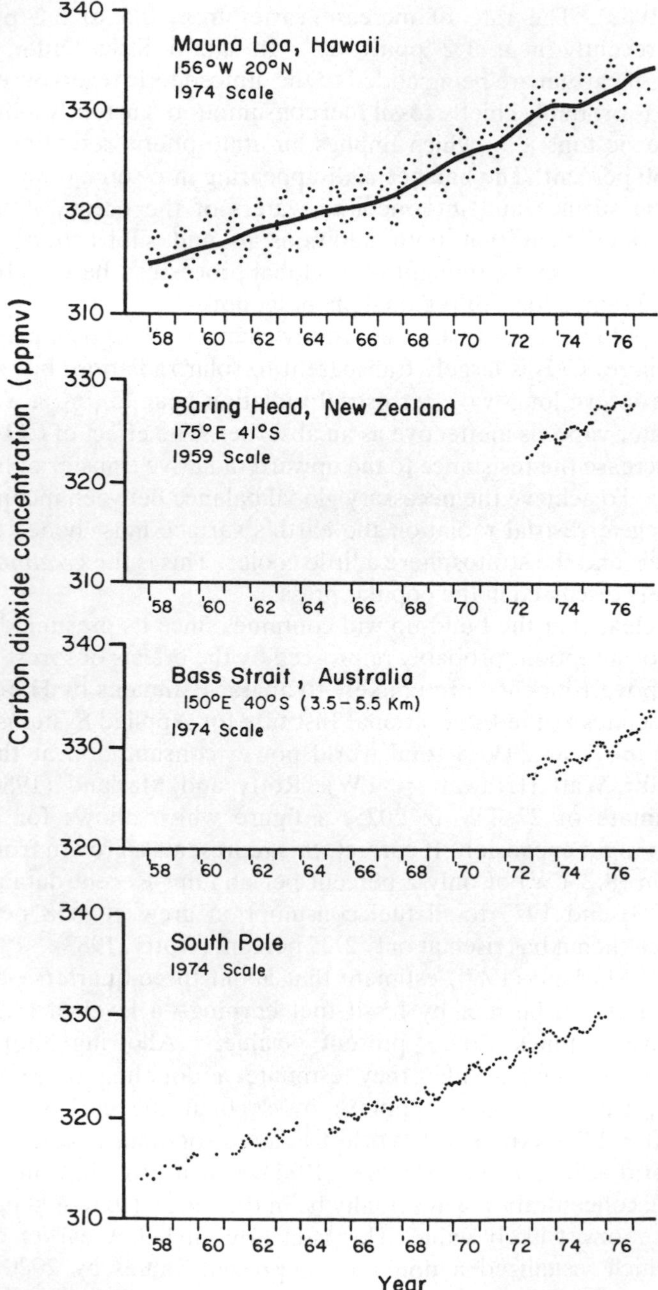
Mauna Loa, Hawaii
156°W 20°N
1974 Scale
Baring Head, New Zealand
175°E 41°S
1959 Scale
Bass Strait, Australia
150°E 40°S (3.5 – 5.5 Km)
1974 Scale
South Pole
1974 Scale
Carbon dioxide concentration (ppmv)
340
330
320
310
'58 '60 '62 '64 '66 '68 '70 '72 '74 '76
Year

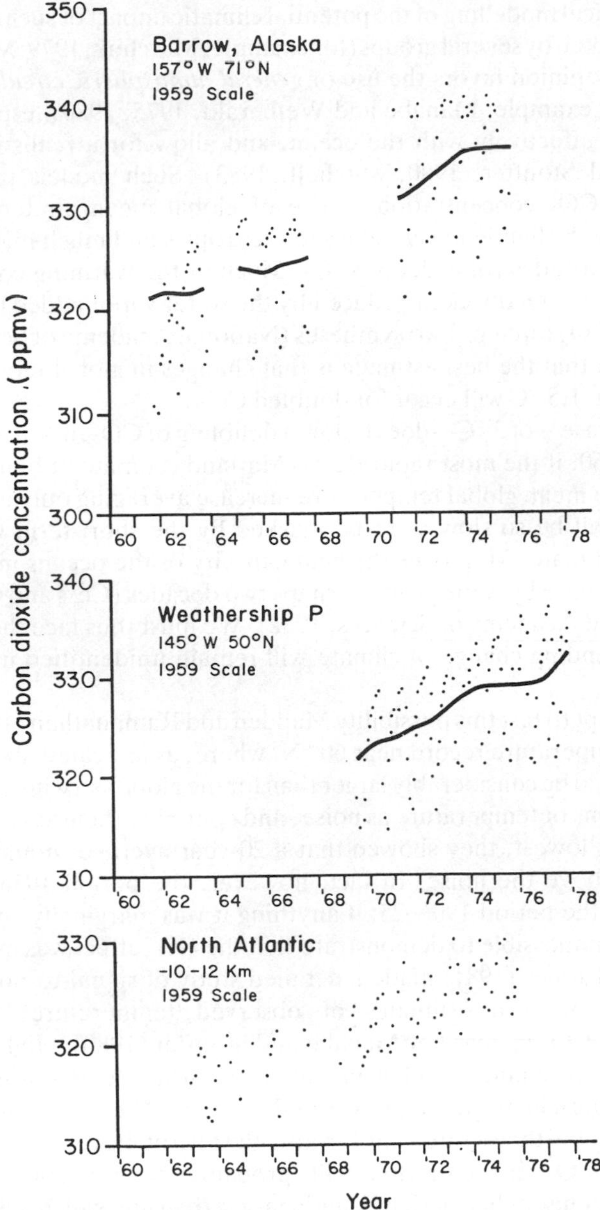

Figure 2.5 Variation of CO_2 concentration (parts per million by volume) at selected stations since 1958. Two slightly different calibration scales are used, but the upward trend is unaffected thereby. A regular seasonal variation is also visible with maximum amplitude in northern hemispheric high latitudes. (Data supplied by World Meteorological Organization)

Mathematical modelling of the potential climatic impact of such increases has been undertaken by several groups (for example, Marchuk, 1979; Mason, 1979). Most expert opinion favors the use of *general atmospheric circulation models* (GCMs) (for example, Manabe and Wetherald, 1975, 1980), especially those that interact effectively with the ocean, and allow for a realistic geography (Manabe and Stouffer, 1980; Mitchell, 1983). Such models predict, for a doubling of CO_2 concentration, a rise of global mean air temperature of 2.0–3.5 °C, with slightly lower values in the tropics and much higher values in north polar latitudes (of order 5–8 °C). Much of the warming comes, in fact, from the positive feedback introduced by the water vapor added to the air as a result of the CO_2 forcing. Two syntheses (National Academy of Sciences, 1979, 1982) suggest that the best estimate is that changes in global temperatures on the order 3° ± 1.5 °C will occur for doubled CO_2.

If this increase—of 3 °C—does follow a doubling of CO_2, it will occur at some date after 2060, if the most rapid Rotty–Marland estimate of build-up occurs. This implies a mean global temperature increase averaging only about 0.04 °C a^{-1}, which will be so slow as to be masked by the short-term variability of the present climate. Moreover the heat capacity of the oceans may delay the expected response by some years, perhaps two decades (Cess and Goldenberg, 1981; National Academy of Sciences, 1982). We must thus face the probability that the impending change of climate will remain unidentified until it is well advanced.

In an attempt to test this possibility Madden and Ramanathan (1980) decided to test the temperature record near 60 °N, where, as indicated above, the CO_2 warming should be considerably larger than for the globe as a whole. Treating all other variations of temperature as noise, and confining themselves to summer, when noise is lowest, they showed that a 20-year average should display the CO_2 signal above the noise. In fact, however, the period 1956–75 was *not* warmer than the period 1906–25; if anything it was marginally colder. Hence they found it impossible to demonstrate that the CO_2 effect was in progress.

Wigley and Jones (1981) made a detailed study of signal-to-noise ratios by comparing their own estimates of observed temperatures with model predictions of CO_2 increase by Manabe and Stouffer (1980). Highest values of the ratio occur in summer, and in annual mean surface temperatures averaged over the northern hemisphere or over midlatitudes. The effects of CO_2...'may not be detectable,' they wrote, 'until around the turn of the century. By this time, atmospheric CO_2 concentration will probably have become sufficiently high...that a climatic change significantly larger than any which has occurred in the past century could be unavoidable.' (Wigley and Jones, 1981, 208).

On the other hand, it is widely held in the Soviet Union that the signal is already visible (Budyko, 1977; Budyko and and Efimova, 1981; Vinnikov and Groisman, 1981). There have also been model simulations incorporating the assumed CO_2 warming with other influences that account quite well for climatic variation over the past century (Hansen *et al.*, 1981; Gilliland, 1982). To assess such contradictions, what is needed is agreement on the statistical tests to be applied (Epstein, 1982) and on rigorous monitoring of the actual unfolding of events in the next few decades (Weller *et al.*, 1983). Given the high importance of the outcome, major effort needs to be expended on such monitoring.

Assuming the CO_2 effect to be real, and to be roughly of the character foreshadowed in this section, we can speculate as to further climatic consequences. The major effect may be on the hydrologic cycle. There have been two complementary approaches to this question:

1. Some of the GCM experiments (notably Manabe and Wetherald, 1980 and Manabe and Stouffer, 1980) show a small overall increase in world precipitation for a doubled or quadrupled CO_2 concentration. A marked decrease in available soil water is foreseen, however, for a midlatitude belt, chiefly because of increased evapotranspiration. The resolution of the models does not allow any statement as to which longitudes (i.e., actual regions) may be affected. World climate is always longitudinally differentiated.
2. In an attempt to provide such longitudinal detail, Kellogg and Schware (1981) compiled a world map of possible soil moisture patterns on a warmer earth. The map synthesized (a) geological experience of distant past epochs, notably the mid-Holocene Altithermal, (b) recent experience of hot and cold years in the northern hemisphere, and (c) the modeling exercises referred to above. The map showed drier conditions over interior North America, Siberia, southeastern South America, the tropical rainforest areas and the polar zones. Wetter conditions occurred in most subtropical countries and west coast temperate areas.

Such foreshadowings show the vital importance in climatic impact studies of the CO_2 effect, acting indirectly via the effect of differential atmospheric heating on the hydrologic cycle. So far the model estimates have withstood a number of objections raised against them (National Academy of Sciences, 1982).

Rising CO_2 concentrations may also have an effect on polar snow and ice distribution. Snow and ice surfaces have a high albedo. If general warming causes their equatorward limit to recede poleward, a significant increase in the absorption of solar radiation occurs, which intensifies the heating. This albedo feedback mechanism is built into the models described above, and accounts for the markedly greater effect of CO_2 heating in high northern latitudes. In the southern hemisphere the feedback is reduced by the regular seasonal break-up of the Antarctic pack-ice (Robock, 1980).

There have been many attempts to estimate the amount of heating needed to break up the permanent pack-ice of the Arctic Ocean, an enormous ice surface that has probably persisted for the past 700,000 years or more. If this ice were removed, or if it were confined to the winter and spring seasons, immense consequences would ensue. Quite apart from the obvious strategic and economic importance of the removal (Hare, 1981b) there is the fact that removal of the polar ice would imply a major reordering of world climate; the present-day pack-ice is a key controlling factor of the earth's radiative balance and general circulation of the atmosphere and oceans.

Manabe and Stouffer (1980) have estimated from their recent GCM experiments that a doubling of CO_2 would fail to remove the pack-ice in summer, but a quadrupling of CO_2 (probable in the twenty-second century) would do so. Thereafter in the average year there would be winter ice only, as at present in Hudson's Bay and on much of the Southern Ocean's Antarctic margin. If this estimate is correct, the next century is likely to see a shrinking effect on the continental snow surfaces, and a trimming of the margins of the polar pack. There would be no clearance of the Arctic Ocean itself, though the ice would certainly become thinner. This view is opposed to that of Parkinson and Kellogg (1979) who found that a doubling of CO_2 concentration would indeed remove pack-ice in summer. Clearly there is a need to bring the assumptions and techniques of the modelers closer together.

Another point concerns the possible effect of a CO_2-induced warming on glacial ice, and hence on sea level. Contrary to public speculation a doubling of CO_2 is unlikely to produce drastic short-term effects on sea level (Barnett, 1982). The predicted temperature rises will probably reduce the extent of alpine glaciers. Sea levels have recently been rising globally at rates generally estimated in the range of 10–15 cm per century, due partly to this process, and partly to expansion of the ocean water. But the main storage of ice on land is in the great continental glaciers of Antarctica and Greenland. The ice cover of Antarctica is very old, and losses of ice are almost entirely by means of ice discharge into the sea, a process that is not very temperature-sensitive at present temperature level. Greenland loses ice half by thermally induced ablation and stream flow and half by ice discharge, mostly through a single ice stream. A considerable increase of the marginal ablation would be likely to follow the CO_2 warming. On the other hand snowfall on the upper and middle slopes of both ice sheets would probably increase, thus altering the entire mass balance. During the next two centuries it seems unlikely that a large net transfer of water to the sea will occur, though the present slow rise will probably accelerate slowly. In the longer term the threat of significant sea level change from net glacial melting remains present. On the scale of many centuries it may well become important.

It is conceivable that catastrophic changes may affect the so-called West Antarctic ice sheet (the part of the sheet north of the constriction between the Weddell and Ross Seas). Much of this ice is afloat (the Ronne and Ross ice shelves) or rests on a rock surface well below present sea level. Mercer (1978), enlarging on a speculation of Hughes (1973), suggested that the West Antarctic ice might disintegrate if a CO_2 warming becomes pronounced. Most glaciologists seem to question this opinion (World Climate Programme, 1981), but agree that should the ice disintegrate sea level would rise by about 5 m worldwide. The last stages of the disintegration might occur within a century (Thomas *et al.*, 1979). Catastrophic surges of the main East Antarctic ice sheet have also been hinted at. Hollin and Barry (1979) have pointed out that such surges are *mechanical* phenomena; only if CO_2 warming leads to increase of snowfall on the ice sheet, and if a warmer ocean trims back the margin of the ice sheet, is a climatically induced surge likely to occur. We conclude that catastrophic rises of sea level due to transfers of glacial ice to the ocean are unlikely, but not impossible. Within the next century it is improbable that major sea level changes will be a significant outcome of the warming.

2.6.2 Other Mechanisms

What other mechanisms may contribute to climatic change in the next two centuries? There is general agreement in the research community that the CO_2 effect will be the most significant influence, even though estimates of its magnitude differ widely. But several other mechanisms have been proposed, some of which may add to the CO_2 effect, and some reduce it. The list includes the effect of changes in the particle and aerosol concentration in the atmosphere, the effect of other infrared absorbing gases, and the possible effect of flip mechanisms in the atmospheric general circulation (the so-called *almost-intransitivity* effect). This last effect is treated in Section 2.7.

There has been much speculation that variable concentrations of atmospheric particles may induce changes of surface temperatures. The primary source of effective particles, chiefly sulphates in the lower stratosphere, is considered to be explosive volcanic eruptions. Distinct coolings were, for example, observed in the wake of the eruptions of Tambora (1815) and Krakatoa (1883). The 1982 eruption of El Chichón in Mexico has yet to show its effect, although the resulting stratospheric dust cloud has been widely observed.

There has been much debate as to the reality of the particle cooling. It has been argued theoretically (with some observational support) that in some circumstances—perhaps dominant—the net effect of aerosol increase may be lower tropospheric and surface warming (Budyko and Vinnikov, 1973, 1976; Chylek and Coakley, 1974; Kellogg *et al.*, 1975). Unfortunately, most of the parameters needed to settle the argument have yet to be routinely observed.

In two papers Bryson and Goodman (1980a,b) calculated atmospheric turbidity since 1880 from measured attenuation of the direct solar beam. Their results showed a prolonged period (1920–60) of low turbidity, corresponding rather closely to a reduced number of large explosive eruptions. They showed a close relationship between northern hemispheric temperature and the turbidity; a clear atmosphere favored high surface temperature. The post-1938 cooling, in this view, resulted from the resumption of more active explosive eruptions. Their model even predicted the upturn since 1965. A rather different analysis by Schneider and Mass (1975) also found that the 1920–40 upturn was explicable in terms of dust loading, but that a reduced solar constant (due to sunspot activity) since 1950 may have contributed to the downturn.

Other recent attempts to include the effect of volcanic dust on global temperature have been undertaken to identify the modifying effect of such dust on the CO_2 signal calculated from models. Hansen *et al.* (1981) estimated the volcanic effect from Lamb's (1970) dust veil index, whereas Gilliland (1982) preferred to use the record of acidity from Greenland ice cores. When the volcanic effect was empirically weighted (in Gilliland's analysis), and variations in solar luminosity were included, a close fit between the CO_2, solar and volcanic forcings and the observed variation of northern hemisphere temperature was obtained.

A factor that appears to reinforce the CO_2 heating is the presence in the atmosphere of numerous other gases that have infrared absorption bands in the 'window' region—the band between 7 and 14 μm wavelength in which the atmosphere is largely transparent to terrestrial radiation. Carbon dioxide and ozone both reduce the transparency in this band, and hence tend to raise surface temperatures. The effects are incorporated into the various models described above, as is the effect of H_2O. But there are other active gases present in currently negligible amounts that are being increased by human activity. Wang *et al.* (1976) showed that a doubling of the concentrations of nitrous oxide (N_2O), methane (CH_4) and ammonia (NH_3) could increase surface temperatures by about 1.1 °C, which is about half the expected CO_2 warming. The extensive use of agricultural fertilizers may indeed lead to increases in the abundance of some of these gases.

More recent synthesis (Chamberlain *et al.*, 1982) has extended the list of such gases, and confirmed their importance by comparison with the effect of CO_2 acting alone. Hansen *et al.* (1981) calculated the equilibrium warming that increases in the chief absorbers may have contributed between 1970 and 1980, using a model that predicted a 3 °C rise for doubled CO_2. They found that CO_2 may have raised global surface temperature by 0.14 °C in this decade, and the other absorbers (methane, nitrous oxide and two chlorofluoromethanes) by 0.10 °C. If this were representative, and if the other absorbers were to increase at significant rates in the next century, the CO_2 effect would clearly be enlarged. Unfortunately we only speculate as to such continued increases for the other absorbers.

2.7 IS SURPRISE POSSIBLE?

The preceding sections showed that the present and future states of the climatic system are only approximately known. There are grounds for supposing, as just shown, that rising CO_2 and trace gas concentrations in the atmosphere may tend to raise global surface temperatures in the next 80 years at an average (but non-linear) rate of 0.06 °C a^{-1}, with maximum effect in high northern latitudes. It seems likely that this warming will affect precipitation and evaporation amounts. The uncertainties in the estimates have been deliberately stressed. Two questions now arise. Does such a change matter to humanity? And are surprise changes possible?

Seen in the perspective of the observed behavior of climate over the past century the above rate of change of temperature is impressive. Between 1880 and 1940 northern hemisphere temperature rose at about 0.01 °C a^{-1}, during the last two decades of the period the rate being about 0.02 °C a^{-1}. From 1938 until 1965 it fell at a comparable rate. The combined heating rate postulated for CO_2 and other infrared absorbers will eventually be much greater. These changes will all be perceived, in a highly variable climate, as changes in the incidence of extremes. A rapid change will get rapt attention, once it begins in earnest. On the other hand human societies have repeatedly absorbed even more rapid local or regional changes. Wittwer (1980) pointed out, for example, that Indiana in the United States had seen a rise of temperature of 2 °C in the past century, and that from 1915 to 1945 the increase was at the rate of 0.1 °C a^{-1}. (See Kates *et al.*, 1984, for other examples of climatic fluctuations in specific climate divisions.)

Changes of climate due to the gradual accumulation in the atmosphere of pollutants, or growth in the concentration of natural constituents like CO_2, are likely to lead to progressive but irregular changes of the sort just described. But more impulsive changes are common in nature, and these may present greater problems for human adaptation. These impulsive changes are often visible in local or regional precipitation records. Habitually they last only a few years, but in a few cases may extend to some decades. There are theoretical grounds for supposing (Lorenz, 1968, 1975) that the general circulation of the atmosphere may flip from one quasi-stable mode of behavior to another with little change of external forcing—although the argument does not justify the conclusion that local flips are of such origin.

Remarkable instances of such flips on the regional scale have recently affected many parts of the world's desert margins, where climate has been highly perturbed in recent decades (Hare, 1977; Charney *et al.*, 1979). In the 1950s, for example, lake levels in East Africa (including Lake Chad) were low, as they had been since the beginning of the century. Early in the 1960s—1961 in the case of Victoria Nyanza, and hence the headwaters of the White

Nile—there occurred an abrupt increase in rainfall over some of the plateaus. Some lake levels rose abruptly to levels not seen since the nineteenth century (Grove *et al.*, 1975; Nicholson, 1980). There was no warning of this increase. High rainfalls persisted for several years in some areas, assisting in the spread of agriculture in the East African highlands (Oguntoyinbo and Odingo, 1979). Desiccation returned after almost a decade of abundance, which misled many African statesmen into believing that a lasting change had occurred.

This East African anomaly was accompanied by a downward trend of rainfall along much of the southern margin of the Sahara. It, too, began in the early 1960s, and even earlier in some West African localities (Bunting *et al.*, 1976; Nicholson, 1980) where, however, rainfall was well above average at the outset. Between 1968 and 1973 the desiccation was of such intensity that the fluctuation became known as the Sahelian drought. In fact it was part of a two- to three-decade-long fluctuation of rainfall over all equatorial Africa (Kraus, 1977; Motha *et al.*, 1980), of which the high rainfall phase of the early 1960s was a part. Fluctuations like this are probably internal to the climatic system. Such unpredicted variations pose a threat of major disruptions to the human economy over large parts of the world.

2.8 URBAN EFFECTS

The foregoing account has stressed general effects that are visible over the entire earth, or over extensive regions. On these large scales climate has an obvious bearing on such activities as agriculture, forestry, transportation and fisheries. One scale has been missed, however: that of the city.

Within the built-up area of the world mankind has contrived, especially within the past century and a half, considerably to influence the local climates. These effects arise from the intensive use of fuels within urban areas, which releases heat and pollutants; from the substitution of concrete, tarmac and other hard, impervious surfaces for natural vegetation and soil; and from interference with the energy and hydrologic cycles. Urban climatology has come of age, and there are now many comprehensive surveys of what has been learned (see, for example, Chandler, 1970; Oke, 1974, 1979; Landsberg, 1981).

Although the anthropogenic modification of urban climates obviously touches the lives of millions of people, the processes involved are subtle and specialized, and are beyond the scope of this chapter. Reference should be made to the reviews cited for details.

It is important to note, however, that the climate of the larger cities may already have been modified by these effects as much as the carbon dioxide build-up is likely to achieve on the global scale for at least half a century. The so-called *urban heat island*, for example, has raised mean annual temperatures in the centers of the large cities, in some cases by over 2 °C, above surrounding open areas. Substantial increases in precipitation over and downwind of cities have also been detected (see Pilgrim, Chapter 13).

2.9 CONCLUSION

What should be the strategy of a project that seeks to understand the impact of variable climate on changing society? Does any clear imperative emerge from this review of the nature of the variation?

The distinction made in the account between internal variability and climatic change is of considerable practical importance. Internal variability is distinguished by two main characteristics: it is as yet largely unpredictable, and may remain so indefinitely; but it is readily identifiable in the climatic record. In other words experience should be the guide to action. Climate was defined above as the expectation of weather. The point was also made that variability is part of the climate itself—and is in many ways the key part for the human economy. Proper measures of variability, and especially of extremes, are hence vital to the description of a climate. Above all the measures must include estimates of the probable duration and frequency of the sort of extremes that cause stress—and the need for adaptive measures—in the economic system.

As the account has shown, climate looks, in the longer perspective, like a very stable system. Since the major changes in precipitation distribution in mid-Holocene times (~ 5000 years ago), climate has been subject only to fairly small changes. If one trusts the crude analyses now available—and this may not be wise—world temperatures have fluctuated within a range less than ± 2 °C, and world rainfall by less than about ± 5 percent. Along the climatic margins—the cold and dry limits for agriculture, for example—these small changes (like the Little Ice Age) have caused considerable stress, but humanity has been well able to cope with them. In fact most professional atmospheric scientists assume that this basic stability will continue; in the absence of any contrary evidence persistence gives the best forecast.

This comfortable assumption is no longer valid if the CO_2 effect and the complementary effect of other infrared absorbers are real. A progressive rise of temperature is predicted on a world basis, with much larger increases in north polar latitudes. Adjustments of precipitation and evaporation are also predicted. These represent changes in climate, but will probably be manifest as sequences of anomalous years, at least for a short time. The eventual rate of change anticipated is much greater than those associated with the fluctuations of the past century. It may be that their effects will be reduced or compounded by other anthropogenic factors—increased particle loads, altered hydrologic cycle, and others. Adaptation of the human economy to such changes ought to be feasible, since they have been anticipated; there is time to innovate. Whether mankind can act rationally and in concert about such questions is, however, open to doubt. The rest of this volume deals with matters that bear on this question.

In brief, the search for better adaptation to climate has two components: a better use of the existing experience of climatic variability, as given by the climatological record; and a search for adaptation to the climatic changes that now seem probable. The clear imperative is to get on with the search.

REFERENCES

Angell, J. K., and Korshover, J. (1978). Global temperature variation, surface-100 mb: An update into 1977. *Monthly Weather Review*, **106**, 755–770.

Barnett, T. P. (1982). *On Possible Changes in Global Sea Levels and Their Potential Causes*. Report TR-001, US Dept. of Energy, Washington, DC.

Bolin, B. (1979). Global ecology and man. In *Proceedings of the World Climate Conference*, pp. 88–111. Publication No. 537, World Meteorological Organization, Geneva.

Bryson, R. A., and Goodman, B. M. (1980a). Volcanic activity and climatic changes. *Science*, **207**, 1041–1044.

Bryson, R. A., and Goodman, B. M. (1980b). The climatic effect of explosive volcanic activity: Analysis of the historical data. MS paper, *Symposium and Workshop on Mount St. Helen's Eruption: Its Atmospheric Effects and Potential Climatic Impact, November 18–21, Washington, DC.*

Budyko, M. I. (1977). *Climatic Changes*. American Geophysical Union, Washington, DC.: 261 pages.

Budyko, M. I., and Efimova, N. A. (1981) The CO_2 effects of climate. *Meteorologiya i Hydrologiya*, **No. 2**, 1–10.

Budyko, M. I., and Vinnikov, K. Ya. (1973). Recent climatic changes. *Meteorologiya i Hydrologiya*, **No. 9**, 3–13.

Budyko, M. I., and Vinnikov, K. Ya. (1976). Global warming. *Meteorologiya i Hydrologiya*, **No. 7**, 16–26.

Bunting, A. H., Dennett, D. M., Elston, J., and Milford, J. R. (1976). Rainfall trends in the West African Sahel. *Quarterly Journal of the Royal Meteorological Society*, **102**, 59–64.

Cess, R. D., and Goldenberg, S. D. (1981). The effect of ocean heat capacity upon global warming due to increasing atmospheric carbon dioxide. *Journal of Geophysical Research*, **86**(C1), 498–502.

Chamberlain, J. W., Foley, H. M., MacDonald, G. J., and Ruderman, M. A. (1982). Climatic effects of minor constituents. In Clark, W. C. (1982), pp. 253–277.

Chandler, T. J. (1970). *Selected Bibliography on Urban Climate*. Publication No. 276, T.P. 155. World Meteorological Organization, Geneva: 383 pages.

Charney, J., Quirk, W. J., Chow, S. H., and Kornfield, J. (1977). A comparative study of the effects of albedo change on drought in semi-arid regions. *Journal of the Atmospheric Sciences*, **34**, 1366–1384.

Chylek, P., and Coakley, J. A., Jr. (1974). Aerosols and climate. *Science*, **183**, 75–77.

Clark, W. C. (Ed.) (1982). *Carbon Dioxide Review, 1982*. Oxford University Press, New York: 469 pages.

Epstein, E. S. (1982). Detecting climatic change. *Journal of Applied Meteorology*, **21**, 1172–1182.

Fedorov, E. K. (1979). Climatic change and human strategy. In *Proceedings of the World Climate Conference*, pp. 15–26. Publication No. 537, World Meteorological Organization, Geneva.

Flach, E. (1981). Human bioclimatology. In Landsberg, H. E. (Ed.) *World Survey of Climatology. Volume 3: General Climatology* **3**, 1–188. Elsevier Scientific Publishing Company, Amsterdam-Oxford-New York.

Gilliland, R. L. (1982). Solar, volcanic and CO_2 forcing of recent climatic changes. *Climatic Change*, **4**, 111–132.

Grove, A. T., Street, F. A., and Goudie, A. S. (1975). Former lake levels and climatic change in the Rift Valley of Southern Ethiopia. *Geographical Journal*, **141**, 177–201.

Häfele, W. (1978). A perspective on energy systems and carbon dioxide. In Williams, J. (Ed.) *Carbon Dioxide, Climate and Society*, pp. 21–34. IIASA Proceedings Series No. 1, Pergamon Press, Oxford.

Haigh, P. A. (1977). *Separating the Effects of Weather and Management on Crop Production*. MS Report to Charles F. Kettering Foundation, Dayton, Ohio: 93 pages.

Hansen, J., Johnson, D., Lacis, A., Lebedeff, S., Lee, P., Rind, D., and Russell, G. (1981). Climatic impact of increasing atmospheric carbon dioxide. *Science*, **213**, 957–966.

Hare, F. K. (1977). Climate and desertification. UN Conference on Desertification Secretariat (Eds.) *Desertification: Its Causes and Consequences*, pp. 63–120. Pergamon Press, Oxford.

Hare, F. K. (1981a). Climate: The neglected factor? *International Journal*, **36**, 371–387.

Hare, F. K. (1981b). Future climate and the Canadian economy. *Climatic Change Seminar Proceedings, Regina, Saskatchewan*, pp. 92–122. Atmospheric Environment Service, Downsview, Ontario.

Hare, F. K., and Sewell, W. R. D. (1985). Awareness of climate. In Burton, I. and Kates, R. W. (Eds.) *Geography, Resources and Environment*. Volume 2: *Essays in Honor of Gilbert F. White*. University of Chicago Press, Chicago, Illinois (forthcoming 1985).

Hastenrath, S. (1978). On modes of tropical circulation and climate anomalies. *Journal of the Atmospheric Sciences*, **35**, 2222–2231.

Hollin, J. T., and Barry, R. G. (1979). Empirical and theoretical evidence concerning the response of the earth's ice and snow cover to a global temperature increase. *Environment International*, **2**, 437–443.

Hughes, T. (1973). Is the West Antarctic ice sheet disintegrating? *Journal of Geophysical Research*, **78**, 7884–7910.

Jones, P. D., Wigley, T. M. L., and Kelly, P. M. (1982). Variations in surface air temperatures. Part 1, Northern Hemisphere, 1881–1980. *Monthly Weather Review*, **110**, 59–70.

Kates, R. W., Changnon, S. A., Jr., Karl, T. R., Riebsame, W., and Easterling, W. E. (1984). *The Climate Impact, Perception, and Adjustment Experiment (CLIMPAX): A Proposal for Collaborative Research*. Climate and Society Research Group, Center for Technology, Environment, and Development, Clark University, Worcester, Massachusetts.

Kellogg, W. W., Coakley, J. A., Jr., and Grams, G. W. (1975). Effect of anthropogenic aerosols on the global climate. In *Proceedings, WMO/IAMAP Symposium on Long-Term Climatic Fluctuations, Norwich, UK.*, pp. 323–330. Publication No. 421, World Meteorological Organization, Geneva.

Kellogg, W. W., and Schware, R. (1981). *Climate Change and Society: Consequences of Increasing Atmospheric Carbon Dioxide*. Westview Press, Boulder, Colorado: 178 pages.

Kraus, E. (1977). Subtropical droughts and cross-equatorial energy transports. *Monthly Weather Review*, **105**, 1009–1018.

Kukla, G. J., Angell, J. K., Korshover, J., Dronia, H., Hoshiai, M., Namias, J., Rodewald, M., Yamamoto, R., and Iwashima, T. (1977). New data on climatic trends. *Nature*, **270**, 573–580.

Lamb, H. H. (1970). Volcanic dust in the atmosphere; with a chronology and assessment of its meteorological significance. *Philosophical Transactions of the Royal Society of London*, **A266**, 425–533.

Landsberg, H. E. (1981). *The Urban Climate*. Academic Press, New York: 275 pages.

Leith, C. E. (1975). The design of a statistical-dynamical climate model and statistical constraints on the predictability of climate. In World Meteorological Organization (1975), pp. 137–141.

Lorenz, E. (1968). Climatic determinism. *Meteorological Monographs*, **8**(30), 1–3.

Lorenz, E. (1975). Climatic predictability. In World Meteorological Organization (1975), pp. 132–136.

Madden, R. A., and Ramanathan, V. (1980). Detecting climatic change due to increasing carbon dioxide. *Science*, **209**, 763–768.

Manabe, S., and Stouffer, R. J. (1980). Sensitivity of a global climate model to an increase of CO_2 concentration in the atmosphere. *Journal of Geophysical Research*, **85**(C10), 5529–5554.

Manabe, S., and Wetherald, R. T. (1975). The effect of doubling the CO_2 concentration on the climate of a general circulation model. *Journal of the Atmospheric Sciences*, **32**, 3–15.

Manabe, S., and Wetherald, R. T. (1980). On the distribution of climatic change resulting from an increase in CO_2 content of the atmosphere. *Journal of the Atmospheric Sciences*, **37**, 99–118.

Marchuk, G. I. (1979). Modelling of climatic changes and the problem of long-range weather forecasting. In *Proceedings of the World Climate Conference*, pp. 132–153. Publication No. 537, World Meteorological Organization, Geneva.

Mason, B. J. (1979). Some results of climate experiments with numerical models. In *Proceedings of the World Climate Conference*, pp. 210–242. Publication No. 537, World Meteorological Organization, Geneva.

McQuigg, J. D. (1979). Climatic variability and agriculture in the temperate regions. In *Proceedings of the World Climate Conference*, pp. 406–425. Publication No. 537, World Meteorological Organization, Geneva.

Mercer, J. H. (1978). Effect of climatic warming on the West Antarctic ice sheet. *Nature*, **271**, 321–325.

Mitchell, J. F. B. (1983). The seasonal response of a general circulation model to changes in CO_2 and sea temperatures. *Quarterly Journal of the Royal Meteorological Society*, **109**, 113–152.

Motha, R. P., Leduc, S. K., Steyaert, L. T., Sakamoto, C. M., and Strommen, N. D. (1980). Precipitation patterns in West Africa. *Monthly Weather Review*, **108**, 1567–1578.

Namias, J. (1976). Negative ocean-air feedback systems over the North Pacific in the transition from warm to cold seasons. *Monthly Weather Review*, **104**, 1107–1121.

National Academy of Sciences (1975). *Understanding Climatic Change*. NAS, Washington, DC: 239 pages.

National Academy of Sciences (1979). *Carbon Dioxide and Climate: A Scientific Assessment*. NAS, Washington, DC: 22 pages.

National Academy of Sciences (1982). *Carbon Dioxide and Climate: A Second Assessment*. NAS, Washington, DC: 72 pages.

Nicholson, S. H. (1980). The nature of rainfall fluctuations in sub-tropical Africa. *Monthly Weather Review*, **108**, 473–487.

Oguntoyinbo, J. A., and Odingo, R. S. (1979). Climatic variability and land use: An African perspective. *Proceedings of the World Climate Conference*, pp. 552–580. Publication No. 537, World Meteorological Organization, Geneva.
Oke, T. R. (1974). *Review of Urban Climatology, 1968–1973*. Technical Note 134, World Meteorological Organization, Geneva: 132 pages.
Oke, T. R. (1979). *Review of Urban Climatology, 1973–1976*. Technical Note 169, World Meteorological Organization, Geneva: 100 pages.
Olson, J. S. (1982). Earth's vegetation and the carbon dioxide question. In Clark, W. C. (1982), pp. 388–398.
Parkinson, C. L., and Kellogg, W. W. (1979). Arctic sea ice decay simulated for a CO_2-induced temperature rise. *Climatic Change*, **2**, 149–162.
Parthasarathy, B., and Mooley, D. A. (1978). Some features of a long homogeneous series of Indian summer rainfall. *Monthly Weather Review*, **106**, 771–781.
Philander, S. G. H. (1983). El Niño southern oscillation phenomenon. *Nature*, **302**, 295–301.
Ratcliffe, R. A. S., Weller, J., and Collison, P. (1978). Variability in the frequency of unusual weather over approximately the last century. *Quarterly Journal of the Royal Meteorological Society*, **104**, 243–255.
Robock, A. (1980). The seasonal cycle of snow cover, sea ice and surface albedo. *Monthly Weather Review*, **108**, 267–285.
Rotty, R. M. (1983). Distribution of and changes in industrial carbon dioxide production. *Journal of Geophysical Research*, **88**(C2), 1301–1308.
Rotty, R. M., and Marland, G. (1980). Constraints on fossil fuel use. *MS Proceedings, Energy/Climate Interactions Workshop, Münster, Federal Republic of Germany, March 3–8, 1980*.
Schneider, S. H., and Mass, C. (1975). Volcanic dust, sunspots and temperature trends. *Science*, **190**, 741–746.
Stockton, C. W., and Meko, D. M. (1983). Drought recurrence in the Great Plains as reconstructed from long-term tree-ring records. *Journal of Climate and Applied Meteorology*, **22**, 17–29.
Thomas, R. H., Sanderson, T. J. O., and Ross, K. E. (1979). Effect of climatic warming on the West Antarctic ice sheet. *Nature*, **277**, 355–358.
Thompson, L. M. (1975). Weather variability, climatic change, and grain production. *Science*, **188**, 535–541.
Van Loon, H., and Williams, J. (1976–1977). The connection between trends of mean temperature and circulation at the surface. *Monthly Weather Review*, **104**, 365–380, 1003–1011 and 1591–1596; (Williams and Van Loon), **105**, 636–647.
Van Loon, H., and Williams, J. (1979). The association between mean temperature and interannual variability. *Monthly Weather Review*, **106**, 1012–1017.
Vinnikov, K. Ya., and Groisman, P. Ya. (1981). The empirical analysis of the CO_2 effects on the present day variations in the mean annual surface air temperature of the Northern Hemisphere. *Meteorologiya i Hydrologiya*, **No. 11**, 21–33.
Wang, W. C., Yung, Y. L., Laus, A. A., Mo, T., and Hansen, J. E. (1976). Greenhouse effects due to man-made perturbations of trace gases. *Science*, **194**, 686–690.
Weller, G., Baker, D. J., Jr., Gates, W. L., MacCracken, M. C., Manabe, S., and Vonder Haar, T. H. (1983). Detection and monitoring of CO_2-induced climate changes. In National Research Council, *Changing Climate (Report of the Carbon Dioxide Assessment Committee)*, pp. 292–382. National Academy Press, Washington, DC.
Weihe, W. (1979). Climate, health and disease. In *Proceedings of the World Climate Conference*, pp. 313–368. Publication No. 537, World Meteorological Organization, Geneva.

White, R. M. (1979). Climate at the millennium. In *Proceedings of the World Climate Conference*, pp. 1–14. Publication No. 537, World Meteorological Organization, Geneva.

Wigley, T. M. L., and Jones, P. D. (1981). Detecting CO_2-induced climatic change. *Nature*, **292**, 205–208.

Wigley, T. M. L., Lough, J. M., and Jones, P. D. (1984). Spatial patterns of precipitation in England and Wales and a revised homogenous England and Wales precipitation series. *Journal of Climatology*, **4** (1, January/February), 1–26.

Wittwer, S. H. (1980). Carbon dioxide and climatic change: An agricultural perspective. *Journal of Soil and Water Conservation*, **35**, 116–120.

Woodwell, G. M. (1982). Earth's vegetation and the carbon dioxide question. In Clark, W. C. (1982), pp. 399–400.

World Climate Programme (1981). *On the Assessment of the Role of* CO_2 *on Climate Variations and their Impact (Villach, Austria, Workshop, November 1980)* WMO, UNEP and ICSU, Geneva: 35 pages.

World Meteorological Organization (1975). *The Physical Basis of Climate and Climate Modelling*. GARP Publication Series No. 16, Geneva: 265 pages.

World Meteorological Organization (1979). *Proceedings of the World Climate Conference*. Publication No. 537, Geneva: 791 pages.

Climate Impact Assessment
Edited by R. W. Kates, J. H. Ausubel and M. Berberian

CHAPTER 3
Research in Climate–Society Interaction

WILLIAM E. RIEBSAME

Department of Geography
University of Colorado,
Boulder, Colorado 80309 USA

3.1 INTRODUCTION

Research on the ways climate, biomes, and society interact is framed by abiding views of the relationship between humans and their environment. Extreme episodes in the early 1970s fostered a resurgence of expressions of human vulnerability to climate, reminiscent of the 'climatic determinism' voiced in the early decades of this century. Simultaneously, recognition of the potential for intentional and unintentional human-induced climate change heightened our awareness of climate as a natural resource that can be managed or mismanaged. This chapter reviews the development of these and other ideas about climate subsumed in four broad categories: climate as setting, climate as determinant, climate as hazard, and climate as natural resource. The discussion considers the basic perspectives and research emphases that tend to be associated with each view.

The classification used here derives from a survey of the literature of climate impacts and appears to embrace and to order that literature reasonably neatly. Clearly, the classification is subjective and alternatives with more or less detail might be developed. Also, theoretical bases for various abiding myths of nature, of which climate is an element, can be proposed. For example, the anthropologists Douglas (1972) and Thompson (1982) explored how nature is characterized by different authors and societal groups. Holling *et al.* (1983) and Kates (1983) have derived taxonomies of natural systems and man–environment interactions based on ecological and geographical perspectives, respectively.

3.2 FOUR VIEWS OF CLIMATE

3.2.1 Climate as Setting

One sense of climate is as the setting for the processes of biophysical and socioeconomic systems. Nature provides the climate, and climate provides the setting. In this view, expressed in the thorough and extensive calculations of climate means and normals, climate is a given and stable element of the terrestrial environment. Much of early scientific climatology, reviewed in detail by Leighly (1949), was an effort to measure, describe, and establish climate as a static, or very slowly varying, element of the environment (see also Chapter 2, this volume).

The view of climate as a stable backdrop to life on earth is strongly evident in the first scientific climate classification schemes. These were based on vegetation assemblages implicitly assumed to have established themselves in equilibrium with the fixed climate (see, for example, Köppen and Geiger, 1930). While a view of climate as strongly repetitive, only gradually changing, with few surprises, has endured to the present, some climatologists (see, for example, Thornthwaite, 1961) have argued for a more 'dynamical' climatology, based on an accounting of flows of energy, mass, and momentum in the climate system. Studies of paleo-climates, polar climates, and climates of other planets also take a broad view of climate as evolutionary setting.

Typical expressions of the present climate as setting are the many regional climatologies produced in the 1940s and 1950s, what Terjung (1976) called studies of 'climate-for-its-own-sake'. Contemporary climatologists, attempting to relate climate more closely to aspects of biophysical and socioeconomic systems, have developed more specifically adapted ways to express climate characteristics. For example, the Growing Degree Day, based on the 'heat unit' idea (see Mather, 1974, 158), relates temperature to a critical base value above which plant growth occurs. Similarly, the Heating Degree Day expresses temperature with regard to a base below which interior space heating is assumed necessary. Such indices are then related to variables like crop yield and energy consumption. Other indices used to describe the climate background in meaningful terms include Palmer's (1965) drought index and various comfort and heat stress indices (see, for example, Belding and Hatch, 1955).

3.2.1.1 Research Implications

A view of climate as setting encourages work aimed at better observation and description of climate variables seen as rigid characteristics of the environment. Regional climatologies continue to appear (see, for example, Nieuwolt, 1977), and these now regularly include discussions of radiation and water balances and synoptic conditions, but they hesitate to draw connections between climate and biological and social systems. The underlying assumption of descriptive climatology is that good description can provide the basis for the finer adjustment of human activities like agriculture, as well as the design and construction of transportation systems and human settlements. The field of applied climatology, as codified in texts such as Griffiths (1976) and Oliver (1973), has developed around the notion that better knowledge of the statistics of climate can inform the design of systems to minimize adverse effects and to maximize positive opportunities of climate. Thus, mapping, sorting and classification are the characteristic activities associated with the view of climate as setting.

3.2.2 Climate as Determinant

The sense of climate as a pervasive, powerful element of the environment forms a basis for climate determinism, the view that climate is a dominating influence in the molding of natural and social systems. The first recorded, concise statements of this view are attributed to Greek and Roman scholars. For example, Hippocrates, in his *Airs, Waters, Places* (see Jones, 1962), cast climate as the determinant of health and disease, and claimed that the microclimates of cities affect the civility and personality of their inhabitants.

Early historians and geographers, blending natural and human scientific exploration and description, lent a scholarly basis to determinist views. They described vegetation, animal, and even human populations as adapted to climatic constraints. This perspective is exemplified by Alexander von Humboldt's *Kosmos* (see Otte, 1849). Although recognizing that humankind's mental powers provide some independence from environmental factors, von Humboldt argued that culture is essentially a product of adaptation to the physical world, a key element of which is climate. In the first decades of the twentieth century, Ellsworth Huntington (1915) extended this view, claiming that climate is all-pervasive in molding social structure, settlement patterns, and human behavior.

Although the contention that climate molds human behavior and development and may actually determine great events fell into disfavor among social scientists after Huntington's time, it occasionally re-emerges in contemporary thought. For example, Carpenter (1968) and Bryson *et al.* (1974) argued that drought caused the decline of Mycenaean Greek civilization during the late Bronze Age. Writing about contemporary disparities in the development of nations, Lambert (1975) and Harrison (1979) attribute to global climate patterns a range of sociotechnical characteristics, from labor productivity to agricultural efficiency. Some modern analysts continue to invoke climate to explain the slow economic development of particular regions, especially the tropics (see, for example, Oury, 1969; Myrdal, 1972). A contemporary natural science approach that has a determinist flavor is found in Chang (1970). Chang explored the temperature dependence of net photosynthesis and concluded tentatively that in warmer climates yields are smaller. A certain fatalism seems to underly the determinist view, a sense that deliberate societal strategies will make little difference in modifying the impacts on society. Instead, it focuses on the importance of how societies are initially endowed with climatic resources, which is largely a matter of chance occasioned by the unequal distribution of energy, water and land mass across the earth's surface.

3.2.2.1 Research Implications

The view of climate as a determinant is evidenced in contemporary research by the search for causal chains that link climate to specific elements or behaviors of biophysical and socioeconomic systems (see Maunder, 1970). This work has concentrated on outcomes, such as crop yields or industrial output, that can be quantified and numerically modeled. A statistical relationship between climate and some element of interest is termed a transfer function. Transfer functions have now been developed for most crops (see Nix, Chapter 5, this volume; Baier, 1977), for water supply (Nováky *et al.*, Chapter 8, this volume; World Meteorological Organization, 1974) and for energy use (Jäger, Chapter 9, this volume; McQuigg, 1975).

The major thrust of research associated with the climate-as-determinant view involves the detailing of climate-crop production linkages. Baier (1977) cites a long history of attempts to correlate crop yields with such elements as precipitation, temperature, soil moisture, and insolation. He argues that much of the work prior to about 1960 was based on shaky methods, but that in recent years several groups, for example, de Wit and colleagues at the Agricultural University of Wageningen, The Netherlands, have advanced the science by applying careful validation techniques. While such approaches are not strictly 'determinist' in the language of statistics, both correlation-based and more physically oriented models, such as the deterministic crop growth approaches reviewed by Baier (1983), are founded on concepts of inescapable cause and effect between climate and crop yields.

Less formal relationships, sometimes simply verbal statements of the result expected in a biophysical or socioeconomic system given some initial climate state, have been described for several other areas. These include fisheries (Kawasaki, Chapter 6, this volume), forestry (Baumgartner, 1979), transportation (Wintle, 1960; Bollay, 1962; Beckwith, 1966; Evans, 1968), recreation (Hentschel, 1964; Clawson, 1966; Paul, 1972; Perry, 1972; Taylor, 1979), and construction (McQuigg and Decker, 1962; Russo, 1966; Musgrave, 1968; Martin, 1970; Maunder *et al.*, 1971). These studies are limited by their exclusive focus on one industry or one economic sector, and they are dated by technological change, but they nevertheless represent qualitative and quantitative attempts to describe causal linkages between climate and human activity.

Several obstacles have hindered the development of reliably applicable causal formulas for climate and impacted activities. Lack of understanding of the specific processes involved limits explicit, deterministic modeling, as does the absence of reliable, long time-series of quantitative data for both selected climate variables and the activities to which they would be linked. Crop yield data are of good quality in some parts of the world, encouraging the development of climate-crop transfer functions. Alternatively, fish populations are difficult to observe and available data series are short, hindering the estimation of quantitative relations between fisheries and climate.

Perhaps the greatest problem for determinist approaches is coping with complexity. Quantitative functions tend to be unidirectional, linear and limited in their ability to represent accurately the effects of simultaneously or abruptly changing factors. Difficulties in quantifying certain kinds of variables and in specifying nonlinear relationships have especially limited the development of reliable formal models of climate and socioeconomic factors such as health, population change, and economic activity. Some recent work, for example, Palutikof's (1983) study of the impact of climate on industrial production in Britain (see Table 4.5, this volume), by virtue of good care in accounting for factors exogenous to the climate–society link, indicates promise for building better climate–society transfer functions in the future.

Often expressed in the scientific vernacular of dependent and independent variables, the contemporary development of climate–society transfer functions maintains the central core of determinist ideas and mostly represents a unidirectional view of cause and effect. While this paradigm is logical with regard to immediate climate–biophysical connections, it is more controversial when applied to connections between climate and human behavior, given the capacity for adjustment and adaptation. As Kates (Chapter 1, this volume) points out, deterministic models are useful simplifying tools that have allowed the explication of several important climate–society linkages, while, of course, not capturing the full complexity of climate–society interaction.

3.2.3 Climate as Hazard

Like the two previously described concepts of climate, the view of climate as a hazard to be suffered, accommodated, or mitigated is long-standing. Climate historians find ample records of extreme climate conditions (see Lamb, 1983), suggesting that the harmful and threatening aspects of climate attract the greatest human attention. For example, after overcoming their early expectations about climate (Kupperman, 1982), the early European inhabitants of North America were impressed by the severe and changeable aspects of that continent's very 'un-European' climate (Ludlum, 1966, 1968). The modern view of climate as natural hazard owes its strength chiefly to droughts that occurred during the first half of this centry. Sub-Saharan Africa experienced extreme climate variability in 1910–15 (perhaps comparable to the 1968–74 'Sahelian' drought), and the North American Great Plains experienced multiyear droughts in the mid–1930s, described in detail by Worster (1979). In the southern hemisphere, parts of Africa south of the equator experienced major drought in 1919 (Tannehill, 1947) and Australia experienced severe droughts in 1902 and 1944–45 (Campbell, 1968).

While such widely reported climate extremes set the stage for the climate-as-hazard view, the accounts of droughts, cold waves, heat waves, and other climate variations were largely anecdotal until the 1930s, when researchers began to investigate systematically the role of natural hazards in human affairs, focusing first on riverine flooding in the United States (White, 1945). During the 1970s this interest evolved into an international research effort involving anthropologists, economists, psychologists, geographers, and sociologists along with geographical scientists. Case studies of a wide range of natural hazards were undertaken, including several atmospheric hazards in both developed and developing countries (White, 1974; see also Heathcote, Chapter 15, this volume).

Extreme climatic episodes during the late 1960s and early 1970s were assessed by some as a switch to a more variable and dangerous climate epoch. This view has been argued recently by Lamb (1983), who provides statistical evidence for a global increase in variability in the 1960s and an impressive list of climate extremes during the two decades 1960–80. Whether or not the global climate system has recently shifted to a more variable state (see Hare, Chapter 2, this volume), this perception has spurred studies of the impacts of extreme events and has been used to amplify the messages of those concerned that society's sensitivity to climate disruption might also be increasing, another of Lamb's arguments.

Several works on climate–society interaction addressed to a broad audience dwell on extremes and emphasize the increasing actual or perceived vulnerability of societies to climate. Roberts and Lansford (1979) present a roster of severe seasons and individual weather events to illustrate climate impacts, especially on global food supply. Case studies of particular climate extremes have ranged from García's (1981) eclectic review of the events of 1972 to assessments of the impacts of recent droughts, heat waves, cold waves, and wet seasons (see, for example, Posey, 1980; Assessment and Information Services Center, 1981).

While the studies of natural hazards have the merit of readily discernible climate–society linkages, the studies of extreme events do not often utilize common assessment methodologies or validation techniques and thus do not provide the comparable information needed to make broader generalizations about climate–society links *in extremis*. Moreover, it is questionable whether one can extrapolate from them to more common, moderate circumstances. The approach also tends to focus on a limited realm of adverse extremes. For example, although studies of extreme failures of grain harvests are common, little or no attention is given to surplus agricultural production situations, which may be equally climate-related and, in some respects, also carry important impacts. The view of climate as hazard is *a priori* rather negative, introducing a bias into the research that may be contrasted with the characteristics of the climate as resource view discussed below.

3.2.3.1 Research Implications

Research in the climate-as-hazard tradition is organized around the approach of 'reasoning from extremes'. By focusing on extreme cases, which initially become known through firsthand reports of the event and its associated impacts, the researcher is virtually assured of relevant new source material and the discovery of climate linkages. While, as mentioned above, one might argue that the climate-as-hazard approach is by its nature limited to exceptional cases, it is also possible that the linkages easily observed in extreme cases also function in some more common 'day-to-day' situations. For example, studies of how Great Plains wheat farmers cope with drought (Warrick, 1980) have illuminated a wide range of strategies inherent in that agricultural system for absorbing many types of climate variations, not only extreme ones.

Using 'reasoning from extremes', natural hazards researchers have exposed issues important to climate impact assessment (see Heathcote, Chapter 15, this volume). First, hazards researchers have attempted to relate the level and distribution of hazard impacts to societal vulnerability across different levels of social integration and technological development (Burton *et al.*, 1978). For example, variations in vulnerability across societies in the developing world were scrutinized by Susman *et al.* (1983), who argued that vulnerability to disruption by natural events is closely tied to the problem of unequal international development and the marginalization of populations forced, by economic and political pressures, to rely on unsuitable lands for subsistence (see also Waddell, 1975; Regan, 1980; Spitz, 1980).

A second contribution of hazard research to climate impacts is focused on the particular mechanisms by which societies adjust to natural events. One collaborative research effort sought to identify adjustments in an international context (White, 1974) by studying 24 hazards, 21 of which were atmospheric in origin. Collective and individual adjustments adopted at local, national, and global levels were described first by a listing of theoretically possible adjustments, which were then compared to adjustments documented in actual field observation. The pattern of differences between theory and actuality led to the development of adjustment models that incorporate limitations in individual and group perception and decision-making (Slovic *et al.*, 1974).

Most generally, natural hazards workers have attempted to develop a perspective that visualizes impacts as the joint product of both geophysical states and levels of social vulnerability, a form of the interactive model described by Kates (Chapter 1, this volume). The underlying bias of hazards research is to see climate and nature as rather capricious and, thus, also to see a need for widespread defensive actions on the part of individuals and social groups. These actions often involve a reduction in the demands placed on climate, for example, through nonstructural increases in the efficiency of water use.

3.2.4 Climate as Natural Resource

Although the notion of climate as resource has a decidedly contemporary tone, especially with regard to wind and solar energy, the view predates the recent burgeoning of climate concern. For example, Landsberg (1946) discussed 'climate as a natural resource' and 'climate as national income', arguing that climate, once properly assessed, could be exploited to enhance human productivity, health, and comfort. During the 1940s and 1950s the notion of climate as resource was used to argue for better record-keeping and analysis. The international weather observation network, oriented toward daily forecasting and aviation safety, was not functioning well as a climate network. By arguing that climate was an important national resource, Landsberg and others urged better climate data management. Thus, an initial focus of the climate as resource theme was on measuring and assessing the climate resource base, similar to efforts in the minerals, energy, water, and forestry resource areas. In this respect, the resource view overlaps with the view of climate as setting.

Where the resource view differs from other perspectives is in its incorporation of attempts to value, allocate, manage, and manipulate climate. These approaches have been apparent in discussions about weather modification, and they are also evident in the development of the discussion of climate as an economic resource.

Resource economists, thinking about climate, have not settled on a single definition. Most resource analysts agree, however, that climate is a nonmarket resource, that is, a resource that is not valued and allocated by price mechanism in private trade. Thus, climate is initially best viewed as a public good. It is essentially a 'free resource' (see Freeman, 1980), available to all economic actors at a given location. It is also a factor in production, and in some respects a common property resource (a resource to which access is open to all and whose utilization might affect all other users).

Although Samuelson, in 1954, provided a definition of public goods that can apply to climate—a good which provides benefits to all users with individual consumption leading to no decrease in any other user's benefits (see also Lovell and Smith, Chapter 12, this volume)— the analytical tools necessary to value and manage climate without recourse to conventional market criteria (which would obviate climate's inherent 'publicness') have been slow to develop. Some momentum was provided by the growth of environmental awareness in the late 1960s, when trends in population, living standards and environmental pollution exposed a wide rift between private and public goals with regard to the environment. Economists began using normative management models to calculate values for selected common property resources like oceanic fisheries (Gordon, 1954) and clean water; public property resources like wilderness; and intangible resources like quiet and scenic views. (See Kneese, 1977, for a broad review of the development and nature of environmental economics.)

During this period, and by virtue of its visibility and the economic questions involved, the issue of air pollution further fostered a view of the atmosphere as an exhaustible natural resource to be valued and managed—and even bought and sold as a receptacle and disperser of industrial wastes (see Ausubel, 1980).

Of course, the resource view of climate does not rest solely on the many definitions emanating from the discipline of economics. Farmers, water resource managers, and even foresters might well consider climate simply as an extension of the commodity resources they manage. Most farmers would reckon climate as a basic resource when it is not being regarded as a hazard. As Ausubel (1980) argued:

> ... climate is matter and energy organized in a certain way. If a climatologist were to say to a farmer that the climate is going to change, the farmer could interpret this to mean that deliveries of matter and energy may be going to change in quantity, time and place, in ways similar to how supplies of fertilizer or gasoline might change. (Ausubel, 1980, 23)

The development of scientific cloud seeding by Langmuir and Schaefer in the mid-1940s (see Schaefer, 1946) also encouraged a natural resource view of weather. A key attribute of natural resources is that they are malleable in some way; that they can be managed for optimum characteristics and yield. Following Langmuir and Schaefer's pioneering work, weather modification was quickly developed and adopted as a resource management tool, although with little or no success. During the 1960s and 1970s, several countries attempted to manage rainfall resources via cloud seeding; operational projects were initiated in Australia, Honduras, the USSR, Israel and the United States, for example (see Hess, 1974, for a review of early weather and climate modification projects). Simultaneously, scientists began to search for economic and decision-making analyses that would determine how best to allocate weather and climate management efforts (see, for example, Sewell, 1966). They turned chiefly to cost-benefit analysis, which had been widely used to assess water development projects (see Prest and Turvey, 1965), thus putting climate (as measured, for example, by average rainfall, frequency of hailstorms, or number of foggy days) on a par with the traditional natural resource of surface water.

The growing recognition that human activity might, unintentionally, change global climate strengthened further its aspect as a natural resource. During the early 1970s, questions about the way an increase in high-altitude, supersonic air transport might affect the climate caught the attention of economists and other social scientists (see Climate Impact Assessment Project, 1975 and Glantz *et al.*, Chapter 22, this volume), who then sought to place social and economic value on climate elements in order to compare the costs of climate change to the benefits of supersonic transportation. Questions surrounding possible socioeconomic impacts of a CO_2-induced climate change also began, in the mid-1970s, to elicit reviews, programs, and analyses aimed at assessing the value of climate, and various scenarios of altered climate, to economic systems (see Clark, 1982). The notion of climate as a fixed constraint weakens in light of the potential consequences of human activities. Although the dominant focus in studies of the impacts of human-induced climatic change has been on negative aspects, the resource view also has led some scientists to consider the potential positive benefits of changed climate (Hare, 1983), and to propose altering human activities in order to manage the climate resource for maximum positive benefit (Kellogg and Schneider, 1974).

3.2.4.1 Research Implications

As noted earlier, research and analysis associated with the natural resource view of climate has evolved slowly, groping for an appropriate theory and perhaps hindered in some circles by the view remaining from the first half of the twentieth century that the background, unmanaged environment has no economic value (Krutilla and Fisher, 1975, 9). The resource of climate is essentially external to traditional economic markets and thus does not lend itself to standard neoclassical economic analysis. The development of welfare economics and the study of public goods has encouraged analysts to account for the effects of economic activity by calculating benefits and costs associated with natural resource utilization outside of the traditional market framework.

When climate is seen as a tangible good, it can theoretically be measured, valued, and managed in optimal ways that increase the derived net public benefit. Questions of climate change and weather modification have been the focus of such research approaches. For example, d'Arge (1979), attempting to assess the costs or benefits associated with anthropogenic climate changes, suggested three approaches for valuing climate:

1. assess the *alternative costs* associated with maintaining current production levels despite the climate change;
2. assess the costs of all production *opportunities* deleted or provided by the climate change; and
3. assess the *willingness to pay* of 'users' of a given climate to maintain or change that climate.

Similar approaches have been applied to traditional natural resources like forests, and other newly recognized ones like wilderness and clean air.

Other approaches used to put a value on climate include calculating the differences it causes in wages and prices at different locations, the value it adds or subtracts from land, the impacts on crop prices, water supply costs and energy demand, and even the health maintenance costs associated with different climates (see, for example, Climate Impact Assessment Project, 1975; Freeman, 1980; Haurin, 1980; and Lovell and Smith, Chapter 12, this volume).

Research under the climate as natural resource theme also focuses on ways that it might be managed and allocated. Curry (1962, 1963) studied how farmers are already managing the climate resource through land valuation, seasonal programming of activities, and support of weather modification projects. Because weather modification represents (in theory) direct management of climate resources, it has elicited research aimed at assessing the value it adds to climate-sensitive production (see Sewell, 1966).

Despite calls for better recognition of the natural resource attributes of climate (see, for example, Taylor, 1974, and White, 1979), this theme remains rather ill-defined and ambiguous. It awaits both further development of natural resources theory and the application of that theory to climate. However, the view of climate as partly influencable by human action is obviously an abiding one, recognizable in early as well as modern culture. What may be changing is that that view is being matched by a capacity to effect the transformations envisioned.

3.3 CONCLUSION

The four views of climate presented are not mutually exclusive. All appear in contemporary thought about climate and its impact on society, and climate certainly partakes of all four attributes. It acts as background, occasionally determines critical outcomes, often presents hazards or triggers disasters, and is a basic resource to human activity. Indeed, the difficulty in seeing the multifaceted role of climate in human affairs may be a key hindrance to progress in our thinking on the climate–society equation. The questions involved demand multiple cultural and disciplinary perspectives, and, as Hare (1982) noted: 'In attempting to grapple with climate impact we (are) shaking the crystal lattices of politics and intellectual clan structure'. While climate research will benefit from the attention of several scientific schools, both physical and social, practitioners may be hesitant to leave their specialized realms and join one other in an integrated effort (Chen, 1981).

One might also argue that it is partly the nature of climate, allowing us so readily to project into it the character of determinant or hazard or resource or setting, that has also hindered our ability to focus research on climate problems. If climate is neither narrowly defined nor the particular specialty of any 'intellectual clan', then who will take responsibility for pushing forward our understanding? This volume illustrates one way: by encouraging broadminded specialists to explore the application of their research tools to climate problems, by looking on climate fairly from many points of view, and by assembling those points of view.

Finally, what does the existence of multiple, enduring perspectives on climate and its impact tell us? Is there an implicit succession or cycle of the kind of themes discussed here? Can we achieve a more profound synthesis or will we largely reaffirm a set of basic biases with more refined arguments? Readers of this book are invited to form their own judgments.

REFERENCES

Assessment and Information Services Center (1981). *The Economic and Social Impacts of the Winter of 1976–77*. National Oceanic and Atmospheric Administration, US Dept. of Commerce, Washington, DC.

Ausubel, J. H. (1980). Economics in the air: An introduction to economic issues of the atmosphere and climate. In Ausubel, J., and Biswas, A. K. (Eds.) *Climatic Constraints and Human Activities*, pp. 13–59. Pergamon Press, Oxford.

Baier, W. (1977). *Crop–Weather Models and Their Use in Yield Assessments*. Technical Note 151. World Meteorological Organization, Geneva.

Baier, W. (1983). Agroclimatic modeling: An overview. In Cusack, D. (Ed.) *Agroclimatic Information for Development: Reviving the Green Revolution*. Westview Press, Boulder, Colorado.

Baumgartner, A. (1979). Climatic variability and forestry. In *Proceedings of the World Climate Conference*, pp. 581–607. World Meteorological Organization, Geneva.

Beckwith, W. B. (1966). Impacts of weather on the airline industry: The value of fog dispersal programs. In Sewell, W. R. D. (Ed.) *Human Dimensions of Weather Modification*. Department of Geography Research Paper 105, University of Chicago, Chicago, Illinois.

Belding, H. S., and Hatch, T. J. (1955). Index for evaluating heat stress in terms of resulting physiological strain. *Heating, Piping and Air Conditioning*, **27**, 129–136.

Bollay, E. (1962). *Economic Impact of Weather Information on Aviation Operations*. Federal Aviation Agency, Washington, DC.

Bryson, R. A., Lamb, H. H., and Donley, D. L. (1974). Drought and the decline of Mycenae. *Antiquity*, **48**, 46.

Burton, I., Kates, R. W., and White, G. F. (1978). *The Environment as Hazard*. Oxford University Press, New York.

Campbell, D. (1968). *Drought: Causes, Effects, Solutions*. F. W. Cheshire, Melbourne, Australia.

Carpenter, R. (1968). *Discontinuity in Greek Civilization*. W. W. Norton, New York.

Chang, J. H. (1970). Potential photosynthesis and crop productivity. *Annals of the Association of American Geographers*, **60**, 92–101.

Chen, R. S. (1981). Interdisciplinary research and integration: The case of CO_2 and climate. *Climatic Change*, **3**, 429–447.

Clark, W. C. (1982). *Carbon Dioxide Review, 1982*. Oxford University Press, New York.

Clawson, M. (1966). The influence of weather on outdoor recreation. In Sewell, W. R. D. (Ed.) *Human Dimensions of Weather Modification*. Department of Geography Research Paper 105, University of Chicago, Chicago, Illinois.

Climate Impact Assessment Project (1975). *Economic and Social Measures of Biological and Climatic Change*. Monograph 6, US Department of Transportation, Washington, DC.

Curry, L. (1962). The climatic resources of intensive grassland farming: The Waikato, New Zealand. *Geographical Review*, **52**, 174–194.

Curry, L. (1963). Regional variation in the seasonal programming of livestock farms in New Zealand. *Economic Geography*, **39**, 95–118.

d'Arge, R. (1979). Climate and economic activity. In *Proceedings of the World Climate Conference*, pp. 652–681. World Meteorological Organization, Geneva.

Douglas, M. (1972). *Natural Symbols: Explorations in Cosmology*. Random House, New York.

Evans, S. H. (1968). Weather routing of ships. *Weather*, **23**, 2–8.

Freeman, A. M., III (1980). The hedonic price technique and the value of climate as a resource. Paper presented at the *Climate and Economics Workshop, April 24–25, 1980, Fort Lauderdale, Florida*. Resources for the Future, Washington, DC.

García, R. V. (1981). *Drought and Man: The 1972 Case History. Vol. 1: Nature Pleads Not Guilty*. Pergamon Press, Oxford.

Gordon, H. S. (1954). The economic theory of a common-property resource: The fishery. *Journal of Political Economy*, **65**, 124–142.

Griffiths, J. F. (1976). *Applied Climatology: An Introduction* (2nd edn). Oxford University Press, Oxford.

Hare, F. K. (1982). Climate: The neglected factor? *The International Journal*, **36**, 371–387.

Hare, F. K. (1983). Future climate and the Canadian economy. In Harrington, C. R. (Ed.) *Climate Change in Canada 3. National Museum of Natural Science Syllogeus Series* **49**, 15–49. National Museums of Canada, Ottawa.

Harrison, P. (1979). The curse of the tropics. *New Scientist*, **22**, 602.

Haurin, D. B. (1980). The regional distribution of population, migration, and climate. *Quarterly Journal of Economics*, **95**, 293–308.

Hentschel, G. (1964). Sports and climate. In Light, S., and Kamenetz, H. L. (Eds.) *Medical Climatology*. Waverly Press, Baltimore, Maryland.

Hess, W. N. (1974). *Weather and Climate Modification*. John Wiley & Sons, New York.

Holling, C. S., Walters, C. J., and Ludwig, D. (1983). *Myths, Time Scales, and Surprise in Ecological Management*. Mimeo. International Institute for Applied Systems Analysis, Laxenburg, Austria.

Huntington, E. (1915). *Civilization and Climate*. Yale University Press, New Haven, Connecticut.

Jones, W. H. S. (1962). *Translation of Hippocrates' Airs, Waters, Places*. Heinemann, London.

Kates, R. W. (1983). Part and apart: Issues in humankind's relationship to the natural world. In Hare, F. K. (Ed.) *The Experiment of Life*. University of Toronto Press, Toronto, Canada.

Kellogg, W. W., and Schneider, S. H. (1974). Climate stabilization: For better or worse? *Science*, **186**, 1163–1172.

Kneese, A. V. (1977). *Economics and the Environment*. Penguin Books, New York.

Köppen, W., and Geiger, R. (Eds.) (1930). *Handbuch der Klimatologie*. Gebrüder Borntraeger, Berlin.

Krutilla, J. V., and Fisher, A. C. (1975). *The Economics of Natural Environments*. Johns Hopkins University Press, Baltimore, Maryland.

Kupperman, K. O. (1982). The puzzle of the American climate in the early colonial period. *American Historical Review*, **87**, 1262–1289.

Lamb, H. H. (1983). *Climate History and the Modern World*. Methuen, London.

Lambert, L. D. (1975). The role of climate in the economic development of nations. *Land Economics*, **47**, 339.

Landsberg, H. (1946). Climate as a natural resource. *The Scientific Monthly*, **63**, 293–298.

Leighly, J. (1949). Climatology since 1800. *Transactions of the American Geophysical Union*, **30**, 658–672.

Ludlum, D. M. (1966). *Early American Winters 1604–1820*. American Meteorological Society, Boston, Massachusetts.

Ludlum, D. M. (1968). *Early American Winters II, 1821–1870*. American Meteorological Society, Boston, Massachusetts.

Martin, D. B. (1970). Construction and seasonality: The new federal program. *Construction Review*, **16**, 4–7.

Mather, J. R. (1974). *Climatology: Fundamentals and Applications*. McGraw-Hill Co., New York.

Maunder, W. J. (1970). *The Value of the Weather*. Methuen, London.

Maunder, W. J., Johnson, R., and McQuigg, J. D. (1971). Study of the effects of weather on road construction: A simulation model. *Monthly Weather Review*, **99**, 939–945.

McQuigg, J. D. (1975). *Economic Impacts of Weather Variability*. Prepared at the Department of Atmospheric Science, University of Missouri, Columbia, Missouri.

McQuigg, J., and Decker, W. (1962). The probability of completion of outdoor work. *Journal of Applied Meteorology*, **1**, 178–182.

Musgrave, J. C. (1968). Measuring the influences of weather on housing starts. *Construction Review*, **14**, 4–7.

Myrdal, G. (1972). *Asian Drama: An Inquiry into the Poverty of Nations*. Random House, New York.

Nieuwolt, S. (1977). *Tropical Climatology: An Introduction to the Climates of the Low Latitudes*. John Wiley & Sons, London.

Oliver, J. E. (1973). *Climate and Man's Environment*. John Wiley & Sons, New York.

Otte, E. C. (1849). *English Translation of Alexander von Humboldt's Kosmos*. Henry G. Bohn, London.

Oury, B. (1969). Weather and economic development. *Finance and Development*, **6**, 24–29.

Palmer, W. C. (1965). *Meteorological Drought*. Research Paper 45, US Weather Bureau, Washington, DC.

Palutikof, J. (1983). The impact of weather and climate on industrial production in Great Britain. *Journal of Climatology*, **3**, 65–79.

Paul, A. H. (1972). Weather and the daily use of outdoor recreation areas in Canada. In Taylor, J. A. (Ed.) *Weather Forecasting for Agriculture and Industry*. David and Charles, Newton Abbot, UK.

Perry, A. H. (1972). Weather, climate, and tourism. *Weather*, **27**, 199–203.

Posey, C. (1980). Heat wave. *Weatherwise*, **33**, 112–116.

Prest, A. R., and Turvey, R. (1965). Cost-benefit analysis: A survey. *The Economic Journal*, **75**, 683–705.

Regan, C. (1980). Underdevelopment, vulnerability and starvation: Ireland, 1845–48. In O'Keefe, P., and Johnson, K. (Eds.) *Environment and Development: Community Perspectives*. Program in International Development, Clark University, Worcester, Massachusetts.

Roberts, W. O., and Lansford, H. (1979). *The Climate Mandate*. W. H. Freeman and Company, San Francisco.

Russo, J. A., Jr. (1966). The economic impact of weather on the construction industry of the United States. *Bulletin of the American Meteorological Society*, **47**, 967–972.

Samuelson, P. A. (1954). The pure theory of public expenditure. *Review of Economics and Statistics*, **36**, 387–389.

Schaefer, V. J. (1946). The production of ice crystals in a cloud of supercooled water droplets. *Science*, **104**, 457–459.

Sewell, W. R. D. (Ed.) (1966). *Human Dimensions of Weather Modification*. Department of Geography Research Paper 105, University of Chicago, Chicago, Illinois.

Slovic, P., Kunreuther, H., and White, G. F. (1974). Decision processes, rationality, and adjustment to natural hazards. In White, G. F. (Ed.) *Natural Hazard: Local, National, Global*. Oxford University Press, New York.

Spitz, P. (1980). Drought and self-provisioning. In Ausubel, J., and Biswas, A. K. (Eds.) *Climatic Constraints and Human Activities*, pp. 125–147. Pergamon Press, Oxford.

Susman, P., O'Keefe, P., and Wisner, B. (1983). Global disasters, a radical interpretation. In Hewitt, K. (Ed.) *Interpretations of Calamity from the Viewpoint of Human Ecology*, pp. 263–283. Allen and Unwin, Winchester, Massachusetts.

Tannehill, I. R. (1947). *Drought: Its Causes and Effects*. Princeton University Press, Princeton, New Jersey.

Taylor, J. A. (Ed.) (1974). *Climatic Resources and Economic Activity*. David and Charles, London.

Taylor, J. A. (1979). *Recreation Weather and Climate*. Sports Council, Aberystwyth, UK.

Terjung, W. H. (1976). Climatology for geographers. *Annals of the Association of American Geographers*, **66**, 199–222.

Thompson, M. (1982). A three-dimensional model. In Douglas, M. (Ed.) *Essays in the Sociology of Perception*. Routledge and Kegan Paul, London.

Thornthwaite, C. W. (1961). The task ahead. *Annals of the Association of American Geographers*, **51**, 345–356.

Waddell, E. (1975). How the Enga cope with frosts: Responses to climatic perturbations in the central highlands of New Guinea. *Human Ecology*, **3**, 249.

Warrick, R. (1980). Drought in the Great Plains: A case study of research on climate and society in the USA. In Ausubel, J., and Biswas, A. K. (Eds.) *Climatic Constraints and Human Activities*, pp. 93–123. Pergamon Press, Oxford.

White, G. F. (1945). *Human Adjustment to Floods*. Department of Geography Research Paper 29, University of Chicago, Chicago, Illinois.
White, G. F. (1974). *Natural Hazards: Local, National, Global*. Oxford University Press, New York.
White, R. M. (1979). Climate at the millennium. In *Proceedings of the World Climate Conference*, pp. 1–11. World Meteorological Organization, Geneva.
Wintle, B. J. (1960). Railways versus the weather. *Weather*, **15**, 137–139.
World Meteorological Organization (1974). *Intercomparison of Conceptual Models Used in Operational Hydrological Forecasting*. Operation Hydrology Report No. 7. WMO, Geneva.
Worster, D. (1979). *Dust Bowl: The Southern Great Plains in the 1930s*. Oxford University Press, New York.

Climate Impact Assessment
Edited by R. W. Kates, J. H. Ausubel and M. Berberian

CHAPTER 4
Identifying Climate Sensitivity

W. J. MAUNDER* AND J. H. AUSUBEL†

**New Zealand Meteorological Service*
P.O. Box 722
Wellington 1, New Zealand

†National Academy of Engineering
2101 Constitution Avenue, N.W.
Washington, DC 20418, USA

4.1 INTRODUCTION

Where does one begin in undertaking specific climate impact studies? Often, such selection is already dictated by circumstances or events; that is, there is a previous commitment or preference to examine the relationship between climate and a particular crop, region or population. However, the questions of what activities to study, in which places, affecting whom also are frequently not predetermined. This paper attempts to assist the impact assessor by surveying which activities, places and populations analysts have found important to study in the past and the methods they have used to identify climate sensitivity.

4.2 SUBJECTS OF EARLIER STUDIES

There are two overlapping sets of studies that constitute implicit, if not explicit, surveys of activities or places that are highly sensitive to climate. The first set consists of studies of the users and value of the information that national weather services generate. The second includes studies connected with integrated assessments, scientific symposia, or governmental and intergovernmental conferences.

Typical of the impact studies related to national weather services are those of Mason (1966) for the United Kingdom and Tolstikov (1968) for the Soviet Union. Maunder (1970) surveyed the value of the weather with examples from many areas. Such studies emphasize the importance of meteorological conditions and information for agriculture and point out their role in a variety of other activities, such as aviation. To illustrate further, the value of meteorological services was discussed at a New Zealand symposium (New Zealand Meteorological Service, 1979). At this event topics examined included the relation of weather and climate to:

- forest fires and forest management,
- agriculture (wool, wheat, potatoes),
- ship operations,
- electricity supply,
- design and siting of power stations, and
- boating and sport.

Although the national meteorological service studies generally focus on weather, rather than climate, they still may be viewed as a first-order approximation of the climate sensitivity of human activities. Clearly, the range of activities impacted by short-term climate variation will be close to that impacted by weather. Table 4.1 presents a summary of annual dollar and percentage losses due to adverse weather in the United States as estimated by Thompson (1977). According to Thompson, agriculture is an order of magnitude more sensitive to weather than any other activity in both relative and absolute terms.

The Assessment and Information Services Center (AISC) (formerly the Center for Environmental Assessment Services, CEAS) of the US Department of Commerce routinely makes climate impact assessments on a weekly and monthly basis for the United States and several other countries and regions of the world. The monthly impact assessments for the United States cover the following eight broad categories of societal activity (Center for Environmental Assessment Services, 1980):

1. *Construction*—including housing starts, commercial and industrial construction, property damage, soil erosion and land use.
2. *Economics and Commerce*—including employment, banking, business, trade, manufacturing and industry.
3. *Energy*—including utilities, supply and consumption of the different energy types and alternative energy sources.

Table 4.1 Losses due to adverse weather in the United States

Activity	Overall losses (US$ × 10^6)	(% of annual gross revenue)
Agriculture	8240.4	(15.5)
Construction	998.0	(1.0)
Manufacturing	597.7	(0.2)
Transportation (rail, highway, and water)	96.3	(0.3)
Aviation (commercial)	92.4	(1.1)
Communications	77.4	(0.3)
Electric power	45.7	(0.2)
Energy (e.g., fossil) fuels	5.1	(0.1)
Other (general public, government, etc.)	2531.8	(2.0)
Totals	12,684.8	

(After Thompson, 1977, 71)

4. *Food and Agriculture*—including food, fiber and orchard crops, forestry and fisheries.
5. *Government and Taxes*—including executive, legislative and judicial branches; federal, regional, state and local.
6. *Recreation and Services*—including travel, vacation activities and sport.
7. *Society*—including fatalities, injuries, health, air pollution, education, crime, and population movements.
8. *Transportation and Communications*—including highway, railroad, air, and mail delivery.

In AISC's estimates of climate sensitivity, agriculture is again dominant, but other sectors, such as energy and transportation, increase their relative position as seasonal factors come into play.

Besides central government groups concerned with the operation and provision of climate-related services, the second major source of information on climate impacts is a rather miscellaneous research literature, much of it stemming from conferences. The agendas of such conferences and the contents of the reports provide a *de facto* judgment about what is sensitive to climate. Table 4.2 compares these individual agendas with one of the weather studies to suggest both the convergence of opinion on climate sensitivity and the idiosyncratic judgments as well. The list includes Maunder's survey (1970), the Climate Impact Assessment Program (CIAP, 1975), the Aspen Institute

Table 4.2 Topics covered in 8 major climate impact studies[a]

Sensitive area[b]	Maunder 1970	CIAP 1975	Aspen Institute 1977	CSIRO 1979	WMO 1979	DOE/AAAS 1980	CEAS 1980	SCOPE 1984
1. Agriculture	X	X		X	X	X	X	X
2. Forests and forestry	X	X		X	X	X	X	
3. Pastoral activities				X		X	X	X
4. Fish and fisheries	X	X			X	X	X	X
5. Ecosystems						X		
6. Environmental conservation								
7. Water supply, demand		X	X	X	X			X
8. Energy supply, demand	X	X	X	X	X		X	X
9. Manufacturing operations, location of plants	X	X		X			X	
10. Offshore operations					X			
11. Mining (extractive industries)				X				
12. Transportation—water, air, rail, highway	X		X	X			X	
13. Construction	X		X	X			X	
14. Materials weathering		X						
15. Esthetic costs		X						
16. Trade	X						X	
17. Public expenditures		X					X	
18. Communications							X	
19. Insurance	X			X				
20. Financial planning and institutions			X				X	
21. Recreation and tourism	X			X			X	
22. Sea level rise, coastal zones						X		
23. Health—mortality, morbidity	X	X			X	X	X	X
24. Migration							X	
25. Social concerns, crime	X	X				X	X	X
26. Military planning and operations								
27. Political systems and institutions	X	X				X		
28. Legal systems and institutions	X					X		

[a] To be checked here, topic must be treated explicitly or extensively.
[b] List includes 2 topics (6 and 26) not covered in studies listed, but covered in other studies.

Conference of 1977 on effects of climatic change (Aspen Institute, 1977), the 1979 Australian Conference on Climate and the Economy (Commonwealth Scientific and Industrial Research Organization [CSIRO], 1979), the 1979 World Climate Conference (World Meteorological Organization [WMO], 1979), the American Association for the Advancement of Science research agenda on effects of CO_2-induced climate change prepared for the US Department of Energy (DOE) (1980), the Assessment and Information Services Center ongoing assessments (CEAS, 1980), and this volume.

This overview suggests a broad consensus on the sensitivity to climate variation of agriculture, water resources, building construction, transportation, and energy activities. In addition, there is repeated concern about impacts on insurance, governmental expenditures, and recreation. The table may suggest more orderliness and conscious selection than exists. Any agenda represents to a certain extent networks of available researchers and not systematically identified topics. Moreover, impact studies and conferences have often focused on geographic regions, especially drought-prone ones (the Sahel, South Asia, northeast Brazil, North American Great Plains) and high latitudes (Parry, 1983), where life is obviously sensitive to climate.

An alternative to the identifying of large sectoral or regional sensitivity is to focus sharply on specific topics where climate impact analysis appears most useful for applications. Topics from a Massachusetts Institute of Technology (1980) conference on 'Climate and Risk' illustrate sensitivities more specifically, for example:

- extreme wind speeds and structural failure risks;
- weather hazard probabilities and the design of nuclear facilities;
- application of climatology to air force and army operational planning;
- impacts and use of climatological information in the hail insurance industry;
- importance of climate and climate forecasting to offshore drilling and production operations in the petroleum industry;
- seasonal climate forecasts and energy management;
- evaluating farming system feasibility and impact using crop growth models and climate data;
- the utilization and impacts of climate information on the development and operations of the Colorado River system; and
- snow management and its economic potential in the Great Plains.

While it is clear from this survey that there have been some attempts to be reasonably comprehensive from a sectoral point of view, few of the efforts have been internally consistent in assessing sectoral impacts. Differing methods and definitions of climate variation have often been employed within a conference or study. Also, the sectoral studies vary in their attention to receptors or exposure units. The largest portion of studies examines impacts at the national or regional level. Other levels may be important, however. For example, municipalities may have their expenditures for education, welfare, police, fire, sanitation, and sewage services affected by climate variation (Sassone, 1975). Households may be sensitive in terms of income and expenditures for food, housing, transportation, clothing and medical care (Crocker *et al.*, 1975) and at the individual level, nutrition (Escudero, Chapter 10, this volume) and migration (Warrick, 1980) may be major indicators of climate sensitivity. Regrettably, few studies have been systematic in their selection of indicators of sensitivity and units of exposure for comparative examination; it could therefore be fruitful to undertake studies of relative sensitivity to a given climate variation stratified in a variety of ways, for example, by urban, suburban and rural families, or by different firms in a particular region, or similar firms in different regions. Table 4.3 suggests a framework for defining sensitivity to climate and illustrates it using several available studies.

An important deficiency in making sensitivity judgments based on studies of the value of weather services or agendas is their dominant relationship to industrialized societies. While individual papers from the World Climate Conference, for example, relate to developing country situations (see Burgos, 1979; Fukui, 1979; Oguntoyinbo and Odingo, 1979), the balance of analysis and methods development has been restricted to industrialized nations. Although there is considerable evidence that primary activities (agriculture, pastoralism, water resources) may be even more sensitive to climate in developing countries than in industrialized countries, careful surveys of sensitivity remain to be done.

4.3 METHODS

There is not likely to be any single best method for making an initial assessment of a society's climate sensitivities. Several early, related steps might be considered. As implied by Section 4.2, these might include:

1. Analysis of uses and users of weather and climatic information in the proposed study area; this would provide an idea of which groups value climate knowledge and, thus, may be exhibiting sensitivity.
2. Review of scientific literature, especially local sources, related to climate impacts; this would lead to an implicit ranking of activities as to their sensitivity to climatic influences.

This section suggests and gives illustrations of several additional analytical methods. These are analyses of communications media and information content, reviews of national income and product accounts, examinations of seasonality, and correlation analysis. These methods all enable one to take readily available data and perform tentative analyses aimed at identifying the climate signal in social and economic activities.

Table 4.3 Sample of impact studies in framework for defining sensitivity to climate

Study	CLIMATE VARIATION				IMPACT LEVELS						INDICATORS USED		
	change	variability	season	extreme	biophysical	social system unit of exposure					quantitative		qualitative
						globe	nation/ region	locality	firm	household	monetary	non-monetary	
CIAP (1975)	X				X	X	X	X	X	X	X		
WMO (1979)		X			X		X	X				X	
DOE/AAAS (1980)	X				X		X						X
Seifert and Kamrany (1974)				X	X			X				X	X
National Defense University (1980)	X				X	X	X				X	X	
García (1981)				X	X	X	X	X			X	X	X
Chambers *et al.* (1981)			X		X			X		X	X	X	X

Studies are placed in framework according to their emphases.

4.3.1 Media Analysis

In many regions perhaps the best guide to weather/climate sensitivity is obtained by a critical analysis of agricultural, economic and business journals, and by a careful appraisal of the area's newspapers. Indeed in many nations—both developed and developing—the media (newspapers, journals, radio and television) often provide the only indication of the importance in a real-time sense of weather and climate.

A media analysis is particularly important in two kinds of situations: first, in data-poor nations or areas, where what little data do exist are available for analysis only several months (and in some cases several years) after the event; and second, in those data-rich nations or regions where decisions are made on a day-to-day (sometimes even hour-to-hour) basis and where the prices of commodities are important. In both cases the financial, economic and agricultural sections of the daily press (or, if available, the more specialized newspapers such as *The Wall Street Journal* or *The Financial Times*) publish valuable information, provided the reports are read with a critical eye.

Probably the most developed use of media-type information (such as that described above) for climate impact assessment is that published by AISC. As a guide, AISC uses the sensitivity of the gross national product (GNP) to widespread anomalous weather as given in Table 4.4. This table was compiled from an analysis of the weather/climate sensitivity of various economic and social sectors in the United States as reported in the *New York Times* during a 10-year period. The AISC survey shows that a major increase in the GNP can result from an unusually hot summer (specifically in GNP elements of personal consumption expenditures on electricity and food at home), and an unusually cold winter (GNP elements of personal consumption expenditures on natural gas, fuel oil and coal; also in net imports), whereas a major decrease in the GNP can be expected as a result of unusually mild conditions (specifically in the GNP elements of personal consumption expenditures on food at home, changes in business inventories and net imports). While the sensitivities shown in Table 4.4 should be regarded as tentative, they do give an indication of how the GNP of an industrialized nation may be affected by widespread anomalous weather conditions.

Specific weather/climate events have also been studied by AISC. For example, during the 1980 summer heat wave and drought, there was a series of reports (see Center for Environmental Assessment Services, 1981) updating mounting economic losses in the United States that were finally estimated at more than $20,000 million. Six months after the last special report on the summer heat wave and drought was issued, official statistics were released confirming these estimates. To compare this impact of weather/climate with previous events, a report was prepared (Center for Environmental Assessment Services, 1982) on the 1976–77 winter in the United States which indicated

Table 4.4 Sensitivity* of Gross National Product elements to widespread anomalous weather

GNP elements	Hot summer	Cold winter	Dry summer	Storm/rain	Snow	Mild
1. Personal consumption expenditures						
(a) Gasoline and oil	–	–	–	–	–	++
(b) Electricity	++	+	?	++	?	–
Natural gas, fuel oil, coal	?	++	?	++	+	–
(c) Furniture and appliances	–	–	–	?	–	++
(d) Food at home	++	+	++	+	++	––
Food away	––	––	?	?	––	++
(e) Apparel	–	+	?	?	–	+
(f) New and used cars	–	–	–	–	–	++
(g) Housing	–	––	?	?	–	++
(h) Transportation	–	–	?	–	–	++
(i) Other	?	?	?	?	?	?
2. Non-residential fixed investment	?	?	?	?	?	?
3. Residential	–	–	–	?	–	++
4. Change in business inventories	+	+	+	+	+	––
5. Net imports	+	++	+	+	+	––
6. Government purchases						
(a) Federal	+	+	+	+	+	–
(b) State and local	+	+	+	+	+	–

*Weather-related changes in consumption: + = increase; ++ = major increase; – = decrease; –– = major decrease.
(Adapted from Center for Environmental Assessment Services, 1980)

that the economic losses during that winter were almost twice those of the 1980 summer heat wave and drought. These media-based studies of a national economy are a useful method of placing specific weather/climate events in perspective. The AISC studies further point out that although last month's climate is history, the measurement of its economic impact is not; short-term losses (for example, suppressed consumption) may be compensated for or magnified by subsequent economic developments, and early indicators may be replaced by more reliable data and information. Further, since weather information is available in real-time, whereas weekly national economic indicators—even in the United States—have a time publication delay of 2 or 3 weeks (or months, depending on the economic parameter), weather-based forecasts of economic activity can be made available 1, 2 or 3 weeks (or months, depending on the economic parameter) *before* the actual production/consumption information is available.

4.3.2 National Income and Product Accounts

A more thorough and orderly process can be a critical reading of national income and product accounts from a climatic point of view. Although when looking at an economy the trained eye may be able to differentiate quickly between sectors of an economy that are weather/climate sensitive and those that are not, systematic approaches can be helpful. As an example, let us consider the climate sensitivity of New Zealand by examining the various components of the economy as they appear in the *New Zealand Official Year Book*. The approach involves an overview of the economy and then a more detailed look at key elements. A few examples will indicate a typical step-by-step process to make an initial assessment of national climate sensitivity.

In calling the roll of activities, let us focus first on transport. Transport, an important component of the economy, should be considered specifically by its various sectors and the climate-sensitive aspects of these sectors. For example, in New Zealand the gross expenditure on railways (using data from the 1981 *Year Book*) was $404 million; this included $26 million (or 6 percent) spent on fuel. On the basis that any change in fuel used—as a result of better weather and climate or information about them for rail transportation—will affect the economy, the fuel expenditure can be said to be weather/climate sensitive. Other aspects of railway expenditures may also be weather/climate sensitive in other countries, but analysis shows that few other sectors of the railway operations in New Zealand are weather/climate sensitive. In the case of shipping, the key weather/climate factors that can be identified are the loading and unloading of containers. In New Zealand, the number of containers moved exceeds 200,000 in a year, and many contain climate-sensitive commodities such as dairy products, wool, meat and fruit. Optimal movement of these perishable commodities is crucial to New Zealand's export competitiveness, and thus it is logical to include container movements in a list of activities that are climate sensitive.

When we look at the energy sector, the key factor is the heavy reliance of New Zealand on imported oil, 50 percent of New Zealand's total energy being supplied by imported oil. In contrast, of the 22 percent of the total energy supplied from primary electricity, a very high 86 percent comes from hydroelectric generation. The energy situation in New Zealand is therefore related mainly to two factors—the need to use imported oil and the natural availability of relatively cheap hydroelectricity. Both factors are sensitive to an adequate supply of water at hydroelectric stations during the critical summer and autumn periods, and the severity of the winter. The latter is a critical determinant of the proportion of electricity to be generated by more expensive oil and gas.

A third area to consider is government expenditure, especially subsidies from public funds. The *New Zealand Official Year Book* lists two relevant subsidies—the first concerning adverse events (such as drought conditions), the second, a fertilizer subsidy. This latter subsidy is in several ways weather/climate sensitive. It reflects concern of the New Zealand Government about farm income variations, in that a subsidy on fertilizers is often given following a relatively poor income (climatic) season, in order to encourage farmers to fertilize their pastures during the following season. A natural response of farmers following a poor season is to spend fewer dollars on fertilizer; the government fertilizer subsidy is in effect a means of smoothing not only the irregularities in the application of fertilizer to New Zealand pastures, but also the climate-induced variations in farmers' incomes.

A second step in a systematic survey of national economic data is to consider both absolute and relative climatic sensitivity through dollar value of production. For example, in New Zealand, wool accounts for 19 percent of the gross value of agricultural production, compared with 3 percent for vegetables. Thus, irrespective of the fact that some aspects of the vegetable sector (such as transport of vegetables to markets) are more weather sensitive than some aspects of the wool industry (such as effects of severe frost on quality of wool), in terms of monetary value the wool sector of New Zealand agriculture is six times more important than the vegetable sector. Similarly, since 62 percent of New Zealand's agricultural income is from pastoral products (that is, wool, dairy products and meat), it is evident that the climatic sensitivities will be strongly related to the state of the nation's pastures. Naturally, in other countries this could be quite different, especially where field crops or horticultural-type crops are the principal agricultural earners.

Finally, in this climatic view of the *New Zealand Official Year Book*, consider overall the components of gross domestic product. Eleven percent of New Zealand's gross domestic product comes from the agricultural sector, and

6 percent from food manufacturing. Other important weather/climate-related market production groups include transport/storage (6 percent), construction (5 percent), energy (3 percent) and wood products/forestry (3 percent). These components, clearly climate sensitive, comprise one-third of New Zealand's gross domestic product. The climate-sensitive sectors of other nations could be—and in many cases are—quite different from those of New Zealand. An analysis similar to that described for New Zealand, using appropriate scans of national 'Year Books' or income and product accounts, should point to those sectors of an economy that are most climate sensitive.

4.3.3 Seasonal Variations

A development of the kind of survey described above is to look for seasonal variations in production and consumption data. It is well recognized that some economic activities follow a recurring seasonal pattern during the year; for example, the retailer is aware that there will be increased business around Christmas or other major holidays, for which to plan purchasing and personnel changes. Similarly, the contractor buys materials and hires additional workers for the increased construction activity that inevitably comes in many countries during the summer months, and the farmer's expenditure rises in the spring and autumn because of planting and harvesting costs. Bankers also recognize these and other patterns of seasonal activity and they plan for an uneven deposit inflow and demand for loans during the year. In the same manner wage earners in industries with high degrees of dependence on seasonal activity realize that their income may not flow evenly during the year. The strongly seasonal character of social welfare in developing countries has also been assessed (Chambers *et al.*, 1981).

Seasonal patterns usually follow a relatively similar pattern from year to year. Procedures have been developed to measure the fluctuations; in many cases a climatic explanation is offered for the variations. For example, economists often use statistical smoothing on weekly, monthly and quarterly data and refer to the results as 'seasonally adjusted', suggesting that the revised figures account for variations in 'seasonal' activities such as Christmas trade, end-of-quarter activities, winter/summer differences, and the weather. Of course, such seasonally adjusted data may have little relationship to weather events as such, but the extent of seasonal adjustments may provide another initial quantitative view of climatic sensitivities over a broad spectrum of activities.

4.3.4 Correlation Analysis

A correlation approach may also suggest sensitivities in the comparison of climate and economic data. For example, in a study of weather and the retail trade, Linden (1962) related—in a simple but telling way–sales of women's winter coats in New York department stores to the average monthly temperature in September and October. Correlations are often pointed out between the behavior of futures markets and the arrival of information about weather and climate.

More detailed correlation analyses can also be made. For example, Maunder (1979) examined climatic data in relation to sheep numbers and wool production. The analysis identified statistically the most important weather/climatic factors and the significant months or combination of months. Palutikof (1983) used multiple regression analysis to evaluate the impact of a severe winter and a hot, dry summer on British industry (Table 4.5), revealing interesting contrasts. For example, severe winters favor utilities (+17.5) and reduce clothing and footwear production (−9.6), while a dry, droughty summer reduces utility performance (−6.4) and increases clothing and footwear performance (+3.0).

In some correlation-type analyses, however, the impact signals are weak, or hidden by larger, non-climatic fluctuations. In such cases it is useful to focus on particularly anomalous climate episodes (for example, major droughts, floods,

Table 4.5 Performance of United Kingdom industries

Industry	Average deviation per month from mean of preceding season	
	1962–63 winter	1975–76 drought
Bricks, cement, etc.	−14.4	−0.9
Timber, furniture	−14.3	−1.8
Clothing and footwear	−9.6	+3.0
Paper, printing and publishing	−5.6	−2.2
Mining and quarrying	−4.7	−2.3
Shipbuilding	−4.3	−0.9
Engineering and electrical goods	−3.7	−2.1
Nonferrous metals	−2.7	−4.3
Drinks and tobacco	−2.4	+4.4
Ferrous metals	−2.3	−10.0
Food	−2.0	+0.7
Metal goods (not elsewhere specified)	−2.0	−4.6
Chemicals	−1.6	+2.1
Leather goods	−0.6	+0.6
Pottery and glass	−0.3	−3.9
Textiles	0.0	−0.6
Vehicles	0.0	−1.3
Coke ovens, oil refining, etc.	+3.3	−0.3
Utilities	+17.5	−6.4

(After Palutikof, 1983)

cold spells and the like) and search for concomitant variations in social and economic activities. If no apparent variation is found, it is safe to assume that the subject under study exhibits little sensitivity to that particular type of climate variation. Such a 'reasoning from extremes' is particularly useful in data-poor areas, and in some cases may be the only reasonable way of quantitatively analyzing climatic sensitivity.

While measures of sensitivity to climate are often economic, other indices are also available. For example, measures can be offered of the numbers of people affected by a climate variation (see Burton *et al.*, 1978; Warrick, 1980) or in terms of dietary levels or patterns of land ownership (see Jodha and Mascarenhas, Chapter 17). The primary appeal of economic indicators, especially monetary ones, is ease of intercomparison. The most extensive effort to assess relative economic sensitivities of different sectors and impacts to climate variation was that of CIAP (1975). CIAP sought to formulate mathematical relationships between long-term climatic change and many of the economic activities mentioned above. The results of one calculation are presented in Table 4.6. While the accuracy of these estimates is questionable (see Ausubel, 1980), the results are surprising and, as such, worth noting. In particular, wage and health effects far surpass others, including agriculture and water resources, in importance. Other studies include those of Eddy *et al.* (1980), who have employed input/output and other economic models to assess the effects of contemporary climate variability on a range of sectors and spatial scales.

4.4 CONCLUSION

The difficulty of identifying climate sensitivities may vary with the scale of activities involved. In many industries, straightforward analysis of firm behavior may reveal climate-sensitive points. Assessment can be more difficult at the national and international levels, where numerous compounding factors play roles in determining resource use and productivity. Extracting the impact of climate variations from the signals of economic growth and decline, business cycles and so forth, is a challenge. Assessing national vulnerability may be especially difficult in developing countries where baseline data are poor and rapid social and economic changes are occurring.

Nonetheless, some general comments about sensitivity may be offered. Most studies focus on agriculture and water, which are clearly directly sensitive. In developing countries a large proportion of the population is engaged in agriculture, and agriculture could well be the most sensitive sector by several measures. In contrast, in developed countries, agriculture typically occupies a small proportion of the population and GNP, so the absolute sensitivity may appear small, although the relative sensitivity may remain high. It is harder to draw conclusions about water. Some studies (for example, Revelle and

Table 4.6 Estimates of economic impacts of a hypothetical global climatic change (−1 °C change in mean annual temperature, no change in precipitation)

Impact studied	Annualized cost—1974 (millions of US dollars)
Corn production (60% of world)	+21
Cotton production (65% of world)	−11
Wheat production (55% of world)	−92
Rice production (85% of world)	−956
Forest production	
(a) US	−661
(b) Canada	−268
(c) USSR (softwood only)	−1383
Douglas fir production (US Pacific Northwest)	−475
Marine resources (world)	−1431
Water resources (2 US river basins)	+2
Health impacts (excluding skin cancer) (world)	−2386
Urban resources (US)	
wages	−3667
residential, commercial and industrial fossil fuel demand	−176 lower bound −232 upper bound
residential and commercial electricity demand	+748
housing, clothing expenditures	−507
public expenditures	−24
esthetic costs	+219

(Source: CIAP, 1975, 1–15 and 1–25)

Waggoner, 1983) suggest this sector is one where impacts are amplified. Beyond agriculture and water, studics typically pick their subjects from about 20 other areas of potential impact. No clear order of importance emerges.

For an initial indication of the sensitivity of activities to climate variation, content analysis of newspapers and other media is usually a helpful beginning. Such analysis can be supplemented with more elaborate surveys (for example, Center for Environmental Assessment Services, 1980) and construction of various indices (for example, Maunder, 1972). Efforts to arrive at more reliable quantitative measures through application of more sophisticated economic models (for example, CIAP, 1975; Eddy *et al.*, 1980) are suggestive but remain subject to criticism (see Glantz *et al.*, Chapter 22; Lovell and Smith, Chapter 12; and Robinson, Chapter 18, in this volume).

Overall, it remains difficult to draw strong conclusions about the importance of various sectors when considered for different units of exposure, levels of development, and categories of climate variation. The Advisory Committee for this volume placed special emphasis on agriculture, pastoralism, fisheries, water resources and energy. Subsequent chapters take these sectors and explore the usefulness of specific methods and perspectives. Despite the reservations expressed in this paper, development of more systematic frameworks for defining sensitivity to climate is certainly possible, and their application will clearly yield useful benefits to research efforts and the operating agencies of governments.

REFERENCES

Aspen Institute (1977). *Living with Climatic Change.* MITRE Corporation, McLean, Virginia.

Ausubel, J. H. (1980). Economics in the air. In Ausubel, J. H., and Biswas, A. K. (Eds.) *Climatic Constraints and Human Activities*, pp. 13–59. Pergamon Press, Oxford.

Burgos, J. J. (1979). Renewable resources and agriculture in Latin America in relation to the stability of climate. In World Meteorological Organization (1979), pp. 525–551.

Burton, I., Kates, R. W., and White, G. F. (1978). *The Environment as Hazard.* Oxford University Press, New York.

Center for Environmental Assessment Services (CEAS) (1980). *Guide to Environmental Impacts on Society.* National Oceanic and Atmospheric Administration, US Department of Commerce, Washington, DC.

Center for Environmental Assessment Services (1981). *Retrospective Evaluation of the 1980 Heatwave and Drought.* National Oceanic and Atmospheric Administration, US Department of Commerce, Washington, DC.

Center for Environmental Assessment Services (1982). *U.S. Economic and Social Impacts of the Record 1976/77 Winter Freeze and Drought.* National Oceanic and Atmospheric Administration, US Department of Commerce, Washington, DC.

Chambers, R., Longhurst, R., and Pacey, A. (1981). *Seasonal Dimensions to Rural Poverty.* Frances Pinter, London.

Climate Impact Assessment Program (CIAP) (1975). *Economic and Social Measures of Biological and Climatic Change.* Monograph 6. US Department of Transportation, Washington, DC.

Commonwealth Scientific and Industrial Research Organization (CSIRO) (1979). *The Impact of Climate on Australian Society and Economy.* CSIRO, Mordialloc, Victoria, Australia 3195.

Crocker, T., Eubanks, L., Horst, R., Jr., and Nakayama, B. (1975). Covariation of climate and household budget expenditures. In CIAP (1975).

Department of Energy (DOE) (1980). *Environmental and Societal Consequences of a Possible CO_2-Induced Climate Change: A Research Agenda*, Volume 1. (A project conducted by the American Association for the Advancement of Science for the US Dept. of Energy.) DOE/EV/10019-01. National Technical Information Service, US Dept. of Commerce, Springfield, Virginia.

Eddy, A., *et al.* (1980). *The Economic Impact of Climate.* Oklahoma Climatological Survey, University of Oklahoma, Norman, Oklahoma, Vols. 1, 2, 3, 4, 5.

Fukui, H. (1979). Climatic variability and agriculture in tropical moist regions. In World Meteorological Organization (1979), pp. 426–474.

García, R. V. (1981). *Drought and Man, the 1972 Case Study.* Vol. 1: *Nature Pleads Not Guilty.* Pergamon Press, Oxford.

Linden, F. (1962). Merchandising weather. *The Conference Board Business Record*, **19**, 15–16.

Mason, B. J. (1966). The role of meteorology in the national economy. *Weather*, **21**, 382–393.

Massachusetts Institute of Technology, Center for Advanced Engineering Study (1980). *Climate and Risk*. MTR-80W322-01. MITRE Corporation, McLean, Virginia.

Maunder, W. J. (1970). *The Value of the Weather*. Methuen, London: 388 pages.

Maunder, W. J. (1972). The formulation of weather indices for use in climatic-economic studies. *New Zealand Geographer*, **28**, 130–150.

Maunder, W. J. (1979). The use of econoclimatic models in national supply forecasting: New Zealand wool production. In *Proceedings of the Tenth New Zealand Geography Conference*, pp. 22–28. New Zealand Geographical Society, Auckland.

National Defense University (1980). *Crop Yields and Climate Change to the Year 2000*, Vol. 1. US Government Printing Office, Washington, DC.

New Zealand Meteorological Service (1979). *Symposium on the Value of Meteorology in Economic Planning*. New Zealand Meteorological Service, Wellington.

Oguntoyinbo, J. S., and Odingo, R. S. (1979). Climatic variability and land use. In World Meteorological Organization (1979), pp. 552–580.

Palutikof, J. (1983). The impact of weather and climate on industrial production in Great Britain. *Journal of Climatology*, **3**, 65–79.

Parry, M. (1983). Climate impact assessment in cold areas. Report of working group on cold margins, *WMO/UNEP Workshop on Impacts of Climate Change, Villach, Austria, September 1983*. International Institute for Applied Systems Analysis, Laxenburg, Austria.

Revelle, R. R., and Waggoner, P. A. (1983). Effects of a carbon dioxide induced climatic change on water supplies in the Western United States. In Carbon Dioxide Assessment Committee, *Changing Climate*. National Academy Press, Washington, DC.

Sassone, P. (1975). Public sector costs of climate change. In CIAP, Monograph 6: *Economic and Social Measures of Climatic Change*. US Department of Transportation, Washington, DC.

Seifert, W. W., and Kamrany, N. M. (1974). *A Framework for Evaluating Long-Term Strategies for the Development of the Sahel-Sudan Region*. Vol. 1, *Summary Report: Project Objectives, Methodologies, and Major Findings*. MIT Center for Policy Alternatives, Cambridge, Massachusetts.

Thompson, J. (1977). In Aspen Institute, *Living with Climatic Change*. MITRE Corporation, McLean, Virginia.

Tolstikov, B. I. (1968). Benefits of meteorological services in the U.S.S.R. In 'Economic Benefits of Meteorology', *World Meteorological Organization Bulletin*, **17**, 181–186.

Warrick, R. A. (1980). Drought in the Great Plains. In Ausubel, J. H., and Biswas, A. K. (Eds.) *Climatic Constraints and Human Activities*, pp. 93–123. Pergamon Press, Oxford.

World Meteorological Organization (WMO) (1979). *Proceedings of the World Climate Conference: A Conference of Experts on Climate and Mankind*. WMO, Geneva.

Part II
Biophysical Impacts

In the simplest of impact models, climate affects places, people and their activities, leading to a set of ordered consequences. The first set of consequences is designated as biophysical impacts, because their causal mechanisms (where known) or transfer functions (where inferred) are in the realm of physical and non-human biological relationships: crops grow, rainfall runs off, cattle graze, fish feed, and buildings cool.

As indicated in the previous chapter on climate sensitivity, all five of the sectoral chapters deal with highly climate-sensitive activities. But they do not exhaust the set of such activities; indeed they only illustrate them. Particularly missing are chapters on unmanaged, natural ecosystems, sensitive industrial activities such as construction and transportation, and service activities such as insurance and recreation. However, the five chapters well illustrate the range of methods available for determining first-order impacts, the comparative precision and validity of these methods, and some common directions for future investigation.

In Chapter 5, Nix suggests five methods for analyzing the impacts of climate on agriculture: trial and error, analogy, correlation, simulation modeling, and systems analysis. This typology serves well to describe not only methods employed in the agricultural sector, but also to compare the methods available in other sectors as well. Agriculture (Chapter 5), Water Resources (Chapter 8) and Energy Resources (Chapter 9) employ the full range of methodologies. The analysis of pastoralism (Chapter 7), as distinct from modern ranching, is with a few exceptions still at the level of analogy and simple correlation. The link between climate and fisheries (Chapter 6), with the possible exception of 'El Niño' types of phenomena, is still trial and error, uncertain analogy and correlation.

Greater predictability of transfer functions is found in physical than biological functions and in small-scale, controllable human activities. Thus it is easier to predict temperature–energy or rainfall–runoff relationships than the climate yields of agriculture, pastoralism or fisheries. Similarly it is easier to predict the energy demand for a building than for a city, or the yield of a field than that of a grazing area or an ocean current. The chapter authors have not limited themselves to first-order impacts. Although concerned with biophysical impacts, all the authors work from a framework in which climate impact relationships are constrained or changed by human action. Thus as Nix (Chapter 5) shows, agricultural models must be confined to a particular crop, place and technology or specifically include changes in technology as a major variable in yield functions. Kawasaki (Chapter 6) explores the still indeterminate issue of whether fluctuations in fishery catch reflect natural (including climate) cause or simply overfishing. A similar issue is posed by LeHouérou relative to the success of human adjustment to drought. Nováky, Pachner, Szesztay and Miller (Chapter 8) argue for the need to consider water-related impacts of climate clearly within the context of socially related management activities. Similarly Jäger (Chapter 9) notes that climate–energy demand relationships are changing rapidly as conservation methods come into widespread use.

Thus a common research question for methodological development is posed by the need, in even the simplest of impact models, to allow for the interaction of climate and society. A second issue is posed most clearly in the analysis of Le Houérou, but is also evident in the chapters on water, agriculture and energy. Le Houérou traces the systematic amplification and dampening of impacts along the causal chain of the impacts model, suggesting that variability in climate is amplified in primary productivity of grazing lands, but in turn is dampened in livestock yield and in human impact. Similarly Chapter 8 shows that fluctuations in rainfall are amplified by fluctuations in streamflow but are dampened by water resource management measures. The systematic comparison of climate–yield–impact ratios for different sectors might well lead to improved understanding of the processes of interaction.

Climate Impact Assessment
Edited by R. W. Kates, J. H. Ausubel and M. Berberian

CHAPTER 5
Agriculture

HENRY A. NIX

Commonwealth Scientific and Industrial Research Organization
Division of Land and Water Resources
Canberra City, A.C.T. 2601
Australia

5.1 INTRODUCTION

Not a single farmer, forester or stockbreeder needs any reminder that production systems are subject to the vagaries of climatic variation and the hazard of extreme weather events. The primary role of climate and weather in conditioning biological and physical processes and in determining much of the structure and function of both natural and man-modified systems should need no further emphasis. Yet, the development, testing and practical application of methods of agroclimatic analysis have not been given a high priority in most agricultural and biological research programs. Why is this so?

While there are intrinsic problems, the major limitations have been extrinsic. Farm, forest and grazing management involve a mix of controlled and uncontrolled variables. Since climate and weather are, for most practical purposes, uncontrolled, research emphasis has been given to those components of the production system that can be subject to some measure of control. Thus, for the past century of active development, agricultural research has been pedocentric and genocentric, that is, major effort has been devoted to soil description, classification and amelioration, and genetic improvement of crops and livestock. Since World War II increasing effort has been devoted to chemical and biological control of pests and pathogens.

In addition, the prevailing strategy of agricultural research and development is based on extensive networks of experimental sites at 'representative' locations, with transfer of results by analogy to 'similar' sites. Thus, for any experiment at such a site the climate and weather are taken as given, i.e., as a common component of all treatments. Such a framework is not conducive to the development of a general understanding of crop/environment interactions.

Intrinsic limitations to the development of methods of agroclimatic analysis have been that:

1. basic understanding of physical and biological processes involved in crop/climate and crop/weather interactions is limited (though now developing rapidly);
2. available modes of analysis lack generality and tend to be descriptive, rather than prescriptive; static rather than dynamic;
3. the necessary climatic data are inadequate, unavailable, or non-existent; and
4. matching sets of crop/soil/weather-management data for widely contrasting environments and production systems are not available.

New approaches and data now becoming available may remove most of these limitations. Even so, the future of agroclimatic analysis and synthesis hinges on convincing demonstrations of their utility, at scales ranging from local to global.

In order to place the more traditional and the newly developing methods of analysis and synthesis in perspective I have used a framework developed elsewhere (Nix, 1968, 1980, 1981). Progress toward an ultimate goal of prediction of probable outcomes (ecological and economic) of any agricultural production system at any location is traced through an evolutionary path of trial and error, transfer by analogy, correlation/regression, analysis of variance, multivariate analysis, and systems analysis. These methods all remain in use and are not mutually exclusive. The implications of each method for climate impact assessment are discussed, and examples of relevant research are offered. The focus is on impacts of climate on agricultural production, with impacts on the food system more generally receiving less emphasis.

5.2 AGROCLIMATIC ANALYSIS: RETROSPECT

5.2.1 Trial and Error

The whole foundation of present food and fiber production rests upon thousands of years of farmer trial and error through hundreds of human generations. Coupled with keen observations of climate/weather/crop/livestock interactions, farmer trial and error has produced most of our domesticated plants and animals, as well as many stable and productive agricultural systems. But the time-scale is daunting and the social cost incalculable, since for each incremental improvement there have been countless failures. Major objectives of agricultural research are to shorten the long time trajectory and reduce the social cost for the discovery and substitution of superior strategies and technologies.

Traditional crop and livestock technologies have evolved such that climate and weather information becomes an integral component of the total system. The often complex timing required to fit operations into the seasonal cycle and to avoid specific weather hazards is codified in ritual and custom. An excellent example is provided by Stanhill (1977), who shows a close correspondence between the timing and sequence of special prayers for rain in the Jewish liturgical calendar, established some 200 years ago, and the present-day dependence of wheat yields in Israel on the critical early part of the growing season. Careful analysis of such ritual and custom can yield valuable information about climatic patterns and weather hazards in regions where instrumental data are lacking. Jodha and Mascarenhas (this volume, Chapter 17) report a variety of traditional indicators of and adjustments to climate variation.

The primary objective of most traditional systems is to minimize year-to-year variation in productivity and, especially, to minimize the risk of total loss of crop and/or livestock. But this stability exacts a price in terms of potential yields foregone. While the traditional-crop cultivars and livestock breeds are hardy and resilient to the impact of climatic variation and weather hazard, they are usually low-yielding and incapable of an economic response to added inputs or new management techniques. So well adapted and finely turned are some of these systems that sudden change in any major component—cultural, biological or physical—can cause massive disruption (see Le Houérou, this volume, Chapter 7). The lesson here is that a thorough understanding of climate/weather/crop/livestock interactions in the existing system is necessary before new or 'improved' technologies can be introduced.

Modern, high-technology agricultural systems usually aim at maximizing the expectation of profit or yield, using crop cultivars and livestock breeds that have been produced to fit specific environments and markets, and inputs of fertilizers, herbicides, pesticides and machine energy. Normally, buffering reserves of food and/or capital permit such systems to adopt higher-risk strategies that lead to much greater production on a time-averaged basis. Acceptance of an occasional crop loss or uneconomic yield may be normal for a capital-intensive farmer, but it is unlikely to be acceptable to a subsistence farmer. These distinctions are important when globally assessing climatic impact on agricultural production. Estimation of risk is an important component of agricultural analysis, but the level of risk acceptance must be taken into account.

What kind of climate impact studies correspond to the history of trial and error in the development of agriculture in its environmental context? Naturally, the studies are primarily cultural, like that of Stanhill mentioned earlier, and historical. An interesting example of historical study is that of Parry (1978), who has documented the location of cultivation limits in Scotland between 1600 and 1800, then traced the shifts of these limits as an expression of climatic (and other) factors. Chapter 14 of this volume describes the methods in detail. Such studies record the trials and errors of agricultural practices in the face of a changing climate or extreme events.

Experiencing a changed climate at a particular location may be analogous to experiencing a new climate at a different place. The movement of people into New World agricultural lands and the process of trial and error, whereby optimistic expectations of the settlers yielded to the realities of different climates, has been documented in North America by Kupperman (1982) for the eastern seabord, Malin (1944) for Kansas, and Hargreaves (1957) for the northern Great Plains. Similar studies can be found for Australia (Heathcote, 1965; Williams, 1974).

The mix of immigrants to new lands created possibilities for trial and error experiments. Malin (1944) cites the key role of German Mennonite farmers who migrated from southern Russia in the 1870s to Kansas, bringing drought- and cold-resistant varieties of hard red winter wheat with them. Over time these experimental efforts became professional and led to systematic trials, the major research device of the agricultural experiment station. With such trials and related selection and breeding, hard red winter wheat was adjusted to climates both warmer and colder and wetter and drier than its Kansas source area (Rosenberg, 1982).

5.2.2 Transfer by Analogy

Evidence suggests that the earliest cultivators recognized that natural vegetation could provide an index of site quality for their crops. Modern man has added a little to this basic concept—the transfer of information by analogy. The basic hypothesis is that all occurrences of a defined class of land will respond similarly to any imposed treatment. Climate, soil, vegetation and multi-attribute land classifications exemplify this approach. These classifications then provide a basis for the selection of 'representative' experimental sites and for the subsequent extrapolation of results to other sites that share similar properties. Most current agrobiological research rests firmly upon such a foundation. Impact assessment can also employ analogues as simulators of future climate change. Analogues are sought that have the characteristics of the simulated or predicted climate, and the crop types or natural vegetation found within such climatic locations serve as the simulated or predicted vegetation.

Such a strategy is logical when the understanding of functional relationships between site attributes and crop response is qualitative or lacking, since this understanding is not required for classification. However, the better such knowledge is, the better will be the choice of attributes and the resultant classification. The thematic maps produced (that is, soil, land, vegetation) form a major component of an information system that is relatively inexpensive and highly portable. Usually agricultural research strategy is based upon climate and/or soil classifications, while forestry and pastoral research may place additional emphasis on vegetation classification.

Major deficiencies of the analogue approach reside in the static, multi-attribute character of the classifications upon which it is based and the implicit assumptions of covariance of attributes within mapped units. Compounding this is the location, season, cultivar and management-specific nature of the results from traditional agronomic experiments. For the individual farmer, the ideal location for an experiment is on his or her farm or as close as possible, since successful extrapolation of results is observed to hinge on proximity to the experimental site. Accordingly, there is continuing sociopolitical pressure to extend the network of research stations and experimental sites.

Given that agricultural research strategy will remain firmly based upon concepts of transfer of information by analogy for some time to come, there is ample scope for improvement and refinement of climate, land and soil classifications and for optimization of experimental networks. Numerical taxonomic analysis and/or pattern analysis (Sneath and Sokal, 1974; Williams, 1976) provide an array of classifications. The more relevant the attributes are, the better is the chance of a useful result. Commonly, the attributes measured represent a compromise between sampling density and site measurement.

Agroclimatic and bioclimatic classifications provide illustrations of the basic problems inherent in the analogue approach. While there is a considerable literature on bioclimatic classification and associated technologies, most of the general, multipurpose classifications are of limited use in solving specific problems. More direct coupling of climate data and specific crop response data offers prospects of more relevant classifications.

Agroclimatological zonation has always been particularly attractive to data-poor developing countries without an extensive network of meteorological and agricultural stations. They construct broad agroclimatological or agroecological zones using available temperature, precipitation, and soil and crop data. One example that illustrates this approach is the work of Jaetzold (Jaetzold and Kutsch, 1982; Jaetzold and Schmidt, 1982–83), who was asked by the Kenyan Ministry of Agriculture to provide a highly differentiated system of agroecological zones to help decide which variety of crops would best utilize the natural potential of a given district of arable land (Jaetzold, 1983). The main zones are given in Table 5.1; the rows are based on annual mean temperature, differentiated where appropriate by maximum or minimum temperature and frost occurrence. The columns are derived from a water balance model that gives the yield probabilities of leading crops (in the cells) and roughly reflects the precipitation/evaporation ratio. This main zonation is further subdivided to give actual varietal suggestions by length of season, soil type, plant density, and the like, samples of which are given in Figure 5.1 and Table 5.2. In effect these varietal suggestions are predictions that these crops will fare well if planted under the given conditions of climate and soils, predictions derived from analogous plantings elsewhere under similar conditions.

Analogue methods are currently in use to predict potential agricultural change under conditions of climatic change, particularly CO_2-induced change. At a continental scale, a characteristic bioclimatic classification is that of Holdridge (1947), based on temperature, precipitation and evaporation. This classification currently is being used to analyze the impacts of a CO_2-induced climate change (Emanuel *et al.*, in press). At other scales of analyses in such data-rich countries as the United States, shifts in the boundaries of the corn belt have been projected for a hypothetical CO_2-induced climate change by Newman (1980) and Blasing and Solomon (1983). The studies on crop yields and climate change by the US National Defense University (1980, described by Glantz *et al.*, this volume, Chapter 22) are a final example of assessment by analogy, in this case drawing on the experience of a group of experts.

5.2.3 Correlation/Regression

Since the pioneering work of Hooker (1907) and, more particularly, Fisher (1924), both simple and multiple regression techniques have become standard practice in the agricultural and biological services, both for exploratory data analysis and for prediction. Such empirical/statistical modes of crop/weather analysis dominate the voluminous literature of agroclimatology. Correlations have been established among a wide range of environmental variables and virtually all aspects of crop production. The utility of many such correlations under the conditions in which they were developed is not doubted, but the majority have limited application. Generally, the simpler the measure used, for example, total annual or seasonal rainfall, or annual mean air temperature, the more location-specific is the relationship.

Data on crop yield, temperature, and precipitation for a given district are the central components in statistical regression models. As reported by Waggoner (1983), the model for a particular crop in a particular area is generally expressed as

$$Y_i = a + b_1 t_i + b_2 X_{2i} + \dots + b_n X_{ni},$$

where

Y_i = estimated yield in the ith year;
a = intercept;
t_i = surrogate for technology in the ith year;
b_1 = coefficient representing the effect of technology in quintals/hectare/year;
b_2 to b_n = coefficients representing the effect in quintals/hectare/unit change in the weather;
X_{2i} to X_{ni} = weather variables such as precipitation, temperature, potential evapotranspiration (PET), and evapotranspiration (ET) in the ith year.

Thus, to quote Waggoner, 'by multiple linear regression the effect of weather, factor by factor, on the yields of crops is distilled from a history of weather and yields, accumulated for decades by faithful observers'. Crop/climate regression models have been reviewed by Baier (1977) and Biswas (1980).

Greatest success with this approach has been achieved where one, or perhaps two, environmental factors dominate crop performance. Thus, in high latitudes, temperatures that determine length of growing season may be most critical; in midlatitude, subhumid to semi-arid regions, rainfall amount may be most critical; and in the humid tropics, with high rainfall and near-continuous cloud cover, solar radiation receipts may be most critical. The use of simple variables, such as monthly mean or seasonal mean precipitation and temperature, is common in large-area studies where the availability of data is restricted and/or computational limitations occur. Examples of successful applications of such analyses using simple data are provided by Thompson (1969a,b; 1970) in his analysis of weather and technology in the production of wheat, corn and soybeans in the United States.

Usually, greater precision and generality are achieved if the primary climatic and/or soil data are transformed into indices that are more closely coupled to system performance. The derivation of crop/water stress indices from water-balance models (c.f. Baier and Robertson, 1968; Mack and Ferguson, 1968; Nix and Fitzpatrick, 1969) has proved successful in the explanation of large-area variation in wheat yields. Subsequent developments in the understanding of crop response to water stress have led to the formulation of empirical/statistical functions for a very wide range of crops (Doorenbos and Kassam, 1979).

Inherent problems of the empirical/statistical approaches based on regression-type analyses relate to explicit assumptions of linearity of response, the interdependence of so-called independent variables, implicit assumptions that correlation implies causation, and, inevitably, the location, season, cultivar and management-specific nature of resultant relationships. With care,

Table 5.1 Agroecological zones of the tropics[1]. Reproduced with permission from Jaetzold (1983).

Main zones / Belts of z.	0 (perihumid)	1 (humid)	2 (subhumid)	3 (semi-humid)	4 (transitional)	5 (semi-arid)	6 (arid)	7 (periarid)
TA Tropical Alpine zones Ann. mean 2–10 °C	Glacier / Mountain swamps	I. Cattle-sheep zone		II. Sheep zone			High altitude deserts	
UH Upper highland zones Ann. mean 10–15 ° Seasonal night frosts	F	Sheep-dairy zone	Pyrethrum-wheat zone	Wheat-barley zone	U. highland ranching zone	* U. H. nomadism zone[4]		
LH Lower highl. zones Ann. mean 15–18 ° M. min. 8–11 ° norm. no frost	o r e	Tea-dairy zone	Wheat/maize[2]-pyrethrum zone	Wheat/(M)[2]-barley zone	Cattle-sheep-barley zone	L. highland ranching zone	* L. H. nomadism zone[4]	
UM Upper midland zones Ann. mean 18–21 ° M. min. 11–14 °	s t	Coffee-tea zone	Main coffee zone	Marginal coffee zone	Sunflower-maize[3] zone	Livestock-sorghum zone	U. midland ranching zone	U. midland nom. zone[4]

LM Lower midland zones Ann. mean 21–24 ° M. min. > 14 °	* Z o n	L. midl. sugarcane zone	Marginal sugarcane zone	L. midland cotton zone	Marginal cotton zone[6]	L. midland livestock-millet zone	L. midland ranching zone	L. midland nom. zone[4]
L Lowland zones **IL** Inner lowland zones Ann. mean > 24 ° Mean max. > 31 °	* e s	*Rice-taro zone	* Lowland sugarcane zone	* Lowland cotton zone	* Groundnut zone	Lowland livestock-millet zone	Lowland ranching zone	Lowland nom. zone[4]
CL Coastal lowl.z.[5] Ann. mean > 24 ° Mean max. < 31 °	*	*Cocoa-oilpalm zone	Lowland sugarcane zone	Coconut-cassava zone	Cashewnut-cass. zone	Lowland livestock-millet zone	Lowland ranching zone	Lowland nom. zone[4]

1. Inner Tropics, different zonation towards the margins. The T for Tropical is left out in the thermal belts of zones (except at TA), because it is only necessary if other climates occur in the same country. The names of potentially leading crops were used to indicate the zones. Of course these crops can also be grown in some other zones, but they are then normally less profitable.
2. Wheat or maize depending on farm scale, topography, a.o.
3. Maize is a good cash crop here, but maize also in LH 1, UM 1–3, LM and L 1–4;
4. Nomadism, semi-nomadism and other forms of shifting grazing
5. An exception because of the vicinity of cold currents are the tropical cold Coastal Lowlands cCL in Peru and Namibia. Ann. mean there between 18 and 24 °C.
6. In unimodal rainfall areas growing periods may be already too short for cotton. Then the zone could be called Lower Midland Sunflower-Maize Zone.

* Not in Kenya

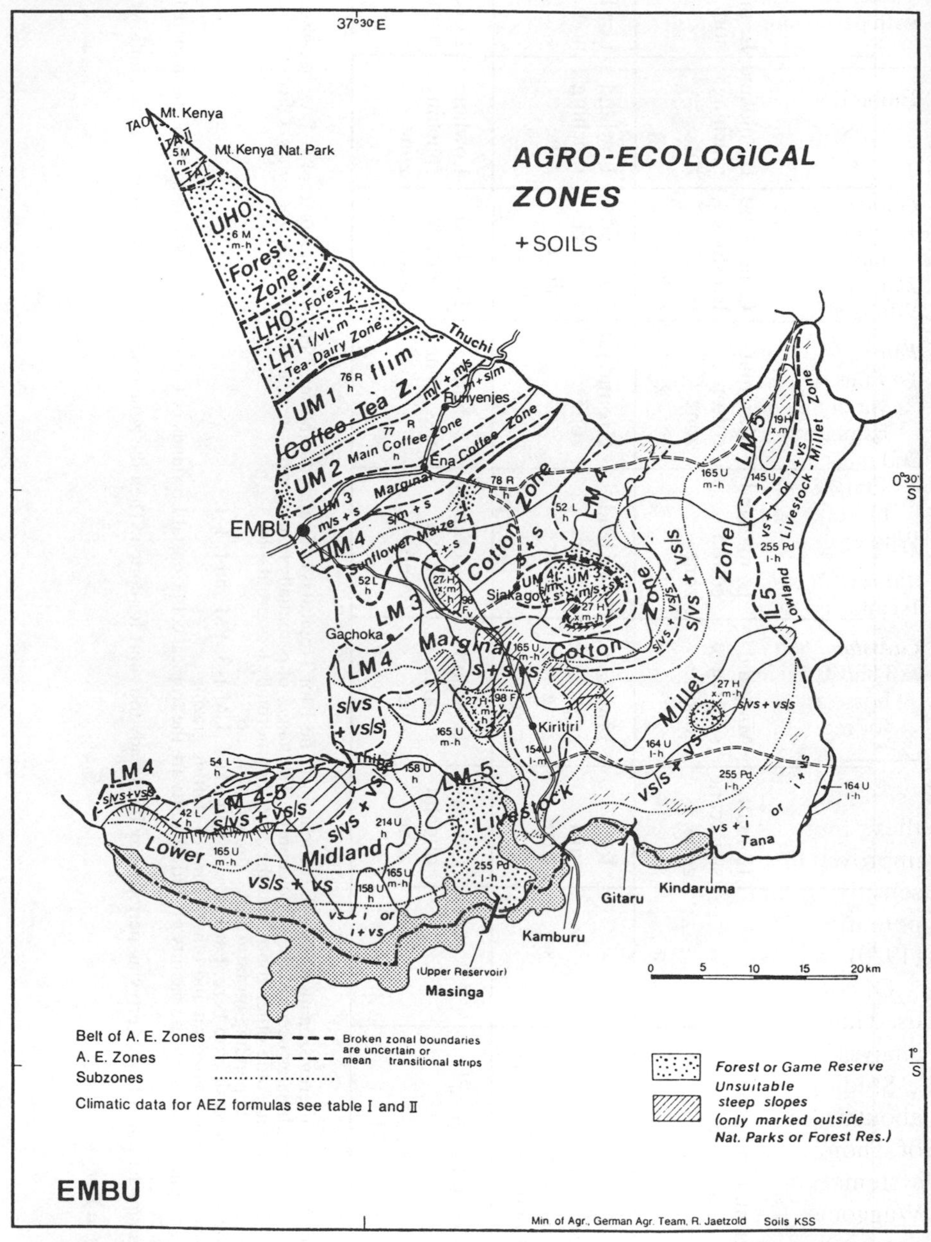

Figure 5.1 Agroecological zones, Embu District, Kenya. Reproduced with permission from Jaetzold (1983)

Table 5.2 Sample of a land use potential for an agroecological subzone. Reproduced with permission from Jaetzold (1983)

Embu District near the margins of rainfed cultivation.

LM 5 = *Lower Midland Livestock-Millet Zone*
vs/s + vs *with a very short to short and a very short cropping season*

Good yield potential
1st rains, start norm. end of March: E. mat. foxtail millet like 1 Se 285 (~60%), e. mat. proso millet like Serere 1; moth beans (~60%)
2nd rains, start norm. end of Oct.: E. mat. proso millet like Serere 1
Whole year: Buffalo gourds (light soils) and Marama beans, Jojoba (in valleys)

Fair yield potential
1st rains: Dwarf sorghum (50–60%), e. mat. bulrush millet (bird rejecting awned var.); bl. and green grams, cowpeas, chickpeas (on h. bl. soils, late pl.), v.e. mat. bambarra groundnuts (on light soils); dwarf sunflower
2nd rains: E. mat. foxtail millet like 1 Se 285, dwarf sorghum (50–60%); dwarf sunflower (40–50%); bl. and green grams, cowpeas, moth beans, chickpeas (on h. bl. soils, late planted)
Whole year: Sisal, castor C-15

Poor yield potential
1st and 2nd rains: Dryland comp. maize

Grassland and forage
> 3 ha/LU on mixed short grass savanna with buffel grass (*Cenchrus ciliaris*) and horsetail grass (*Chloris roxburghiana*) predominant; saltbush best palatable shrub for reestablishing pasture on overgrazed and eroded places.

---

these limitations can be minimized, and new techniques offer prospects of improved linear and nonlinear estimation procedures. Thus, Katz (1979), in a sensitivity analysis of statistical crop/weather models, demonstrated the potential value of ridge regression techniques as developed by Marquardt (1970) and Marquardt and Snee (1975).

Of course, regression techniques form part of the basic armory of methods used in the newer systems-based research strategies, both for exploratory data analysis and to quantify functional relationships between variables.

Studies employing regression have been developed for a variety of questions about agriculture in relation to climate variability and change. As in the studies of Thompson cited above, a major objective has been the identification of systematic patterns of yield variation. For example, Newman (1978) and Waggoner (1979) have examined whether there has been an increase or decrease in crop yield variability between the 1930s and the present and whether there is a correlation in behavior between major grain-producing regions. In general, they find relative variability has decreased somewhat, while variability expressed as absolute yield has increased. Michaels (1982a) has examined the same question related to new high-yielding (green revolution) varieties of wheat in Mexico and India, using as a control US winter wheat. Contrary to US trends, the Mexican and Indian wheats do display increased variability.

Regression-centered studies were extremely popular during the 1970s as nations and regions tried to assess their vulnerability to climatic variability. Good examples of national studies are found in McKay and Allsopp (1977) for Canada, National Research Council (1976) for the United States, and Takahashi and Yoshino (1978) for Japan and southeast Asia. For the World Climate Conference, Oguntoyinbo and Odingo (1979) explored relations among agriculture, climatic variability and various other causal factors in Africa, while Mattei (1979), McQuigg (1979), and Fukui (1979) prepared perspectives on the semi-arid regions, temperate regions and humid tropics, respectively. All report on regression and correlation analyses quite extensively.

A recent application of a regression approach to impacts of long-term climatic change is found in Waggoner (1983). Waggoner explored what the impacts of an annual average warming of 1 °C and decrease of 10 percent in rainfall might be on wheat, corn and soybean production in the US Middle West, under assumptions of no adjustment or adaptation. While such studies are obviously tentative, they may give a sense of the order of magnitude of prospective impacts. These empirical/statistical models, like those that relate crop yields to atmospheric pressure patterns (Steyaert *et al.*, 1978; Michaels, 1982b) may give sound results, providing that climate change does not induce atmospheric patterns that lie outside the range for which the yield regressions have been developed. This was recently illustrated by Santer (1984), using the work of Hanus (1978), whose models for winter wheat usually predict actual West German yields within 4 percent, except for the anomalous year 1976. In that year precipitation in the key month of June was a quarter to a half of the long-term average for most of West Germany, effectively outside the range from which the regressions were developed. Applying scenarios derived from two general circulation models of CO_2-induced change to the Hanus models evidenced similar problems, leading Santer to conclude that simple multiple linear regression appraisal may be basically unsuitable for making any credible assessment of the impacts of climatic change on crop yields (Santer, 1984).

5.2.4 Multivariate Techniques

These techniques represent a logical mathematical development of regression-based analysis in that they permit simultaneous analysis of sets of interacting variables; multivariate analysis, factor analysis, principal component analysis, and principal coordinate analysis are examples. While of principal value for data reduction and in exploratory data analysis, they can provide a basis for useful prediction and partial explanation. However, the derived principal component values, eigenvalues, or factor scores remain static representations of dynamic interactive processes, and their validity is strictly limited to the data domains used to develop them. While it is possible to determine the relative contributions of separate variables to the derived composite axes, it is often difficult to provide any physical explanation for their operation.

Sophisticated applications of multivariate approaches are found in the work of Monteith (1972, 1977a,b, 1978a,b). Monteith (1972) developed a simple model of crop growth for tropical crops, including climatic factors, and has modified it for application in other climatic zones. For example, Monteith (1977b) employed the model to estimate the ultimate limit set by climate on the productivity of British farms, defining efficiency of crop production in thermodynamic terms as the ratio of energy output (carbohydrate) to energy input (solar radiation). Temperature and water supply were found to be the main climatic constraints on efficiency. Over most of Britain, where radiation and thermal climates are uniform, Monteith found rainfall to be the main discriminant of yield between regions.

Another example of climate impact studies employing multivariate analysis is found in the discussion of Gifford (1979) and Monteith (1978b), who have debated maximum growth rates for crops like maize, sorghum, millet and sugarcane (the so-called C_4 crops), that grow well *only* in warm climates, and potatoes, sugar beets, rice, kale and cassava, growing either in warm or cool climates. Such studies come close to attempts at full simulation of plant behavior. The simulations, discussed below, are based more on hypothesized causal relations than on observed historical relationships, but the distinction is often blurred.

5.2.5 Systems Analysis

Very simply, systems analysis is concerned with the resolution of any complex system into a number of simpler components and the identification of important linkages between them. Subsequent synthesis in the form of a symbolic representation (diagrams, flow charts) that is formalized as a set of logical statements or algorithms and mathematical formulae (program) leads to the construction of a model of the real system. This model is then tested, using appropriate data inputs, and revised and developed to the point where it successfully simulates the behavior of the real system. Beyond this point the model can be used to predict probable outcomes of experiments performed on the system. For many complex systems simulation-based experiments represent the only possible means of examining perturbations to the system. Climate and weather systems and biological systems are examples of complex systems that can benefit from the systems approach. Figure 5.2 offers a generic

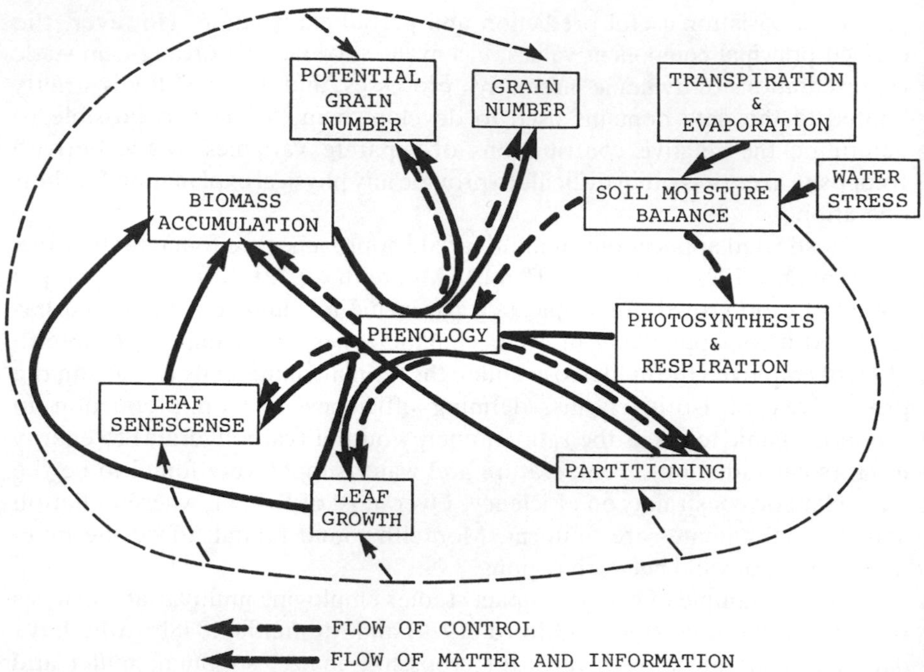

Figure 5.2 Plant process model diagram. Source: Hodges, T. (1984), personal communication

representation of a plant-level simulation model combining the physics of the water balance with the physiology of the growing plant. Simulation models based on such process models have been used both in the United States and Europe to simulate the behavior of plants in response to a CO_2-induced climate change (personal communication, Sakamoto, 1984). The effects of a climate change potentially induced by CO_2 (+1 °C temperature/−10% precipitation) is shown for North Dakota spring wheat in Figure 5.3 from Waggoner (1983). The warmer and drier climate shifts the frequency of yields simulated from actual 1949–80 weather to yields two quintals lower, from a median of 8.5 to 6.5 quintals per hectare of yield. Using the simpler and more general Briggs plant biomass model for Europe, Santer (1984) has simulated two sets of general circulation model (GCM) climate results for a doubling and quadrupling of atmospheric carbon dioxide. He concludes that a simple crop/weather simulation model may be more suitable for the purposes of agricultural impact analysis than the linear regression models frequently used in such studies.

The process of formal analysis of any complex system yields other benefits in that it formalizes what is known about the system, identifies the more important components and processes and significant constraints on perform-

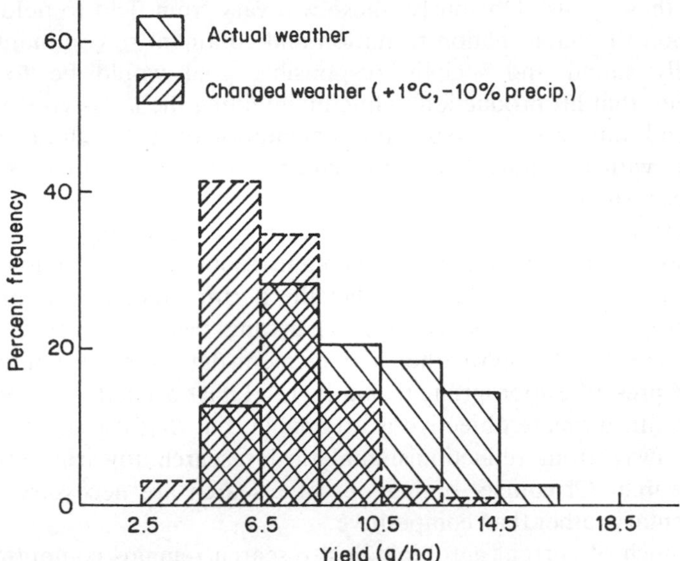

Figure 5.3 North Dakota simulated spring wheat yield (1949–80). Reproduced with permission from Waggoner (1983, 404)

ance, and provides a suitable vehicle for testing alternative management strategies and the 'what if' questions posed by scenarios of global climatic change. Despite the proliferation of systems terminology and its ready incorporation into program titles and statements of research objectives, few agrobiological research strategies are truly systems-based. Essentially, this is because a systems-research strategy demands interdisciplinary teamwork that transgresses the boundaries of long-established discipline and subject-specialist groups and competes with them for scarce resources. As in any research team, leadership is critical, but special problems relate to the maintenance of subject-specialist skills of team members, peer recognition, and the allocation of rewards and responsibilities. Most successful systems-research groups have evolved around a nucleus of a small group of committed scientists rather than through the imposition of a systems-research structure by higher authority. Glantz *et al.* (Chapter 22, this volume) discuss the achievements and difficulties of several systems-oriented studies of climate impacts.

5.3 AGROCLIMATIC SYNTHESIS: PROSPECT

Past, present and, predictably, future modes of agroclimatic analysis and synthesis are inextricably linked with the broader goals of agricultural, pastoral and forestry development and land and water resource management. What, then, are these goals? Obviously, these will vary from field to field, farm to farm, region to region, nation to nation and continent to continent. But an ecologically sound and socially responsible goal would be to develop technologies that lift production, while maintaining the long-term stability of the land and water resource base. Implementation of such technologies rests, ultimately, with the individual land manager, whether profit maximizer or subsistence farmer.

The central problem, then, is how to prescribe a technology that is relevant to the land, labor, capital and management resources of the individual land manager. If it were possible to predict the performance of any production system (crop, pasture/livestock, forest) at any location given a specified minimum set of soil/crop/weather-management data, then it would become possible to prescribe appropriate technologies. I have argued (Nix, 1968, 1976, 1981) that this ultimate objective is attainable, but that it requires a shift in emphasis away from reductionist/analytical research towards holistic/synthetic research. Of course, both types of research are necessary; they are complementary rather than competitive.

While much of current agrobiological research remains committed to the statistical differentiation of treatment effects on specific crops and cultivars, and on specific soils at specific locations in specific seasons, there are encouraging signs that greater efforts are being directed toward solution of the general problem. Of necessity, the earliest attempts at modeling agricultural systems were concerned with specific processes and/or specific products within restricted environments. Now, wide ranges of fully operational models of crop, livestock and forest systems are available or are under active development. Some, at least, are capable of general application across a spectrum of crops and environments. An example is 'EPIC', the Erosion-Productivity Impact Calculator (Williams *et al.*, 1983a,b).

Active development of simulation modeling and access to sophisticated computer technology are not in themselves, however, sufficient evidence of adoption of a systems-research strategy that aims at prescriptive rather than descriptive solutions. If, as stated, the primary objective is to develop models capable of providing both general solutions and location-specific prescriptions, then the focus must be on the balanced development of two interactive components—(a) the models, and (b) the data base. The models should be fully accessible and operational at all times, yet capable of continuing improvement in logical structure and function. The use of subsystem modules facilitates flexibility. The data base contains only those physical, biological, social and economic data that are specified as necessary for implementation of the model. Balanced development of both components is critical. Without access to prescribed input and test data, models cannot be implemented. On the other hand, the acquisition of data for which no specific requirement exists can be wasteful of scarce resources.

5.3.1 Model Development

Many would agree with Passioura (1973) that models should be simple; most would agree that they should be testable and capable of improvement after testing. Following the last two decades of development of models of ever-increasing complexity, there are signs of real efforts to return to simpler and more general crop models. In many cases, when it comes to implementation of the model in the real world, this is no more than a response to the reality of data limitations. In other cases the ability to simplify and generalize functional relationships has been a consequence of real improvements in understanding through basic research. Increasingly, models are being constructed using modular concepts that permit rapid updating and transfer of modules between models. Thus, it is possible to recognize common modules for, say, the water balances processes in many different models.

Van Dyne and Abramsky (1975) reviewed agricultural systems models then extant in some detail. They defined agricultural systems broadly and included those renewable natural resource systems and artificial biological resource systems that are directly under the control of man. They tabulated characteristics (structure; mathematical techniques; time-step; number of driving variables, processes and parameters; programming language; availability of flow charts and source code; stated objective and actual use and comments) for 90 papers and referenced 66 others in modeling applications to dairying and beef feedlotting, grazing sheep, irrigation, pest control, crop growth and yield, forestry, fisheries, wildlife management and agricultural economies. A general conclusion was that most models were poorly documented and thus not readily transferable (see also Robinson's discussion in Chapter 18). Many models provided a useful learning experience for the developers, but few become fully operational in day-to-day planning and management. It is hoped that this situation is changing, but Van Dyne and Abramsky's recommendations on documentation bear repeating:

1. specify objectives,
2. specify the hypotheses or assumptions used,
3. specify the general mathematical form of the model,
4. list the specific driving variables,
5. list the state variables used,
6. show a diagram of the model structure,
7. list equations or functional relationships,
8. list the computer code,
9. provide adequate comment statements,
10. report model deficiencies and limitations,
11. provide test samples of data used, and
12. provide output of a test case.

Although new scientific journals are catering to the specialized requirements of agricultural systems models, few can afford to publish lengthy computer program listings. Full documentation along the above lines as part of the program listing issued to potential users on microfiche could be one solution.

Various attempts have been made to classify the wide variety of models in use and under development, ranging from basic purpose (for example, descriptive *vs* prescriptive, simulation *vs* optimization) to type of formulation (for example, static *vs* dynamic, deterministic *vs* probabilistic). Van Dyne and Abramsky (1975) in their comprehensive review found that most agricultural system models were deterministic; the same would be true today. Stochasticity is introduced through the use of historical, synthesis or real-time weather and other environmental data. Ideally, the output should be expressed as a probability function.

Agricultural production systems of all kinds and at all scales draw upon three basic types of resources. Natural resources include the physical (for example, climate, terrain and soil) and the biological (for example, crop and livestock gene pools). Capital resources include items of plant and machinery, draught animals, fertilizers, herbicides, pesticides and so on. Human resources, the vital component, include all aspects of labor and management. Many models have concentrated on the biological and physical processes that determine crop growth, development and yield and/or animal production, and others have focused on optimization procedures that explore economic and/or social objectives. Few, as yet, have attempted to encompass the whole spectrum of interactions and relational processes. Robinson (Chapter 18, this volume) discusses modeling attempts to place the agricultural system in a global context.

An interesting systems perspective, though not formalized in a mathematical model, is that of Anderson (1979). Anderson, employing a general framework encompassing farm, regional, industrial, sectoral and national levels, has examined impacts of climate variations on Australian agriculture. Anderson begins with behavioral assumptions at the farm level, assuming first that satisfaction in farming depends on both personal characteristics and the probability distribution of financial performance. Financial performance depends in turn on allocative decisions, institutions and governmental interventions, prices and yields. The last two are, of course, in turn dependent on climate and a variety of other factors.

As Anderson (1979) reports, the effects of extreme climatic regimes have been sketched as unambiguously severe in affected regions. At high levels of geographic aggregation, impacts are often less severe because of the 'cancelling out' of good and bad experiences (Waggoner, 1979). McIntyre (1973) estimated that gross rural output in Australia in the drought years 1965–66 and 1967–68 fell by 11 percent and 15 percent, respectively, below smoothed trend values. There is debate over the multiplier effects of drought on the national economy, some results suggesting amplification and others dampening.

The Erosion-Productivity Impact Calculator (EPIC) model referred to above is one of the most comprehensive agricultural system models now operational, and it is undergoing continued development. It addresses a vital socioeconomic and biophysical issue—soil erosion—at national levels, but it could have addressed climate change and variation equally well. EPIC has been developed by a large modeling team formed within the Agricultural Research Service of the United States Department of Agriculture. The brief given was to determine the relationship between erosion and productivity for the whole range of soil and crop interactions in the United States. Williams *et al.* (1983a) provide a detailed description of the model, which has physically based components for simulating erosion and plant growth and economic components for assessing the cost of erosion, determining optimal management strategies, and so on. The physical components include weather simulation, hydrology, erosion/sedimentation, nutrient cycling, plant growth, tillage and soil temperature.

5.3.2 Data Base Development

For every model it is necessary to identify the minimum data set necessary for its successful development, validation and subsequent implementation. The early stages of conceptualization, flow-charting, and programming usually rely on the current understanding of process, function and interaction as established by literature review. Any testing and validation relies on readily available data. Many modeling exercises never progress beyond this stage. Those that do soon realize that data limitations become the major constraint on further progress. This is particularly so when further development and testing require balanced and matching site/crop/weather-management data from a range of contrasting environments. Ideally, specific field, laboratory and/or controlled environment experiments would be designed to acquire the necessary minimum data set in the shortest possible time with the least expenditure. In reality, most model building and testing in this phase have had to rely on a fortuitous assembly of data from a range of sources. Few existing experiments, unfortunately, can satisfy even the barest minimum requirement for balanced and matching sets of soil/crop/weather-management data.

Proposals that standardized data sets be collected from experiments are not new. A number of national and international research programs have adopted this principle. What is new about the concept of *minimum* data sets is that it arises directly from the adoption of a systems-research strategy. Each model specifies the minimum data set necessary for its development, valid action and implementation. Review of existing operational models of agricultural production systems in general, and of crop systems in particular, indicates that many data items are common to all minimum data sets. In fact, a remarkable convergence in specification is under way. All share the need for primary weather data; terrain and soil attributes that modulate water balance and nutrient cycling; plant growth attributes that specify growth and development responses and yield accumulation processes; and management inputs that modify any of the foregoing. While it is possible to erect a hierarchy of minimum data sets with increasing range, precision, accuracy and frequency of measurement, at each level emphasis remains on a balanced monitoring of the whole crop system (Nix, 1979).

A new international program, International Benchmark Site Network for Agrotechnology Transfer (IBSNAT), was initiated by the US Agency for International Development in 1983 with the objective of developing a systems-based strategy for agrotechnology transfer. Through collaborative experiments in a large number of tropical and subtropical countries, standardized minimum data sets are to be collected for major food and cash crops. Development of operational models of these crops will proceed in parallel. The field experimenter will have access to models that will run on his or her own data, and the modelers will have access to standard minimum data sets from a wide range of contrasting environments. Useful and profitable interactions between the two groups can be expected to follow. Provided that the specific minimum data sets are monitored, no particular restrictions on experimental design need apply.

The IBSNAT program should help to remove constraints that currently hamper the further development and testing of operational models of crop systems. But what of the application of these developed models to real-world problems on an extensive scale? Such implementation requires that the appropriate model be coupled to a specified data base. Usually this will involve a climatic data file and either standardized or actual terrain, soil, crop and management variables. Obviously, the full flowering of these new technologies will depend on the coupling of an appropriate model to a resources data base that provides the necessary climate, terrain, soil and management data at the required scale and level of resolution.

New technologies currently available and under active development offer prospects of economical and efficient computer storage of geocoded data. Automatic digitizing equipment can process and store contour maps as arrays of grid points. Sophisticated surface fitting algorithms (see, for example, Wahba and Wendelberger, 1980) can be used to reconstruct the contoured surface, but more importantly, to estimate slope, aspect, elevation, position in landscape, water concentration, and overland flow trajectories and other important terrain-related attributes. Combined with similar algorithms that permit estimation of long-term mean climatic data at any point (Williams, 1969; Williams *et al.*, 1980; Hutchinson *et al.*, 1984), the stage is set for the derivation of necessary input data for models at levels of resolution that are useful.

The generation of stochastic sets of daily weather data based on statistical analysis of historical weather sequences and correlations between individual weather elements has been shown to be practicable. Initial attention focused on daily rainfall (see, for example, Nicks, 1974), but this has been extended to include other important elements such as maximum and minimum temperature and daily solar radiation (Richardson, 1981; Larsen and Pense, 1982). More recently still, Richardson (1984) has demonstrated that interpolation procedures can be used to generate satisfactory sequences of daily weather data at any geocoded point. The implications of this new technology for systems-based research and use of models in agrotechnology transfer are profound.

Technically, there are no reasons why these techniques should not be extended to generate stochastic sequences of daily weather data for much of the surface of the planet. On land, the value of such a data base for land evaluation, planning and management would be incalculable. Perturbation of the basic equations could be used to generate notional climates in some postulated climatic change and coupled to crop models to provide 'what if' answers to speculative questions.

5.4 CONCLUSION

All forms of crop/climate analysis have been used in land evaluation for specific forms of production; for analysis of crop/environment interactions in relation to crop adaptation; for crop monitoring and yield forecasting; for development and testing of new and modified management strategies and tactics; for risk assessment; for pest and disease assessment, management and control; for research and development strategy; and for developing our understanding of complex agricultural production systems.

Progress towards an ultimate goal of prediction of probable outcomes (physical, biological, ecological, economic) of any specified management treatment on any production system at any location has followed an evolutionary path from simple observation, trial and error, transfer by analogy, regression, multivariate techniques, analysis of variance through to systems analysis.

Over time each of the methods of modeling crop systems has accounted for more of the variance in crop yields (Nix, 1983). As shown in Figure 5.4, simple observation may be associated with 15 percent of the explained variance, trial and error field tests about 40 percent, transfer by analogy, 50 percent, and correlation and regression, 65 percent. The first generation (0) of systems and simulation explains about 80 percent of the variance, and future generations (1, 2, 3) of these models might bring it to 90 percent.

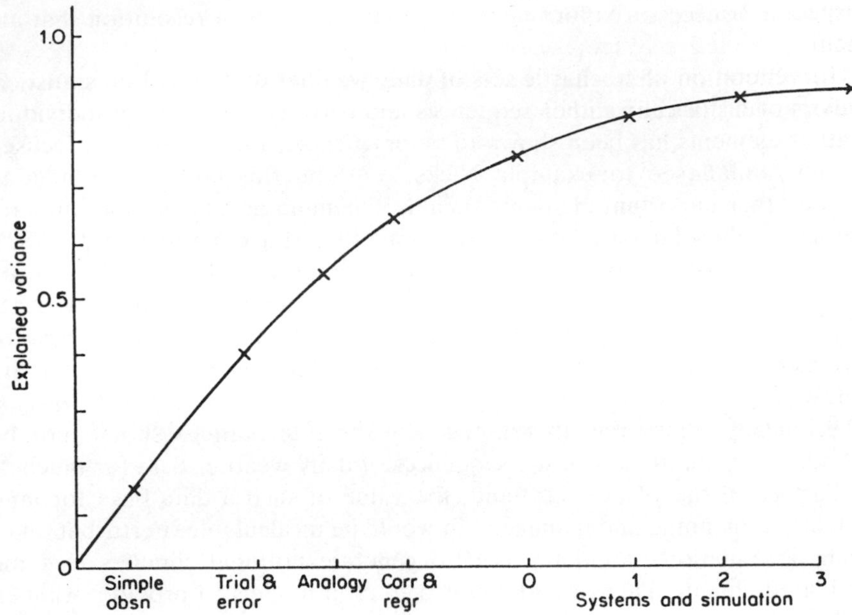

Figure 5.4 Modeling crop system. (Source: Sakamoto, 1983, modified after Nix, 1983)

It must be emphasized that these methods did not evolve in direct linear succession and are not mutually exclusive. Each can be expected to continue to play a role in climate impact assessment. Obviously, however, systems analysis and simulation techniques offer the best prospects of understanding and managing the complex interactions between climate and agricultural production systems.

In recent years the possibility of secular climatic change has focused attention on climate-related human activities and, in particular, agricultural production. While most scientists grapple with the problems of predicting crop response to weather variations within existing climate, a few have been so bold as to make predictions about global food production in relation to postulated climatic change. The empirical/statistical modes of analysis used generally do not inspire confidence in the results.

Agricultural production system models now operational and currently under development offer prospects of both general solutions and location-specific prescriptions. Their impact on all aspects of agricultural research, development and production will be profound. Because climate and weather provide primary inputs and forcing functions, these models offer the most direct means of climate impact assessment. General application of the new generation models, however, will depend on the availability of the minimum data sets that are specified as input. Accordingly, we can expect to see the development of dynamic interactive systems of environmental data banks coupled to appropriate simulation models that will be used for all aspects of agricultural development, planning and management.

REFERENCES

Anderson, J. R. (1979). Impacts of climatic variability on Australian agriculture. In *The Impact of Climate on Australian Society and Economy*. Commonwealth Scientific and Industrial Research Organization (CSIRO), P.O.B. 77, Mordialloc, Victoria, Australia.

Baier, W. (1977). *Crop-weather Models and Their Use in Yield Assessment.* WMO No. 458, Technical Note 151. World Meteorological Organization, Geneva: 48 pages.

Baier, W., and Robertson, G. W. (1968). The performance of soil moisture estimates as compared with the direct use of climatological data for estimating crop yields. *Agricultural Meteorology*, **5**, 16–31.

Biswas, A. K. (1980). Crop-climate models: A review of the state of the art. In Ausubel, J., and Biswas, A. K. (Eds.) *Climatic Constraints and Human Activities*, pp. 75–92. Pergamon Press, Oxford.

Blasing, T. J., and Solomon, A. M. (1983). *Response of the North American Corn Belt to Climate Warming*. Prepared for the US Dept. of Energy. National Technical Information Service, US Dept. of Commerce, Springfield, Virginia.

Doorenbos, J., and Kassam, A. H. (1979). *Yield Response to Water.* FAO Irrigation and Drainage Paper. Food and Agriculture Organization, Rome: 193 pages.

Emanuel, W. R., Shugart, H. H., and Stevenson, M. P. (in press). Climatic change and the broad-scale distribution of terrestrial ecosystems complexes. *Climatic Change* (accepted for publication May 1984).

Fisher, R. A. (1924). The influence of rainfall distribution on the yield of wheat at Rothamstead. *Royal Society of London Philosophical Transactions*, **Series B213**, 89–142.

Fukui, H. (1979). Climatic variability and agriculture in tropical moist regions. In *Proceedings of the World Climate Conference*, pp. 426–474. World Meteorological Organization, Geneva.

Gifford, R. M. (1979) *Australian Journal of Plant Physiology*, **1**, 107.

Hanus, H. (1978). *Forecasting of Crop Yields from Meteorological Data in the EC Countries*. Statistical Office of the European Communities, Agricultural Statistical Studies No. 21, Luxembourg.

Hargreaves, M. W. M. (1957). *Dry Farming in the Northern Great Plains 1900–1925*. Harvard University Press, Cambridge, Massachusetts.

Heathcote, R. L. (1965). *Back of Bourke*. Melbourne University Press, Carlton, Victoria, Australia.

Holdridge, L. R. (1947). Determination of world plant formations from simple climatic data. *Science*, **105**, 367–368.

Hooker, N. H. (1907). Correlation of weather and crops. *Royal Statistical Society Journal*, **70**, 1–42.

Hutchinson, M. F., Booth, T. H., McMahon, J. P., and Nix, H. A. (1984). Estimating monthly mean values of daily total solar radiation for Australia. *Solar Energy* (in press).

Jaetzold, R. (1983). Agro-ecological zonation in Kenya. *Network for Environment & Development*, **3**, (3, December), 1–4. Available from International Development

Program, Center for Technology, Environment, and Development, Clark University, Worcester, Massachusetts.

Jaetzold, R., and Kutsch, H. (1982). Agro-ecological zones of the tropics, with a sample from Kenya. *Der Tropenlandwirt*, **83**, (2, April), 15–34.

Jaetzold, R., and Schmidt, H. (Eds.) (1982–83). *Farm Management Handbook of Kenya*. Volume 2: *Natural Conditions and Farm Management Information*. Part A: West Kenya (1982), Part B: Central Kenya (1983), Part C: East Kenya (1983). Ministry of Agriculture, Nairobi.

Katz, R. W. (1979). Sensitivity analysis of statistical crop-weather models. *Agricultural Meteorology*, **20**, 291–300.

Kupperman, K. O. (1982). The puzzle of the American climate in the early colonial period. *The American Historical Review*, **87** (5, December), 1262–1289.

Larsen, G. A., and Pense, R. B. (1982). Stochastic simulation of daily climatic data for agronomic models. *Agronomy Journal*, **74**, 510–514.

Mack, A. R., and Ferguson, W. S. (1968). A moisture stress index for wheat by means of a modulated soil moisture budget. *Canadian Journal of Plant Science*, **48**, 535–543.

Malin, J. C. (1944). *Winter Wheat in the Golden Belt of Kansas: A Study in Adaptation to Subhumid Geographical Environment*. University of Kansas Press, Lawrence, Kansas.

Marquardt, D. W. (1970). Generalised inverses, ridge regression, biased linear estimation and nonlinear estimation. *Technometrics*, **12**, 591–612.

Marquardt, D.W., and Snee, R. D. (1975). Ridge regression in practice. *The American Statistician*, **29**, 3–20.

Mattei, F. (1979). Climatic variability in agriculture in the semi-arid tropics. In *Proceedings of the World Climate Conference*, pp. 475–509. World Meteorological Organization, Geneva.

McIntyre, A. J. (1973). Effects of drought in the economy. In Lovett, J. V. (Ed.) *The Environmental, Economic and Social Significance of Drought*. Angus and Robertson, Sydney.

McKay, G. A., and Allsopp, T. (1977). Climate and climate variability. In *Climatic Variability in Relation to Agricultural Productivity and Practices*. Committe on Agrometeorology, Department of Agriculture Research Branch, Ottawa, Canada.

McQuigg, J. D. (1979). Climatic variability and agriculture in the temperate regions. In *Proceedings of the World Climate Conference*, pp. 406–425. World Meteorological Organization, Geneva.

Michaels, P. J. (1982a). The response of the 'green revolution' to climatic variability. *Climatic Change*, **4**(3), 255–271.

Michaels, P. J. (1982b). Atmospheric pressure patterns, climatic change and winter wheat yields in North America. *Geoforum*, **13**(2), 263–273.

Monteith, J. L. (1972). Solar radiation and productivity in tropical ecosystems. *Journal of Applied Ecology*, **9**, 747–766.

Monteith, J. L. (1977a). Climate. In De T. Alvim, P. (Ed.) *Ecophysiology of Tropical Crops*. Academic Press, New York and London.

Monteith, J. L. (1977b). Climate and the efficiency of crop production in Britain. *Royal Society of London Philosophical Transactions*, **Series B281**, 277–294.

Monteith, J. L. (1978a). Models and measurement in crop climatology. *International Society of Soil Science, Canada*, 385–398.

Monteith, J. L. (1978b). Reassessment of maximum growth rates for C_3 and C_4 crops. *Exploratory Agriculture*, **14**, 1–5.

National Research Council (1976). *Climate and Food: Climate Fluctuation and US Agricultural Production*. Committee on Climate and Weather Fluctuations and Agricultural Production, National Academy of Sciences, Washington, DC.

Newman, J. E. (1978). Drought impacts on American agricultural productivity. In Rosenberg, H. J. (Ed.) *North American Droughts*. Westview Press, Boulder, Colorado.

Newman, J. E. (1980). Climate change impacts on the growing season of the North America 'cornbelt'. *Biometeorology*, **7**(2), 128–142.

Nicks, A. D. (1974). Stochastic generation of the occurrence, pattern and location of maximum amounts of daily rainfall. In *Proceedings of Symposium on Statistical Hydrology, Tucson, Arizona*, pp. 154–171 (Miscellaneous Publication No. 1275).

Nix, H. A. (1968). The assessment of biological productivity. In Stewart, G. A. (Ed.) *Land Evaluation*. Macmillan, Melbourne, Australia.

Nix, H. A. (1976). Climate and crop productivity in Australia. In *Rice and Climate*. IRRI, Los Banos, Philippines.

Nix, H. A. (1979). Agroclimatic analogues in transfer of technology. In *Development and Transfer of Technology for Rainfed Agriculture, Proceedings of an International Symposium, Hyderabad, India*. ICRISAT, Hyderabad.

Nix, H. A. (1980). Strategies for crop research. *Proceedings of the Agronomic Society of New Zealand*, **10**, 107–110.

Nix, H. A. (1981). Simplified simulation models based on specified minimum data sets: The CROPEVAL concept. In Berg, A. (Ed.) *Application of Remote Sensing to Agricultural Production Forecasting*. Balkema, Rotterdam.

Nix, H. A. (1983). Minimum data sets for agrotechnology transfer. Presented at the *ICRISAT-IBSNAT-SMSS International Conference on Minimum Data Sets for Agrotechnology Transfer, Hyderabad, India, March 1983*. Conference Proceedings published by IBSNAT, Honolulu (forthcoming 1984).

Nix, H. A., and Fitzpatrick, E. A. (1969). An index of crop water stress related to wheat and grain sorghum yields. *Agricultural Meteorology*, **6**, 321–337.

Oguntoyinbo, J. S., and Odingo, R. S. (1979). Climatic variability and land use. In *Proceedings of the World Climate Conference*, pp. 552–580. World Meteorological Organization, Geneva.

Parry, M. L. (1978). *Climatic Change, Agriculture and Settlement*. William Dawson & Sons, Folkestone, UK.

Passioura, J. P. (1973). Sense and nonsense in crop simulation. *Journal of the Australian Institute of Agricultural Science*, **39**, 181–183.

Richardson, C. W. (1984). Stochastic simulation of daily precipitation, temperature, and solar radiation. *Water Resources Research*, **17**(1), 182–190.

Richardson, C. W. (1984). *WGEN: A Model for Generating Daily Weather Variables*. USDA, ARS-8. US Dept. of Agriculture, Agricultural Research Service, Washington, DC: 83 pages (forthcoming 1984).

Rosenberg, N. J. (1982). The increasing CO_2 concentration in the atmosphere and its implications on agricultural productivity II. Effects through CO_2-induced climatic change. *Climatic Change*, **4** (3), 239–254.

Sakamoto, C. M. (1983). New generation of crop models. Lecture presented at the *Agroclimatic Technology Program for Southeast Asia, April 11–15, 1983, Bangkok, Thailand*.

Santer, B. (1984). Impacts on the agricultural sector. In Meinl, H., and Bach, W., Socioeconomic Impacts of Climatic Changes Due to a Doubling of Atmospheric CO_2 Content. Contract No. CL1-063-D, Commission of the European Communities, Brussels.

Sneath, P. H. A., and R. R. Sokal (1974). *Numerical Taxonomy*. W. H. Freeman and Co., San Francisco, California.

Stanhill, G. (1977). Quantifying weather-crop relations. In Landsberg, J. J., and Cutting, C. V. (Eds.) *Environmental Effects on Crop Physiology*. Academic Press, New York.

Steyaert, L. T., Leduc, S. K., and McQuigg, J. D. (1978). Atmospheric pressure and wheat yield modeling. *Agricultural Meteorology*, **19**, 23–24.

Takahashi, K., and Yoshino, M. M. (Eds.) (1978). *Climatic Change and Food Production*. University of Tokyo Press, Tokyo.

Thompson, M. L. (1969a). Weather and technology in the production of wheat in the United States. *Journal of Soil and Water Conservation*, **24**, 219–224.

Thompson, L. M. (1969b). Weather and technology in the production of corn in the US corn belt. *Agronomy Journal*, **61**, 453–456.

Thompson, L. M. (1970). Weather and technology in the production of soybeans in the central United States. *Agronomy Journal*, **62**, 232–236.

Van Dyne, G. M., and Abramsky, Z. (1975). Agricultural systems models and modelling: An overview. In Dalton, G. E. (Ed.) *Study of Agricultural Systems*. Applied Science Publishers, London.

Waggoner, P. E. (1979). Variability of annual wheat yields since 1909 and among nations. *Journal of Agricultural Meteorology*, **20**, 42.

Waggoner, P. E. (1983). Agriculture and a climate changed by more carbon dioxide. In National Research Council, *Changing Climate: Report of the Carbon Dioxide Assessment Committee*, pp. 383–418. National Academy Press, Washington, DC.

Wahba, G., and Wendelberger, J. (1980). Some new mathematical methods for variational objective analysis using splines and cross validation. *Monthly Weather Review*, **108**, 1122–1143.

Williams, G. D. V. (1969). Applying estimated temperature normals to the zonation of the Canadian Great Plains for wheat. *Canadian Journal of Soil Science*, **49**, 263–276.

Williams, G. D. V., McKenzie, J. S., and Sheppard, M. I. (1980). Mesolscale agroclimatic resource mapping by computer, an example for the Peace River region of Canada. *Agricultural Meteorology*, **21**, 93–109.

Williams, J. R., Dyke, P. T., and Jones, C. A. (1983a). EPIC: A model for assessing the effects of erosion on soil productivity. In *Proceedings of the Third International Conference on the State-of-the-Art in Ecological Modelling*.

Williams, J. R., Renard, K. G., and Dyke, P. T. (1983b). EPIC: A new method for assessing erosion's effect on soil productivity. *Journal of Soil and Water Conservation*, **38** (5), 381–383.

Williams, M. (1974). *The Making of the South Australian Landscape*. Academic Press, London.

Williams, W. T. (Ed.) (1976). *Pattern Analysis in Agricultural Science*. Elsevier/CSIRO, New York.

Climate Impact Assessment
Edited by R. W. Kates, J. H. Ausubel and M. Berberian

CHAPTER 6
Fisheries

TSUYOSHI KAWASAKI

Faculty of Agriculture
Tohoku University
Sendai 980, Japan

6.1 THE NATURE OF THE FISHERIES RESOURCE

A natural resource may be defined as a substance which can be extracted from the earth (including the ocean and atmosphere) and usefully transformed by human labor. Natural resources are generally classified in two categories. One category, exemplified by mineral resources, is 'non-renewable'; on relevant time-scales the stock of the resource decreases irreversibly by the amount that is removed. The second category, exemplified by living resources, is 'renewable'; stocks may recover readily after harvesting by man. Living resources may further be classified according to the time required for renewal. Forest resources, for example, tend towards one side of the recovery time-scale while fishery resources tend to the other.

Although many species of fishery resources are capable of rapid recovery following exploitation (for example, herring, sardines, scallops, oysters), some may be very slow to recover (for example, whales and seals). This rate of recovery may also be significantly lengthened if the intensity of harvest is higher than some species-specific optimum.

The renewable nature of fisheries and the inherent spatial and temporal variation in stock size are primary characteristics of fisheries. It is the primary goal of fisheries scientists to understand and ultimately predict the timing and extent of this variation. Spatial and temporal variation in stock size are species-specific characteristics and require knowledge of the species life history such as reproduction, growth, etc. Predicting fluctuations in abundance is further complicated because the oceanic milieu is neither static nor homogeneous in its composition. Furthermore, since the ocean and atmosphere are coupled, both the distribution and abundance of fishery resources are capable of being modified by climate.

6.2 THE CLIMATE/OCEAN/FISHERIES LINKAGE

6.2.1 General Description

Climate exerts an as yet incompletely understood linkage with fisheries. The linkage, however, occurs through the transfer of energy from the atmosphere to an ocean surface layer of variable thickness. The depth of this layer varies with latitude, distance from shore, and seaon but is generally less than 100 meters and may be considerably shallower in the presence of a pycnocline, or density gradient. Below 100 meters less than 1 percent of sunlight penetrates even in the clearest ocean water and wind-generated currents, with the exception of western boundary currents like the Kuroshio or Gulf Stream, are seldom important. In the surface layer or euphotic zone, however, several components of climate, including solar radiation, wind and temperature, may impact fisheries. These variables are of course correlated, but their oceanographic implications and hence their impact on fisheries differ.

Solar radiation influences the ocean through heating and evaporation at the air/sea interface and by providing light energy for photosynthesis throughout the surface layer. Biological production in the sea originates with primary production, a photosynthetic process whereby chlorophyll-containing phytoplankton generate organic matter. In this process, phytoplankton take in dissolved phosphorus and nitrogen-containing compounds, nutrients necessary for growth. The phytoplankton are eaten by zooplankton, or by some clupeids, such as sardines, and shellfish. Zooplankton generally are utilized by small fish, which in turn are consumed by larger ones.

The bodies of organisms and feces are then subject to bacterial decomposition or autolysis. The remains, being denser than sea water, slowly sink or are turbulently mixed downward, resulting in a flux of nutrients from the euphotic zone. Unless compensating upward transport of nutrients occurs, the euphotic zone will become impoverished. Turbulent mixing in the arctic and shallow margins of temperate oceans, as well as upwelling, are primarily responsible for the return of nutrients to the surface layer.

Vertical mixing occurs primarily during winter when solar radiation is low, vertical density gradients weak, and winds are strong. Mixing occurs through turbulence and the depth to which it occurs depends on wind speed, direction and duration. A useful index of turbulent mixing in fisheries studies is simply to use the cube of the wind speed which can be employed in exploratory analysis with fisheries parameters through correlation or regression analyses (Murray *et al.*, 1983). Turbulent mixing results in the redistribution of nutrients which, with the vernal warming due to increasing incident radiation, leads to a flowering of phytoplankton termed the spring bloom. The spring bloom is primarily a high latitude process which is followed by the propagation of animals of higher trophic levels. Tropical areas are generally less productive than higher latitudes. This lower productivity is partly explained by the greater incident radiation and hence deeper euphotic zone, which is not mixed with deeper nutrient-rich waters, and partly by lower inputs of continentally derived nutrients via rivers.

In addition to turbulent mixing, wind also impacts fisheries through wind-generated surface currents which may result in egg and larval transport away from nursery areas and hence, fluctuations in year-class strength (Murray *et al.*, 1983). Wind-generated surface currents may also control productivity through upwelling. This is a process of water motion whereby cold, deeper waters move upward toward the surface as a result of winds displacing warm surface water. While upwelling may occur anywhere, it is most common along the western coasts of continents. The movement of surface water in response to wind is described by a simple mathematical model and is referred to as Ekman transport. In this model, if the wind blows in the same direction for a sufficient duration, say a few days, it will set a surface layer in motion. This surface current is approximately 2 percent of the wind speed and to the right of the wind direction in the northern hemisphere (to the left in the southern hemisphere). The movement of this layer sets in motion the layer below it which in turn sets in motion a deeper layer. The speed of each layer is slower than the layer above it and the direction is displaced further to the right (northern hemisphere) or left (southern hemisphere) than the layer above. For modeling egg and larval transport occurring in the upper few meters of the water column, however, it is preferable to use the modified transport model of Stolzenbach *et al.* (1977) in which current speed is 3 percent of the wind speed and 15° off the wind direction. However, Ekman transport involves a deeper water column (basement is usually taken as the depth at which the currrent speed decreases to 4 percent of the surface speed) and the net transport is 45° to the right or left of the wind direction. Therefore if the wind is blowing approximately parallel to the shore, the surface water will be displaced seaward and replaced by deeper (down to 200 meters), colder nutrient-rich waters. As a result, upwelled waters are highly productive of fish and other organisms. The most marked coastal upwellings are found off western North America, South America, Africa and Australia. The upwelling along the boundary of the Peruvian Current maintains the large stocks of plankton-feeding fish such as the Peruvian anchovy, *Engraulis ringens* (Paulik, 1981). Diverging currents also cause upwelling. Where adjacent surface waters flow away from each other, deeper water must also rise. Such divergence and associated upwelling are found around Antarctica, along the Aleutian Island Chain, along the equator, and along the northern margin of the Equatorial Countercurrent.

6.2.2 El Niño

One of the most dramatic impacts of climate on fisheries occurs every few years off the Peruvian coast. This phenomenon occurs during the southern hemisphere summer months of December to March, when a transequatorial flow of warm low-salinity water originating from the equatorial countercurrent displaces the northbound Peruvian current. Called El Niño, or Christ Child, because of its arrival during the Christmas season, this intrusion of high-temperature and nutrient-poor water 5–10 degrees further south than normal results in mass mortalities of fish, as well as increased evaporation and, consequently, increased precipitation on the adjacent land. Dead fish are washed ashore where they decay; the subsequent formation of H_2S, when combined with sea fog, may even blacken the paint of ships. Sea birds feeding on fish die of starvation and the decline in bird population results in reduced production of guano, the basis of the fertilizer industry.

In 1972, El Niño was very intense; total average production in the euphotic zone during this event was 0.39 $gC/M^2/day$, three times lower than the long-term average off southern Peru. The catch of Peruvian anchovy, which had been 13 million metric tons (19 percent of the world catch in 1970) declined dramatically, while other species increased notably during the following year. In 1972 jack mackerel *Trachurus murphyi*, until then accounting for under 2 percent of the catch in northern Chilean fisheries, suddenly increased to 10 percent. Dramatic increases since 1972 have also occurred in the catch of the Chilean sardine *Sardinops sagax*, paralleling the rise of jack mackerel and accompanying the decline in anchovy (Figure 6.1). The decline in anchovy catch may be explained by intense exploitation as well as by ecological disruption from El Niño in 1972–73 (Caviedes, 1981).

An important feature of El Niño is that it may be predictable several months in advance in the larger context of the Southern Oscillation, one of the major irregular, periodic fluctuations of atmopheric circulation. Wright (1978), Thompson (1981) and Ramage and Hori (1981) discuss causal

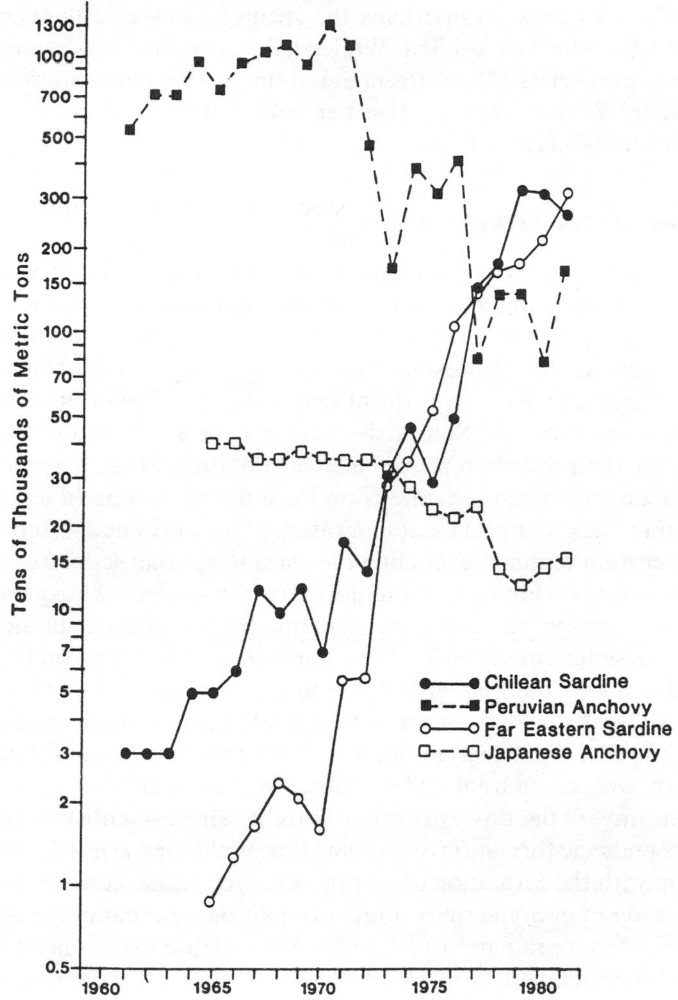

Figure 6.1 Interannual variation in catches of sardine and anchovy

relationships concerning the Southern Oscillation and El Niño. Basin-wide atmospheric and oceanic anomalies appear to be as important as local winds in creating El Niño. Large-scale changes in wind fields far out in the Pacific are of particular interest. The trade winds were unusually strong in 1970 and 1971, but were unusually weak in 1972. The eastward flowing North Equatorial Countercurrent intensified and the South Equatorial Current weakened during the 1972 El Niño. It appears that warm water accumulated in the eastern Pacific, deepening the usually shallow thermocline and covering the colder upwelled waters (Wooster and Guillén, 1974).

The 1982–83 El Niño was perhaps the strongest event of the century and has certainly been the best studied. While analyses are still underway, it is clear that this was not a classical occurrence, but one with a pattern different from earlier events of the century (Barber and Chavez, 1983; Cane, 1983; Rasmusson and Wallace, 1983).

6.2.3 Current Meanders and Oceanic Rings

Other examples of ocean/climate phenomena with important implications for fisheries are the variation in western boundary current axes and the formation of eddies or rings from these currents. It has long been observed that the stream axis of currents like the Kuroshio, Gulf Stream and East Australian current meander. These meanders in turn influence the circulation of water on the continental slope and shelf. Since fish stocks may be associated with particular water masses, their distributions will also be modified. If the current meander grows large enough it can separate from the current as either a warm-core or cold-core ring, depending on the temperature of the entrained water relative to that of the current remnant encircling the core. Rings that separate seaward of the current rotate cyclonically in the northern hemisphere (anticyclonically in the southern hemisphere); those that separate landward rotate in the opposite direction. These features are 100–300 kilometers in diameter, rotate at several knots, and maintain their integrity for as long as several years. During a ring's life they drift and exchange nutrients, biota and energy with the adjacent water mass. Backus *et al.* (1981) have reviewed the dynamics of cyclonic Gulf Stream rings which are comparable with those of the Kuroshio and with the anticyclonic rings of the East Australian current. These scientists point out that to preserve mass the formation of one ring type (cold- or warm-core) must be in equilibrium with the formation of an opposite type. Based on this assumption and their study of cyclonic rings, they estimate that warm-core rings transfer $3–20 \times 10^{15}$ grams of salt and 10^{21} calories of heat from the Sargasso Sea to the continental slope annually. They further estimate that cold-core rings transfer 5×10^{11} grams of carbon as living organic matter from the slope water to the Sargasso Sea, increasing productivity in 10 percent of the Sargasso Sea by 50 percent. The importance of rings to fisheries lies only partly in their role in determining the distribution of oceanic properties. Rings have also been implicated in the destruction of deep water crab pots in the Baltimore Canyon and larger catches of swordfish have been reported from ring boundaries than from adjacent waters (Chamberlain, personal communication).

The cause of meanders and rings is poorly understood; however, Teramoto (1981) discusses the Kuroshio meandering as a local manifestation of a large-scale fluctuation in the ocean/atmosphere system. Temperature changes around Oshima Island south of Tokyo may be caused by the meandering of the Kuroshio, and this change is inversely correlated with change in the surface velocity of the Kuroshio. Changes in ocean surface temperature produce changes in heat transmission that are connected with movement in the lower atmosphere. This interchange between the lower atmosphere and oceanic circulation is depicted in Figure 6.2 as a feedback process.

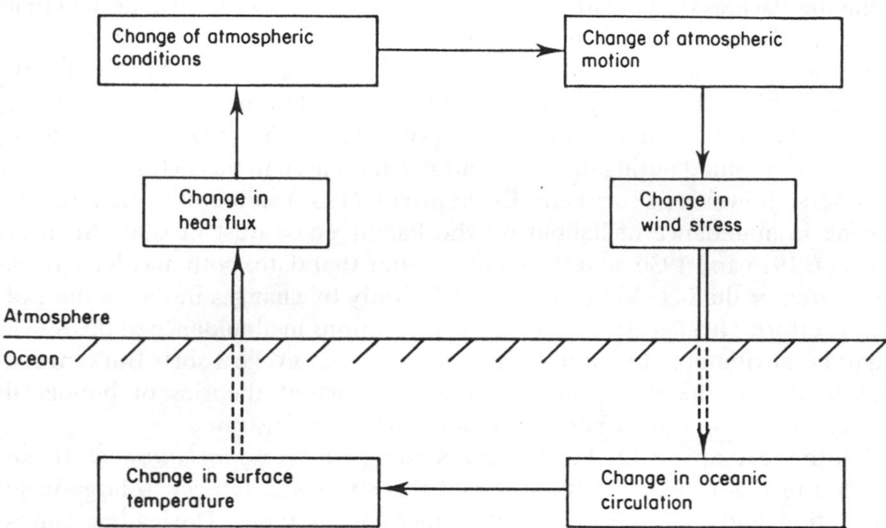

Figure 6.2 Large-scale feedback processes of the ocean/atmosphere system. (Source: Teramoto, 1981, 55)

While links between ocean and atmosphere are easily recognized on local or regional scales, there may also be identifiable distant linkages. So-called 'teleconnections' exhibit significant relation to variations in widely separated areas of the ocean and atmosphere. Teleconnections that impact fisheries have been reviewed by Wyrtki (1973) and Teramoto (1981); for example: precipitation and cloudiness on Canton Island (3 °S, 172 °W) and Ocean Island (1 °S, 169 °E) in the central equatorial Pacific and the difference between water levels along the northernmost and southernmost areas of the North Equatorial Countercurrent (an index of the current); the preceding difference in water levels and El Niño; the difference between water levels and longitudinal components of the velocity of geostrophic winds on a 700 mb plane along 30 °N latitude. Statistical analysis suggests that these teleconnections are lagged by 3–8 months. This same series of teleconnections also seems to be associated with the variation of the Kuroshio. Changes in living resources which might be connected with the Kuroshio meandering will be discussed later.

6.3 NATURAL CAUSE OR OVERFISHING?

There is continuing controversy about the cause of fluctuations in many stocks, since natural (including climatic) causes, or overfishing, or both, can result in a major change in a given fishery. An example illustrating the often equivocal interpretations possible is that surrounding the halibut *Hippoglossus stenolepis* along the Pacific coast of North America. In the early 1950s it was argued that there was an approximately reciprocal relationship between the catch-per-unit of fishing gear and the number of sets of units of gear (Thompson, 1952). Fishing effort had been controlled based on this relation, and it was claimed that regulation of fishing effort was responsible for the recovery of catches, which had declined during the 1930s but had increased in the 1940s.

Others, however, disagreed. Burkenroad (1948) contended that in the decline in abundance of halibut on the Pacific coast west of Cape Spencer between 1915 and 1930, and the increase after that date, both may have been much greater than could be accounted for only by changes in the amount of fishing effort. He suggested that major fluctuations in abundance of this stock might be attributable to natural causes of a regular cyclical sort. Burkenroad thus held that the desirability of applying current theories of biological management to marine fisheries remained to be demonstrated.

Another example is the Far Eastern sardine *Sardinops melanosticta.* It can be seen in Figure 6.3 that the catch of this species is subject to large-scale fluctuation, and this has been recorded since ancient times. The leading causes of the fluctuation have been attributed to both environmental factors and overfishing.

The catch of this sardine fell abruptly from a peak of 2,730,00 tons in 1937 to 9000 tons in 1965, a drastic reduction in population size (Figure 6.3). Nakai (1962) noted that in an area south of the Kil Peninsula a cold water mass had developed during the years 1938–45. He argued that this anomaly of the Kuroshio Current caused mass mortality of the early life stages of this sardine resulting in the depletion of the adult sardine population.

In the early 1970s sardine abundance again began to rise and the catch increased rapidly, resulting in a take of 3,610,000 tons in 1981 (Figure 6.3). The fisheries of the Republic of Korea and the USSR were also high. Kondo *et al.* (1976) explained this increase by a combination of environmental factors. In the spring of 1972 the Kuroshio had shifted its pathway along southern Honshu, Japan, from a meandering to a non-meandering pattern, a condition favoring the distribution of copepod nauplii, a small crustacean fed upon by larval sardines. The increased supply of food for larval sardines enabled the 1972 sardine year-class to recover from the earlier stock reductions.

While some scientists maintain that the ups and downs of stock level of the Far Eastern sardine have resulted from changes of food conditions of larvae due to the shift of the Kuroshio, others emphasize the role of human intervention. Cushing (1975) points out that in 1930 there were 700 purse seiners for the Japanese sardine, 1000 in 1940. Peak catches occurred in 1936, but subsequently fell dramatically with the failure of the 1938–41 year-classes. The collapse in the early 1940s was followed by a period of decline in which Japanese fishermen concentrated on fish in their first summer of life, and a further decline in catch followed a period of high and sustained effort. While Nakai associated this failure in recruitment with an anomaly in the flow of the Kuroshio Current in the years 1938–45, Cushing disputes the suggested environmental source of recruitment failure, emphasizing that fishing effort was high on prespawning fish. Murphy (1977) also attributes the population collapse to intensive exploitation. He criticizes the environmental hypothesis on the grounds of the apparent lack of a consistent cause/effect mechanism for these years. Environmental arguments have been advanced to explain most fish population collapses, but usually a different causal mechanism has been ascribed to each collapse. Murphy points out that the source of, and difficulty with, these environmental hypotheses is that the ocean climate is highly variable; thus, it is possible to find 'significant' shifts or non-shifts in ocean climate associated with almost any biological event one wishes to specify. Murphy concedes that poor year-classes are observable throughout the record, but he contends that they are associated with crashes only when the population is also heavily fished.

Certainly one must be careful in ascribing a change in a stock to a specific oceanic variation. As Murphy (1977) shows, it is not difficult to find an oceanic phenomenon which matches a specific change in a stock. Nakai (1962) and Kondo *et al.* (1976) explained the fluctuation in abundance of the Far Eastern sardine by shifts of the Kuroshio axis, more specifically, the occurrence and disappearance of a cold eddy south of Honshu. However, when the data are examined closely, it is apparent that it was 1934 when the long-lived cold eddy occurred, whereas stock decline began in 1938. Although recent recovery of the sardine population was promoted by a strong 1972 year-class, the origin of the recovery existed before 1970. In spite of the frequent occurrence of anomalies of the Kuroshio between 1945 and 1964, corresponding changes of sardine stock were not seen.

Advocates of the explanation that fluctuation in stocks lies in fishing intensity also face difficulties. The Far Eastern sardine has been known to fluctuate widely with an irregular cycle of decades to a century, and since ancient times the variation in catch has been associated with the history of success and failure of fishing villages. Such large-scale fluctuations in stock in days past, when fishing effort was very small in comparison with this century, are strong evidence that stock fluctuations can be caused by environmental change.

The solution to this problem does not lie exclusively in the alternatives of environment or fishing. It is an oversimplification to seek the cause of stock changes only in a shift of the Kuroshio axis or solely from the effects of fishing. What is important is that we evolve both theory and methods which consider all sources of variation comprehensively. An effort in this direction is offered in the next section.

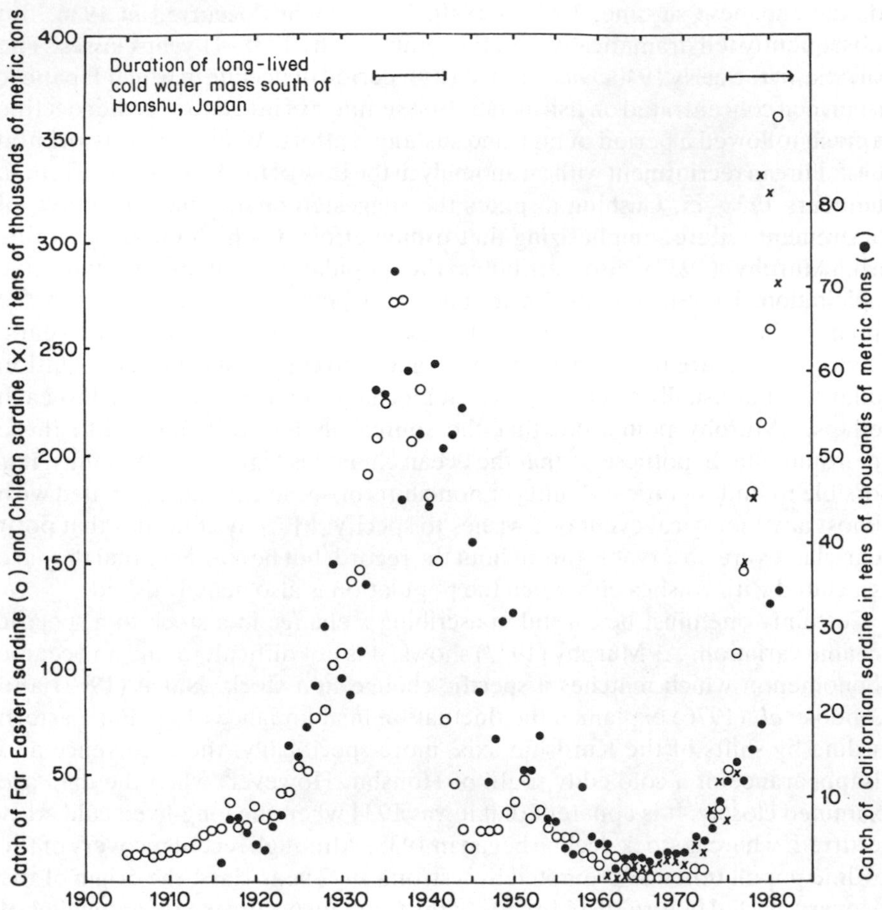

Figure 6.3 Large-scale variations in the catch of three sardine species, Far Eastern, Californian and Chilean

6.4 SPECIES-SPECIFIC PATTERNS OF FLUCTUATION IN NUMBER

6.4.1 General Description and a Methodology

Patterns of fluctuation in the number of marine teleosts (all fish except sharks and rays) can be assigned to two broad types.

Type I: unstable and unpredictable
- Subtype IA: irregular and short-term, e.g. Pacific saury (*Cololabis saira*)
- Subtype IB: large-scale and cyclical, e.g. sardines (genus *Sardinops*)

Type II: stable and predictable, e.g. scombroids (tuna, mackerel, etc.)

Major ecological features of these types are elaborated in Kawasaki (1980).

The above patterns seem to reflect the characteristics of environmental variation in the sea. Examination of the cycle of production of algae and herbivores may illuminate some of the underlying explanation. Cushing (1975) represents production cycles in different regions diagrammatically (Figure 6.4). He notes a single peak in the arctic midsummer, a double peak in temperate waters, and only minor oscillations in the tropics. Seasonal differences increase with latitude, as does the delay period between the production of algae and grazing by herbivores.

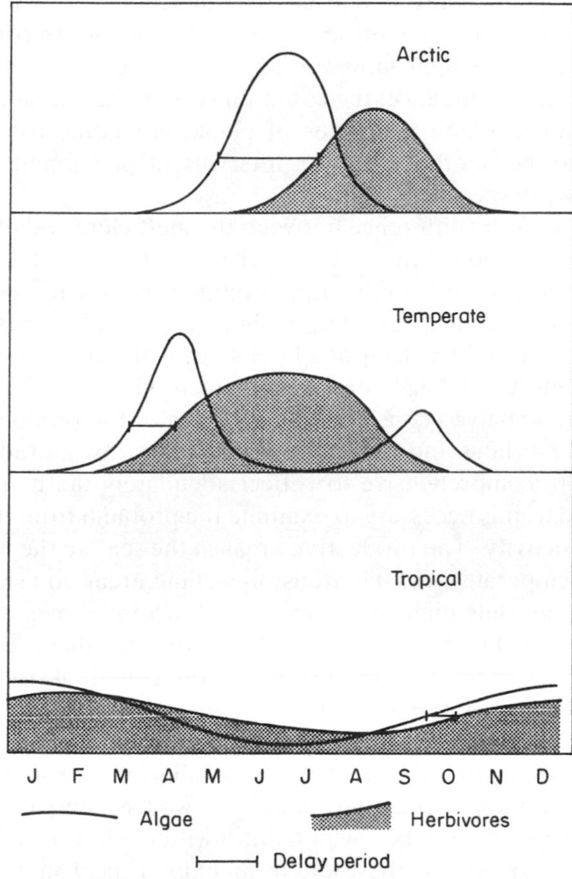

Figure 6.4 Diagrammatic representation of production cycles in different latitudinal zones. Reproduced by permission of Cambridge University Press from Cushing (1975)

Because the tropical cycle is of low amplitude and is continuous throughout the year in a deep euphotic layer, Cushing considers it an efficient cycle. Food is available throughout the year, and both growth and reproduction in the marine animals are probably continuous. In temperate and high-latitude systems, a large quantity of material is transferred from one trophic level to the next at low efficiency, whereas in the quasi-steady-state tropical system less material is transferred more efficiently. Cushing estimates that the transfer coefficients are three times higher in the 'poor' areas than in the 'rich' areas.

The different cycles correspond to different degrees of stability at higher trophic levels. Cushing notes that a stable community is diverse, that is, there are many species, each few in numbers, such as those found in the deep tropical ocean where the quasi-steady-state cycle of production is found. Because the cycle is of low amplitude, numbers are more or less stable. In contrast, in the high-amplitude cycle of high latitudes and upwelling areas, there are few species, each of high abundance; the community is neither stable, nor diverse. In the case of upwelling, a few types of plankton-feeding fish alternate in dominance with one another at short intervals, a phenomenon known as 'alternation between species'.

Cushing attributes the difference between the inefficient and efficient areas largely to the delay period in the cycle, which depends on the onset of grazing. In the open tropical ocean, production is limited by grazing, possibly at all stages of the cycles, whereas in the upwelling areas and in temperate waters production is controlled by grazing at a later stage in the cycle, by which time a considerable quantity of living material has been produced.

While Cushing's analysis considers the problem of the production cycle in the sea for high latitudes and upwelling areas versus low latitudes, its scope needs to be more comprehensive to reflect adequately the marine environment. In particular, it is necessary to examine the problem from the viewpoint of primary productivity. The productive areas in the sea are the high latitudes (subarctic and temperate), coastal areas, upwelling areas and surface layers. These areas are not only highly productive and rich in biomass, but they are also highly variable. Food is readily available for animals of higher trophic levels in these areas. In contrast, low latitudes (tropics and subtropics), oceanic areas (in particular the central parts of the oceans), and subsurface and bottom layers are not only lower in productivity and biomass, but are also considerably more stable environments. Obtaining food is laborious for animals in these areas. This situation holds true interannually as well as annually. Contrasting patterns of fluctuation in number, which differ with life history stages, seem to have evolved in response to these environments. Type I matches the high productivity environment and Type II the lower productivity environment.

Type I, in turn, is divided into Subtypes IA and IB, corresponding to specific temporal scales of environmental variation in the sea. The observed variation in subarctic and temperate seas is considered to result from short-term variation, with intervals of one to a few years, and tends to occur in a relatively small area. Large-scale variations, with cycles of several decades to centuries, tend to occur in ocean-wide areas. Each may drive the variation in abundance of a variety of species of fish in an area. Fish species of Subtype IA respond more closely to short-term variation in the environment, whereas those of Subtype IB respond more closely to large-scale variation. As shown in Figure 6.5, the Pacific saury, Far Eastern sardine, and scombroids, such as chub mackerel *Pneumatophorus japonicus* and albacore *Thunnus alalunga*, reveal the fluctuation patterns of Subtypes IA and IB and Type II, respectively.

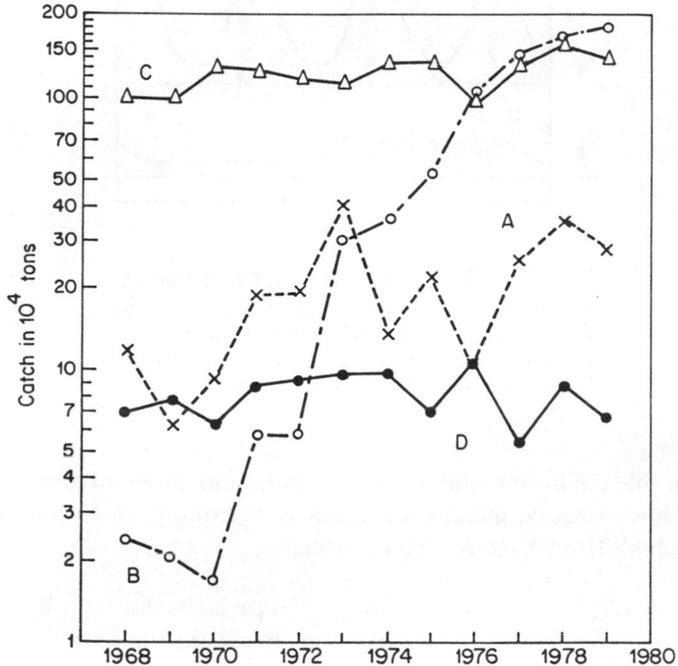

Figure 6.5 Interannual variations in the catch of Pacific saury (A), Far Eastern sardine (B), chub mackerel (C), and albacore (D), caught in Japanese fisheries. (Source: Kawasaki, 1982, 211)

The variations of the three types are represented diagrammatically in Figure 6.6. The figure employs an assumption common in population biology, that variation in stock size can be represented by a logistic model. Expressing the carrying capacity of the environment K (equivalent to maximum stock size)

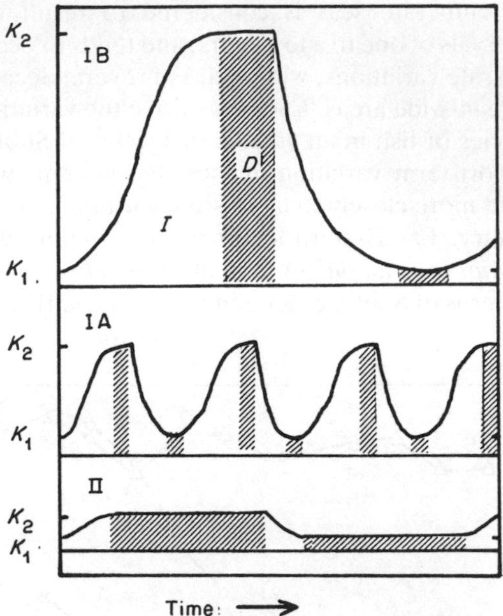

Figure 6.6 Three types of patterns of fluctuation in number of marine teleosts. K_2, high K; K_1, low K; D, period when density-dependent are dominant factors; I, period when density-independent are dominant factors

under favorable environmental conditions as K_2 and under adverse conditions as K_1, the stock size, N, increases when K goes from K_1 to K_2 and decreases when K declines from K_2 to K_1. More formally,

$$dN/dt = rN(K_2 - N)/(K_2 - K_1) \quad \text{represents the increasing phase} \quad (K_1 \rightarrow K_2)$$

$$dN/dt = rN(K_1 - N)/(K_2 - K_1) \quad \text{represents the decreasing phase} \quad (K_2 \rightarrow K_1)$$

$$K_1 \leqslant N \leqslant K_2$$

Now, suppose that the stock is fished. Then,

$$dN/dt = rN(K_2 - N)/(K_2 - K_1) - aXN \quad \text{represents the increasing phase}$$

$$dN/dt = rN(K_1 - N)/(K_2 - K_1) - aXN \quad \text{represents the decreasing phase}$$

where X denotes fishing effort, and fishing is carried out at a rate proportional to stock density per unit effort.

As seen in Figure 6.6, the values r and $K_2 - K_1$ are species-specific attributes and vary between species, even if they inhabit a common area. In this figure I (unshaded area) denotes periods when density-independent factors (environmental factors) are the main influence, while D (shaded area) denotes periods when density-dependent factors (biological factors) are the main influence. The duration of I is long and D is short for Type I species, indicating that they are apt to be subject to environmental change. In contrast, the duration of I is short and D is long for Type II species, indicating that they tend to be more influenced by overfishing. Far Eastern sardine and halibut are examples of Types IB and II respectively.

6.4.2 Interannual Variation: Pacific Saury (Subtype IA)

The catch of Pacific saury around Japan has been characterized by short-spaced, interannual fluctuation (as shown in Figure 6.5). The fluctuation in catch is thought to reflect a change in stock size, and it is believed that variation of oceanic conditions, especially the pathway of the Kuroshio Extension east of Japan, has caused the fluctuation in stock (Fukushima, 1979). As shown in Table 6.1, the shift of the Kuroshio has given rise to a change of the Oyashio, the southbound cold current east of Japan, eventually resulting in changes in the distribution, size and catch of fish. This pattern of short-term fluctuation depends on the life history features of saury, which is short-lived (2 years), reproduces early, has a high intrinsic rate of natural increase r and feeds on zooplankton. For more detailed explanation, see Kawasaki, 1980.)

6.4.3 Long-term and Phenomenal Variation: Sardines (Subtype IB)

Figure 6.3 displays the large-scale fluctuations in catch exhibited by three species of the genus *Sardinops*: Far Eastern sardine, California sardine, and Chilean sardine. While these species are geographically distinct, occurring in the northwest, northeast and southeast parts of the Pacific Ocean respectively, their fluctuations are in phase with one another. This marked phenomenon suggests that variations in stocks of the three species are governed by some common Pacific-wide oceanic variation. If we accept the explanation in Section 6.2 about large-scale ocean changes as valid, it is understandable why the three sardine species have shown almost identical trends of variation in stock size. A long-lived A-type cold water mass associated with Kuroshio meandering may be regarded as the local manifestation of large-scale variation, such as occurred in 1934–43 and has been occurring since 1975 to the present, a period when sardine stocks have been most abundant (see Figure 6.3).

Table 6.1 Relation between oceanic conditions and distribution, size, and catch of the Pacific saury

Period northernmost path of the Kuroshio	Trend of the Oyashio	Major fishing area	Dominant size in catch	Amount of catch (Typical years)
38°00′N	The 1st branch develops	Coastal Long. 142°E–145°E	Medium-sized fish (27–28 cm)	200,000 (1950–53)
37°30′N	The 1st and 2nd branches develop	Coastal and nearshore Long. 142°E–146°E	Bimode of large and medium-sized fish (30 cm and 27–28 cm)	400,000 (1955–59)
37°00′N	The 1st branch is weak, and the 2nd branch develops	Offshore Long. 144°E–148°E	The same as above but some change in the medium sized fish (30 cm and 26–27 cm)	300,000 (1960–63)
36°30′N	The 1st branch is weak, and the 2nd and 3rd branches develop	Offshore Long. 145°E–149°E	Large in even years or medium in odd years for 1964–67, or large in odd years or medium in even years for 1972–75	200,000 (1964–67 = 1972–75)
36°00′N	The 1st and 2nd branches are weak, and the 3rd branch develops	Offshore Long. 145°E–151°E	Principally small fish (24–25 cm)	100,000 (1968–71)

Source: Fukushima, 1979, 4.

Soutar and Isaacs (1974) studied temporal change in the scale-deposition rate of the California sardine in cores from anaerobic sediments in the Santa Barbara Basin off northern California. Radiometric methods were used to estimate the age of the laminae, and scale-deposition rates were obtained. Comparison of estimates of the sardine population derived from solution of a fishery catch equation and the scale-deposition rate in the Santa Barbara Basin indicate a parallel trend. Deposition rates of the sardine scale were very low in 1865–90 and 1940–70, suggesting long-term variation. Such large-scale fluctuation exhibited by sardines is a function of their ecological characteristics. They are long-lived, fast-growing herbivores living in coastal areas, and these features mean that they are especially subject to changes in oceanic conditions. To summarize, large-scale fluctuation of *Sardinops* appears to be caused by global-scale environmental variation and depends on whether the sardines are able to utilize a large quantity of phytoplankton or not.

6.4.4 Variation Caused by Man: Southern Bluefin Tuna (Type II)

Southern bluefin tuna *Thunnus maccoyii* is a Type II species distributed over the relatively stable waters of the deep oceans of the southern hemisphere. Fluctuation of stock size is density-dependent, and stock size changes little under unexploited conditions. Figure 6.7 shows a relation between the catch by longline and the number of hooks used. The hooked-rate as an index of stock density had been approximately 4 percent from 1954 to 1962, decreasing as fishing intensified, and has fallen to about 0.5 percent since 1975. This is a typical example where heavy fishing resulted in a decline in stock. While the variation in this stock is caused primarily by overfishing, there remains a need to understand the variation with the species-specific pattern of environmental fluctuation in mind.

6.5 CHANGES IN STOCK BASED ON INTERSPECIES RELATIONS

Changes in stock are brought about not only by environmental and human factors, but by relations between species as well. For example, it has been found that the catch of species of *Sardinops* (sardines) is inversely correlated with the catch of *Engraulis* (anchovies). This has been seen in different areas of the world oceans and may be explained by the partly overlapping ecological niches of sardine and anchovy (see Figure 6.1). The former feeds on both phytoplankton and zooplankton, while the latter depends mainly on zooplankton; both inhabit coastal areas. Murphy (1977), using the results of egg and larval surveys, maintained that as the sardine population off California diminished, the populations of northern California anchovies increased. The anchovy population lagged far enough behind the sardine that Murphy concluded it was filling the void left by the sardine, rather than causing the sardine population to decline.

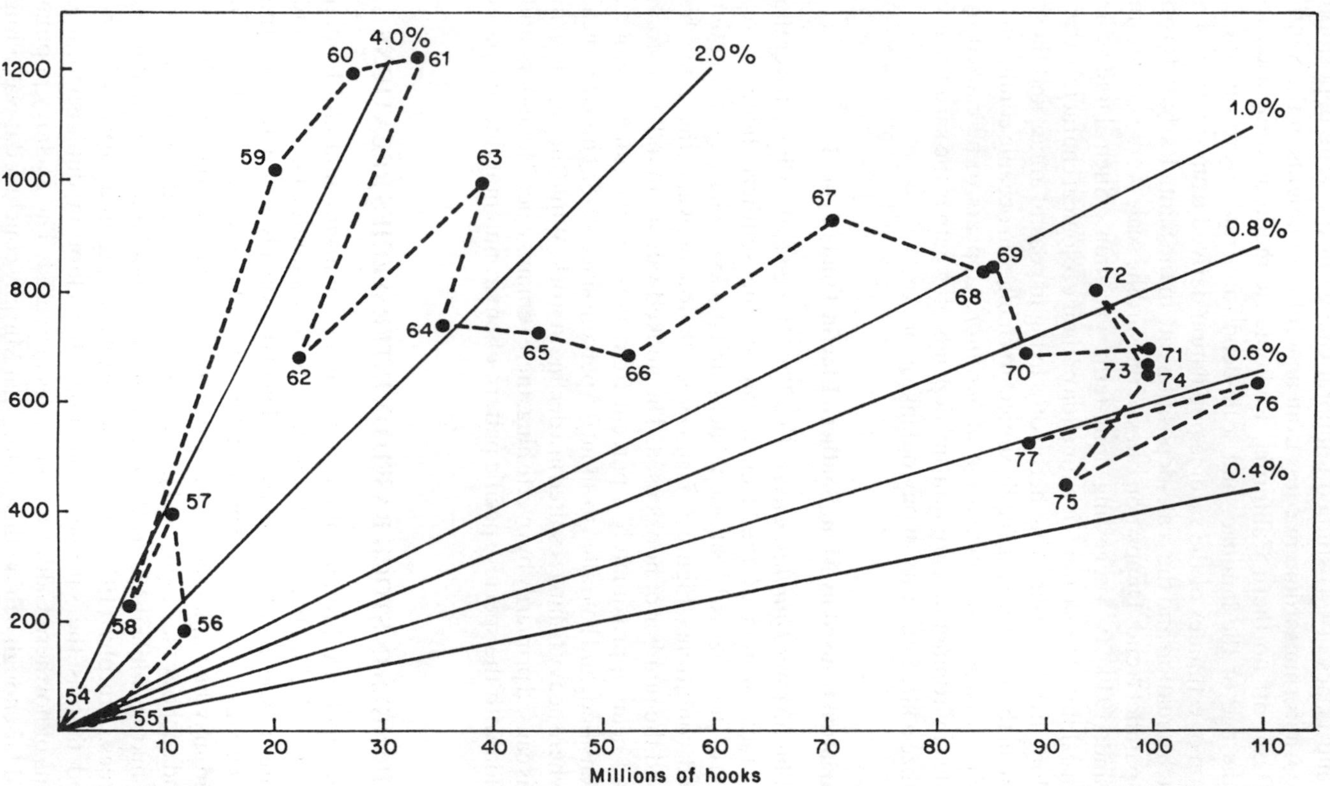

Figure 6.7 Relation between fishing effort and catch of southern bluefin tuna in a longline fishery. Unbroken lines show catch-per-100 hooks as an index of population density. (Source: Shingu and Hisada, 1980, Figure 2)

6.6 VARIATION OF FISH SPECIES AND FISH BIOMASS IN AN AREA THROUGH SUSTAINED ENVIRONMENTAL CHANGE

Variation in ocean biota attributable to sustained shifts in ocean climate may be illustrated by the events of the so-called Russell cycle, discovered by Sir Frederick Russell (Cushing and Dickson, 1976). A northward spread of warm water species occurred in the western English Channel, with warming in the north Atlantic from the 1920s to the 1940s, and was reversed some 30 years later during the early 1970s. The first event was the decline in recruitment of the Plymouth herring stock (*Clupea harengus*), which began with the 1925 year-class. In 1931, the last recorded year-class entered the fishery, which subsequently collapsed in 1936 or 1937. In the summer of 1926, pilchard eggs *Sardina pilchardus* were recorded in considerable numbers and remained so until 1960. Between 1925 and 1935, the winter phosphorus concentration declined by one-third. In the autumn of 1931, macroplankton (plankton in the 0.2–2.0 mm size class) declined by a factor of four and the numbers of summer-spawned fish larvae decreased.

When the catches in 1919–22 are compared with those in 1944–52, one of the northern species, ling *Molva molva*, virtually disappeared, while boarfish *Capros aper* decreased by an order of magnitude. During the same period, southern species like horse mackerel *Trachurus trachurus*, hake *Merlucius merlucius*, red mullet *Mullus surmuletus*, and red bandfish *Cepola rubescens*, all increased by an order of magnitude. Phytoplankton in the 1920s resembled

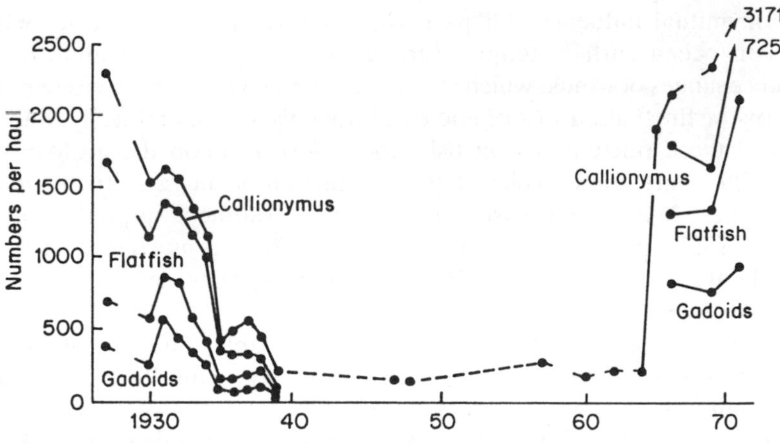

Figure 6.8 The Russell cycle in monthly averages of the planktonic stages of teleostean fish, excluding clupeids, between 1924 and 1971 in the vicinity of Plymouth, England. Reproduced by permission of Cambridge University Press from Russell (1973)

those in the 1970s in species composition, whereas in the intermediate years a number of southerly species were found. Furthermore, in 1965, the numbers of spring-spawned fish larvae (generally northerly species) increased again by an order of magnitude. In 1970, the macroplankton increased for the first time since the autumn of 1930, and in the following winter the phosphorus concentration rose to the pre-1930 level. This cycle is summarized in Figure 6.8.

There are three events common to the two changes in the 1930s and the 1970s. In 1930 macroplankton and winter phosphorus declined, and the spring spawners decreased 5 years later. In 1965 the spring spawners increased in numbers, to be followed by increases in macroplankton and winter phosphorus in 1970. Thus, the later events are the mirror image of the earlier ones; if the decline of spring spawners in 1935 indicated the end of the earlier sequence of events, then their recovery signalled the start of the later ones.

It should be noted that alternation between herring and pilchard has occurred before. A summer pilchard period occurred in the first half of the seventeenth century, the early years of the eighteenth century, and in the first 30 years of the nineteenth century (Cushing and Dickson, 1976).

6.7 CONCLUSION

The linkage between climate and fisheries must be considered in two jointly coupled relationships. One coupling occurs in the relationship between climate and ocean, and the other in the relationship between ocean and fisheries.

The relationship between the atmosphere and the sea is a feedback system, a system of mutual influence. Of particular importance are processes where changes in ocean surface temperature cause changes in the atmosphere, especially changes of winds, which in turn cause changes in ocean circulation.

Patterns of fluctuation of oceanic conditions vary geographically, and the impacts of these fluctuations on fish stocks depend upon the ecology of a species. There are three typical types of fluctuation in recruitment in the marine fishes. While some fish fluctuate in recruitment unpredictably or irregularly (Subtype IA), recruitment of other species is characterized by large-scale fluctuation (Subtype IB). For some species, stock size is less changeable (Type II).

The Type I species (variable and unpredictable) are distributed in environments where sea conditions change widely—upwelling areas, high latitudes, and the surface layer. Species of Subtype IA are matched to short-term fluctuations of the ocean environment and include saury and sand eels. Species of Subtype IB correspond to large-scale fluctuations, and herrings and sardines exemplify the category. Fish of Type II inhabit stable environments, such as tropical areas and bottom or subsurface layers, fluctuate slightly, and tend to be more subject to overfishing than to environment. Flatfishes and tunas are typical of this type.

It may remain controversial whether changes in particular stocks are more attributable to natural causes, like climate, or to overfishing. Characteristic debates have been held on several species, for example, herring, sardine, and Pacific halibut. While the former two taxa, belonging to Subtype IB, are less apt to be directly subject to drastic impacts from fishing, a stock of the latter species, typical of Type II, tends to be quite sensitive to fishing effort.

Mankind owes its existence in large part to the seas. Living resources of the seas continue to be a valuable source of food. For this reason and because of their intrinsic natural value, an increased understanding of the behavior of fish stocks must be sought. Understanding the relationship between climate and fisheries can be an important part of sustaining fisheries as a renewable resource.

REFERENCES

Backus, R. H., Flierl, G. R., Kester, D. R., Olson, D. B., Richardson, P. L., Vastano, A. C., Wiebe, P. H., and Wormuth, J. H. (1981). Gulf Stream cold-core rings: Their physics, chemistry and biology. *Science*, **212**, 1091–1100.

Barber, R. F., and Chavez, F. P. (1983). Biological consequences of El Niño. *Science*, **222**, 1203–1210.

Burkenroad, M. D. (1948). Fluctuation in abundance of Pacific halibut. *Bulletin of the Bingham Oceanographic Collection*, **11**(4), 84–85.

Cane, M. A. (1983). Oceanographic events during El Niño. *Science*, **222**, 1189–1195.

Caviedes, C. N. (1981). The impact of El Niño on the development of the Chilean fisheries. In Glantz, M. H., and Thompson, J. D. (Eds.) *Resource Management and Environmental Uncertainty, Wiley Series in Advances in Environmental Science and Technology* **11**, pp. 351–368. John Wiley & Sons, New York.

Cushing, D. H. (1975). *Marine Ecology and Fisheries*. Cambridge University Press, Cambridge, UK: 278 pages.

Cushing, D. H. (1982). *Climate and Fisheries.*.Academic Press, London: 373 pages.

Cushing, D. H., and Dickson, R. R. (1976). The biological response in the sea to climatic changes. In Russell, F. S., and Yonge, M. (Eds.) *Advances in Marine Biology* **14**, pp. 1–122. Academic Press, London.

Fukushima, S. (1979). Synoptic analysis of migration and fishing conditions of saury in the Northwest Pacific Ocean. *Bulletin of the Tohoku Regional Fisheries Research Laboratory*, **41**, 1–70 (in Japanese).

Guillén, G. O., and Calienes, Z. R. (1981). Biological productivity and El Niño. In Glantz, M. H., and Thompson, J. D. (Eds.) *Resource Management and Environmental Uncertainty, Wiley Series in Advances in Environmental Science and Technology* **11**, pp. 255–282. John Wiley & Sons, New York.

Kawai, H. (1972). Hydrography of the Kuroshio and Oyashio. In Iwashita, M., *et al.* (Eds.) *Kaiyo-butsuri (Physical Oceanography), Fundamental Series of Marine Sciences* **2**, pp. 129–328. Tokai University Press, Tokyo (in Japanese).

Kawasaki, T. (1980). Fundamental relations among the selections of life history in the marine teleosts. *Bulletin of the Japanese Society of Scientific Fisheries*, **46**, 289–293.

Kawasaki, T. (1982). *Pelagic Fish Populations, Fisheries Series*, **9**. Koseisha-Koseikaku, Tokyo (in Japanese): 327 pages.

Kawasaki, T. (1983). Some problems in estimating the fisheries productivity. *Bulletin of the Japanese Society of Fisheries Oceanograph*, **42**, 75–77 (in Japanese).

Kondo, K., Hori, Y., and Hiramoto, K. (1976). Ecology and resources of the Far Eastern sardine. *Suisan-Kenkyu-Sosho (Fisheries Research Series)*, **30**, Suisan-Shigen-Hogo-Kyokai, Tokyo (in Japanese): 68 pages.

Murphy, G. I. (1977). Clupeids, In Gulland, J. A. (Ed.) *Fish Population Dynamics*, pp. 283–308. John Wiley & Sons, New York.

Murray, T., Le Duc, S., and Ingham, M. (1983). Impact of climate on early life stages of Atlantic mackerel *Scomber scombrus*, L.: An application of meteorological data to a fishery problem. *Journal of Climate and Applied Meteorology*, **22**, 57–68.

Nakai, Z. (1962). Studies relevant to mechanisms underlying the fluctuation in the catch of the Japanese sardine, *Sardinops melanosticta* (Temminck & Schlegel), *Japanese Journal of Ichthyology*, **9**, 1–115.

Paulik, G. J. (1981). Anchovies, birds, and fishermen in the Peru current. In Glantz, M. H., and Thompson, J. D. (Eds.) *Resource Management and Environmental Uncertainty, Wiley Series in Advances in Environmental Science and Technology* **11**, pp. 35–79. John Wiley & Sons, New York.

Ramage, C. S., and Hori, A. M. (1981). Meteorological aspects of El Niño. *Monthly Weather Review*, **109**, 1827–1835.

Rasmusson, E. M., and Wallace, J. M. (1983). Meteorological aspects of the El Niño/Southern Oscillation. *Science*, **222**,1195–1202.

Russell, F. S. (1973). A summary of the observations of the occurrence of planktonic stages of fish off Plymouth, 1924–1972. *Journal of the Marine Biological Association of the United Kingdom*, **53**, 347–355.

Shingu, C., and Hisada, K. (1980). Recent status of the southern bluefin tuna stock. Paper presented to the *Conference of Japanese and Australian Researchers on the Southern Bluefin Tuna held in Tokyo on August 11, 1980*.

Soutar, A., and Isaacs, J. D. (1974). Abundance of pelagic fish during the 19th and 20th centuries as recorded in anerobic sediment off the Californias. *Fisheries Bulletin of the US Fish and Wildlife Service*, **72**, 257–273.

Stolzenbach, K. D., Madsen, O. S., Adams, E. E., Pollock, A. M., and Cooper, C. K. (1977). *A Review and Evaluation of Basic Techniques for Predicting the Behavior of Surface Oil Slicks*. Report No. 222, Department of Civil Engineering, Massachusetts Institute of Technology, Cambridge, Massachusetts.

Teramoto, T. (1981). Long-term variation of the Kuroshio. *Kisho-kenkyu Note (Meteorological Research Note)*, **141**, 45–58 (in Japanese).

Thompson, J. D. (1981). Climate, upwelling and biological productivity. In Glantz, M. H., and Thompson, J. D. (Eds.) *Resource Management and Environmental Uncertainty, Wiley Series in Advances in Environmental Science and Technology* **11**, pp. 13–33. John Wiley & Sons, New York.

Thompson, W. F. (1952). Conditions of stocks of halibut in the Pacific. *Journal du Conseil*, **18**, 141–166.

Thompson, W. F., and Bell, F. H. (1934). Biological statistics of the Pacific halibut fishery (2): Effect of changes in intensity upon total yield per unit of gear. *Report of the International Fisheries Commission*, **8**, 3–49.

Wooster, W. S., and Guillén, O. (1974). Characteristics of El Niño in 1972. *Journal of Marine Research*, **32**, 387–404.

Wright, P. B. (1978). The Southern Oscillation. In Pittock, A. B., Frakes, L. A., Jenssen, D., Peterson, J. A., and Zillman, J. W. (Eds.) *Climatic Change and Variability: A Southern Perspective*, pp. 180–185. Cambridge University Press, Cambridge, UK.

Wyrtki, K. (1973). Teleconnections in the Equatorial Pacific Ocean. *Science*, **180**, 66–88.

Climate Impact Assessment
Edited by R. W. Kates, J. H. Ausubel and M. Berberian

CHAPTER 7
Pastoralism

HENRI N. LE HOUÉROU

CNRS/CEPE
Louis Emberger
B.P. 5051
Montpellier Cedex 34033, France

7.1 INTRODUCTION

Pastoralism is understood in the present paper as the unsettled and non-commercial husbandry of domestic animals. Thus it differs from sedentary subsistence husbandry systems and profit-oriented transhumant operation and from commercial ranching, although the same term may be used to describe these other forms of animal rearing (such as pastoralism in Australia). Pastoralism thus described is a way of life and livelihood for an estimated 60–70 million people, mainly in Africa and in Asia.

Studying the impacts of climate on pastoralism is complicated by the fact that pastoralism is essentially—but not solely—a form of adaptation of human societies to hazards and hardships induced, and imposed on them, by climatic constraints. The wandering husbandry of domesticated animals as a response of rural societies to climatic variability or limitations for human subsistence is about as old as irrigated agriculture, which first developed in the Middle East and then in Egypt some 5000–10,000 years ago. Runoff farming also developed in the Middle East and then in North Africa 2500–3000 years ago. Evidence of the latter does not go further back than 4000–5000 years (Bonte *et al.*, 1981), while the first references to domesticated animals (which may not necessarily have been nomadic) date back to 8000–12,000 years ago for sheep and goats, 6000–8000 for cattle and 4000–5000 for camels (Zeuner, 1963).

7.1.1 Pastoral Systems

The impacts of climate on pastoralism, as a managed ecosystem that evolved as an adaptation to climate, vary in important but often subtle ways by the type of pastoralism practiced and the various management techniques adopted. Pastoral types of animal production systems may take many forms based on the degree of nomadism, its regularity and the type of animals kept. (See Bernard and Lacroix, 1906; Capot-Rey, 1953; de Planhol and Cabouret, 1961–72; Bataillon, 1963; Johnson, 1969; Monod, 1973, 1975; Bonte *et al.*, 1981 for a thorough analysis of these forms.)

One may first distinguish (Monod, 1973, 1975) among true *nomad pastoralism* where the whole household moves with the herd or flock, *transhumant pastoralism* where only the herder moves with the stock, and *semi-pastoralism* where cultivation (usually of cereals) constitutes a significant segment of activity.* There are intermediate forms where the whole family may move with the herd or with a part of it for shorter or longer periods. The movements may also be *periodic*, pendulum-like (that is, linked to climatic seasonality), or *aperiodic* (that is, depending on random climatic vagaries in time and space). In addition, one can further differentiate systems according to the type of stock which is kept: cattle, one- or two-humped camels, sheep and

* In oil-rich countries, new forms of 'semi-pastoralism' have developed based on herds managed in a pastoral-like way by expatriate waged shepherds from similar cultural origin in low-income neighboring countries; this new type of management, due to very high internal meat prices, employs costly management practices such as the trucking of water for stock and intensive supplementary feeding of grain and other subsidized concentrates. This type of husbandry is not included under the term of pastoralism described herein.

goats, horses, reindeer and even yak or llama. These characteristics may be combined with various geographic situations. One may thus differentiate between transhumant pastoralists and periodic nomads, those who having their homeland in the mountains spend the cold seasons in the plains from those who having their homeland in the plains move to the mountain areas for the dry hot season.

Within the various forms of nomadism, one may further distinguish among those who are almost sedentary like the Masai of Kenya and Tanzania; those who move over short distances within a few thousand square kilometers, such as the Tuareg of the bend of the Niger River in Mali, Niger and Upper Volta (Barral, 1967, 1977), and most East African pastoralists; from those who may move over very long distances (1000–2000 km), such as the Reguibat Lgouacem Moors of Mauritania (Cauneille, 1950; Bonnet-Dupeyron, 1951); the Kabbabish of the Republic of Sudan (Asad, 1970); or the Al Murrad of Saudi Arabia. There are also fundamental differences between those who are territorially based, such as the Moors and Tuaregs (Rodd, 1926; Nicolas, 1950; Nicolaisen, 1963; Bonte, 1973; Bernus, 1981), and those who have no permanent base, such as the nomadic Fulani, who consequently move over continental distances during the course of centuries and even within decades. The latter might be regarded as the true pastoralists since they are tied only to their herds and have had no real homeland throughout history (Stenning, 1959; Dupire, 1962, 1970; Gallais, 1967, 1975, 1977).

Finally, one can differentiate among those who have other significant means of subsistence (for example, hunting, fishing, cultivation, trading, caravaning, wage labor, raiding, handicraft industries, and/or religious activities) while practicing pastoralism as an important source of livelihood. There are even those who live as settled farmers, traders, shopkeepers or civil servants and own flocks or herds which are managed in a pastoral way by waged or share-owning shepherds, within a nomadic or transhumant production system. The latter case is no longer subsistence husbandry but it is not yet commercial husbandry, either.

7.1.2 Non-climate-related Pastoral Movement

At first sight it would seem that there is a clear linkage between pastoralism and climate, since most pastoral areas are subject to arid and semi-arid climates and therefore pastoralism would be the usual way to avoid the effects of seasonal drought on livestock in traditional societies. This is obviously grossly true, but there are, as ever, a number of exceptions.

First, pastoralism exists in a wide array of climatic conditions, from desert to subhumid, under temperate continental, mediterranean, tropical and equatorial climates (not to speak of the tundra); under various rainfall and temperature regimes in many regions of the world: central Asia, middle Asia, southwest Asia, northern Africa, the Sahelian and Sudanian zones of West Africa, arid and semi-arid East Africa, and southern Africa. It also used to exist (in the Middle Ages and until the beginning of the present century) under the temperate climates of western Europe (Braudel, 1949).

Second, there are many areas within the arid zone where settled livestock husbandry is the rule, including traditional societies in Africa and Asia, not to mention the ranching system of production developed in America, Australia and southern Africa. Conversely, there are many semi-arid areas where pastoralism is not found either because of livestock health constraints such as trypanosomiasis or human health problems such as onchocerciasis (river blindness).

Third, nomadism, or transhumance, or herd movement, is not necessarily triggered by drought or climatic constraints. On the contrary, in some instances movement is initiated by floodings and the subsequent pullulation of flies, mosquitoes, ticks, worms and parasites of all kinds, affecting herds and herders alike. This is the case, for example, in the inner delta of the Niger River, the Sudd and the Bahr El Ghazal, and the lower Chad Basin, although in these cases there is an indirect linkage with climate on the catchment basins. In other cases movement may be caused by the search for more comfort, towards warmer winter or cooler summer areas (Barth, 1961).

In most cases, however, movement is motivated either by the quest for better pastures, the exhaustion of the range, or the shortage of drinking water. The flocks of Provence in southeast France used to move for centuries to their summer lush 'alpages' in the Alps Mountains (the word Alp, incidentally, means summer montane pasture in the local dialect), and come back to the Mediterranean shores with the first late summer frosts on the highlands and the first autumn rains in the lowlands. The same thing occurred between the Languedoc lowland garrigues and the Cevennes highlands. The Waled Nail and the Larbaa of Algeria spend the winter season in their home country on the steppic ranges of the northern fringe of the Sahara and move for the summer months to the stubble fields at the northern limit of the highlands, 300–400 km further north, thus travelling 600–800 km every year (Bernard and Lacroix, 1906; Sagne, 1950; Capot-Rey, 1953; Cauneille, 1950, 1968; Bataillon, 1963). The migration of the Delta Fulani of Mali is regulated by the flooding of the Niger River and by the filling up of water ponds along the 250-km trail to the Nema of southeast Mauritania, where they move for the rainy season, while the return trip is governed only by the desiccation of the ponds along the trail (Gallais, 1967, 1975; Le Houérou and Wilson, 1978.)

Fourth, the degree of adaptation of ethnic groups of pastoralists to drought varies greatly. Thus in central Mali the Moors continue to utilize throughout most of the dry season ranges that have been abandoned by the Fulani at the end of the rainy season, and by the Tuareg a little later (Le Houérou and Wilson, 1978). In other words some pastoralists move only when they are compelled to, either by the lack of pasturage or the shortage of water, while others move long before the resources are exhausted in order to find better forage or cleaner water elsewhere, leaving the land to less demanding groups.

Thus 'pastoralism' is not a simple system; there are virtually as many kinds of pastoralisms as there are ethnic groups practicing it. Even within the main ethnic groups in Africa, such as the Tuaregs, Fulani, Teda, Nilotic or Hamitic, there are very great differences in habitats, life styles, water, land and herd management practices, and social, cultural and religious values and beliefs. Strategies of adaptation to climatic variability and other hazards thus may be extremely different from one group to another, and therefore the impact of climate is necessarily quite variable according to the ecological conditions and the sociocultural adaptation strategies which are considered. The impact of climate should thus not be seen *per se*, in isolation, but within a sociocultural framework. This makes it extremely difficult to generalize and synthesize in a meaningful way as to the impact of climate at the societal level. Thus, I will limit myself to methods and findings related to primary and secondary resource bases from which pastoralism derives its mere existence—land, water, pasture, stock—which seem amenable to some valid generalization. In the ensuing sections I shall examine the link between climate variability and change and biological production in grazing lands, then the linkage between grazing land production and animals, and then I shall examine the many varied links between animals and human sustenance. In addition, I shall describe human modifications affecting each of these linkages, and conclude with some remarks on the future of pastoralism.

7.2 THE RELATIONSHIP BETWEEN CLIMATE AND THE PRIMARY PRODUCTION OF GRAZING LANDS

7.2.1 Climatic Factors

The climatic factors affecting primary production parameters are essentially rainfall (that is, water availability for plant growth) and energy flow (that is, temperature and potential rates of photosynthesis and transpiration). Temperature may strongly affect seasonal plant growth in temperate or montane climates, but it does not seem to affect greatly total annual production in pastoral zones except, of course, in truly cold climates such as in central Asia, the tundra, or high mountains. Plant production is usually affected when daily maximum temperatures drop to +10 °C, or below, for substantial periods of time, the zero growth being usually in the vicinity of +5 °C. In montane pastoral areas seasonal variations in temperature from year to year thus may strongly affect migration patterns in time and space, either in hastening or delaying migration as a result of early cold, mild winters, early or late springs, and the like.

There is a consensus among scientists that rainfall and evaporation are the overriding climatic factors that affect primary production in most pastoral zones; other climatic variables are often omitted in primary production predictive models (Rosenzweig, 1968; Van Keulen, 1975; De Vries *et al.*, 1978; Young and Wilson, 1978; Floret *et al.*, 1982). The predictive value of combining effective rains (that is, water that is actually infiltrated) and actual evapotranspiration is quite good. The only difficulty is then to determine how much rain has actually infiltrated and how much is being evapotranspired. This is, in principle, very simple to measure, but, in fact, is extremely difficult to integrate over large geographic areas, since it is strongly affected by many local factors (vegetation cover and structure, topography, soil characteristics and rainfall distribution in space and time). Since, for a given type of climate, yearly variations in evaporation are very small as compared to variations in rainfall, and since in arid and semi-arid zones annual evapotranspiration is always much higher than rainfall, annual rainfall is often used as the basic climate factor to relate to primary production over large areas.

7.2.2 Estimating Primary Production

There are three major methods for estimating the actual primary production of large geographic areas: ground appraisal at simple sites, aircraft and satellite remote sensing, and indirect appraisal through measurements of secondary livestock production.

There are many measurement systems employed in appraising the primary production of grazing lands. These include the use of biomass (gross, net or consumable), vegetation survey and mapping, range evaluation, and the like. These in turn are often presented as measures of carrying capacity for typical animals converted into an arbitrary standard animal unit, based on equivalents in metabolic weights (kg 0.75). For example:

300 kg Lwt zebu cow = 72.08 $\text{kg}^{0.75}$
30 kg Lwt goat = 12.82 $\text{kg}^{0.75}$

Therefore one 300 kg Lwt cow = 72.08/12.82 = 5.47 times a 30 kg Lwt goat.

Because of the rarity of long-term series of measurements of both rainfall and range production, attempts have been made to mitigate the absence of ground data by using remote sensing. Low altitude systematic reconnaissance flights are routinely used for aerial livestock and wildlife censuses, for instance for the Ecological Monitoring of Rangelands in Kenya (KREMU) and Senegal (Inventory and Monitoring of Sahelian Pastoral Ecosystems); the same method has been used by private contractors for the governments of Sudan, Somalia, Kenya, Ethiopia, and others, either for livestock and wildlife censuses, or range appraisal, or both. They could thus help, in the course of each season, in the evaluation of annual production and better localization of productive and unproductive areas. The cost of this method is high and the extension to large pastoral areas in developing countries will certainly not expand in the foreseeable future because the benefit-cost ratio may be low and because of the 'benign neglect' attitude of many governments towards their pastoralists and their range resource. Remote sensing of biomass and production, using airborne reflectivity sensors, is not reliable (IPAL, 1981); but such sensors may be used to measure 'greenness' of vegetation, which is more tied to the feed quality of the range than to the amount of fodder available.

The use of satellite imagery would, in principle, make it possible to know the pasture production and its distribution over time and space at the end of each rainy season. This would be of great advantage in areas like the Sahelian zone where the rainy season lasts only 1–4 months, beginning and ending at fairly regular dates every year.

It would thus become possible to know in advance the forage availability in any given area for the 8–11 coming months of each dry season and so to plan range and stock management accordingly (Le Houérou, 1972), but this may remain in the domain of 'range management fiction' for many years to come!

Experiments which lasted 4 years over a large zone in southern Tunisia (ARZOTU Project) showed, however, that satellite images, in order to be used in a reliable way for biomass assessment, need important and costly means of ground control (Long *et al.*, 1978), but further technical progress may improve the feasibility of this technique in the future.

Finally, methods of indirect assessment (deducing primary production from secondary production in fully stocked areas) are used in climatically comparable areas under ranching systems in America and Australia, but there are no similar statistical data available on range production in pastoral areas of Africa and Asia. There are no reliable statistical data on animal numbers and stocking rates either; the number of animals slaughtered or sold is not known; milk production is almost totally unknown. Thus there is no way of deducing primary production from secondary production.

7.2.3 Rainfall–Primary Production Relationships

Many authors have tried to relate rainfall and primary production or rainfall and carrying capacity (Lomasson, 1947; Walter and Volk, 1954; Coupland, 1958; Le Houérou, 1958, 1962, 1963, 1964, 1965, 1969, 1971, 1975, 1977a, 1982, 1984a,b; Sneva and Hyder, 1962; Condon, 1968; Braun, 1973; Cassidy, 1973; Tadmor *et al.*, 1974; Breman, 1975; Philipson, 1975; Bille, 1977; Pratt and Gwynne, 1977; De Vries *et al.*, 1978; Floret and Pontanier, 1978, 1982; Cornet, 1981; Cornet and Rambal, 1981; Grouzis and Sicot, 1981; Floret *et al.*, 1982).

7.2.3.1 Annual Rainfall

When average annual rainfall is compared to average annual primary production in a given ecological and geographic zone, the correlations are quite satisfactory ('*r*' > 0.8, '*P*' < 0.01), especially in arid and semi-arid zones (Le Houérou, 1982, 1984a,b). But there are very great differences in productivity levels, as might be expected, depending on vegetation types and on the kind of management which is applied to them. The value of mean annual precipitation as a predictive indicator of average primary production is therefore limited to large geographic areas having identical climates, similar vegetation types, and comparable range management practices. This is, for instance, the case for broad ecological zones such as the Mediterranean arid and semi-arid zones of the Near East and North Africa, the Sahelian and Sudanian zones of northern Africa, the East African arid and semi-arid zones, and the Miombo of southern and eastern Africa, where climates, vegetation patterns and range management practices are rather homogenous—in their diversity—within each zone.

Mean annual precipitation or the actual amount of annual rain is of no value in predicting forage yield in a given site for a particular year. The latter depends more on actual rainfall distribution over the annual cycle than on the amount that falls, all other conditions being equal (Le Houérou and Hoste, 1977; Cornet and Rambal, 1981; Floret *et al.*, 1982; Le Houérou, 1982) as described in Section 7.2.3.3.

Accurate prediction (±10%) of actual yields in a given site for a particular growing season can be done only by knowing actually infiltrated rains and the rate of evapotranspiration, as shown by many scientists (Briggs and Shantz, 1913; De Witt, 1958; Hanks *et al.*, 1969; Floret and Pontanier, 1982). But this method can hardly be used for large geographic areas such as pastoral zones because it postulates the use of many data which are available only for a very small number of sites, such as: actual rainfall distribution in time and space, rainfall intensities, runoff rates, infiltration rates, evaporation from soil surface, fertility status of the soil, vegetation type, range condition. The integration of all these data over large geographic pastoral areas is an almost impossible task at the present time. As a matter of fact all predictive models and their validation are based on data from one site or a small number of sites, and cannot therefore be extended to large geographic areas having different characteristics.

At the present time the easiest, if not the best, assessment of range production in pastoral zones can be based on statistics of annual rainfall, which are available and fairly reliable, along with site experiments and surveys providing production figures (often excluding browse) that can be related to rainfall data. One can thus determine a rain use efficiency (RUE) factor which is the number of kilograms of above-ground dry matter produced per

Table 7.1 Rain use efficiency in arid and semi-arid grazing lands

No.	Zones	Average RUE (kg DM mm^{-1} ha^{-1} y^{-1})	Range
1	Mediterranean	2.8–3.8	1–10
2	Mediterranean	3.2	1– 6
3	Sahelo-Sudanian	2.6–3.6	1–10
4	Sahelo-Sudanian	3.3	–
5	Sahelo-Sudanian	3.0	2–11
6	Sahelo-Sudanian	3.5	2– 5
7	Sahelo-Sudanian	2.2	–
8	East Africa	6.0	3–12
9	East Africa	3.2	1.5– 5.0
10	N. Kenya	4.5	2.2– 9.6
11	USA	4.2	–
12	Australia	1.5	0.7– 3.3

[1] Detailed surveys over some 30 million hectares of rangelands in Tunisia, Algeria, Morocco, Libya, Syria, Israel, Egypt, Italy, Spain, France and Greece (Le Houérou, 1962–1982). Lower figure refers to southern Tunisia, higher figure from compilation of other surveys and experiments.
[2] Experimental data from seven range types of southern Tunisia over seven consecutive years, in range conditions somewhat better than the average of the area (Floret and Pontanier, 1982).
[3] Compilation of a number of surveys over some 100 million hectares in Mauritania, Senegal, Mali, Upper Volta, Niger, Nigeria, Chad, Ivory Coast and the Republic of Sudan over a period of some 15 years. Lower figure excludes browse, higher figure includes it. (Le Houérou and Hoste, 1977; Le Houérou, 1982.)
[4] Experimental work in three sites of northern Senegal over a decade, in good range condition (Cornet, 1981).
[5] Experimental data collected over 7 years in one site of northern Senegal in range conditions well above the average Sahelian situation (Bille, 1977).
[6] Experimental data from a dozen sites over a period of 4 years in central Mali in range condition above the average Sahel situation (Hiernaux *et al.*, 1980).
[7] Experimental data from many sites in northern Upper Volta over 4 years, in range condition below the average Sahelian situation (Grouzis and Sicot, 1981).
[8] Experimental data gathered over a period of 2 years in the Serengeti National Park, Tanzania, under range conditions far above the average situation in East Africa (Braun, 1973).
[9] Integrated global estimation for the rangelands of East Africa (Pratt and Gwynne, 1977).
[10] Experimental and survey data from the arid zone of northern Kenya over 4 years (Lamprey and Yussuf, 1981).
[11] Synthesis of the data available on the arid and semi-arid zones of the United States. This figure does not refer to pastoral areas; it is given here for the sake of comparison (Szarec, 1979).
[12] Data shown are computed from a study by Condon (1968) on the carrying capacities of the rangelands of the southwest of New South Wales. The figures given show clearly the low fertility status of the Australian rangelands. RUE increases from 0.7 under 120 mm of rainfall to 3.3 under 500 mm. But the figures given by Condon are concerned with livestock density in range zones; they are therefore somewhat lower than if extracted from carrying capacities *strictu sensu*.

millimeter of annual rain over one hectare in one year (Le Houérou, 1982, 1984a). Calculated RUE values are shown in Table 7.1 and evidence good agreement between survey data and experimental findings, and rather amazing consistency across various ecological zones when large geographic areas are concerned.

In the Mediterranean region and intertropical Africa, the overall order of magnitude of RUE is 3–4 kg DM mm^{-1} y^{-1} (Le Houérou, 1982). This figure, based on a large number of surveys over many million hectares, has been confirmed by long-term experimental data both in the Mediterranean zone and in the Sahel (De Vries *et al.*, 1978; Hiernaux *et al.*, 1980; Cornet, 1981; Grouzis and Sicot, 1981; Floret and Pontanier, 1982). These estimates also accord well with figures for the United States and East Africa, with the exception of the Serengeti Plains. Indeed, the major exception is for Australia (with the above-mentioned reservations), and overall the figures from various ecological zones are amazingly consistent.

At the continental level, rainfall distribution over seasons, in different ecological zones, does not seem to influence total productivity greatly over the whole annual cycle, contrary to what might be expected. The RUE factor is almost similar under comparable management practices under the winter rains regimes of North Africa and the Near East as under the tropical summer rains regimes of the Sahelian and Sudanian zones, as well as under the bimodal equatorial regime of East Africa.

But the quality of the production along the annual cycle is totally different in the three cases. The Mediterranean arid zone, having a much longer potential growing season, with fair probabilities of rain occurrence over 5–7 months per year, produces green forage for longer periods than under the monomodal tropical regime of the Sahel, where the growing season lasts 1–4 months only. The equatorial bimodal rainfall regime of East Africa allows for two growing seasons per year and green forage is thus available for 1–3 months twice each year. The protein content (that is, the feed value) of the forage is thus kept at desirable levels for much longer periods every year in the Mediterranean and in East Africa, as compared to the Sahel.

Range types and composition are also very different in the three cases. While dwarf shrubs are dominant in the Mediterranean and perennial grasses in East Africa, the Sahel ranges are dominated by annual grasses of very low feed value outside the growth period. Moreover, there may be compensation in the Mediterranean and East Africa between the annual growing seasons. If rains fail for one season they may be adequate 3–4 months later, whereas in the dry tropics such as the Sahelian or the Sudanian zone, seasonal rain failure means total drought for at least one year, since there are 8–11 months of continuous dry season.

Within a given rangeland type RUE may vary considerably with range condition, as we shall see in Section 7.3. Productivity may currently be decreased by a factor of 3 or even 5 to 1 or sometimes up to 10 to 1 for a given range type in a given site by mismanagement practices such as prolonged heavy overstocking. Conversely, RUE can often be increased by a factor of 3 to 5, using various techniques and heavy inputs such as water conservation techniques, fertilizers or reseeding. These are, however, usually not economically or socially feasible in the pastoral zone.

In general, RUE increases with the total biomass and plant cover present, as water losses from direct evaporation through soil surface are inversely related to these parameters. RUE also seems to increase with average rainfall up to an optimum and then decreases due to other limiting factors such as soil fertility or water-logging, as suggested by the research of Floret and Pontanier (1978) in Tunisia, Cornet (1981) in Senegal and De Vries *et al.* (1978) in Mali. Other experiments, however (Tadmor *et al.*, 1972a), suggest that RUE remains constant for a wide spectrum of water applications. The subject deserves further investigation.

7.2.3.2 Annual Variability

It is a well known fact that in all arid and semi-arid zones of the world, rainfall variability increases with aridity. South of the Sahara, for instance, the coefficient of variation* for annual rainfall varies from 0.40 at the edge of the desert to 0.15 in the rain forest, with values between 0.25 and 0.40 in the pastoral area (Le Houérou and Popov, 1981). In the Mediterranean North Africa it varies from 0.50 to 0.60 in the Northern Sahara, to 0.20 in the higher rainfall coastal zones (Le Houérou, 1959).

It is less well known that variability in annual primary production is greater than variability in the amount of annual rains. If we compare the ratio of maximum to minimum recorded yields to the ratio of maximum to minimum rainfall (MY/my ÷ MR/mr) as published by several scientists for various arid and semi-arid zones, we shall find values similar to those shown in Table 7.2.

The use of maximum/minimum ratios as an indicator of variability is

Table 7.2 Rainfall to yields variability ratios

Area	Max rain ratio $\frac{\text{Max}}{\text{min}}$	Max yield ratio $\frac{\text{Max}}{\text{min}}$	$\frac{\text{Yr}}{\text{Rr}}$	No. of years	Reference
USA (various sites)	3.48	6.55	1.88	7–20	Cook and Sims, 1975
S. Tunisia	3.75	6.95	1.85	7	Floret and Pontanier, 1982
N. Senegal	13.6	21.3	1.57	7	Bille, 1977

* $V = \sigma/P$ where σ = standard deviation of annual rainfall and P = mean annual precipitation.

debatable. It surely is not, in theory, the best possible indicator. The ratio between coefficients of variation is a better indicator $(\sigma/\bar{x})$. A worldwide study of arid and semi-arid rangelands by the present writer (Le Houérou, 1984b) over 77 series of coupled data on annual primary production/annual rainfall totalling 835 years, i.e. an average 12 years per series (3–43), concluded as follows:

1. the ratio vY/vR averaged 1.50 (s.e. = 0.07);
2. the ratio MY/my ÷ MR/mr averaged 1.81;

where: vy = coefficient of variation of annual primary production $(\sigma/\bar{x})$;
vR = coefficient of variation of annual rainfall;
MY = maximum annual yield recorded within each series;
my = minimum annual yield recorded within each series;
MR = maximum annual precipitation recorded within each series;
mr = minimum annual precipitation recorded within each series.

The 'variation in variability', however, is high. In some 10 percent of the cases variability in production is lower than in rainfall (depressions, water tables, higher rainfall areas 7600 mm). In the same 835 years over the same 77 sites RUE averaged 4.0 kg DM mm^{-1} ha^{-1} y^{-1}, again with large variations mainly due to range condition, that is, essentially as a result of management practices. The latter may totally hide the effect of climatic aridity. The study concluded that rain use efficiency depends more on management practices than on climatic aridity; the RUE factor is therefore a good criterion for evaluating ecosystem condition, health and productivity.

The consistency between the three sets of data in Table 7.2 is unexpectedly good and one may say that variability in annual production, as measured by maximum/minimum ratios, is expected to be 150–200 percent of the variability in annual precipitation. It would seem that a similar situation prevails in arid Australia; Young and Wilson (1978) wrote: 'when the rainfall received in a particular year is half the median, forage growth may be reduced to one quarter.' This means a variability in production 50 percent greater than the variability in rainfall, as found in our study (Le Houérou, 1984b).

7.2.3.3 Seasonal Distribution

Seasonality of rainfall does not seem to have a strong effect on total primary production when expressed in dry matter. The distribution of rain within the rainy season, on the contrary, has a very strong effect on production. It is a well known fact that the highest production is not necessarily the consequence of the highest rain. Moderate rains with regular distribution patterns and moderate intensities are the most efficient. Modelers have designed ideal distribution patterns (Van Keulen, 1975; De Vries *et al.*, 1978; Floret *et al.*, 1982) but, as the number of possible situations of rainfall events over a given rainy season is almost infinite, it is extremely difficult to quantify the effect of rain distribution on primary range production as has been done for crops. A simple and crude criterion used by several authors has been the number of rainy days. The use of this criterion does not improve the predictability of production as drawn from the amount of rain fallen. Several authors have also attempted to define 'useful rain'; there are as many empirical and more or less arbitrary definitions of 'useful rains' as the number of authors. The principle is to discard all precipitation beneath a certain amount per day or per week or per month. The subject is poorly documented as far as range production in pastoral areas in relation to 'useful rains' is concerned.

In higher rainfall areas such as the southern Sudanian zone of West Africa, production is linked to the length of the rainy season rather than to the amount that falls (Boudet, 1975). In otherwise similar regions, areas with 7 or 8 months of rainy season and, say, 1200 mm of precipitation, will have a higher production than areas having 1500 mm falling in 5 or 6 months. But these are hardly pastoral zones.

7.3 MODIFICATION OF CLIMATE–PRIMARY PRODUCTION RELATIONSHIP BY MANAGEMENT PRACTICES

The effect of climate on pastoralism cannot be validly considered in isolation but should be examined within a socioeconomic framework. Many surveys and experimental studies have shown that vegetation response to climate depends as much on its status and condition as on climate variables.

7.3.1 Management Practices

It has been mentioned above that RUE may vary by more than 300 percent within a given plant community according to plant cover and biomass. The decrease of RUE with reduced plant cover as a result of overgrazing, wood-cutting, and inappropriate cultivation is one of the main mechanisms of desertization, as shown by many scientists (Le Houérou, 1959, 1962, 1968, 1971, 1975; Floret and Le Floch, 1979; Floret and Pontanier, 1982). All management practices which tend to reduce plant cover and biomass will therefore decrease RUE and, conversely, management practices tending to maintain plant cover and biomass will also increase the productivity of the ecosystem. The effect of different practices—no grazing, heavy grazing and overstocking—on plant cover, production and RUE for two North African plant communities is shown in Table 7.3.

The absence of grazing or undergrazing because of insecurity (as in northern Kenya, eastern and southern Ethiopia) or because of the lack of permanent water (as in many areas of the Sahel) can be a local problem. But underused

Table 7.3 Effect of management practices on plant cover, primary production and RUE

Site	Rainfall	No grazing			Heavy grazing			Overstocking		
		Plant cover (%)	Production (kg DM ha^{-1} y^{-1})	RUE (kg DM mm^{-1})	Plant cover (%)	Production (kg DM ha^{-1} y^{-1})	RUE (kg DM mm^{-1})	Plant cover (%)	Production (kg DM ha^{-1} y^{-1})	RUE cover (kg DM mm^{-1})
Hodna Basin, Algeria[1]	200	25	1044	5.22	5	425	2.13	3	257	1.29
Southern Tunisia[2]	314	25	1069	3.40	8	615	1.96	4	415	1.32

[1] *Salsola vermiculata-Anabasis oropediorum* community (silty gypsic soil, Hodna Basin, Algeria [Le Houérou, 1971]).
[2] *Rhantherium suaveolens-Stipa lagascae* community (coarse sandy soil, southern Tunisia [Bourges *et al.*, 1975]).

areas probably represent less than 1 percent of the total pastoral zones of Africa and Asia and therefore, generally speaking, are not a relevant factor.

The presence of edible trees and shrubs increases considerably the stability of production in range ecosystems, thus buffering and damping out the effects of climatic variability. This fact is due to various characteristics of trees and shrubs, including the ability to draw water from many meters down into deep soil layers (up to 50–100 m in some cases, but routinely 10–30 m in arid zones) not accessible to the shallow root systems of grasses and herbs, and to make use of out-of-season rains. In addition, trees and shrubs can remain green throughout the dry season, and many species can carry their green leaves, phyllodes, or cladodes for 18–24 months, accumulating forage from two or more growing seasons. These leaves contain high contents of protein, carotene and minerals, whereas dried-out grasses are virtually void of digestible protein and carotene. Some trees, such as *Acacia* and *Prosopis*, also produce highly nutritive fruits that may be stored and marketed.

The value of shrubs and trees was particularly exemplified during the 1968–73 drought in the Sahel where browsers (camels and goats) managed to survive throughout the drought while grazers (sheep and cattle) perished of starvation in great numbers. The conservation and maintenance of multi-story range ecosystems thus extends the usable primary production. Unfortunately, in many pastoral areas, tree and shrub cover is in rather rapid regression due to overexploitation (browsing, lopping, wood-cutting), which in turn renders them more sensitive to the effects of drought. Additionally, bush fires are held responsible for the annual destruction of 80 million hectares of pastures in the semi-arid and subhumid savannas of Africa (Wickens, 1968), thus diminishing the grazable acreage and increasing the pressure of livestock on the unburnt areas.

Overgrazing for long periods of time in arid and semi-arid zones may lead to sharp evolution in pastoral groups (Lamprey, 1983). The grazing lands of the Samburu in northern Kenya have thus become progressively unable to sustain cattle and Samburus are now shifting from their traditional cattle pastoralism to camels and small stock, mainly goats. Similar recent revolutions have been reported in other East African pastoralist groups (Dyson-Hudson, personal communication).

Thus conservation methods are essentially the continuous adaptation of stocking rates and kinds of livestock to carrying capacities and the nature of the range; the utilization of grazing systems compatible with the regeneration of vegetation; and, in some cases, improvement techniques such as ripping, subsoiling, scarifying, pitting, contour benching and water spreading, which all tend to increase the water intake of the soil and therefore of RUE. These may be combined with oversowing or planting of shrubs which, in addition, aim at improving the use of water by the utilization of more productive plant material.

All these practices, by increasing water intake by the soil, tend to smooth off the effects of irregular rainfall and increase rain efficiency.

7.3.2 Trends in Management

The impact of climate variability on rangeland ecosystems in pastoral areas may be enhanced or reduced by human actions (management practices). The trend in such practices unfortunately appears to be in the direction of exacerbating climate variability impacts. The sharp increase of livestock numbers in Africa since the 1950s, due mainly to efficient vaccination campaigns and the increase of the human population (be it sedentary or nomadic), continuously augments the pressure on ecosystems. It should be stressed that the increase in human pastoral populations is light as compared to the case of settled farmers, even when the farmers are former nomads such as the Buzu Tuaregs or the Farfaru Fulani, the rate of demographic increment being 1.0–1.5 percent versus 2.5–3.5 percent for the farmers in West Africa (Bernus, 1981). The surface of land under cereal cultivation has doubled in the Sahel of Africa for the past 20 years, at the expense of rangelands and of the pastoralists, as shown in several local surveys in Niger, Mali and Senegal; the same holds true for many parts of the pastoral zone of East Africa (ILCA, 1978). Boreholes discharging great quantities of water without any control of stock numbers have contributed to attract large numbers of stock (15,000–20,000 tropical livestock units* around some boreholes in the dry season in the pastoral zones of Niger [Bernus, 1971]), thus destroying all vegetation in a radius of 20–30 km. Wood-cutting affects large areas around cities and villages in the pastoral area; in some cases one has to go many kilometers from towns in search of firewood, and it may cost the town dweller as much to boil the pot as to fill it.

One of the conclusions of a detailed study of a Sahelian ecosystem in northern Senegal is that it would take two to three decades for the vegetation to return to its predrought conditions after the 1968–73 episode, in a situation of total protection (Bille, 1978); in the present conditions of exploitation the impact of the Sahelian 1968–73 drought will never be cancelled or absorbed. It has also been concluded that the continuation of the present type of exploitation of the Sahel would render most of the pastoral area unfit for animal nutrition for 6–9 months of dry season every year within the next 50 years or so, owing to the elimination of the woody plants which at present supply most of the protein and carotene in stock diets during the dry season (Le Houérou, 1980a).

The situation is less dramatic in East Africa, for the time being, but it is still worsening rather fast (Lamprey, 1983). In northern Africa, however, the situation is improving because supplementation with concentrate feeds is taking

* TLU = conventional mature head of zebu cattle weighing 250 kg $\simeq$ 5 sheep = 6 goats = 2 asses = 0.8 camel, metabolic weight equivalents.

place. In many pastoral areas around 30 percent of animal nutrition needs presently are met in this way—a significant achievement considering that supplementary feeding was hardly known, let alone practiced, in these regions only 10 years ago (Le Houérou and Aly, 1981).

The present impact of man in pastoral areas—overstocking, wood-cutting, bush fires, 'wild' water development, expansion of inappropriate cultivation, and ever-increasing animal numbers—all intensify the pressure on grazed ecosystems. These, in turn, become less resilient and less able to 'absorb' climatic droughts, which they used to do under the light exploitation conditions of past centuries. If this chaotic situation imposed onto the pastoralists persists—and there is unfortunately little doubt that it will—one can predict only a gloomy future for pastoralism. Range ecosystems will become more and more sensitive to periodic droughts (which are a constant characteristic of the arid and semi-arid pastoral zones). Pastoralism will either have to disappear quietly or evolve towards other systems of animal production, with increased reliance on supplementary feeding of grain and concentrates for longer and longer periods.

7.4 PASTORAL LIVESTOCK PRODUCTION

Secondary production from livestock in pastoral systems is very low by modern animal husbandry standards. Birth rates in pastoral cattle barely reach 60 percent (50–65), with 5–15 percent abortions and stillbirths; 40 percent of the calves (30–45) die in the first year and about 5 percent per year subsequently, save disease outbreaks or drought (Doutresoulle, 1947; Coulomb, 1971, 1972; Wilson and Clarke, 1975; SEDES, 1975; ILCA, 1978; OMBEVI, 1978; Republic of Niger, 1981; Diallo and Wagenaar, 1981; Wilson *et al.*, 1981; Wilson and Wagenaar, 1983). In the East African pastoral zone, predation by carnivores (spotted hyenas, lions, jackals) may reach up to 10 percent (Kruuk, 1980). Output (offtake and herd increment) is thus of the order of 8–12 percent in cattle, 20–25 percent in small stock, and 7–8 percent in camels.

Net annual body weight gain of African zebu cattle is usually around 50 kg, so that mature size is reached only at 5–6 years of age. In lambs and kids daily gains are 50–100 g up to 6 months and 100–300 g for camel calves (IPAL, 1981). These low performances are due to feed restriction, since milk is shared on about a 50 percent basis between the pastoralist and the calf, and to the fact that these breeds have been selected for centuries on the criterion of survival rather than on production. Production, in addition, is seriously hindered by parasitism and diseases. Conversely, modern productive stock could possibly not survive in the ecological conditions in which pastoral stock do. In spite of this, research has shown that such 'local breeds' (such as Boran, Gobra, Azawak cattle, Awasi, Somali, Persian, Fat Tail Barbary,

Karakul sheep) are amenable to considerable improvement in their performance when they are properly managed.

It should be kept in mind that meat production is not—with the exception of sheep—the primary objective of African pastoralists, but rather milk production, and, occasionally in East Africa, blood production. Milk production varies with ecological conditions and breeds. It is of the order of 300–600 liters per annum for cattle, 80–100 liters for goats, 50–70 liters for sheep, and 1500–2000 liters for camels; about half of these quantities are left for the calves, kids and lambs and the other half levied by the pastoralist (Dahl and Hjort, 1976; ILCA, 1978; IPAL, 1981).

7.4.1 Climate–Livestock Relationships

Livestock numbers are strongly influenced by climatic variability, but unlike primary production, they dampen rather than amplify climatic variability. Le Houérou (1962) showed that, over a period of 25 years (1936–60) in southern Tunisia, the ratio of maximum annual number over the minimum was 3.7 for sheep and goats while the ratio of maximum to minimum rainfall over the same period was 6.8, whereas camels' and donkeys' numbers were little affected by rainfall variations (±20 percent). The variabilities ratio was thus $6.8 \div 3.7 = 1.84$. In Agadez, Niger, the ratios for 1968–72 were the following: cattle, 4.6; sheep and goats, 3.0; and camels, 2.0; with rainfall, 4.0 (Bernus, 1981). Variability in stock numbers is thus 25–50 percent smaller than variability in rainfall, contrary to what happens for primary production. In addition, there is a time lag of 1–2 years in the correlation between rainfall and stock numbers (Le Houérou, 1962); therefore a 1-year drought is no catastrophe. Serious problems arise when two or more dry years occur in succession, as happened in the Sahel in 1910–15, 1940–44 and 1969–73 (Bernus and Savonnet, 1973) and in the north of the Sahara in 1920–25, 1944–48, and 1959–61 (Le Houérou, 1962, 1968, 1979).

During such a severe multiyear drought, 30–70 percent of the herds may die. It is reckoned that 50–70 percent of small stock died of starvation in the North African pastoral zone during the 1944–48 drought, while the overall losses in the Sahel during the 1968–73 drought are estimated at 30–50 percent for the whole region. (The low was in Senegal at 20 percent, and the high in Mauritania at 70–80 percent.)

The performance of the surviving livestock is also considerably reduced in terms of growth, of milk and meat production, and in terms of reproductive rate. The consequences of reduced nutrition and fertility in connection with drought or disease on the future performance of the herd has been studied for cattle by Dahl and Hjort (1976), via simulation modeling. Similar work remains to be done for small ruminants, but some data for sheep are available. According to Israeli workers it would seem that the reproductive performance of Awasi sheep is not affected until normal body weight is reduced by 30 percent or more (Tadmor, personal communication); our own results on Fat Tail Barbary sheep in Libya are in support of that statement (Le Houérou and Dumancic, unpublished).

A 50 percent reduction in pastoral production potential probably could be ascribed to the consequences of climatic variability, now that major disease outbreaks have been virtually curbed. It has been calculated in Tunisia that the interannual stabilization of the national flock, the variability of which formerly reached 55 percent (+29 to −26 percent of the mean), would increase its net productivity by 25 percent (Le Houérou and Froment, 1966). At the level of the pastoral zone, where the variability is over 110 percent (+50 to −60 percent of the mean), one would therefore expect the loss in productivity due to climatic variability to be more than 50 percent of the potential, all other management conditions (including health) remaining equal. This figure is confirmed when examining offtake from herds and flocks in the pastoral zone. These offtakes are hardly half of what they are known potentially to be under proper management (essentially feeding) conditions.

7.4.2 Modification of Climate–Livestock Relationships by Drought Adjustments

The high variability in primary production is effectively dampened by the drought-resistant breeds of livestock raised by pastoralists as described above, by the mix of livestock held, and by various sociocultural relationships in addition to the basic nomadic movement.

The strategy of keeping mixed herds—cattle and small ruminants or camels and small stock—is a drought insurance strategy, as herd build-up in small ruminants is much faster than in large stock; save disease or accident a couple of cows will produce 10 offspring in 5 years while two ewes will produce 32 and two goats 130, in the same time span (Lundholm, 1976). At the same time, keeping grazers (cattle and sheep) and browsers (camels and goats) increases the risk-sharing of drought.

Other strategies of general use among pastoralists in the African ecocultural environment include the complex system of leasing or borrowing-lending—gifts of animals among kinsfolk, relatives and friends—creating a complicated network of duties and rights, of gifts and claims, and of reciprocity. Intermarriages between ethnic groups occupying different ecological niches (such as the Samburu, Rendille and Turkana) and various political alliances, such as among the Çoffs of North Africa, also tend to reduce the risks (drought and disease) by spreading the herds over as large an area as possible under as varied ecological conditions as possible. Under such expedients the possession of the larger number of animals—or the strategy of debtors and clients—are the best possible drought insurances.

Other tactics, such as range exclosures and deferred grazing, were practiced for centuries in some Bedouin pastoral societies in Arabia (Hema) and northern Africa (Gdal) as part of a drought-evading strategy. Some practices, such as the sophisticated and labor-intensive cattle-watering system of the Boran of South Ethiopia, in conjunction with a form of rotational-deferred grazing through the organized seasonal use of water resources, also tended (in addition to the age-class social structure) to keep stock and human numbers under control (Helland, personal communication).

Past methods also involved supplementary support and included the possession or control of oases and grain fields cultivated by slaves or dependent farmers for their pastoralist masters (Tuaregs, Moors). Other strategies were raiding and warfare, the major goal of which was sometimes the replacement of flocks or herds decimated by disease or drought and hence, word for word, a matter of life and death.

Today, pastoralist adaptation goes beyond the system itself and includes settlement as farmers of the pastoral population in excess of the carrying capacity, wage labor, village shop-keeping and trade, government employment (army, police), and emigration to cities.

7.5 LIVESTOCK AND HUMAN POPULATION RELATIONSHIPS

7.5.1 Livestock and Human Ratios

Large differences in the ratio of livestock to human beings occur in different pastoral systems. These are shown for various African systems in Table 7.4. They range between 3900 kg of live weight per person among the Masai to 400 kg for the Ifora Tuaregs. The groups of the lower part of Table 7.4, particularly the Ifora Tuaregs, are destitute pastoralists who managed to survive during the Sahel drought only because of food assistance; several thousand of this group sought refuge and assistance in Niamey and Tamanrasset; probably many died during the 1968–73 drought. Their near neighbors on the table experienced similar conditions in 1979–82. The Masai, on the contrary, appear as 'rich' pastoralists, living as they do under the much better ecological conditions of semi-arid to subhumid climates with lush grass savannas (at least as compared to the conditions in which the Saharan pastoralists live).

The comparison of pastoralists keeping similar species in approximately similar proportions, may, however, be considered valid, for instance Masai, Samburu, Boran, Fulani, Karimojong, Baggara, who are primarily cattle pastoralists, as opposed to Somali, Rendille, Kabbabish, Tuaregs, Afar, Moors, Al Murrah, who are primarily cameleers, and to Turkana, Middle Eastern and North African Bedouins, who are essentially small stock owners. Overall the limit of survival, in the absence of cereal cultivation, seems to be around 800–1000 kg of live weight per person (Brown, 1971; Lamprey, 1983).

Table 7.4 Livestock to human population ratios

Pastoral ethnic groups	kg Lwt of stock per person*	References
Kenya Masai	3900	Allan, 1965
Tanzania Masai	3500	Allan, 1965
Somali	1500	Allan, 1965
Turkana	1300	Allan, 1965
Baggara	1300	Allan, 1965
North African Bedouins	800–1500	Le Houérou, 1962
Bororo (woodabe) Fulani	1175	Dupire, 1970
Kel Dinnik Tuaregs	1075	Bernus, 1981
Rendille	1020	Lusigi, 1981
Jie	880	Allan, 1965
Kel Air Tuaregs	625	Bernus, 1981
Karimojong	600	Dyson-Hudson, 1966
Kel Ifora Tuaregs	400	Swift, 1975

* The amount of live weight per person has been calculated from various units: large stock units (500 kg), standard stock units (SSU = 450 kg), livestock units (200 kg), tropical livestock units (TLU = UBT = 250 kg), ovine units (40 kg small ruminants), as used by various authors. These ratios should be considered with caution as livestock equivalence factors are only approximate—more precise comparisons would emerge from the use of metabolic weights (wt 0.75)—and as their value in terms of survival strategy are quite different. But there is a zonation in the kinds and breeds of animals which are kept; this zonation is based on drought risk or, to put it in a different way, in rainfall belts or ecoclimatic zones (Le Houérou and Popov, 1981).

Thus the meaning of the ratio may differ substantially with the terms of trade between grain and livestock products. Fewer and fewer pastoralists live on a purely animal products diet; more and more include grain in their food, particularly in the dry season. The Kel Dinnik Tuareg pastoralists of Niger, for example, depend on millet grain for 52 percent of their annual energy consumption (Bernus, 1981). Grain is either cultivated by the pastoralists themselves, or bought or bartered against livestock products. In the latter cases the terms of trade may depend considerably on year, season, region and country. The subsistence and survival ratios thus may vary substantially in time and space.

Sensitivity to climatic variability is inversely related to the livestock/humans ratio; this fact explains the constant trend of pastoralists to maximize stock numbers as a form of drought insurance. As stock productivity is related to amount of rain and thus inversely related to climatic variability, one would expect the ratio of livestock to humans to increase with aridity; it would seem rather that the reverse actually occurs. The reasons for this paradoxical situation remain nebulous to the author; the only explanation proffered is that overstocking, overpopulation and the increasing destitution of pastoralists in the drier zones has stretched the system to its limit of elasticity.

7.5.2 Economic Costs of Climate Variability

According to Food and Agriculture Organization (FAO) statistics, Africa had in 1979 some 526 million head of livestock (200 million FAO Standard Animal Units*), that is, an increase of 75 percent over the last 30 years. More than 400 million heads (or 150 million SAU) are still kept under more or less traditional pastoral or semi-pastoral production systems, in the author's estimation. Asia (without the USSR and China), on the other hand, harbors another 200 million heads (50 million FAO SAU) of pastoral stock. The world's number of pastoral stock is thus more than 600 million heads, or 20 percent of the total livestock population of the globe (pigs excepted).

Given the very low output and offtake rates and the poor individual production performances of pastoral stock, one may reckon that not more than one-third of its genetic production potential is able to manifest itself. The loss in production from the potential of pastoral stock thus may be grossly ascribed to climate hazards in conjunction with poor management of range and herds for about 50 percent, while the balance may be ascribed to diseases (in conjunction with poor nutrition), parasitism and predation.

If we accept the assumption that production† could, in principle, be tripled to meet genetic potential (and many experiments suggest such an assumption, indeed) the elimination of climatic hazards alone would increase production by some 100–150 percent in the 600 million heads of pastoral stock. If the pastoral stock of the world numbers about 200 million SAU with a market value of some $200 per unit, the global capital value is $40 billion; assuming an average actual output‡ of, say, 15 percent, we reach an annual production value of $6 billion against a potential of $18 billion; the loss in productivity due to climatically related causes would be of the order of $6 billion per annum. These figures are, naturally, nothing more than very crude orders of magnitude; they do, however, tend to show that the impact of climate variability on pastoralism is important.

The absolute variability in pastoral output, however, does not depend only on the absolute variability of rains or on the intensity of drought. It is still affected more by the duration of drought. A drought of 1-year duration is usually more or less well 'absorbed' by the system. Catastrophes occur when two or more 'dry years' occur in sequence.

* FAO's SAU = 1.1 camel = 1.0 horse, mule and buffalo, = 0.8 cattle and ass, = 0.1 sheep and goat.

† Production = animals slaughtered + herd increment + animal products and services (milk, blood, hides, hair, wool, draught power, transport, etc.)

‡ Average offtake ratios plus herd increment: camels 7 percent; cattle 10 percent; shoats 25 percent; equines 10 percent.

7.6 CONCLUSION

The linkage between climate and pastoral stock is, as we have seen, complex and affected by many variables which may be studied in various ways. This range of methods can be summarized by way of a simplified impacts analysis chain, in which a variable climate is reflected in the variation in primary productivity of biomass and edible plants. This in turn leads to fluctuation in livestock numbers, which in turn affects human numbers and wellbeing. A wide range of relatively standardized methods are available to study each element in the chain and the links between them, and some of these are shown in Table 7.5.

The impact of climate variability differs, of course, and is affected primarily by:

1. ecological conditions, in particular the nature of the vegetation, and the condition and trend of the rangeland;
2. adaptive strategies of the pastoralists to minimize the impact of drought.

In general, variability in primary production is much greater than variability in rainfall. Variability in stock numbers is somewhat smaller than variability in rainfall, but variability in secondary production is very poorly documented; it would seem to be of the same order of magnitude as the former. And the ratio of livestock to human beings is inverse to rainfall. Thus overall, variability in climate is amplified in primary production, but dampened in the production of livestock and in the sustenance of human beings. This is primarily due to various adaptive strategies.

Although these adaptive strategies are extremely diversified, they have a high cost in the long-term survival of pastoral systems. In pastoral systems the general tendency is to maximize stock numbers and their geographical distribution in order to reduce the risks. This strategy inflicts disastrous consequences on the environment.

Pastoral societies in past centuries were based on traditional land tenure systems which have been upset in the last few decades by sharp political and economic changes, whereby power has moved from the hands of the pastoralists to the settlers. The traditional systems, in which warfare and raiding were important components, were more or less in equilibrium with their fragile environment, but much lighter densities of both human and livestock populations prevailed in those times. Furthermore, both human and livestock populations were periodically swept out by disease outbreaks (the rinderpest epizootic of the 1880s in East Africa, for instance).

As human and stock populations increased as a result of imposed peace and of large prophylactic measures against human and animal diseases, the importance of climatic variability on pastoral production increased correlatively. The main constraint to pastoral husbandry thus shifted from disease

Table 7.5 Climate impact analysis of pastoralism

	Climate variability →		Primary production →		Secondary production →		Sustenance
Indicators	Rainfall, evapotranspiration		Production—biomass, edible plants		Production—livestock		Human wellbeing
Methods	1. Statistical analysis of rainfall records 2. Available moisture indication	Study of statistical relationships between rainfall and primary production and carrying capacity	1. Range survey 2. Measurement of primary production 3. Remote sensing 4. Low altitude reconnaissance flights	Analysis of livestock feeding patterns	1. Analysis of production parameters 2. Herd demography surveys 3. Livestock and wildlife censuses 4. Low altitude reconnaissance flights (animal numbers and seasonal distribution)	1. Livestock/people numbers ratios 2. People/crop aggregate ratios 3. Land supporting capacity surveys (humans/livestock)	1. Population census, births, mortality data 2. Nutrition surveys 3. Migration estimates 4. Household economy surveys

outbreaks to periodic food and water shortages, and hence to climatic variability.

The impact of climatic variability in turn is intensified by the depletion of rangelands, which are less and less productive and less and less able to 'absorb' prolonged drought events. The continuing expansion of cultivation in pastoral areas increases still more the stress on pure pastoralists. The usual response of pastoral societies is sedentarization and progressive shifting from pastoralism to agropastoralism and subsistence farming, a tendency generally encouraged—at least implicitly—by governments, since no serious large-scale integrated attempts to organize and modernize pastoralism have ever been made by any government in any country, to the author's knowledge.

There are, nevertheless, some islands of stability where pastoralists have more or less managed to keep their traditional system of control on human and stock populations through various sociocultural mechanisms and balanced water and land use systems; an outstanding example of this are the Boran of southern Ethiopia. How much longer will such systems last? Pastoralists themselves are evolving very fast, their values are changing, their expectations and needs increase as they become more and more aware of the outside world and of opportunities for easier conditions of life. One can only predict a point of rupture between decreasing resources, increasing populations, and the growing needs and expectations of pastoralists.

Indeed such a point may already have been reached. Table 7.6 shows the present situation as regards the population-supporting capacities in the arid and semi-arid zones of Africa. The study, made by a large multiorganizational, interdisciplinary team over several years, shows very clearly that arid (100–400 mm) and semi-arid (400–600 mm) zones of Africa already are supporting more

Table 7.6 Population-supporting capacities in arid and semi-arid zones of Africa

Ecoclimatic zone	Length of growing season (days)	Average present population density (person/km^{-2})	Potential population-supporting capacities	
			Under low inputs	Under intermediate inputs
Desert	0	0.03	0.01*	0.02†
Arid	1–74	0.06	0.03*	0.04†
Semi-arid	75–119	0.16	0.07*	0.20
Dry subhumid	120–179	0.20	0.32*	1.54

* Present-day intensity above potential under low input.
† Under intermediate input.
Source: Le Houérou and Popov (1981), computed from a study by Kassam and Higgins (1980).

people than they possibly could on a long-term sustained basis under low input conditions. Under intermediate inputs only the dry subhumid zone (600–800 mm) within the pastoral area could sustain more people—about seven times more than the present population. These prospects leave little future for the expansion of pastoralism.

ACKNOWLEDGMENT

The kind assistance of Bob Kates, Edmond Bernus and Francesco DiCastri in the preparation of this paper is acknowledged with gratitude.

REFERENCES

Allan, W. (1965). *The African Husbandman.* Oliver and Boyd, Edinburgh: 505 pages.

Asad, T. (1970). *The Kabbabish Arabs: Power, Authority and Consent in a Nomadic Tribe.* Hurst, London: 263 pages.

Barral, H. (1967). Les populations d'éleveurs et les problèmes pastoraux dans le Nord Est de la Haute-Volta. *Cah. ORSTOM, Sces. Hum.*, **IV**, 1, 3–30.

Barral, H. (1977). Les Populations Nomades de l'Oudalan et Leur Espace Pastoral. *Trav. Doc. ORSTOM*, **77**. ORSTOM, Paris: 119 pages.

Barth, F. (1961). *Nomads of South Persia: The Basseri Tribe of the Khamseh Confederacy*. University Press, Oslo: 159 pages.

Bataillon, C. (Ed.) (1963). *Nomades et Nomadisme au Sahara.* Arid Zone Res. UNESCO, Paris.

Bernard, A., and Lacroix, N. (1906). *L'Évolution du Nomadisme en Algérie.* Challamel, Paris: 342 pages.

Bernus, E. (1971). Possibilities and limits of pastoral watering plans in the Nigerian Sahel. *Seminar on Nomadism, FAO, Cairo* (mimeo: 13 pages).

Bernus, E. (1981). *Touaregs Nigériens: Unité Culturelle et Diversité Régionale d'un Peuple Pasteur.* ORSTOM, Paris: 508 pages.

Bernus, E., and Savonnet, G. (1973). *Les Problèmes de la Sécheresse dans l'Afrique de l'Ouest.* Presence Africaine, Paris.

Bille, J. C. (1977). *Étude de la Production Primaire Nette d'un Ecosystème Sahélien.* Trav. and doc., ORSTOM, Paris: 82 pages.

Bille, J. D. (1978). Woody forage species in the Sahel: Their biology and use. In *Proceedings of the First International Rangelands Congress*, pp. 392–395. Society for Range Management, Denver, Colorado.

Bonnet-Dupeyron, F. (1951). *Atlas Pastoral pour la Mauritanie et le Sénégal.* ORSTOM, Paris: 37 pages.

Bonte, P. (1973). L'élevage et le commerce du bétail dans l'Ader-Doutchi Majiya. *Et. Niger*, **23**. CNRSH, Niamey.

Bonte, P., Bourgeot, A., Digeard, J. P., and Lefebure (1981). Pastoral economies and societies in tropical grazing ecosystems. In *UNEP-FAO-UNESCO, Res. on Nat. Res.*, **XVI**, 260–302. UNESCO, Paris.

Boudet, G. (1975). *Manuel sur les Pâturages Tropicaux.* IEMVT, Maisons Alfort: 254 pages.

Bourges, J., Floret, C., and Pontanier, R. (1975). Étude d'une toposéquence type du

Sudtunisien. Djebel Dissa: Les sols, bilan hydrique, érosion, végétation. *Étude Speciale No. 93*, Div. des Sols, Min. Agric., Tunis.

Braudel, F. (1949). *La Méditerranée et le Monde Méditerranéen à l'Époque de Philippe II*. A. Colin, Paris: 589 pages.

Braun, M. H. M. (1973). Primary production in the Serengeti. *IBP, Symp., Lamto, Ann. Univ. of Abidjan*, **6**(2), 171–188.

Breman, H. (1975). Maximum carrying capacity of Malian grasslands. *Proceed. Sem. Eval. and Mapp. Trop. Afr. Rang.*, pp. 249–256. ILCA, Addis Ababa.

Briggs, L. J., and Shantz, H. L. (1913). *The Water Requirements of Plants, I: Investigations in the Great Plains in 1910 and 1911*. US Dept. of Agric. Bur. of PI Ind. Bull. 284. Washington, DC.

Brown, L. H. (1971). The biology of pastoral man as a factor in conservation. *Biol Cons.*, **3**, 93–100.

Capot-Rey, R. (1953). *Le Sahara Français*. PUF, Paris: 564 pages.

Cassidy, J. T. (1973). The effect of rainfall, soil moisture and harvesting intensity on grass production in two rangeland sites in Kenya. *E. Afr. Agr. For. J.*, **89**, 26–36.

Cauneille, A. (1950). Les nomades Reguibat. *Trav. Inst. Rech. Sah.*, **6**, 83–100. Alger.

Cauneille, A. (1968). *Les Chambaa (Leur Nomadisme): Évolution de la Tribu durant l'Administration Française*. Centre Rech. Nord-Afr., CNRS, Aix en Provence: 317 pages.

Condon, R. W. (1968). Estimation of grazing capacity on arid grazing lands. In *CSIRO Symposium on Land Evaluation*, pp. 112–124. Canberra.

Cook, C. W., and Sims, P. L. (1975). Drought and its relationship to dynamics of primary productivity and production of grazing animals. *Symp. on Eval. and Mapp. of Trop. Afr. Ranglds.*, pp. 163–170. ILCA, Addis Ababa.

Cornet, A. (1981). *Le Bilan Hydrique et son Rôle dans la Production de la Strate Herbacée de Quelques Phytocoenose au Sénégal*. Thèse Dr.-Ing. Fac. Sces. Montpellier, Mém. ORSTOM, Paris (mimeo: 353 pages).

Cornet, A., and Rambal, S. (1981). *Simulation de l'Utilisation de l'Eau par une Photocoenose de la Zone Sahélienne au Sénégal: Test de Deux Modèles*. Informatique et Biosphère, Paris.

Coulomb, J. (1971). Zone de modernisation pastorale du Niger. *Économie du Troupeau*, IEMVT, Maisons-Alfort.

Coulomb, J. (1972). Projet de développement de l'élevage dans la région de Mopti. *Étude du Troupeau*. IEMVT, Maisons-Alfort.

Coupland, R. T. (1958). The effects of fluctuations in weather upon the grassland of the Great Plains. *Bot. Review*, **24**, 273–281.

Dahl, G., and Hjort, A. (1976). *Having Herds: Pastoral Herd Growth and Household Economy*. Department of Social Anthropology, University of Stockholm: 335 pages.

de Planhol, X., and Cabouret, M. (1961–72). Nomades et pasteurs. *Rev. Geogr. de l'Est*: I–XII.

De Vries, P. F. W. T., Krul, J. M., and Van Keulen, H. (1978). Productivity of Sahelian rangelands in relation to the availability of nitrogen and phosphorus from the soil. In *Workshop on Nitrogen Cycling in West African Ecosystems*. IITA, Ibadan, Nigeria.

De Witt, C. T. (1958). *Transpiration and Crop Yields*. Wageningen.

Diallo, A., and Wagenaar, K. (1981). Livestock productivity and nutrition in the pastoral systems associated with the flood plains. In Wilson, R. T., De Lecuw, P. N., and De Haan, C. (Eds.) *Systems Research in the Arid Zones of Mali: Preliminary Results*. Research Report No. 5, ILCA, Addis Ababa.

Doutresoulle, G. (1947). *L'élevage en Afrique Occidentale Française*. Larose, Paris: 298 pages.
Dupire, M. (1962). *Peul Nomades, Étude Descriptive de Woodabe du Sahel Nigerien*. Trav. et Mem., Inst. Ethnol., LXIV, Paris: 336 pages.
Dupire, M. (1970). *Organisation Sociale des Peul: Étude d'Ethnographie Comparée*. Rech. en Sces. Hum., No. 32. Plon, Paris: 625 pages.
Dyson-Hudson, N. (1966). *Karimojong Politics*. Clarendon Press, Oxford: 280 pages.
Floret, C., and Le Floch, E. (1979). *Recherche et Développement des Parcours du Centre-Sud Tunisien*. Rapp. techn. AG: DP/Tun/69/001. FAO, Rome: 195 pages.
Floret, C., and Pontanier, R. (1978). *Relations Climat-Sol-Végétation dans Quelques Formations Végétales Spontanées du Sud Tunisien*. ORSTOM, Paris/CNRS. Montpellier: 112 pages.
Floret, C., and Pontanier, R. (1982). *L'aridité en Tunisie Présaharienne: Climat, Sol, Végétation et Aménagement*. Thèse D. Sc., Univ. Sces. Techn. Languedoc, Mém. Orstom, sous Presse, Paris: 580 pages.
Floret, C., Pontanier, R., and Rambal, S. (1982). Measurement and modelling of primary production and water use in a south Tunisia steppe. *J. Arid. Envir.*, **5**(1), 77–90.
Gallais, J. (1967). *Le Delta Intérieur du Niger: Étude Géographique Régionale* (2 vols.). IFAN, Dakar: 625 pages.
Gallais, J. (1975). *Pasteurs et Paysans du Gourma: La Condition Sahélienne*. CEGET, Bordeaux: 239 pages.
Gallais, J. (Ed.) (1977). *Stratégies Pastorales et Agricoles des Sahéliens durant la Sécheresse 1969–74*. Trav. Doc. No. 30, CEGET, Bordeaux: 231 pages.
Grouzis, M., and Sicot, M. (1981). *Pluviométrie et Production des Pâturages Naturels Sahéliens: Étude Méthodologique et Application à l'estimation de la Production Fréquentielle du Bassin Versaut de la Mare d'Oursi, Haute Volta*. ORSTOM, Ouagadougou (multigr).
Hanks, R. J., Gardner, H. R., and Florian, R. L. (1969). Plant growth–evapotranspiration relations for several crops in the Central Great Plains. *Agr. J.*, **61**, 30–34.
Hardin, G. (1968). The tragedy of the commons. *Science*, **162**, 1243–1248.
Hiernaux, P., Cisse, M. I., and Diarra, L. (1978, 1979, 1980). *Rapports Annuels d'Activité*. Section Écologie, CIPEA, Bamako.
ILCA (1978). See Le Houérou and Wilson (1978).
IPAL (1981). See Lusigi (1981).
Johnson, D. L. (1969). *The Nature of Nomadism: A Comparative Study of Pastoral Migrations in Southwestern Asia and Northern Africa*. Department of Geography Research Paper, University of Chicago, Chicago, Illinois.
Kassam, A. H., and Higgins, G. M. (1980). Report on the second FAO/UNFPA expert consultation on land resources for population of the future. AGLS/FAO, Rome.
Kruuk, H. (1980). *The Effects of Large Carnivores on Livestock and Animal Husbandry in Marsabit District, Kenya*. Techn. Rep. E-4, IPAL, UNESCO, Nairobi.
Lamprey, H. F. (1983). Pastoralism yesterday and today: The over-grazing problem. In Bourlière, F. (Ed.) *Tropical Savannas*, pp. 643–666 (Vol. 13 in *Ecosystems of the World*). Elsevier, Amsterdam.
Lamprey, H. F., and Yussuf, H. (1981). Pastoralism and desert encroachment in northern Kenya. *Ambio*, **10** (2–3), 131–134.
Le Houérou, H. N. (1958). Écologie, phytosociologie et productivité de l'olivier en Tunisie méridionale. *Bull. Carte phytogéogr., B*, **IV**, 1: 7–77. CNRS, Paris.

Le Houérou, H. N. (1959). *Recherches Écologiques et Floristiques sur la Végétation de la Tunisie Méridionale.* Mem. H.S. No. 8, Inst. Rech. Sah. Univ. Alger: 510 pages.

Le Houérou, H. N. (1962). *Les Pâturages Naturels de la Tunisie Aride et Désertique.* Inst. Sces. Econ. Appl. Paris-Tunis: 120 pages.

Le Houérou, H. N. (1963). Méthodes d'inventaire de la végétation naturelle et leurs relations avec la productivité et l'utilisation des herbages. *Seventh FAO Working Party on Mediterranean Pastures, Madrid, September 1963.* AGP, FAO, Rome.

Le Houérou, H. N. (1964). The grazing lands of the Mediterranean Basin and their improvement. *Goat Raising Seminar, FOD/AGP/AGA.* FAO, Rome.

Le Houérou, H. N. (1965). *Improvement of Natural Pastures and Fodder Resources.* Report to the Government of Libya. Exp. Progr. of Techn. Ass. Rep. No. 1979. AGP, FAO, Rome.

Le Houérou, H. N. (1968). La désertisation du Sahara septentrional et des steppes limitraphes. I.B.P./C.T., Coll. Hammamet; *Ann. Algér. de Géogr.*, **6**, 2–27. Alger.

Le Houérou, H. N. (1969). La végétation de la Tunisie steppique. *Ann. Inst. Nat. Rech. Agron.*, **4**, **5**, Tunis: 624 pages.

Le Houérou, H. N. (1971). An assessment of primary and secondary production of the arid grazing lands ecosystems of North Africa. In *International Symposium on Ecophysiological Foundations of Ecosystems Productivity in Arid Zones*, pp. 168–172. Nauka, Leningrad.

Le Houérou, H. N. (1972). *Le Développement Agricole et Pastoral de l'Irhazer d'Agadès.* AGP, NER/71/001. FAO, Rome.

Le Houérou, H. N. (1975). The rangelands of North Africa: Typology, yield, productivity and development. *Proceed. Sem. on Eval. and Mapp. of Trop. Afr. Rangldls.*, pp. 41–55. ILCA, Addis Ababa.

Le Houérou, H. N. (1977a). The grasslands of Africa: Classification, production, evolution and development outlook. *Proceed. XII Grassland Congress, Leipzig*, Vol. I, pp. 99–116. Academie Verlag, Berlin.

Le Houérou, H. N. (1977b). The scapegoat. *Ceres*, **56**, 14–18.

Le Houérou, H. N. (1977c). Biological recovery versus desertization. *Economic Geography*, **53** (4), 413–420.

Le Houérou, H. N. (1979). Écologie et désertisation en Afrique. *Trav. Inst. Géographie*, **39–40**, 5–26. Reims.

Le Houérou, H. N. (1980a). Browse in the Sahelian and Sudanian zones of Africa. In Le Houérou, H. N. (Ed.) *Browse in Africa*, pp. 83–103. ILCA, Addis Ababa.

Le Houérou, H. N. (1980b). The rangelands of the Sahel. *J. Rge. Mgmt.*, **33** (1), 41–46.

Le Houérou, H. N. (1982). Prediction of range production from weather records in Africa. *Techn. Conf. on Climate in Africa.* WMO, Geneva.

Le Houérou, H. N. (1984a). Rain use efficiency: A unifying concept in arid land ecology. *J. Arid Envir.*, **7**(2), 1–35.

Le Houérou, H. N. (1984b). Towards a probabilistic approach to arid rangelands development. *Proc. 2nd Internat. Rglds. Congr., Adelaide, Australia* (in press).

Le Houérou, H. N., and Aly, I. M. (1981). *Perspective Study on the Libyan Agriculture: The Rangeland Sector.* FAO, Tripoli: 85 pages (mimeo).

Le Houérou, H. N., Claudin, J., and Haywood, M. (1974). *Étude Phytoécologique du Hodna.* AGP, FAO, Rome: 150 pages.

Le Houérou, H. N., and Froment, D. (1966). Définition d'une doctrine pastorale pour la Tunisie steppique. *Bull. Ec. Nat. Sup. Agron.*, **10–12**, 72–152. Tunis.

Le Houérou, H. N., and Hoste, C. (1977). Rangeland production and annual rainfall relations in the Mediterranean Basin and in the African Sahelian and Sudanian Zones. *J. Rge. Mgt.*, **30**(3), 181–189.

Le Houérou, H. N., and Popov, G. F. (1981). *An Ecoclimatic Classification of Intertropical Africa.* AGPE, FAO, Rome: 40 pages.

Le Houérou, H. N., and Wilson, R. T. (Eds.) (1978). *Study of the Traditional Livestock Production Systems in Central Mali (Sahel and Niger Internal Delta).* ILCA, Addis Ababa: 430 pages (mimeo).

Lomasson, T. (1947). *Development in Range Management: The Influence of Rainfall on the Prosperity of Eastern Montana, 1878–1946.* Bull. No. 7, US Forest Service.

Long, G., Lacaze, B., Debussche, G., Le Floch, E., and Pontanier, R. (1978). *Contribution à l'Analyse Écologique des Zones Arides de Tunisie avec l'Aide des Données de la Teledétection Spatiale.* CNRS/CEPE, Montpellier.

Lundholm, B. (1976). Domestic animals in arid ecosystems in Rapp. In Le Houérou, H. N., and Lundholm, B. (Eds.) *Can Desert Encroachment be Stopped*, 29–142.

Lusigi, W. J. (1981). *Combatting Desertification and Rehabilitating Degraded Production Systems in Northern Kenya.* IPAL, UNESCO, Nairobi: 141 pages.

Monod, T. (1973). *Les Déserts.* Horizons de France, Paris: 247 pages.

Monod, T. (Ed.) (1975). *Pastoralism in Tropical Africa.* Oxford University Press, London: 502 pages.

Nicolaisen, J. (1963). *Ecology and Culture of the Pastoral Tuareg.* Nat. Mus. Skift, Copenhagen: 548 pages.

Nicolas, F. (1950). *Tamesna, les Ioullemmeden de l'Est ou Touareg Kel Dinnik.* Imp. Nat., Paris: 270 pages.

OMBEVI (1978). *Enquête sur les Effets de la Sécheresse au Mali.* OMBEVI, Bamako.

Philipson, J. (1975). Rainfall, primary production and carrying capacity of Tsavo National Park, (East) Kenya. *East Afr. Wildlife Journ.*, **13**, 171–201.

Pratt, D. J., and Gwynne, M. D. (1977). *Rangeland Management and Ecology in East Africa.* Hodder and Stoughton, London: 310 pages.

Republic of Niger, Ministry of Livestock Production (1981). *Resultats Enquêtes Zootechniques 1980–81.* Cellule Suivi-Evaluation Centre Est.

Rodd, F. J. R. (1926). *People of the Veil.* Macmillan, London: 520 pages.

Rosenzweig, M. L. (1968). Net primary productivity of terrestrial communities prediction from climatological data. *Amer. Nat.*, **102**, 67–74.

Sagne, J. (1950). *L'Algérie Pastorale.* Alger: 267 pages.

SEDES (1975). *Approvisionnement en Viande de l'Afrique de l'Ouest*, Vols. I–IV. SEDES, Paris.

Sneva, F. A., and Hyder, D. N. (1962). *Forecasting Range Herbage Production in Eastern Oregon.* Agric. Exp. Stn., Bull. 588, Oregon State University, Corvallis, Oregon.

Stenning, D. J. (1959). *Savannah nomads: A Study of the Woodabe Pastoral Fulani of Western Bornu Province.* Oxford University Press, London: 267 pages.

Swift, J. (1975). Pastoral nomadism as a form of land use: The Tuaregs of the Adrar N'foras. In Monod, T. (Ed). *Pastoralism in Tropical Africa.* Oxford University Press, London.

Szarec, S. R. (1979). Primary production in four North American desert communities: Indices of efficiency. *J. Arid. Env.*, **2**, 187–209.

Tadmor, N. H., Evenari, M., and Shanan, L. (1972a). Primary production of pasture plants as a function of water use. *Proc. Symp. on Ecophysiol. Found. of Ecosystems Productiv. in Arid Zone*, pp. 151–156. Nauka, Leningrad.

Tadmor, N. H., Eyal, E., and Benjamin, R. (1972b). Primary and secondary production of arid grasslands. *Proc. Symp. on Ecophysiol. Found of Ecosystems Productiv. in Arid Zone*, pp. 173–176. Nauka, Leningrad.

Tadmor, N. H., Eyal, E., and Benjamin, R. W. (1974). Plant and sheep production on semi-arid annual grassland in Israel. *J. Rge. Mgmt.*, **27**, 427–432.

Van Keulen, H. (1975). *Simulation of Water Use and Herbage Growth in Arid Regions.* Centre of Agric. Public and Doc., Wageningen: 176 pages.

Walter, II., and Volk, O. N. (1954). *Osunlagen des Weider Wirtshaft in Sud West Africa.* Ulmer, Stuttgart: 218 pages.

Wickens, G. E. (1968). *Savanna Development in Sudan: Plant Ecology.* UNDP/SF/SUD/25. AGP, FAO, Rome: 82 pages.

Wilson, A. J., Dolan, R., and Olahu, W. M. (1981). *Important Camel Diseases in Selected Areas in Kenya.* Techn. Rep. E.-6, UNESCO, IPAL, Nairobi: 100 pages.

Wilson, R. T. (1980). *Livestock Production in Central Mali: Structure of the Herds and Flocks and Some Related Demographic Parameters.* CIPEA, Bamako.

Wilson, R. T., and Clarke, S. E. (1975). Studies on the livestock of southern Darfur, Sudan. II. Production traits in cattle. *Trop. An. Hlth Prod.*, **8**, 47–57.

Wilson, R. T., and Wagenaar, K. T. (1983). Enquête préliminaire sur la démographie des troupeaux et sur la reproduction chez les animaux domestiques dans la zone du projet 'Gestion des Patûrages et Élevage de la République du Niger'. *Progr. Doc. AZ80, CIPEA, Bamako*: 95 pages.

Young, M. D., and Wilson, A. D. (1978). The influence of climatic variability on Australian society's economy. *Phillip Island Conference, Victoria, 27–30 November, 1978.* CSIRO, Canberra.

Zeuner, F. (1963). *The History of Domesticated Animals.* Hutchinson, London: 560 pages.

Climate Impact Assessment
Edited by R. W. Kates, J. H. Ausubel and M. Berberian

CHAPTER 8
Water Resources

BÉLA NOVÁKY,* CSABA PACHNER,† KÁROLY SZESZTAY*
AND DAVID MILLER‡

**Institute for Water Management*
Alkotmány u. 29
Budapest, 1054 Hungary

†West Transdanubian Water Authority
Vörösmarty u. 2
Szombathely, 9700 Hungary

‡Department of Geological Sciences
The University of Wisconsin-Milwaukee
Milwaukee, Wisconsin 53201, USA

8.1 INTRODUCTION

Water is an indispensable element of life, and water resources are highly sensitive to climate variability and change. Traditionally, 'trial and error' has been one of the most basic approaches in the evolution of human responses to climate. In fact, the first large-scale social responses to water-related climatic impacts took form as the fluvial civilizations of antiquity and were based to a large extent on the successes and failures of many small village communities during the preceding millennia (Teclaff, 1967; Mumford, 1967 and 1970). Computerized simulation models and many other impact assessment techniques are essentially also based on the trial and error principle, and assessment is still needed to supplement learning through historic experience for the following major reasons:

1. The accelerated social and economic changes of the present age tend to create situations for which little or no historic experience is at hand. Water supply problems of large metropolitan agglomerations and industrial centers, or flood problems of rapidly growing or changing communities, are situations in which climate variability or change would affect water resources in new ways.
2. Rapid advances in science and technology not only trigger economic and social changes, they also offer new strategies and tools for coping with climatic impact, for which again little historic experience can be found. High-efficiency drilling and pumping equipment, new materials for cheap and durable pipelines, chemicals treating soil surfaces for water harvesting, high-efficiency machines and materials for the construction of dams, and canals and tunnels for large-scale and long-distance water transport are examples of new technologies available for human responses to climatic impacts (Ackerman and Löf, 1959). To explore potential applications of new technologies in adaptive responses to climate should be one of the main water-related directions of the World Climate Programme.
3. Recent technology has also reduced apparent demands for water, which often turn out to be illusory. Industrial conservation and recycling of water is widely practiced, especially when pricing incentives come into play. Some cities are renovating the great volumes of wastewater they generate, to make it available for industrial or agricultural use. Every drought brings forth new ideas that reduce urban demands for water (Meier, 1977); the 1976–78 drought, for example, resulted in many household and institutional changes in the cities of central California. Future impacts of climatic change on the supply areas of these cities will be cushioned by these proven means of reducing water use.

   The often inordinate 'demands' for irrigation water, especially where it is subsidized, can be reduced by better recharge and conveyance practices, drip or sprinkler application of water, better timing of irrigations, and improvements in crop management including genetic changes and better knowledge of true water needs. These can be illustrated for California

(Davenport and Hagan, 1981), in East Asia (VanderMeer, 1968) and in the North American Great Plains, where many technological practices can reduce the impact of drought (Rosenberg, 1978). Many water-demand reductions have been proven in such semi-arid lands as Israel and Australia: water harvesting, better on-site retention, and so on (Thames and Fischer, 1981). Where climatic change might increase rainfall and raise the level of groundwater, hydraulic management can be applied, as in the Netherlands, Finland, and even in the low-energy agriculture of Meso-America (Wilken, 1969).

4. A fourth major reason why historic experience of responses to climate must be supplemented by systematic assessment of impacts lies in the processes of climate formation. Certain types of land use (such as large-scale drainage, soil-conserving practices and irrigation) significantly alter the factors of climate formation, such as the radiation and heat balance at the land surface, as well as soil moisture and atmospheric humidity. Through this they weaken the validity of historic experience for the selection of the appropriate human responses.

This chapter deals with the 'climate–water resources–water management–society' pathway of an assessment of the impact of climate on water. Section 8.2 reviews the elements for the 'climate–water resources' relation and Section 8.3 does the same for the 'water resources–water management–society and economy' sequences. Section 8.4 addresses issues of assessment integrated over the whole sequence, with emphasis on various types of societal and technological settings for water-related policy analysis and adjustments.

Complicating the human modification of the natural hydrologic system is the fact that many modifications take place unintentionally, and a wide range of human activities, including all types of land use, may intervene in the natural hydrologic processes (precipitation, infiltration, storage and movement of soil moisture, surface and subsurface runoff, recharge of groundwater and evapotranspiration).

Figure 8.1 attempts to capture some of this complexity, identifying not only the central pathway of this chapter, 'climate–water resources–water management–economy–society', as marked by the thick arrow lines, but also the feedback relationship with water as an element of the climate system and with the many human-induced modifications of the natural cycle.

8.2 THE IMPACT OF CLIMATE ON WATER RESOURCES

Water resources are essentially the products of climate (Voeikov, 1884), significantly influenced, however, by land factors. Figure 8.2 provides a structured scheme of the 'climate–water resources' link of Figure 8.1 with inclusion of the land elements. This section is focused on two key features of

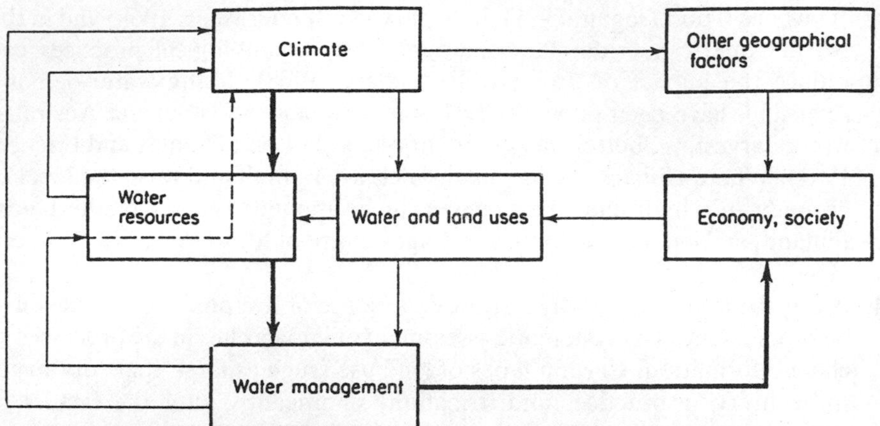

Figure 8.1 Conceptual scheme of water-related climate impact assessment

that linkage: the specific parameters and roles of climate in continuously redistributing the earth's water resources, and the present state of knowledge in quantifying these relations under various sets of natural and man-made conditions.

8.2.1 Climate and Water

Water in the different domains of the earth displays different rates of turnover and so reflects climatic fluctuations occurring at different time-scales. For example, a brief extension of the rainless period between summer rainstorms can bring about a large increase in the number of days with low soil moisture, in which crops suffer moisture stress.

Longer fluctuations may reduce the level of upper groundwater and the base flow in streams, on which many urban and industrial uses depend. Such a drought may be exacerbated by societal factors, as occurred in Pennsylvania in 1980–81 (Perkey *et al.*, 1983). Still longer fluctuations in climate change the level of large lakes and affect navigation, hydropower production, and riparian access, as in the Laurentian Great Lakes (Phillips and McCulloch, 1972). Very long fluctuations affect vegetation cover and even soil; allied with short-sighted practices of land management, they may result in desertification (Biswas and Biswas, 1980; Kovda, 1980).

8.2.1.1 Land Factors

Land factors of morphology, soil, and plant cover, as shown in Figure 8.2, play an important role in mediating the impacts of climate fluctuations on the hydrosphere. These factors determine the storage of water on the surface or in

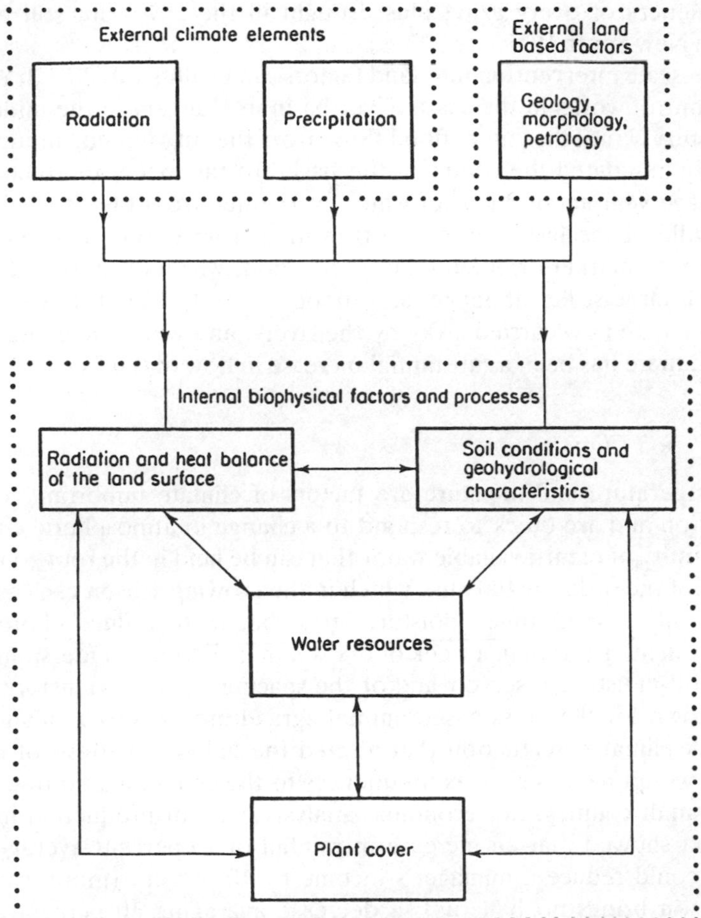

Figure 8.2 Conceptual scheme of the climate–water resources relationship

the soil, percolation to groundwater, evaporation, and runoff. Their role is particularly significant over short periods of time in humid areas; their effects in arid and semi-arid lands are long-lasting. L'vovich (1969, 142) finds that changes in land-surface management in the Dnieper Basin are likely to decrease storm flow in rivers by 9 mm annually and increase base flow by 1 mm, for a net loss of 8 mm. In arid and semi-arid areas the impact of land factors on evaporation can exceed those of climatic variation (Sokolowsky 1968). These impacts can be expressed in one or another index of aridity (Thornthwaite, 1948; Szesztay, 1965; Budyko, 1974, 324–335; Mather, 1974, 112–120). The process of desertification, now serious in many parts of the globe, is a nearly irreversible change in vegetation, land utilization and even soil resources; the consequences of overgrazing plus drought in the 1890s are still evident in western New South Wales.

Large-scale intervention into land factors can be illustrated by drainage and flood-control works in the Tisza River basin in Hungary in the middle of the last century. Until that time, flood flows from the surrounding mountains had regularly inundated the center of the basin, to the extent of 16,000 km^2. In an average year about 2 km^3 of water evaporated from the flooded areas. In the middle of the last century the transition from flood-recession land and water use to market-oriented grain production, which evaporates less water, required large-scale drainage and flood control. Flood flows from the mountains are now carried away by the rivers, and while the climate has not changed there has been a substantial increase in flow of the Tisza.

8.2.1.2 Soil Climate

Soil temperature and moisture are factors of climate important to primary production and are quick to respond to a change in atmospheric circulation. The quantity of plant-available water that can be held in the root zone of most crops is of the order of 100 mm, which in the growing season can sustain crop growth only a short time; moisture stress begins to reduce photosynthetic production after less than a week of dry weather. Farmers understand the role of the soil-moisture reservoir and of the spacing between rainstorms, as has been shown for the grass-based animal agriculture of New Zealand (Curry, 1962). A climatic fluctuation that altered the habitual pattern of rainstorm spacing would have serious consequences to the economy and trade balance of this small country. An economic analysis of wool production in western Australia showed that an increase in rainfall of 10 percent averaged over a decade could reduce a manager's income by 10 percent (more water in the wet season brings no benefits); a decrease averaging 10 percent could cut farm income by nearly two-thirds and double the risk (Arnold and Galbraith, 1978).

Agriculture in most regions is closely attuned to the frequency of days of soil-moisture deficit (Mather, 1974, 207–213; Zur and Jones, 1981). Any climatic fluctuation that would increase the number of stress days would have an immediate impact on crop yield, whether of corn (Dale and Shaw, 1965) or pulpwood growth (Bassett, 1964), up to the point of complete loss, as in the North American middle west in the 1930s.

8.2.1.3 Groundwater

Climatic fluctuations that persist over long periods affect first the shallow groundwater resource, hence domestic wells and the base flow in small streams used for irrigation. These changes in the water table cause wells to go dry or at the least necessitate the lowering of pumps, may require the hauling of water for livestock, and impair the habitat of aquatic life.

Longer fluctuations have an impact on deep aquifers, reducing water pressure, permitting compaction and resulting land subsidence, and sometimes the intrusion of saline water. The effects observed on a local and regional scale as a result of overpumping in many places give an indication of the potential effects of prolonged drought or other reductions in aquifer recharge. For example, introduction of fall plowing and other practices in the central Chernozem area of the Soviet Union have the potential to increase recharge by reducing surface runoff (Grin, 1965; L'vovich, 1969, 142). In northwestern Russia, a possible climate change by the year 2020 might increase streamflow (Budyko, 1982, 243) and implies a rise in shallow groundwater and increased swamping of forest. Examples from East Africa show the effects on catchment yields, which are in part groundwater outflows, when rain forest is cut down to plant tea or when bamboo forests are replaced by pines (Pereira, 1973). Groundwater storage, a useful cushion over short fluctuations in climate, is vulnerable to long fluctuations.

8.2.1.4 Streamflow

River water, which has a relatively short turnover period and is a major source of fresh water, has great importance for humankind. The classification of climate from the point of view of surface flow can conveniently be based on precipitation and potential evapotranspiration, which can be tied to solar radiation or air temperature (Thornthwaite, 1948; Szesztay, 1965; Mather, 1974, 112–122). On the basis of these parameters, combined into an aridity index based on the ratios of evapotranspiration to precipitation, nine types of climate are specified (Figure 8.3). Water flows on the surface in four of the nine types (about 62 percent of the 149 million km^2 total land area), and is frozen in polar ice and glaciers over 12 percent of the land area; in 26 percent, there are deserts and semideserts without permanent surface water. Land areas with different types of surface water have varied historically as the climate of the Earth has changed.

Fluctuations of the Danube River's discharges during the period 1948–68 can be related to the fluctuations of climatic elements, specifically, to the fluctuations of ocean surface temperatures in the northern part of the Atlantic ocean (Nováky, 1981). For small catchments or low flows, the variability of surface flow is amplified, because runoff is a residual of precipitation and evaporation, and its variability surpasses the variability of precipitation, particularly in areas with little runoff (Schaake and Kaczmarek, 1979). This is illustrated in Figure 8.4 for selected catchment areas in the Tisza River basin (Nováky, 1981).

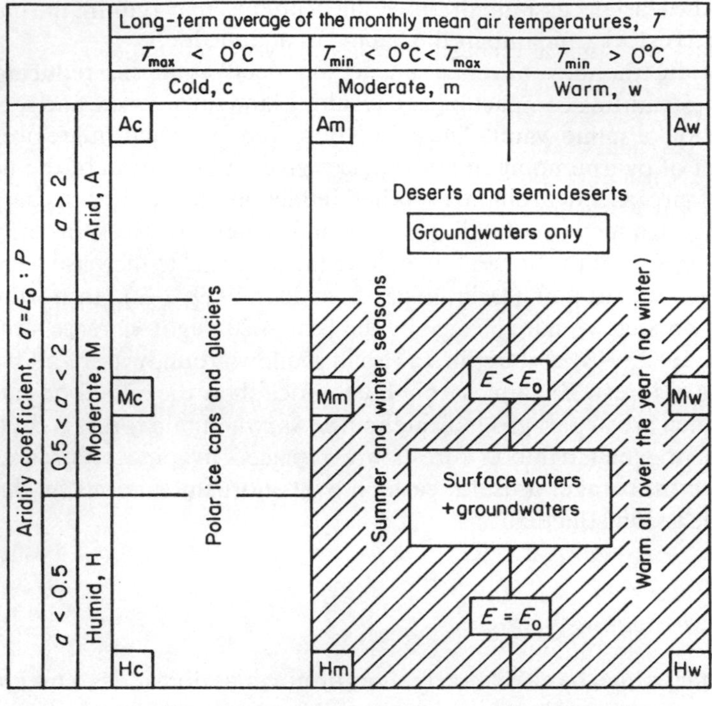

Figure 8.3 Classification of land surfaces according to climate. (Source: Szesztay, 1965)

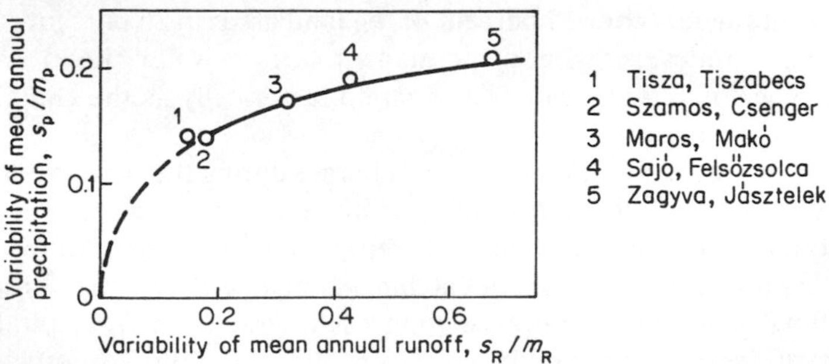

Figure 8.4 Relationship between the variability of mean annual precipitation and the variability of mean annual runoff in the Tisza River basin

8.2.1.5 Lakes

The impact of climatic changes can bc analyzed particularly well in those elements that have a character of storage and accumulate climatic impacts over long periods, such as deep groundwater and lakes. The water level of the Caspian Sea has decreased since the middle of the past century. Water withdrawals for irrigation for water supply played an important role in this decrease, but it was also the result of climatic changes in the drainage area. Winter precipitation decreased, and summer temperatures increased, which led to an increase in evaporation and to a decrease in the flow of the Volga River. The change in water level of the Caspian Sea followed the change of climatic elements with a time lag of 15 years (Klige, 1978).

8.2.2 Quantifying Climate–Water Resource Relationships

In quantifying the climate–water resources relationship, transfer functions are used to transform climatic characteristics into water resources. Transfer functions are classified by Schaake and Kaczmarek (1979) into three categories: statistical, analytical and numerical. The theoretical base becomes more complete in progressing from the statistical through the analytical to the numerical models.

Statistical transfer functions are, for example, the empirical relationships between proxy information on fluctuations of climate (such as tree-ring indices or glacial activity) and water resources. Analytical transfer functions are based on simplified physical principles, such as the balance between climate elements (for example, precipitation, evaporation) and water resources (runoff, change of storage in soil moisture, etc.). Numerical transfer functions are based on conceptual hydrological models, which allow for more detailed physical considerations than the analytical functions but also require digital computers for their application.

As Schaake and Kaczmarek (1979) comment, the application of any transfer function is limited by three main technical factors:

1. the inherent accuracy of the model,
2. the degree to which model parameters depend upon the climatic conditions for which the model was calibrated,
3. the accuracy of the input data.

The climate–water resources relationship must be based on characteristics of climate aggregated over the long term, especially precipitation, its mean annual and seasonal values, monthly values and dispersions. The third point above must be kept in mind: many climatic measurements are of dubious accuracy, especially in recent decades, and rainfall data are notoriously defective (United Nations, 1972, Chapter 2; Mather, 1974, 51–56, 100; World Meteorological Organization, 1975, 13, 23).

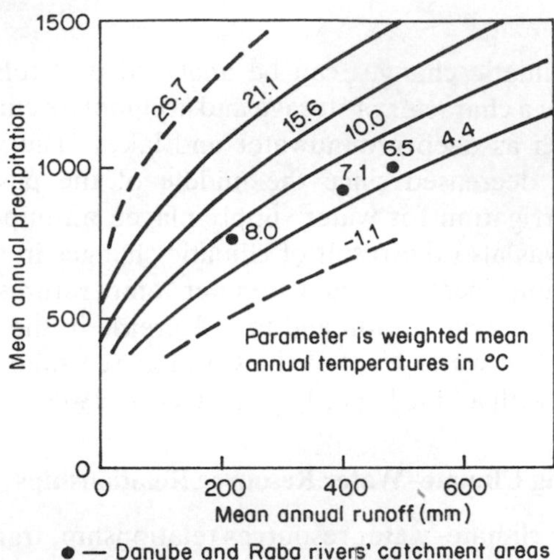

Figure 8.5 Relationship of annual runoff to precipitation and temperature. (After Langbein, 1949)

The relationship of mean annual streamflow to precipitation and temperature was evaluated by Langbein (1949) for regions of the United States (Figure 8.5), and a tentative analysis of data from the Political and Economic Atlas of the World (edited by the Hungarian Cartographical Institute, 1974) suggests that this relationship can be applied to other regions of the world. Mean annual streamflow, precipitation, and air temperature in the drainage area of the Danube River and its tributary, the Raba River, are well in line with the relationship elaborated by Langbein.

Another model relates to Lake Balaton, the largest lake in Hungary, with a surface area of about 600 km^2. The precipitation on the surface is 630 mm, the inflow from the drainage area is 880 mm, and the total supply is 1510 mm. This total supplies 870 mm evaporation and 640 mm regulated outflow. The balance of these elements is presented in Figure 8.6 (after Szesztay, 1960). Suppose a change of climate around the lake such that it would be similar to the present-day climate in the middle of the Tisza River basin; for example, suppose an increase in temperature by 0.5 °C and a decrease in annual precipitation by 5 percent. These relatively slight changes in climate would result in a significant change in the life of the lake: evaporation from the lake surface would consume most of the precipitation and inflow, and outflow would decrease to a tenth of its present value. Lake Balaton would become nearly a closed inland lake, with a water surface smaller than that of today. The renewal of the lake's water would be slower, which would have an effect on water quality and biological regime.

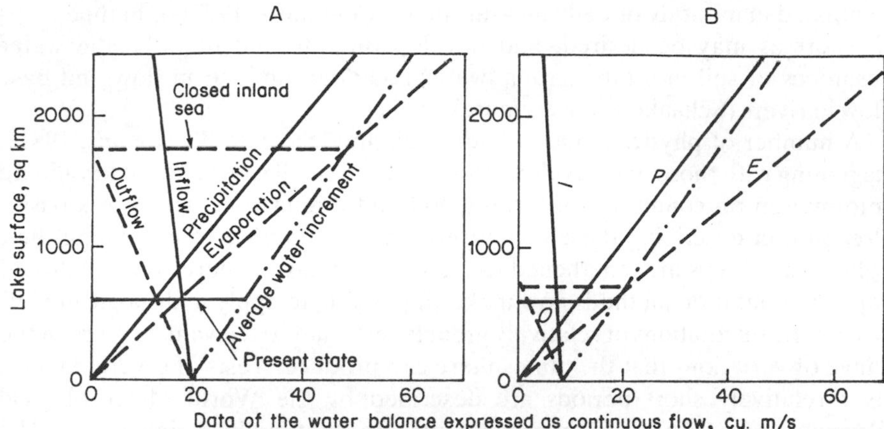

Figure 8.6 Water balance of Lake Balaton, plotted against the lake surface. A, under present climatic conditions (after Szesztay, 1960) and B, under simulated climatic conditions

Two complicating factors should be noted: the role of water itself in the formation of climate, and the human modifications in water fluxes, especially those at the land surface.

Water is an internal and almost ubiquitous factor in the processes that form weather and climate, because it stores, transports, delivers and redistributes energy in many ways and at many scales. Water is not an external parameter of climate formation but rather plays an important role in the biophysical impacts of climatic changes, as well as in the responses of climate to biophysical changes. It is therefore a key element in the assessment of every human-caused climatic change, and water management is a prospective tool for influencing climate formation. In fact, the impact of water management on certain elements of the local climate is frequently quite rapid and obvious (as in the case of large-scale drainage or irrigation), whereas impacts along the 'climate–water management' line usually remain slow and indirect. The impact of water on climate belongs, however, to the climate research sector of the World Climate Programme and lies essentially outside the scope of the present paper.* Nonetheless, these and land-related feedbacks complicate the quantification of climate–water resources relationships.

Both the feedback relationships of water in climate processes and the effects of human activity on water fluxes can be incorporated in models of the climate–water relation. These models accept inputs of climate or weather data,

* In order to acknowledge this linkage, major aspects of the 'water management–climate' pathway have been reviewed in one of the preparatory papers of this study program (Szesztay, 1981).

expressed in monthly or daily measurements (Willmott, 1977) or in time-steps as short as may be desired, and develop outputs that describe the water resources of soil moisture, groundwater recharge, and storm flow and base flow in rivers (Schaake and Kaczmarek, 1979).

A number of physically based hydrologic models exist (Peck *et al.*, 1981), beginning with those that develop flash floods on small rivers, useful to validate information on conditions of terrain, soil and drainage networks of a basin. Precision in describing these conditions lends confidence to estimating their role as time-steps are lengthened to 6 hours to a day or more. Availability of rapid computation methods now makes it possible to apply hydrologic models to climatic fluctuations of relatively great length, and so to evaluate more of the range of variations that the atmosphere can produce. Tests of several models over relatively short periods are described by the World Meteorological Organization (1975), and improvements are continually being made. The SSARR model (US Corps of Engineers, 1975), for example, was verified in the upper basin of the Missouri River (Cundy and Brooks, 1981).

Nemec and Schaake ran the Sacramento soil-moisture accounting model (Burnash *et al.*, 1973) at 6-hour time-steps over periods of 12 years for a river basin in Texas, defining 16 parameters that describe the upper and lower soil zones and percolation from them. The calibrated model was then run under different postulated values of rainfall and air temperature (as a means of incorporating the energy input that drives evapotranspiration), and produced the probable streamflow, the reservoir storage required to obtain a specific degree of river regulation that will produce yield equal to 0.2 of mean annual streamflow, and reservoir yield. A model for a humid river basin was similarly calibrated and run, giving comparative responses of that basin to an increase or decrease in rainfall. In both basins, a change of 0.01 in rainfall produces approximately 0.02 change in reliable yield from reservoirs (Nemec and Schaake, 1982).

Stochastic models are sometimes used to evaluate the probable range of variation in future streamflow, using analyses of the statistical properties of the past record. These records, however, are even more limited than those of rainfall, and evaluation of extremes is risky, whether these be design floods or prolonged low flows. Stochastic models of rainfall, usually the most important input into such hydrologic models as the Sacramento or SSARR, also help define some of the range of fluctuations in atmospheric deliveries of water, although the fact that every year rainfall records are broken by the dozen by new extremes reminds us that nature has a great potential to surprise us. The concepts of probable maximum precipitation, having a base in physical hydrometeorology, and probable minimum flow, using geological factors, are useful in designing water resource developments. However, in using all these hypothetical constructions it must be kept in mind that longer-range variations in climate are hardly being adequately sampled by the available records.

Rainfall itself is a crude output of some of the larger models of the atmospheric circulation, such as those run to foreshadow possible effects of increased atmospheric carbon dioxide. Only the largest models attempt to produce longitudinal differences in the zonally averaged outputs, so regional information is often approximate. The value of general circulation models for regional hydrologies is therefore not clear at the present time; yet the critical impacts of climate upon water resources are those that occur at regional scales, like the Colorado River basin in the southwestern United States. Partly as a consequence of relying on unrepresentative streamflow data, this relatively small river has been over-allocated, and societal institutions are overtaxed to manage present-day flow resources (National Research Council, 1968; Dracup, 1977; Peterson and Crawford, 1978). The low quality of the river water exacerbates the management problem, and has given rise to further hydrological modelling efforts (Clyde *et al.*, 1976).

The effects of changes in land management in a river basin can be evaluated by changing basic parameters in hydrologic models, which focus on processes at the surface of the earth and the immediately underlying soil layers. Properties of the surface, soil, and vegetation, which with over-use may be degraded hydrologically, can be entered in the models to evaluate the consequences of abuse of the land. Good practices that maximize the infiltration of rainwater into the soil and minimize storm flow can be expected to reduce the impact of a climatic fluctuation on the societal value of the water and biological productivity of a region. Moreover, it should prove possible to evaluate the economic benefit of good management practices that improve the status of soil moisture and photosynthetic production, the stability of groundwater and streamflow, and regulated storage as resources of a region.

8.3 THE SOCIETAL CONTEXT OF WATER-RELATED IMPACTS OF CLIMATE CHANGE

Given a transfer function between climate variation and water resources, it is then necessary to transform the physical quantity and availability of water into economic and social values. This can be done by identifying the socially significant attributes and factors of the hydrogeographical endowments and hydrological processes of a given area and the major means of water management in satisfying society's demands for water-related services, taking into account human as well as climate impacts.

8.3.1 Use and Purpose in Water Resource Development

A climate-induced increase or decrease of water resources takes on value only in terms of the actual or potential benefits and hazards to humans. A classification of these attributes in terms of water-resource use and purpose is given in

Table 8.1 Water resources use, purpose, and evaluation methods

A Direct utility and safety needs		B Utilization of the hydrological potential	
Needs of population and settlements	Needs connected with production and services	Economic utilization of the natural resource	Role of hydrogeographic endowments in social welfare
– level of drainage and of protection against floods and excess waters (3) – water for drinking and other domestic uses (3) – water quality-related health safety and the disposal of waste waters and wastes (3)	– soil moisture in crop production (1) – soil moisture in forest production (1) – water-related conditions of soil conservation (3) – hydropower (1) – navigation (1) – needs for drainage and damage prevention (flood, erosion) of industries and transportation systems (1) – disposal of used waters and wastes (1) – water-related safety conditions of mining (1) – water needs of industry, agriculture and services (3)	– yields of waters as production sites (fish, reed, etc.) (3) – refuse and waste processing capacity of waters (3) – medical waters (3) – gravel and mineral stock of channels (1) – flood recession land use (3)	– water-related recreation (2) – shaping the environment of settlements (2) – cultural and landscape aesthetic values (2) – impacts on life styles (2)

Evaluation possibilities: (1) Analytical methods
(2) Based on social policy criteria
(3) Joint consideration of economic and social criteria

After Orlóci and [illegible], 1981

Table 8.1, in which 'utility and safety' that are directly related to human habitat and production are separated from 'hydrological potentials', which are less directly associated.

The industrialization of recent centuries was usually accompanied by a shift from hydrologic potentials toward growing interest in the groups of direct utility and safety. In Hungary, for example, during the period from the eleventh to the eighteenth centuries, economic stability was largely based on a traditional system of flood-recession land use along the Danube and Tisza Rivers and their major tributaries. In the sophisticated and productive system of land use, annual flooding was not prevented but was rather promoted and regulated in order to achieve high yields and a variety of foodstuffs (fish, cattle, poultry, fruits, grain, vegetables, honey) and to provide power and transportation by watermills and inland waterways (Andrásfalvy, 1981). The gradual replacement of this traditional economy by market-oriented grain production and industrialization during the eighteenth and nineteenth centuries required large-scale drainage and flood-control works affecting more than a third of the present area of the country.

For each of the socially significant attributes of water resources listed in Table 8.1 there exist approaches and methods by which the nature and extent of social interest can be described and quantified. These include the conventional methodology of economic analysis (see the items marked by '1'), evaluations based on verbal descriptions and social policy criteria (items marked by '2'), or on a joint consideration of the previous two approaches (items marked by '3'). The suggestions of Table 8.1 are tentative and much will depend on the availability of data and on other local conditions in any given case. Commonly, the analytically assessable attributes of water belong to the group of direct utility and safety needs connected with production activities, and the socially significant attributes are assessed by verbal descriptions and social policy criteria.

8.3.2 Water Management Techniques

For each preference with regard to the socially significant attributes of water there exist specific methods and technologies through which the demands for water and water-related services are satisfied. In a narrower sense the technologies applied in satisfying water-related demands are summarized under the term 'water management' and they include the 12 groups of activities listed in Figure 8.7, with the indication of their linkages to the four groups of social demands and interests of Table 8.1.

8.3.3 Climate–Water Management Sensitivity

Major water management activities are variously affected by climate events, depending on their time-scale: within-year weather, yearly fluctuations,

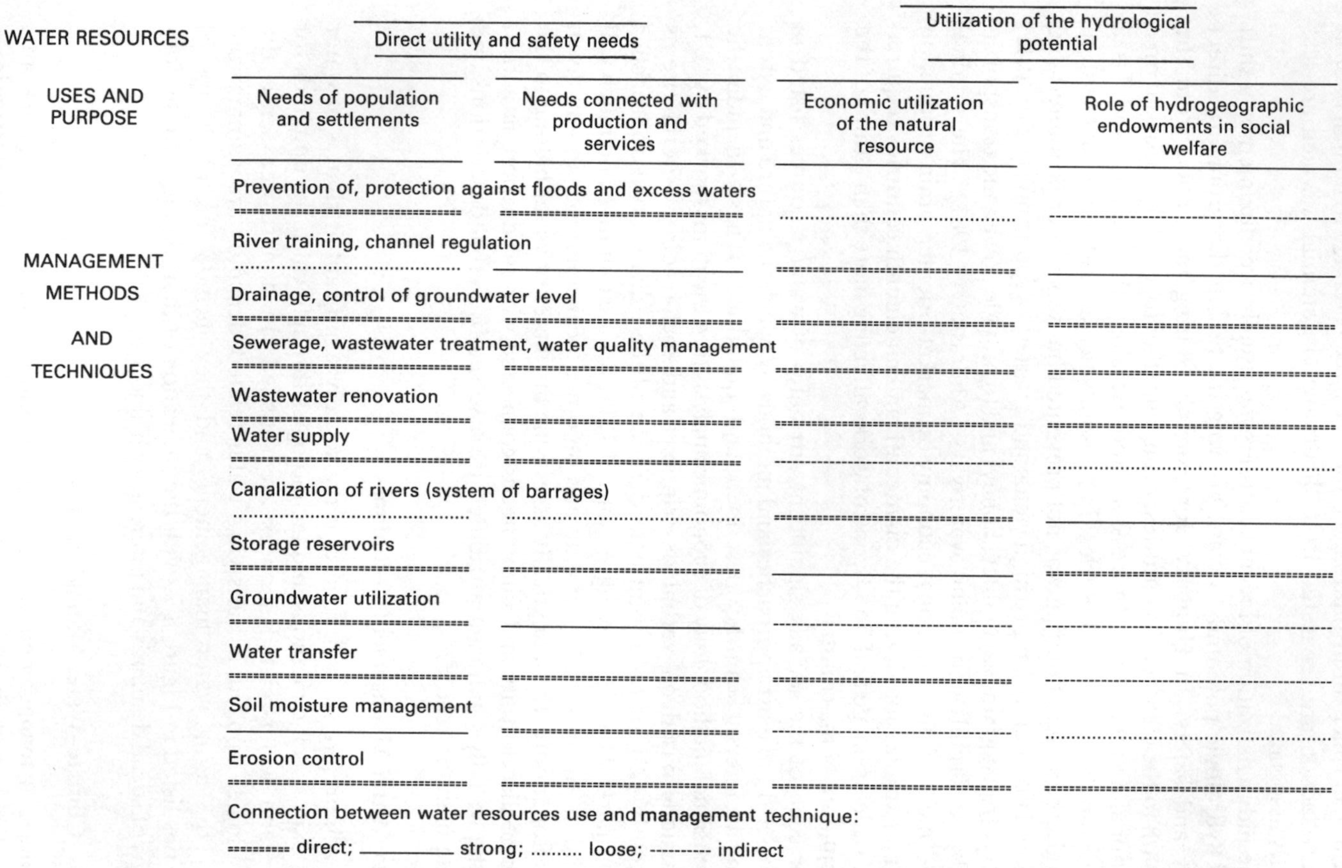

WATER RESOURCES	Direct utility and safety needs		Utilization of the hydrological potential	
USES AND PURPOSE	Needs of population and settlements	Needs connected with production and services	Economic utilization of the natural resource	Role of hydrogeographic endowments in social welfare
MANAGEMENT METHODS AND TECHNIQUES				
Prevention of, protection against floods and excess waters	direct	direct	loose	indirect
River training, channel regulation	loose	strong	direct	strong
Drainage, control of groundwater level	direct	direct	direct	direct
Sewerage, wastewater treatment, water quality management	direct	direct	direct	direct
Wastewater renovation	direct	direct	direct	direct
Water supply	direct	direct	indirect	loose
Canalization of rivers (system of barrages)	loose	loose	direct	strong
Storage reservoirs	direct	direct	strong	direct
Groundwater utilization	direct	strong	indirect	indirect
Water transfer	direct	direct	direct	indirect
Soil moisture management	strong	direct	indirect	loose
Erosion control	direct	direct	direct	direct

Connection between water resources use and management technique:

======== direct; ________ strong; .... loose; --------- indirect

Figure 8.7 Water resources, purpose and management methods

Table 8.2 Sensitivity of water management to climatic events

Management methods and techniques	Sensitivity to climatic events Within-year	Annual	Multiyear	Century
Protection against floods	X	X		
River training	X	X	X	
Drainage		X	X	X
Water quality management	X	X	X	X
Wastewater renovation		X	X	
Water supply		X	X	X
River canalization (dams)		X	X	
Storage reservoirs		X	X	X
Groundwater utilization		X	X	X
Water transfer		X	X	X
Soil-moisture management	X			
Erosion control	X			

multiyear variations, and century or longer changes. Each management activity can be evaluated as to its sensitivity (Table 8.2) and its reliability (National Research Council, 1977), as well as its ability to recover after a failure and the likely consequence of a failure (Cohon, 1982).

8.3.4 Human Activity–Water Resource Sensitivity

The socially significant attributes of the water resources of a given region are determined and influenced not only by climate and other environmental factors; they may also be altered and affected to a considerable extent by human impact upon the environment. Water-related climatic impacts can be assessed and evaluated only if they are large in comparison to hydrologic changes caused by humans, and if the climatic and the human impacts can be reasonably separated. For this reason the assessment of human-caused hydrologic changes should go hand in hand with the assessment of water-related climatic impacts. Figure 8.8 offers a structural scheme and a few indicative examples for such an assessment.

It is obvious that water management activities, that is, water use and regulation, always have impact on hydrologic processes, but perhaps less obvious that land uses also alter the hydrologic regime, and that these alterations frequently exceed those caused by water management activities (as, for example, in the case of large-scale mining operations, chemicalized agricultural land use, or toxic metals in industrial wastewater entering a lake). In order to arrive at a definite conclusion with regard to the social significance of human-caused hydrologic changes, the sequence of impact assessment

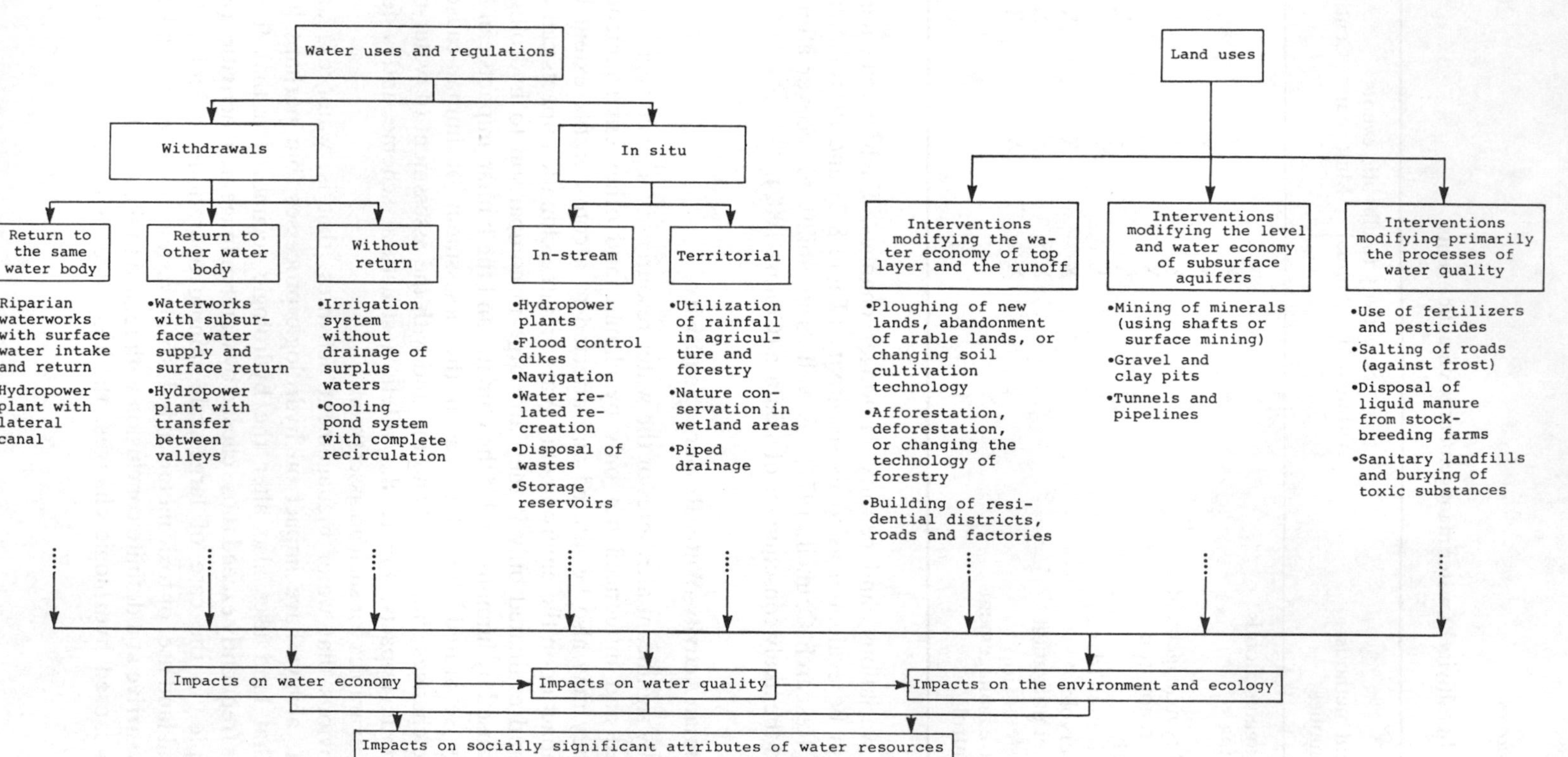

Figure 8.8 Impact of human activities on hydrological processes and their feedbacks on society (structural scheme with indicative examples. (After Orlóci and Szesztay, 1981)

indicated at the bottom of Figure 8.8 is important. Changes in the societally significant attributes of the region's water resources constitute the concluding phase of the assessment procedure.

8.4 INTEGRATED ASSESSMENT OF WATER-RELATED CLIMATIC IMPACTS

8.4.1 Defining Assessment Objectives

While climate as a key element in hydrology for water management is often studied, integrated climate impact assessment is rare. A beginning point for such assessment is to select and define a few specific assessment objectives.

To look at a country's (region's) water management in its entirety and its historical evolution as a specific human response to climate could be a sound point of departure. A general survey could identify characteristic levels and turning points in water management, and compare them with corresponding levels and trends in the region's climatic and social conditions. Policy-oriented global reviews on major issues of water management (Falkenmark and Lindh, 1976; United Nations, 1976; Szesztay, 1982) could help in the formulation of questions that should be asked, and analytical studies on related topics (Kates, 1981) could provide guidance on methodological approaches that could be applied in such regionwide surveys.

Settings in which relatively small changes in climate might trigger substantial consequences in water resources and water management deserve particular attention. Shallow lakes can dry out or reappear under the cumulative effects of relatively small changes in aridity. In a cold climate, the snow line as well as river and lake ice are affected by relatively small consecutive fluctuations in winter temperatures. Revelle and Waggoner (1983) have shown that warmer air temperatures and a slight decrease in precipitation would probably severely reduce both the quantity and quality of water resources in the western United States, and that similar effects can be expected in many water-short regions elsewhere in the world.

In formulating assessment projects, priority generally should be given to regions and situations where relatively small changes in the water-resources regime might produce significant consequences in water management and its societal implications (regions where withdrawals are close to the dependable river flow resources, densely populated or intensively cultivated flood plains).

8.4.2 Assessing Climatic Impacts by Matrices

After having defined the scope and objectives of the assessment program in the light of current issues of water management planning and policies, the implementation of the program should proceed. Three interrelated phases of implementation can be distinguished:

1. identification of the particular attributes of the water management system that are sensitive to climatic impacts (Section 8.3);
2. identification of specific elements of climate that affect the system indices (Section 8.2.1);
3. formulation of the relation between climatic parameters and water management factors in terms of impact functions (Section 8.2.2).

These steps can be brought together in two illustrations: the metropolitan water supply system of the northeastern United States and an analysis of floods.

Based on a thoughtful effort to investigate how sensitive are the large metropolitan water supply systems of the northeastern United States to climatic change, Schwarz (1977) prepared Table 8.3. Of nine attributes of the systems which are judged to respond significantly to a change of climate, Nos. 1–4 relate to the 'climate–water resources' part of the impact scheme and express changes in the quantity and quality of water available for supply, and Nos. 5–9 relate to the 'water resources–water management–society' part and describe technical, economic and managerial aspects of system operation. Five climate fluctuations are confronted in the table with these nine system attributes.

Table 8.4 contains a similar matrix of the impacts of climatic change on flood hazards under various hydrologic conditions and managerial situations. Four flood-hazard situations are shown in this table against four variations of climate. The matrix emphasizes the fact that the relevant parameters of climate differ, even in the same group of water management activities, with the size and composition of the system.

Flood-hazard simulation is the one major field of water management in which a sound basis and relatively rich experience are available, mainly as a result of the work of a group under White at the University of Chicago and his later Natural Hazards group at the University of Colorado.

8.5 CONCLUSION

In this chapter, the authors have tried to describe some of the manifold ways in which fluctuations in atmospheric circulation and climate might alter the water resources of a river basin, region or nation. These fluctuations occur at many time-scales and have correspondingly diverse impacts on water resources: impacts on the resources of soil moisture and storm flow occur at short time-scales; those on groundwater, base flow in rivers, and the level of large lakes occur at long time-scales, represent a different kind of alteration in the circulation of the atmosphere, and are immune to short-period fluctuations.

Table 8.3 Speculative water supply impact matrix of climatic change

Attributes of water supply systems	Parameters of climatic change				
	A Decrease in mean streamflow	B Increase in variance of streamflow	C Increase in skew of streamflow	D Increase in persistence of streamflow	E Speed with which change occurs
1. Yield from unregulated streams	Some effects, but likely not very large except if change in mean is large or combined with other changes	Severe effects; however, generally short term	Significant effects because number of days of low flow increase relative to few high flow periods	Significant effects more through duration of low flows than severity	Not applicable
2. Yield from reservoirs	Significant to severe effects particularly if reservoirs develop a high percentage of the average flow	Medium to no effects depending on the size of the reservoir in relation to drainage area; larger reservoirs will suffer smaller effects	Medium to no effects depending on the size of the reservoir in relation to drainage area; larger reservoirs will suffer smaller effects	Significant to severe effects especially if reservoir long-term storage is limited	Not applicable
3. Yield from groundwater	Significant in the long run, especially if draft on aquifer is near average recharge	Little if any significance	Little if any significance	Effects severe and of long duration	Not applicable

4. Quality of raw water	Probably insignificant effects except where large reservoirs are drawn to very low levels	Generally no effects except possible increase in turbidity during high flows	Little if any significance	Little if any significance	Not applicable
5. System reliability	Some effects, other than effects accounted for under 1–4	Some reduction due to constant change in flows in addition to effects under 1–4	Little or none, other than effects under 1–4	Little or none, other than effects under 1–4	Sudden changes severely affect reliability, slow ones less or not at all
6. Effectiveness of intersystem and interbasin connections	No change	Increased effectiveness if variance increases	Little effect	Reduced efficiency of interconnections because long droughts are usually also widespread	No change
7. Magnitude and control of demand	No significant effect	No significant effect; often reccurring short-term restrictions may reduce their effectiveness	No significant effect	No significant effect; emergency restrictions likely to become less effective over long droughts	Significant and visible effects, relatively fast changes could force major steps toward conservation and demand control

8. Cost of operation of water system	No significant effects except for additional construction that might eventually ensue to alleviate long-term shortages	Possible increase due to turbidity, increased pumping between systems if applicable; possible additional reservoir construction	No significant effects likely	No significant effects except search for new sources	No effects
9. Pressure on and ability of the water system to respond to change	Pressure for expansion would be created if shortages occur repeatedly; ability to respond would not be affected by hydrologic event	Pressure for expansion would be created, but rapid return to normal may for some time inhibit expansion	Pressure for expansion would be created if shortages occur repeatedly; ability to respond would not be affected by hydrologic event	Pressure for expansion would mount over time and increase likelihood of action; however, long high flow periods may inhibit development	Sudden or relatively near future changes could increase action; long-term changes (20 years+) even if known would likely be ignored by existing institutions

Reproduced with permission from Schwarz, 1977, 116–117.

Table 8.4 Speculative flood hazard impact matrix of climatic change

Attributes of flood hazard management systems		Parameters of climatic change			
		Increase in short-term peak intensity or rainstorms	Increase in average intensity or duration of rainstorms	Increase in average intensity or duration of the snow melting period	Increase in persistence of multiannual cycles without exceptional or catastrophic floods
Small urban or rural catchment areas	with flood retention reservoirs	Slight impact on reservoir operation	Significant revision of reservoir design and operation, or increase of flood hazard	No or little impact	Slight to high increase of flood hazard due to unwarranted intensification of land use in the risk area, and to insufficient maintenance of flood control installation and services
	without flood retention reservoirs	Slight to medium increase of flood hazard	Medium to significant increase of flood hazard		
Large river basins	with flood recessive land and water use	No impact	No or little impact	Change in land use pattern with no or very little damage	
	with dike system along the major streams			Significant revision of dike system design and operation, or very substantial flood losses	

The impact on soil moisture, groundwater, and storm and base flow resources of a change in water or energy delivered to a river basin can be evaluated by several kinds of models. Particularly useful is the conceptual hydrologic model that reconstitutes basin hydrology under changing weather at short time-steps and collects the data into periods of years or decades, as appropriate. These changes in the resources of soil moisture adequacy, streamflow, and groundwater can then be assessed in terms of possible ameliorative or coping technology and management practices.

In order to select these technologies and practices in a socially desirable way, all those properties of the hydrogeographical endowments and hydrologic processes that are of actual or potential benefits or hazards to man within the given region should be assessed. For the purposes of analytical evaluation these socially significant properties should then be tied to water-related climatic parameters via impact matrices or other tools of correlative description.

REFERENCES

Ackerman, E. A., and Löf, G. O. G. (1959). *Technology in American Water Development*. Johns Hopkins Press, Baltimore: 710 pages.

Andrásfalvy, B. (1981). *Flood Recessive Land and Water Use along the Danube River in Hungary*. Consultant report for the Institute for Water Management, Budapest: 54 pages (in Hungarian).

Arnold, G. W., and Galbraith, K. A. (1978). Cultural and economic aspects. Case study one: Climatic changes and agriculture in Western Australia. In Pittock, A. B., Frakes, L. A., Jenssen, D., Peterson, J. A., and Zillman, J. W. (Eds.) *Climatic Change and Variability: A Southern Perspective*, pp. 297–300. Cambridge University Press, New York.

Bassett, J. R. (1964). Tree growth as affected by soil moisture availability. *Proceedings of the Soil Science Society of America*, **28**, 436–438.

Biswas, M. R., and Biswas, A. K. (Eds.) (1980). *Desertification*. Pergamon Press, Oxford, UK: 523 pages.

Budyko, M. I. (1974). *Climate and Life*. Translation by D. H. Miller of *Klimat i Zhizn'*, Gidrometeiozdat, Leningrad (1971). Academic Press, New York.

Budyko, M. I. (1982). *The Earth's Climate: Past and Future*. Academic Press, New York: 307 pages.

Burnash, R. J. C., Ferral, R. L., and McGuire, R. A. (1973). *A Generalized Streamflow Simulation System: Conceptual Modeling for Digital Computers*. US National Weather Service and California Department of Water Resources, Joint Federal-State River Forecast Center, Sacramento, California: 204 pages.

Clyde, C. G., Falkenborg, D. H., and Riley, J. P. (1976). *Colorado River Basin Modeling Studies*. Utah Water Resource Laboratory, Utah State University, Logan, Utah: 616 pages.

Cochrane, H. C., and Howe, C. W. (1976). A decision model for adjusting to natural hazard events with application to urban snow storms. *Review of Economics and Statistics*, February, pp. 50–58.

Cohon, J. L. (1982). Risk and uncertainty in water resources management. *Water Resources Research*, **18**(1), 1.

Cundy, T. W., and Brooks, K. N. (1981). Calibrating and verifying the SSARR model—Missouri River watersheds study. *Water Resources Bulletin*, **17**, 775–782.

Curry, L. (1962). The climatic resources of intensive grassland farming: The Waikato, New Zealand. *Geographical Review*, **52**, 174–194.

Dale, R. F., and Shaw, R. H. (1965). The climatology of soil moisture, atmospheric evaporative demand, and resulting moisture stress days for corn at Ames, Iowa. *Journal of Applied Meteorology*, **4**, 661–667.

Davenport, D. C., and Hagan, R. M. (1981). Agricultural water conservation in simplified perspective. *California Agriculture*, **35**(11–12), 7–10.

Dracup, J. A. (1977). Impact on the Colorado River Basin and Southwest water supply. In National Research Council, *Climate, Climatic Change, and Water Supply*, pp. 121–132. National Academy of Sciences, Washington, DC.

Falkenmark, M., and Lindh, G. (1976). *Water for a Starving World*. Westview Press, Boulder, Colorado: 204 pages.

Grin, A. M. (1965). *Dinamika Vodnogo Balansa Tsentral'no-Chernozemnogo Raiona*. Nauka, Moscow: 147 pages.

Hungarian Cartographical Institute (1974). *Political and Economic Atlas of the World*. Budapest: 384 pages (in Hungarian).

Károlyi, Zs. (1981). *The History of Flood Recessive Land Use in Hungary*. Consultant report for the Institute of Water Management, Budapest: 73 pages (in Hungarian).

Kates, R. W. (1981). *Drought Impact in the Sahelian-Sudanic Zone of West Africa: A Comparative Analysis of 1910–15 and 1968–74*. CENTED, Clark University, Worcester, Massachusetts: 92 pages.

Klige, R. K. (1978). Some problems of global water balance. In *Problems of Hydrology*, Institute of Water Problems of the Academy of Sciences of USSR, pp. 36–50. Nauka, Moscow (in Russian).

Kovda, V. A. (1980). *Land Aridization and Drought Control*. Westview Press, Boulder, Colorado: 277 pages.

Langbein, W. B. (1949). *Annual Runoff in the United States*. US Geological Survey, Circular 52.

L'vovich, M. I. (1969). *Vodnye Resursy Budushchego*. Izdat. Prosveschenie, Moscow: 174 pages.

Mather, J. R. (1974). *Climatology: Fundamentals and Applications*. McGraw-Hill, New York: 412 pages.

Meier, W. L., Jr. (1977). Identification of economic and societal impacts of water shortages. In National Research Council, *Climate, Climatic Change, and Water Supply*, pp. 85–95. National Academy of Sciences, Washington, DC.

Mumford, L. (1967 and 1970). *The Myth of the Machine*, Vols. 1 and 2. Harvest Books, New York.

National Research Council (1968). *Water and Choice in the Colorado Basin: An Example of Alternatives in Water Management*. Committee on Water, G. F. White, Chair. National Academy of Sciences, Washington, DC.

National Research Council (1977). *Climate, Climatic Change, and Water Supply*. Panel on Water and Climate, J. R. Wallis, Chair. National Academy of Sciences, Washington, DC: 132 pages.

Nemec, J., and Schaake, J. C. (1982). Sensitivity of water resource systems to climate variation. *Hydrologic Sciences*, **27**(2), 327–343.

Nováky, B. (1981). *Influences of Climatic Changes on the Hydrosphere*. Paper prepared for the Toronto meeting of the SCOPE/ISCIS programme, September: 20 pages.

Orlóci, I., and Szesztay, K. (1981). *Recent Trends in the Assessment of Water Resources and Demands*. Institute for Water Management, Budapest: 17 pages.

Peck, E. L., Keefer, T. N., and Johnsen, E. R. (1981). *Strategies for Using Remotely Sensed Data in Hydrologic Models*, NASA-CR-66729. National Aeronautics and Space Administration, Greenbelt, Maryland.

Pereira, H. C. (1973). *Land Use and Water Resources in Temperate and Tropical Climates*. Cambridge University Press, New York.

Perkey, D. J., Young, K. N., and Kreitzberg, C. W. (1983). The 1980–81 drought in Eastern Pennsylvania. *American Meteorological Society Bulletin*, **64** (2, February), 140–147.

Peterson, D. F., and Crawford, A. B. (Eds.) (1978). *Values and Choices in the Development of the Colorado River Basin*. University of Arizona Press, Tucson, Arizona: 337 pages.

Phillips, D. W., and McCulloch, J. A. W. (1972). *The Climate of the Great Lakes Basin*. Canada, Atmospheric Environment Service, Climatic Studies 20: 42 pages.

Revelle, R. R., and Waggoner, P. E. (1983). Effects of a carbon dioxide-induced climatic change on water supplies in the western United States. In *Climate Change (Report of the Carbon Dioxide Assessment Committee)*, pp. 419–432. National Academy Press, Washington, DC.

Rosenberg, N. J. (Ed.) (1978). *North American Droughts*. AAAS Selected Symposium Series, 15. Westview Press, Boulder, Colorado.

Schaake, J. C., and Kaczmarek, Z. (1979). Climate variability and the design and operation of water resource systems. *World Climate Conference*, Overview Paper 12, WMO, Geneva: 23 pages.

Schwarz, H. E. (1977). Climatic change and water supply: How sensitive is the Northeast? Chapter 7 in National Research Council, *Climate, Climatic Change and Water Supply*. National Academy of Sciences, Washington, DC.

Sokolowsky, D. L. (1968). *River Flow*. Gidrometeoizdat, Leningrad (in Russian).

Szesztay, K. (1960). *Water Balance Survey of Lakes and River Basins in Hungary*. Publication of the International Association of Scientific Hydrology No. 51, General Assembly of Helsinki.

Szesztay, K. (1965). *Some Aspects of Hydrological Network Design with Special Regard to Mountainous Areas*. Publication of the International Association of Scientific Hydrology No. 68, Symposium of Quebec.

Szesztay, K. (1981). *The Role of Water in the Climate Formation Process*. Paper prepared for the Toronto meeting of the SCOPE/ISCIS programme, September: 16 pages.

Szesztay, K. (1982). River basin development and water management. *Water Quality Bulletin (WHO)*, **7** (4, October), 152–162.

Teclaff, L. A. (1967). *The River Basin in History and Law*. Martinus Nijhoff, The Hague: 228 pages.

Thames, J. L., and Fischer, J. N. (1981). Management of water resources in arid lands. In Goodall, D. W., and Perry, R. A. (Eds.) *Arid-land Ecosystems: Structure, Functioning and Management*, Vol. 2, pp. 519–547. Cambridge University Press, New York.

Thornthwaite, C. W. (1948). An approach toward a rational classification of climate. *Geographical Review*, **38**, 55–94.

United Nations, Economic Committee for Asia and the Far East (1972). *Water Resource Project Planning*. Water Resource Series 41: 220 pages.

United Nations (1976). *River Basin Development Policies and Planning*. United Nations Publication Sales No.: E.77.II.A.4. New York-Budapest, two volumes.

U.S. Corps of Engineers, North Pacific Division (1975). *Program Description and User Manual for SSARR Model: Streamflow Synthesis and Reservoir Regulation*. Portland, Oregon: 188 pages.

VanderMeer, C. (1968). Changing water control in a Taiwanese rice-field irrigation system. *Annals of the Association of American Geographers*, **58**, 720–748. Reprinted in Coward, E. W., Jr. (Ed.) (1980), *Irrigation and Agricultural Development in Asia*, pp. 225–262. Cornell University Press, Ithaca, New York.

Voeikov, A. I. (1884). *The Climates of the Earth, Particularly of Russia*. Reedited by Academy of Sciences of USSR (1948) (in Russian).

Wilken, G. C. (1969). Drained-field agriculture: An intensive farming system in Tlaxcala, Mexico. *Geographical Review*, **59**, 215–241.

Willmott, C. J. (1977). WATBUG: A Fortran IV algorithm for calculating the climatic water budget. *Publications in Climatology*, **30** (2): 55 pages.

World Meteorological Organization (1975). *Intercomparison of Conceptual Models Used in Operational Hydrological Forecasting*. Report No. 429. WMO, Geneva: 172 pages.

Zur, B., and Jones, J. W. (1981). A model for the water relations, photosynthesis, and expansive growth of crops. *Water Resources Research*, **17**, 311–320.

Climate Impact Assessment
Edited by R. W. Kates, J. H. Ausubel and M. Berberian

CHAPTER 9
Energy Resources

Jill Jäger
Fridtjof-Nansen-Strasse 1
D-7500 Karlsruhe 41
Federal Republic of Germany

9.1 INTRODUCTION

The impacts of climate and weather on energy supply and demand have received increasing attention in recent years, especially since a number of severe winters in the northern hemisphere middle latitudes have highlighted mankind's vulnerability to climatic variability. Cold winters have increased the demand for energy and have also led to disruptions of supply. At the same time, interest has been growing with regard to future energy supply and the possibilities of using renewable sources of energy, especially solar energy, instead of nonrenewable, especially fossil fuel, resources. The renewable energy sources tend to be dependent upon climatic elements such as solar radiation, wind, rainfall and cloudiness.

This chapter is divided into two main parts: one considering the impact of climate on energy demand and one on the impact on supply. Studies on energy demand have concentrated mostly on the demand for energy for space heating and cooling. This emphasis is a reflection of the fact that a considerable proportion of the energy demand in industrialized countries is for space conditioning and this demand tends to be more climate sensitive than, say, the demand for energy for transportation. McKay and Allsopp (1980) state that over one-third of the energy consumed in industrialized North America and about 50 percent of that consumed in Europe (Denmark and Great Britain) is used to overcome the direct or indirect consequences of climate.

McKay and Allsopp point out that although the influence of climate on energy demand is most evident in the case of space heating, significant demands also occur in agriculture, transportation, and in outdoor industries such as construction and forestry. The latter uses, however, are small in comparison to the demand for energy for space conditioning. Pimentel (1981) estimates that America and Europe currently use about 17 percent of their total energy for their food systems. About 6 percent is used directly for agricultural production. In developing countries, Pimentel estimates that the amount of energy used in the food system is 30–60 percent.

There are three basic methods that have been used so far to study the impacts of weather, or climate, on energy demand. These three methods are discussed in Section 9.2. The first method is that of the case study, in which the impacts of particular climatic anomalies are documented. The other two methods involve models. One set of models is referred to as physical models. These models consider the actual heat losses from a building, or a number of buildings, and compute the changes of these heat losses as a function of changes in climatic variables such as outside temperature and windspeed. Physical models are useful for studying the detailed response of individual structures, but the amount of input information that is required makes them inappropriate for studies of the impact of a cold winter on the demand for energy for the space heating of an entire country. In such a case, it is necessary to aggregate data and use statistical samples.

The second set of models is referred to as statistical or empirical models. Usually they involve regression analyses, with some expression of energy demand as the dependent variable and some climatic variable, usually temperature or degree days, as one of the independent variables in the regression equation.

Section 9.3 considers the impact of climate on energy supply, mainly on renewable sources of energy. Basically, there are two methods for studying these impacts. The first method involves data acquisition and analysis. The main aim of such studies is to provide a detailed description of the availability of renewable energy resources such as solar energy, wind energy, and hydropower and to provide information on their likely variability over time as a result of climatic variability. The second method is the development of computer models to simulate the performance of renewable energy technologies in different climate zones. These models have been useful, for example, in showing the interplay between solar energy availability and heating requirements as a function of climate.

Section 9.4 discusses energy supply and demand in developing countries. Attention must be devoted to traditional, non-commercial sources of energy in the developing countries. Studies have been made of the potential role of renewable energy sources, with some studies suggesting only a limited role in the short term. Increasing pressure on traditional, non-commercial energy sources in developing countries has the end result of deforestation. Studies must consider the multiple uses of land for energy and agriculture, both of which are influenced by climate.

9.2 THE IMPACT OF CLIMATE ON ENERGY DEMAND

9.2.1 Case Studies of Climatic Events

Relatively few studies have been made of the impact of climate on energy demand. A brief description of the impact of the cold winter of 1947 and the more prolonged and severe winter of 1963 on British fuel supply and demand was given by Burroughs (1978).

Living with Climatic Change (Beltzner, 1976) presented a series of illustrative examples, or scenarios, as a guide for concrete investigations of climate sensitivity. The purpose was to present to the planner possible sequences of climatic events which are, in some sense, representative of the type of stress which climatic variations place on the social and economic structure of North America. Two considerations prompted the study to select real periods of past climates as models for the future. First, the situations are inherently credible: what has occurred can occur again. Second, it was felt that only real data contain the complex richness of detail that characterizes the atmosphere and is essential for planning in the real world. Therefore, periods of past weather that placed stress on society were studied. It is pointed out that well-documented instances are necessary for credible scenarios, so that only instrumentally observed events after 1880 were considered.

The scenarios that affected the energy sector were:

variability (1895–1905),
Midwest drought (1933–37),

energy (1935–36),
Mexican drought (1937–44),
variability (1950–58),
eastern urban drought (1961–66),
sea ice (1964–65 and 1971–72),
snowfall (1970–74).

The variability scenario (1895–1905) was characterized by: a cool climate; a wet period in the northwestern Great Plains; sustained drought in the Pacific Northwest; extreme cold in the Gulf States; heat waves in California and the Midwest; and East Coast and Great Lakes storms. Stress on the economy and society was not continuous but varied in type, time and place. The authors point out that a cold winter followed by a scorching summer as in 1899–1900 could today overtax the energy supplies and systems upon which society has become increasingly dependent.

In a brief description of the impacts of the cold winter of 1976–77 in eastern Canada, Won (1980) points out that secondary climatic impact on energy consumption involves the disruption or expenditure of energy resources due to extremes in weather. For example, many communities had expended their yearly budgets for snow removal before midwinter and disruptions in transportation occurred often. Ice storms that resulted in the collapse of transmission towers and lines caused extended power interruptions.

Quirk and Moriarty (1980) discussed the impact of the 1976–77 winter on the United States. The winter brought continued drought to the western United States and cold weather to the eastern United States. Quirk (1981a) indicates that the winter of 1976–77 had 11 percent more population-weighted heating degree days than in the preceding year, which implied an increase in the demand for heating fuel of about the equivalent of 350 million barrels of oil. The drought on the West Coast caused reductions in hydropower, resulting in an additional demand for another 50 million barrels of fuel to produce electricity. These anomalies are illustrated in Figure 9.1 from Quirk (1981a).

An extension of the case study approach is the development of climatic scenarios. The scenarios are based on recurrent patterns of anomaly identified by principal components analysis. These can be used to postulate future situations and their impact (Quirk, 1981b). As Quirk points out, the scenarios must contain information on the geographical pattern, amplitude and timing of typical climate anomalies. Data on climatic anomalies are obtainable for the land areas of the temperate latitudes of the northern hemisphere for the last 80 years, and the use of eigenvector analysis helps to find the minimum number of patterns representing the large part of the variations. Diaz and Fulbright (1981) have used eigenvector analysis to find the empirical orthogonal functions describing temperature deviations from

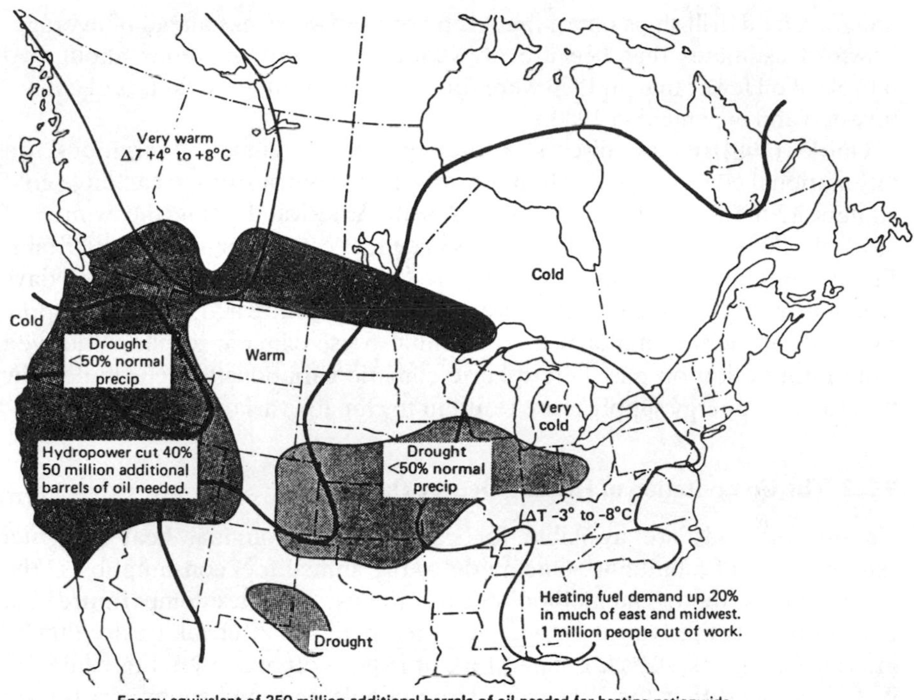

Figure 9.1 Climatic impacts of the winter of 1976–77 in North America. Reproduced by permission of the American Meteorological Society from Quirk, *Bulletin of the AMS*, **62**, 623–631 (1981)

normal in the United States. They found that only three patterns were needed to explain 86 percent of the variance in winter temperature for the period 1894–1978.

Quirk (1981b) outlines a methodology for deriving climate scenarios. It is necessary first to obtain monthly averaged data for all the heating season months. Statistical analysis (asymptotic singular decomposition) is then required to find the most common patterns. Then data on the variation in amplitude of these patterns for each month of a heating season are used to give information on the timing and amplitude of the climatic anomalies. A scenario thus consists of a month-by-month description of the pattern and amplitude of the temperature anomaly of the heating season. The probability of the occurrence of each scenario could also be given.

Lawford (1981) has documented the climate events in Canada in 1980 that significantly affected the demand for and supply of energy, the exploration for new reserves, and the production of hydroelectricity. He points out that movements of the pack ice in the Beaufort Sea were unusual in the spring and summer of 1980. The pack ice closed in on the Arctic shoreline in September and forced oil drill ships out of the Arctic one or two weeks ahead of average. Lawford estimates that because of weather and ice conditions about two barrels of oil fewer than in 1979 were found for every hundred dollars of capital invested and expended in 1980.

Quirk (1981b) has discussed the impact of climatic variations on international energy supply. He points out that western Europe can have cold winters at the same time as eastern North America. During the winter of 1978–79, the United States, Canada, Sweden, West Germany, Great Britain, France, and northern Italy all had about 8 percent more heating degree days than normal. It is clear, therefore, that those concerned with oil supply shortfalls at the international level could also use climatic scenarios derived from information on past hemispheric climatic variations to determine how best to make energy supply more resilient to climatic variations.

9.2.2 The Computation of Heating Degree Days

Various methods are available for calculating the annual heat and fuel requirements of a building. One guide to the annual fuel consumption is the degree day method. The method requires the use of a base temperature, T_b, that represents the typical indoor mean temperature after taking account of internal heat gains. A base temperature of 18 °C is often chosen. For example, if it is assumed that the heat gains from people and appliances in a house, together with solar energy gains through windows, would raise the temperature by 2 °C, heating is required in order to keep the temperature of the house at 20 °C (a typical value) if the outside temperature falls below 18 °C (T_b). For a well-insulated house the base temperature should be lower, say 12 °C, because the internal heat gains would contribute a proportionately large part of the daily heating load. For a day on which heating is required, the number of heating degree days (DD) equals the difference between the average temperature on that day and the base temperature (T_b). For example, a day during which the average temperature is 13 °C has five degree days, if T_b is 18 °C. The number of heating degree days for the whole heating season can be added and used as a guide to the annual heating requirement. In a similar manner it is possible to calculate the number of cooling degree days required by assuming a base temperature above which cooling is necessary to maintain a particular temperature level in a house.

There are several reasons why degree day figures are useful. First, they are cumulative so that the degree day total for a period is proportional to the total heating load for that period. Second, the relationship between degree days and fuel consumption is usually assumed to be linear. That is, it is assumed that if the heating degree day total is doubled, the fuel consumption is doubled. It has, however, been pointed out (Jäger, 1981) that this assumption is an approximation because fuel consumption also depends on the efficiency of the heating system, which depends on the operation frequency of the system. Fuel consumption also depends on insulation, construction, building exposure, life styles, and so forth.

If the seasonal degree day totals in different locations are compared, the relative amounts of fuel consumption can be estimated. Table 9.1 shows the degree day totals for base temperatures of 18 °C and 12 °C for a number of European locations. It can be estimated that the annual fuel consumption in a building in Lerwick (Shetland Islands, United Kingdom, 3940 degree days, T_b = 18 °C) would be about 1.8 times as much as the fuel consumption in a similar building in Toulouse (southern France, 2210 degree days, T_b = 18 °C).

Mitchell *et al.* (1973) used the degree day concept in a study of the extent to which the United States national total demand for heating fuels is dependent

Table 9.1 Degree days to base temperatures of 18 °C, 12 °C, 17 °C and 19 °C for different locations in Europe

Latitude band °N	Place		Degree days base		
		18 °C	12 °C	17 °C	19 °C
59–61	Lerwick Shetlands	3940	1880		
55–57	Glasgow Scotland	3370	1520		
53–55	Dublin Ireland	3156	1317		
	Hamburg FR Germany				3350
51–53	Kew (London)	2780	1200		
	Valentia Ireland	2786	961		
49–51	Uccle (Brussels)	2580			
	Lille France	3062	1378		
	Reims France	3010	1396		
	Nürnberg FR Germany				3370
47–49	Brest France	2653	899		
	Strasbourg France	3061	1900		
	Freiburg FR Germany				3050
	Munich FR Germany				3730
45–47	Limoges France	2820	1231		
	Milan Italy	2350		2120	
43–45	Montélimar France	2233	938		
	Toulouse France	2210	842		
	Montpellier France	1875	665		
	Genoa Italy	1494		1270	
41–43	Ajaccio Corsica	1866	405		
	Rome Italy	1570		1350	
39–41	Naples Italy	1355		1142	
37–39	Messina Italy	806		623	

Values are not given for all countries because calculation methods are not entirely consistent and sometimes correction factors are added.
After Jäger, 1981, 47.

on the weather. A probability analysis was made of the nationwide variability of seasonal total heating degree days, based on a long series of temperature data and information on the geographical distribution of heating fuel demand. The authors first calculated the seasonal total heating degree days for each of the 48 conterminous states of the United States and for each of the 42 heating seasons from 1931–32 to 1972–73. The heating degree day totals were then averaged together into a nationally averaged heating degree day total for each of the 42 heating seasons. Five different weighting procedures were used, based on the contribution of each state to the national total demand for fuel in each of five categories: all fuels, gas, oil, electricity, and liquefied petroleum gas (chiefly propane). The series of 42 nationally averaged heating degree day totals for each of the five fuel categories was then treated as a direct measure of the relative variations of total national heating fuel demand in that fuel category, for the assumption of a constant economy. Each series was examined for evidence of systematic trends. Lastly, the 42 values in each series were treated as random samples from populations of such data. This provided the basis for constructing appropriate statistical models for the assessment of probabilities of extreme fuel demand in an arbitrarily chosen heating season. Using this approach, Mitchell *et al.* (1973) were able to determine the influence of weather on heating fuel demand in terms that were independent of the long-term growth of demand attributable to economic, demographic and technological trends.

The results of the analysis made by Mitchell *et al.* showed, for example, that in only one year out of one hundred years should one expect the national total demand for heating oil to exceed the long-term average demand (for constant economy) by as much as 10.6 percent. Similarly, it was found that the demand for heating oil can be expected to exceed its average demand (for constant economy) by at least 3 percent on an average of one heating season in five.

Mitchell *et al.* noted that in a situation where the national total heating fuel demand is higher than average, it is quite likely that excess demand would be found to center on one section of the nation where the problem is severe, while near-average or even below-average demands would be found in other areas. Therefore, the fuel demand and degree day series were computed for nine regions. As expected, the probable extreme deviations (when expressed as percentage deviations from average regional demands) were found to be somewhat larger than those for the nation as a whole, especially in the southern and Pacific states.

Figure 9.2 shows the heating degree days (base temperature 65 °F) accumulated for each heating season (defined as October–March) for the United States and weighted by population (Diaz and Quayle, 1980). The anomalously cold winter of 1976–77 is quite distinct, as is the decline in the heating degree day total between 1900 and about 1940 and the subsequent increase.

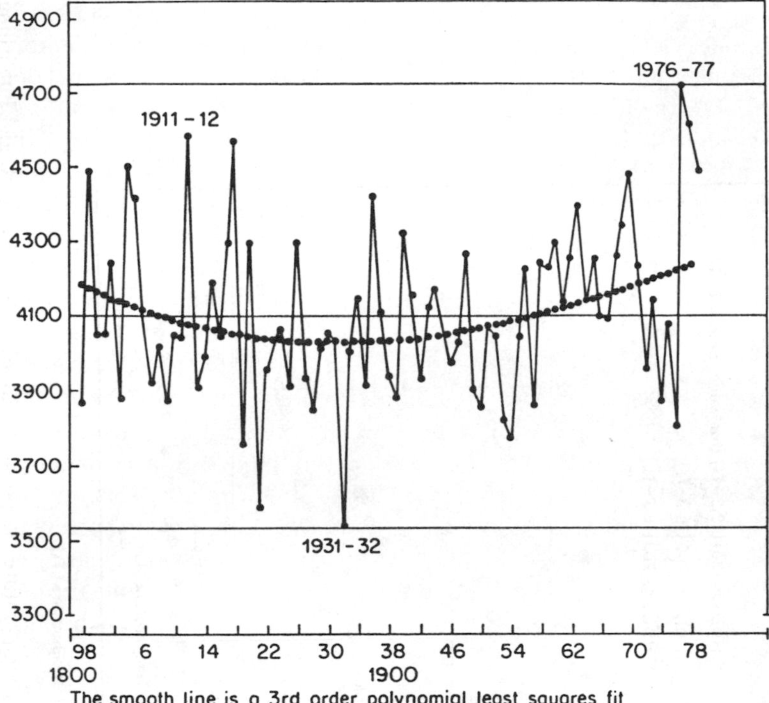

Figure 9.2 Total October–March heating degree days (65 °F base) for the United States weighted by population for 1898–1978. Reproduced by permission of the American Meteorological Society from Diaz and Quayle, *Monthly Weather Review*, **108**, 687–699 (1980)

Ahti (1975) cites Robinson (1974) in pointing out that the degree day method is not appropriate for estimating energy consumption by air conditioning for the following reasons:

- the system energy demand does not necessarily increase linearly with a temperature increase,
- the system energy demand is affected by relative humidity,
- the system energy demand is affected by solar radiation.

McKay and Allsopp (1980) showed the relationship between heating degree days and mean temperature presented in Figure 9.3. On the basis of the known relationship between mean annual air temperature and population distribution, and assuming that the heating requirement approaches zero at 18 °C and 35 percent of the total energy use at 0 °C (based on US and Canadian rates), McKay and Allsopp calculated that the greatest energy use occurs in the regions where the annual air temperature is between 5 °C and 15 °C with a maximum at 10 °C. They conclude that a one-degree change in annual

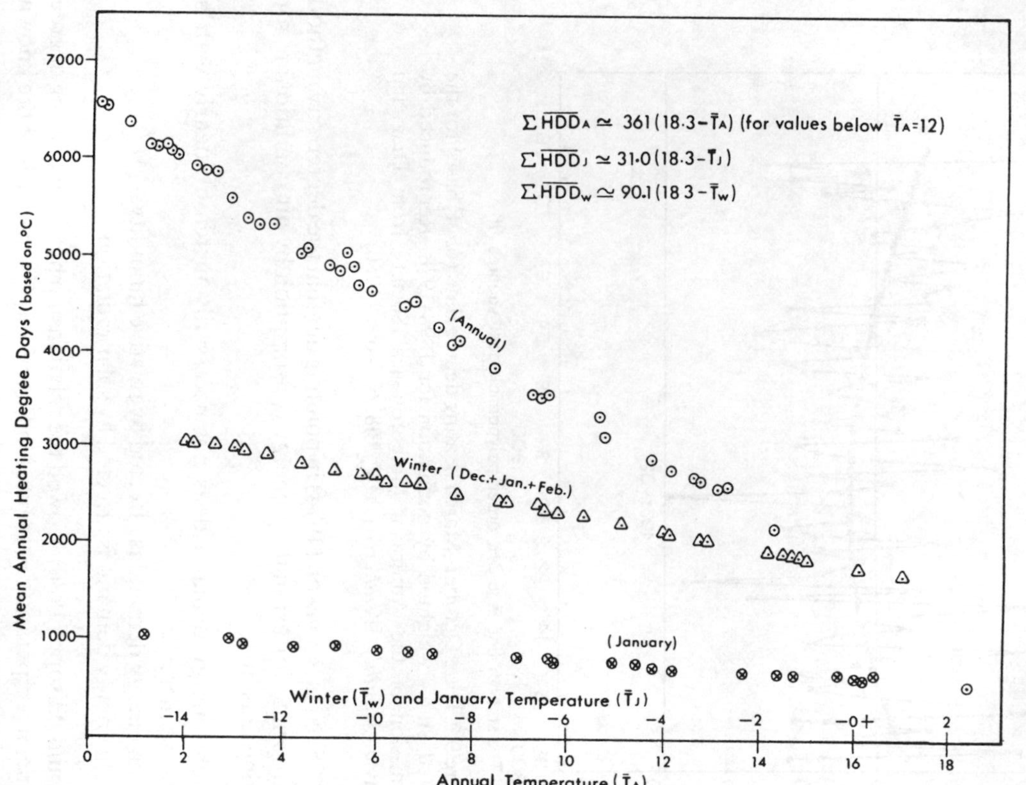

Figure 9.3 Relationships between heating degree days and mean temperature. Reproduced by permission of D. Reidel Publishing Company from McKay and Allsopp (1980)

temperature would alter space heating energy demand for people living in the present intensive energy use area by roughly 10 percent. National cumulative abnormal heating and cooling costs for the United States are published, along with information on heating and cooling degree days, by the US Department of Commerce (National Oceanic and Atmospheric Administration, ongoing).

Cohen (1981) has designed what he believes to be a more realistic climatic index of long-term residential energy consumption and shows how this index is better than heating degree days. The index includes non-temperature elements and provides a more complete representation of climate for use in multivariate energy demand models. A statistical approach was used to relate seasonal frequencies of daily upper air circulation patterns to energy consumption. Principal component analysis was used to identify the significant upper air (500 mb) flow types. Simple linear regression analysis and Pearson product-moment correlation analysis were used to examine separately the statistical relationship between seasonal energy consumption and the two variables: seasonal frequencies of upper air circulation types and seasonal heating degree day totals for each state in the United States. The energy data base consisted of annual per-household consumption of natural gas for each state from 1960 to 1978. Cohen concluded that the 500 mb circulation type frequencies are a better climatic index of long-term residential natural gas consumption than heating degree days.

9.2.3 Computing the Heating Requirements of a Building

In order to calculate the heat losses of a building a number of factors must be considered:

- the coefficients of heat transmission of all parts of the building through which heat can be lost (walls, windows, doors, etc.);
- the outer surface area of these parts of the building;
- the temperature difference between the interior and exterior of the building;
- the ventilation losses.

The transmission heat loss from a building is dependent on the mean coefficient of transmission of the building, the total exterior surface area, and the temperature difference from the interior to the exterior. The ventilation heat loss depends on the airflow and the specific heat of the air. Not only does the outdoor temperature continually change during the day, but indoor conditions also vary due to changes in ventilation, absorption of solar radiation by the building, solar radiation coming through windows, and heat emissions from people and appliances. The most accurate models for the determination of heating requirements take account of the dynamic effects arising from the thermal storage capacity of the building components and heat transfer in these components as a function of time.

Jäger (1981) has used a detailed computer model for the calculation of heating requirements. The model was based on guidelines set up by the Federal Republic of Germany for the dimensioning of heating systems. The different components of the energy balance of a building (such as transmission and ventilation losses and heat gains from occupants) were calculated for each hour. The model used hourly meteorological data (temperature, direct and diffuse radiation). This method allows a more exact calculation of the annual heating requirements than can be made with the degree day method, which can consider the heat gains from the sun only approximately through a reduction of the base temperature. The model did not, however, take into account the dynamic effects mentioned above. For the calculation of the heating requirements a reference house was considered. Two sets of house insulations were studied. The first set consisted of those required by present regulations in the Federal Republic of Germany. The second set of insulation standards was based on the more stringent present standards in Denmark. The annual heating requirement for the two house types was calculated using meteorological data from two locations, Copenhagen in Denmark and Carpentras in southern France. The computed annual heating requirements are shown in Table 9.2. For both house types the heating requirement in southern France was about 63 percent of that in Denmark. In both locations the heating requirement in the better-insulated house was about 50 percent of that in the house built according to lower standards.

Table 9.2 Computed annual heating requirement for two types of reference houses in two European locations

Annual heating requirement (kWh)	German standard	Danish standard
Copenhagen, Denmark	30 420	15 350
Carpentras, France	19 280	9750

Source: Jäger, 1981, 49.

Hörster (1980) has reviewed the methods of calculation of energy requirements of buildings. He points out that a number of mathematical models exist which are based on first principles. These models calculate the magnitude and distribution of interior temperature and the relative humidity in individual rooms. The models require much computer time; it is therefore necessary to develop simplified models. As Hörster points out, the simplified methods use the most important physical factors that determine the heating requirements.

Hörster distinguishes between dynamic and stationary models. The simplified stationary models do not take into account the energy flows within the building. These flows are especially important when the insulation standard is high.

9.2.4 Climate–Energy Use Models

9.2.4.1 A Buildings Model

In a series of reports, Reiter *et al.* (1976, 1978, 1979, 1980, 1981) have discussed the results of a long-term project to study the impact of weather variations on energy demand for space heating. The authors decided that statistical models relating weather variables and energy consumption, especially models based on historical data, have a number of limitations. These are accurate only if physical structures, use patterns, and comfort levels remain constant. Therefore, Reiter *et al.* (1976) decided to base their model on physical features which could be derived from basic heat transfer relationships.

The authors point out that a model cannot be developed for each individual building in a region, but in the United States there is a remarkable degree of thermal similarity among the vast majority of residential, commercial and industrial high- or low-rise buildings constructed within fairly distinct time periods. It was concluded that buildings could be grouped into thermally equivalent classifications and aggregated. The physical model therefore consists of:

1. a generalized computer program for predicting the space-conditioning energy requirements of selected building classifications,
2. a set of modules generated from the above program, with each module representing a particular building type, use and vintage classification.

The generalized computer program computes transmission losses and infiltration losses as outlined in Section 9.2.3.

The meteorological input to the model consists of insolation or percent of cloudiness (amounts, types, thicknesses and altitudes), windspeed, and ambient temperature (air and ground). A second requirement is a census of the buildings that comprise the population region to be studied by types, usages, ages, numbers, sizes, construction characteristics, materials, shading and sheltering from sun and wind, energy sources, heating and cooling systems, internal heat loads and locations. The model also provides for the selection of thermostatic settings, for the introduction of multipliers to characterize infiltration rates, and for the variation of the physical (thermal) characteristics of structures for known or assumed building uses and occupant habit patterns. This latter part of the model is referred to by the authors as the adaptive portion of the model, and it is suggested that the policy-maker could use it to introduce a variety of alternative scenarios into the physical model in order to investigate their relative impacts on energy consumption for space conditioning.

Reiter *et al.* (1976) applied the overall model to predict the daily gas consumption for Greeley, Colorado, during the period 1 December 1975 through 29 February 1976. The model predicted the mean daily energy consumption to within 8 percent, with a standard deviation of 5 percent. The model underpredicted during cold periods and overpredicted during warm periods. The authors investigated several possible causes for this, including the variations in furnace efficiency caused by more frequent switching between on- and off-cycles during cold periods, and the converse during warm periods.

Reiter *et al.* (1978) also reported on refinements of the energy model. They point out that the model developed in the previous report was a micromodel, in which the system is decomposed as much as possible. The information needed by a micromodel is extensive. Their first study had, by means of an exhaustive survey, obtained detailed information on physical variables as well as the social behavior of the building occupants. A refinement of this procedure, in which a less detailed building census was generated by means of statistical sampling schemes and procedures, was adopted in 1978. The second approach to the development of a model for energy demand was a macromodel. Macromodels use socioeconomic data that are available in census reports and other data available within the community. Reiter *et al.* indicate that micro- and macromodels both have advantages and disadvantages, and the best approach is probably a combination of the two model types. The micromodel was used to compute the energy consumption for the winter of 1976–77 for Greeley, Colorado, and the results were 99.9 percent of the actual consumption. The model was also applied to Cheyenne, Wyoming, and it predicted 97.8 percent of the actual consumption during January, February and March 1977. The slightly larger error was explained as being due to the fact that no detailed building census was taken in Cheyenne, but a statistical sampling technique was employed instead.

Reiter *et al.* (1979) extended the model of space heating demand by adding a set of hypothesis-testing procedures to measure how the model results compare with the actual energy consumption of a community. These testing procedures enable the model to detect changes in habit patterns and also, by updating, to cope with the evolution of a community. To improve the model without increasing the amount of input data required, a time-series description was developed to complement the original physical model. The time-series description was based on the residual (observed minus predicted) energy consumption and its correlation with various meteorological input variables in time sequence. A description of this type was developed for Greeley, Colorado, and it improved the estimation of energy consumption from a root mean square error of 9.1 percent to 5.8 percent for the 1975–76 winter season.

In addition, Reiter *et al.* (1979) described the development of a second model, referred to as a statistical reference model. This model used the same heuristic algorithm that was used in the physical model for identification of the coefficients of the heat transfer equations used to model individual buildings within a certain typical structure category. However, the statistical reference model does not use actual building information. Instead, it uses the meteorological input and the actual response of the community in terms of energy consumption to derive a single high-order equation that can be used to model the response of the entire community. The model was developed for estimating the performance confidence interval, which was used to show which parts of the model output are acceptable, and indicated when the real community was changing in complexion over time in contrast to the earlier identified physical model assumptions. The authors suggest that the statistical reference model has limitations, largely because it is based on coefficients and does not explicitly include the physics of processes involved in the need for energy consumption for space heating. Thus the model cannot be used to answer questions about the effects of behavioral, structural or other changes. On the other hand, the authors indicate that the model could be useful for assessing the energy use of communities with a small amount of data.

Reiter *et al.* (1980) also report on the development of a model for use in studying the effects of optimal utilization of weather and climate information on energy systems design and operation. To test a preliminary model, the authors adopted a scenario in which a solar energy installation backed up by a resistance heat source should be optimized and the requirements of heat storage for various given climatic factors would be studied. The authors point out that in such a scenario various trade-off decisions concerning the number or size of solar panels, the capacity of the heat storage device, and requirements for auxiliary energy under various cost configurations can be made.

9.2.4.2 A Fossil Fuels Model

Nelson (1976) has examined the influence of climate on the demand for fossil fuels in residential and commercial space heating. Climatic variability was considered in terms of heating degree days. The demand for fossil fuels was considered to be the total oil, natural gas, and coal consumed in the residential and commercial sector in a cross-section of the states of the United States of America in 1971.

It is pointed out by Nelson (1976) that studies of the demand for energy in the residential and commercial sectors can be divided into short-run models and long-run models. The short-run demand for a particular fuel refers to the demand for fuel when the stock of heating appliances is held constant. Nelson suggests that the short-run per-customer demand can best be studied by using time-series data that control for the prices of substitute fuels and heating appliances. In contrast, long-run demand for particular fuels involves variations in both the stock of appliances and the usage of that stock. Studies of the long-run per capita demand require cross-sectional data including the prices of substitute fuels and heating appliances.

Nelson assumes that for each state the total residential and commercial demand for fossil fuel energy is a function of:

– the number of customers,
– the price of fossil fuel energy,
– income,
– and several non-economic variables including climate.

The number of customers per state using a particular heating appliance is assumed to be a function of:

– population size,
– the price of fossil fuel energy,
– prices of substitute fuels,
– income,
– and several economic variables including climate.

Nelson found that

1. Population has a significant and positive effect on energy demand.
2. Income has a significant and positive effect on energy demand when it is added to the regression equation, but when the urbanization variable is added, the income variable is no longer significant. Nelson suggests that this may be due to the high correlation between these two variables.
3. The price of fossil fuels has no effect on the total fuel demand.
4. The electricity price variable has a positive and significant effect on the total fuel demand.
5. Degree days have a positive and significant effect on total fuel demand. A 10 percent increase in degree days would increase total fuel demand by about 5 percent.
6. The two most important variables in explaining changes in total fuel demand were found to be population and degree days.

Because population was found to have a significant impact on energy demand, Nelson rederived the model in terms of the per capita demand for fossil fuel. With per capita fuel consumption as the dependent variable, Nelson found that

1. Income was not significant and became negative when the urbanization variable was added.
2. The price of fossil fuels, the electricity price variable, and degree days all retained their significance and relative magnitude.
3. Climate (degree days) and the price of electricity are most important for per capita demand changes.

Nelson has discussed in detail the possible causes of the fact that a 10 percent increase in degree days would increase total fuel demand only by about 5 percent. He points out that the model does not control for interstate differences in housing insulation and construction standards. Because homes in northern climates are better insulated, the degree day variable will be biased downwards if estimated with cross-sectional data. Also, Nelson points out that the dependent variables examined in this study included some non-space heating uses of fossil fuels. It is estimated roughly that space heating accounts for 80 percent of the gas, oil and coal consumed in the residential and commercial sectors. Since non-space heating uses would not always vary with outdoor temperature, the change in degree days would not entirely determine the change in fuel demand.

Nelson describes two applications of the empirical model: an examination of the expected crude petroleum savings due to lower thermostat settings and higher prices for fuel oil, and a prediction of increased energy consumption if supersonic air transportation were to reduce mean annual global surface temperatures. It was calculated that a 6 °F reduction in thermostat settings would reduce per capita demand by 13.65 percent if all else were constant. For a price increase of 10 cents per gallon relative to November–December 1973 prices, the saving was computed to be 8 percent of 1973 demand. It was calculated that an 8.2 percent increase in degree days in the United States would increase the per capita residential and commercial demand for fossil fuels by 4.1 percent.

9.2.4.3 An All-electric Commercial Buildings Model

Crocker (1976) has examined the impact of climatic variations on electricity demand in commercial buildings, using detailed histories of month-by-month electricity consumption and meteorological variables during a period of a year for about 80 all-electric commercial buildings throughout the United States. The estimates made by Crocker are considered to be long-run because it is assumed that building insulation and heating and cooking equipment are selected on the basis of expected climate.

The dependent variable used in the ordinary least-squares cross-sectional regressions was kilowatt-hours consumed in a given month. The variables used in the empirical analysis were: heating degrees, cooling degrees, average electricity price, total connected load, designed heat loss, designed heat gain, apartments, churches, stores and offices, light manufacturing, and motels.

The ordinary least-squares regressions discussed by Crocker are multiplicative, according to a statistical function used to transform combinations of electricity and other inputs into a variety of outputs. The year was partitioned so that only summer and winter months were considered. The results of the March, June, and September regressions indicated that there was substantially less response to variations in heating degree days (March) or cooling degree days (June and September) during these months than for other months for which regression results were reported. Crocker concluded that these relatively low elasticities were consistent with a relatively small absolute change in quantity demanded with respect to a one-unit absolute change in degree days. Over the observed temperature ranges for the months in question, the electricity consumption for heating and cooling purposes was relatively unresponsive to changes in degree days. Crocker suggests that this behavior seems particularly likely during the spring and autumn months in the temperate climates where the buildings used in the study were located. The results also indicated that for the buildings in question there exists for temperatures below 75 °F or 80 °F a uniformity in the responsiveness of electricity consumption to variations in climate.

9.2.4.4 A Real-time Data Model

Warren and LeDuc (1980) discussed the need to assess the impact of weather on the economy, in particular, on energy use and price. The authors point out that models to estimate the impact of weather on energy demand can be classified according to the energy demand or load that they are intended to estimate. At the lower end of the range there are models for estimating the next day's load, using information such as weather effects and the latest load behavior. Most of these models are utility-specific and integrating them into a model at a regional or national level would require the aggregation of data for many separate utilities.

Warren and LeDuc (1980) describe a model used for estimating natural gas consumption in the United States. Two demand functions for residential consumers were formulated: one for customers who do not use gas for space heating, and a second for space heating customers. The relationship for consumers who do use gas for heating is a function of base consumers, price, and heating degree days. A heating degree day index was formulated to account for the spatial distribution of customers in the division and also for a temporal reporting lag. The average rates of consumption of natural gas in nine regions that together covered the contiguous United States were estimated. The authors conclude that weather alone does not explain all the variations in energy consumption. For example, price and seasonal adjustments were necessary in the above model to explain the variations.

9.3 THE IMPACT OF CLIMATE ON ENERGY SUPPLY

The impact of climate on conventional energy supplies has not been studied in detail. The range of impacts is generally known, however. For instance, McKay and Allsopp (1980) state that exploration and transportation phases are climatically sensitive, especially at high latitudes and at sea. Likewise, it is

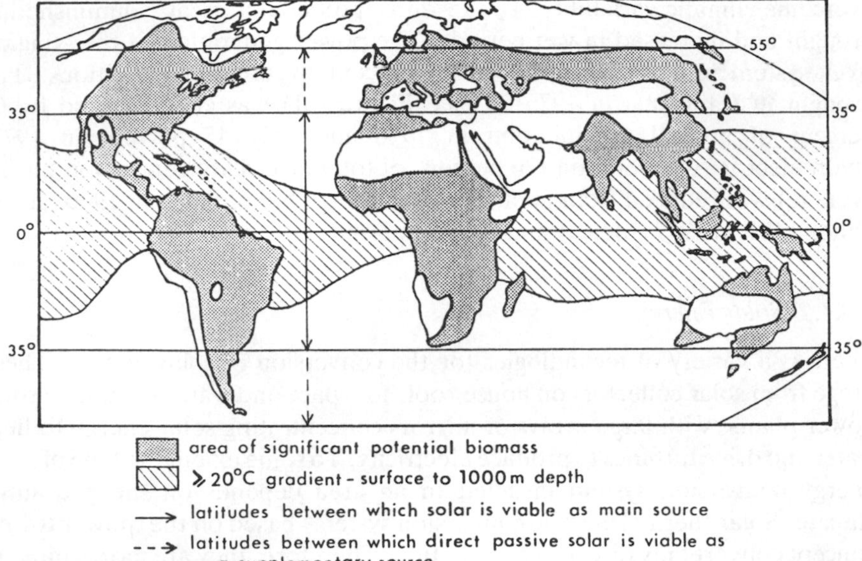

Figure 9.4 Areas favorable for exploitation of solar, biomass and ocean thermal energy. Reproduced by permission of D. Reidel Publishing Company from McKay and Allsopp (1980)

pointed out that climate must be taken into account in the selection of optimum shipping routes, ports and harbors. Strategies, economics, engineering and environmental aspects of offshore drilling, and the development of coastal facilities are also found to be climatically sensitive. McKay and Allsopp (1980) also point out that climatic elements such as air temperature, humidity, solar radiation, wind, and precipitation must be considered in the design and operation of storage systems.

Most impact studies concerning renewable energy sources are at the resource assessment stage with little study, yet, of the variability of the supply due to climatic variability. But as McKay and Allsopp (1980) have indicated, climate dictates the potential supply of renewable energy sources. They provide a general analysis of this supply, showing the influence of geography (Figure 9.4). It is suggested that solar energy systems are most promising between 35 °N and 35 °S. Significant hydropower and biomass production also occur in this zone. Ocean thermal energy is most obtainable in the tropical latitudes, wave energy in ice-free waters, and wind energy is ubiquitous (McKay and Allsopp, 1980).

9.3.1 Renewable Energy Supply

9.3.1.1 Hydropower

The impact of climate on hydropower supplies has been discussed by McKay and Allsopp on a quantitative basis. They point out that storage must be provided to overcome climatic variability. The resource potential is greatly diminished in drought and increased in wet periods. The power generation of the Niagara river system in 1964 was reduced 26 percent by drought conditions. The drought in California in 1976–77 (see Figure 9.1) was characterized by 65 percent of normal precipitation in 1976 and only 45 percent in 1977. Hydroelectricity generation (20 percent of total electricity generation on the average) was reduced from about 33 billion to 16 billion kWh in 1976 and 13 billion kWh in 1977 (McKay and Allsopp, 1980).

9.3.1.2 Solar Energy

There is a variety of technologies for the conversion of solar energy. These range from solar collectors on house roofs for space and water heating to solar power plants, with large arrays of mirrors concentrating solar energy to heat water and drive turbines to produce electricity. To some extent the type of solar energy conversion system installed in an area depends on the prevailing climate. Solar thermal electric conversion systems based on the 'power tower' concept convert only direct solar radiation. Therefore, they are more suited to the dry desert climate regions, where the proportion of direct radiation is high, than to the cloudier middle latitudes, where the proportion of direct radiation is much smaller, especially in the winter months. Insolation resource assessment programs (e.g., Riches and Koomanoff, 1978; Commission of the European Communities, 1979) have been established to collect, record and archive solar radiation data. Analyses of temporal and spatial distributions of solar radiations are reported for various countries in the energy journals (e.g. Balling and Vojtesak, 1983).

The use of solar energy for domestic space and water heating depends on a number of climatic factors, as has been discussed in detail by Jäger (1981). Not only does the solar energy collection depend on climatic variables, but the heat losses in a house are also climatically sensitive, as described in Section 9.2. The design of a solar house should minimize heat losses in winter, while maximizing the heat gains and also ensuring that the house is acceptably comfortable in summer. Information on microclimatic conditions (shading, sheltering, etc.) as well as on the macroclimate is necessary for the design of a solar-heated house.

Jäger (1981) has emphasized the importance of making a balanced assessment of all factors when making an evaluation of the economic viability of solar heating systems. He points out that if one were to consider only the distribution of incoming solar radiation in Europe, one might draw the conclusion that solar heating would be more economically attractive in the southern locations where the solar radiation availability is high. However, the length of time over which heating is required is relatively short in the south because of the relatively higher temperatures associated with lower cloudiness, more sunshine and less wind. A longer operating season in less climatically favorable regions could give an installed solar space heating system greater economy.

9.3.1.3 Wind Energy

The impact of climate on wind energy resources has also been considered. The distribution of wind energy varies markedly in space and time and a knowledge of these distributions is necessary so that the selection and siting of wind energy conversion systems can be effective. Hardy and Walton (1977) have listed the meteorological properties needed for wind energy conversion system studies as:

- areas of smoothly accelerated or enhanced winds,
- locations of flow separation zones,
- mean hourly wind velocity distributions,
- characteristics of local turbulence and gusts,
- occurrences of extreme winds and calms,
- vertical profiles of wind velocity as a function of atmospheric stability,
- frequency of severe thunderstorms, lightning, hail, icing, tornados, or hurricanes,
- presence of salt spray or blowing dust.

Data of these types rarely exist in many areas that are considered most appropriate for wind energy systems development. For an evaluation of the continental scale wind energy resource, wind observations from the standard meteorological stations are available, although many of these observations are made at airports, which are generally not located in the more windy areas. Traditional wind data are also difficult to interpret in mountains and hilly and coastal regions.

Wind energy development involves three stages: initial evaluations, site selection and assessment, and machine design and performance. The initial evaluations generally can be made using standard meteorological observations. Data for site selection and assessment are more difficult to obtain because the density of measurements at scales less than 100 km is poor.

Wind variations over long periods of time must also be considered in planning systems. Hardy and Walton (1977) report that long-term (10-year) variations in wind energy conversion system output have been estimated from standard meteorological data from 15 locations in the United States. Power output was estimated by the hour over the 10-year period at each site and averaged. It was found that in addition to expected seasonal variations, significant year-to-year variations also occurred. Interannual variations of about 25 percent of the long-term mean were estimated at most locations. It was also estimated that minimum energy costs could be achieved at all sites with only a modest amount of energy storage to buffer changes in output.

Hardy and Walton (1977) show that in areas without enough observational data, the following steps are needed:

- collection of meteorological observations from multiple sources,
- use of a numerical model to simulate the windfield over terrain,
- statistical analysis of regional wind velocity patterns,
- coordinated application of the field measurement and statistical and numerical modeling effors.

A study incorporating these aspects has been conducted for the island of Oahu, Hawaii, by Hardy (1977a,b).

A methodology to provide accurate estimates of wind power at potential sites from data that are already available has been developed by Bhumralkar *et al.* (1980). The model, which is three-dimensional and based on physical equations, incorporates the effect of underlying terrain and uses available conventional wind data from selected nearby meteorological stations. The program is essentially an objective analysis that interpolates values of wind from observations at irregularly spaced stations.

Another approach to resource assessment is the investigation of joint wind/solar availability (Kahn, 1979; Aspliden, 1981). In this case, one investigates whether the wind blows when the sun does not shine and vice versa. Kahn shows the results of a study on joint wind/solar availability for a single site in Texas. At this location, the wind energy available at two heights was considered. It was found that the larger amount of wind energy available at the higher level produced excess power that was available only in the spring, whereas a smoother aggregate of solar and wind energy was available using wind energy conversion at a lower height. Kahn points out that it is difficult to generalize from such results and it is also necessary to look at smoothing effects, such as geographical dispersion. The relationship between aggregate energy availability and geographical dispersion has been investigated more extensively for wind energy conversion than for solar energy conversion (Kahn, 1979). For example, Kahn cites a study for West Germany which showed that at one site there was no wind energy output for 1500 hours per year. When three sites were considered there was no output for several hundred hours per year, and with 12 sites there was no output for less than 20 hours.

9.3.1.4 Biomass

Biomass is another renewable energy source that is influenced by climate. The overall prospects for the development of the use of biomass as a source of energy in Europe have been evaluated by Palz and Chartier (1980). They point out that the main climatic factors controlling plant growth in Europe are temperature and the length of the growing season, expressed as periods above

5 °C or numbers of frost-free days, and these factors rather than total solar energy received determine the units of productivity. The climate varies sufficiently within the countries of the European Community so that a number of geographic zones can be outlined in which different opportunities exist and different approaches will be necessary if biomass is to be produced specifically for energy. After regional climate, the factors affecting plant productivity are: local climate, soil, topography, and land use. Palz and Chartier (1980) have summarized the characteristics of a number of lowland and upland zones and indicated the most favorable biomass production opportunities arising in them. The main divisions were based on precipitation deficits and winter isotherms. It was concluded that the potentially available source of biomass energy which can be generated without effecting major changes in land use in the European Community represents 15 percent, in gross energy terms, of the projected demand by the Community in 1985.

9.3.2 Model Studies of Solar Energy System Performance

When detailed data assessments or established monitoring of solar heating systems are not practicable, models of solar heating systems and a reference house are useful for assessing the potential for solar energy systems in various climatic zones. The models usually simulate the performance of a solar system over a complete year. There are three major model types for simulating solar systems: finite-element models, component models, and approximative models. Jäger (1981) has used a component model in which the different components of a solar system are presented by individual differential equations and heat transfer between the components is considered. Weather data on an hour-to-hour basis are required in this model type. It has been found that the models are able to make performance evaluations with satisfactory accuracy when all model input data (climate and technical system characteristics) are well known.

Jäger (1981) has computed the performance of solar systems with 40 m^2 of single-glazed, non-selective collectors and 80 liters of store volume per m^2 of collector area. The solar systems were assumed to be built into a house insulated according to high Danish standards. The same house and solar system were considered for each country in the European Community. In this way the influence of climate on performance of the system could be isolated. The disadvantage of this approach is that the chosen reference system may not represent the optimum situation with respect to component dimensions and configuration of the system. Consequently, for some countries, the given results could underestimate the ultimately achievable solar system performance. The performance results were determined by using hour-by-hour simulation in the cases where suitably detailed climatic data were available (Belgium, Denmark, France, the Federal Republic of Germany and

Ireland). For those countries in the European Community where such climatic data were not accessible, the reference system performance was extrapolated from information in the literature or from data for adjacent countries with a comparable climate. Except for France, only one reference system was defined for each country. The comparison of the heating requirements in Table 9.3 suggests that the reference locations in Belgium, Ireland, Luxembourg and the United Kingdom have about comparable annual heating requirements. These countries all have a maritime climate with relatively warm and cloudy weather in winter. The contribution which the reference solar system makes to the heating supply ('solar energy share') is also comparable, except for Ireland where more favorable results were obtained from simulation. Jäger suggests that the better performance in Ireland is a result of a better matching of the heating requirement and solar energy supply, since in the maritime climate of Ireland the summers are still cool enough to require some heating that can be supplied by the more available solar energy in that season. Further east, towards a more continental climate, in central France and the Federal Republic of Germany and in the more northern latitude of Denmark, the heating requirements and solar energy share increase. In the Mediterranean areas of France and Italy a higher solar share can be obtained. However, since heating requirements decrease, the amount of fuel that can be substituted by solar energy does not increase in proportion to the solar energy share.

9.4 ENERGY SUPPLY AND DEMAND IN DEVELOPING COUNTRIES

In comparison to the numerous studies described above, relatively little attention has been given to the impact of climate on energy supply and demand in developing countries. The role and potential of renewable sources of energy were discussed, however, at the United Nations Conference on New and Renewable Sources of Energy in Nairobi in 1981. Shakow *et al.* (1981) have pointed out that in eastern and southern Africa, for example, only 5–25 percent of the total energy budget is derived from commercial fuel. They add that if draught animals, human power and passive solar power were included, this figure would be substantially reduced. Commercial fuel, therefore, is not the dominant part of the total energy budgets of oil-importing developing countries. Shakow *et al.* also point out that in eastern and southern Africa 50–60 percent of all commercial energy is used in transport and related activities, 20–30 percent is used in commercial and industrial production, and only 5–10 percent is used in agriculture. Therefore, attention must be devoted to traditional, non-commercial sources of energy in the developing countries. In eastern and southern Africa, for example, household energy consumption is about 75–80 percent of the total energy consumption. Eighty percent of this energy is used mainly for cooking, with firewood, charcoal, dung and crop residues accounting for most of the energy supply in rural areas (Shakow *et al.*, 1981).

Table 9.3 Performance and fuel saving characteristics of reference solar energy systems in countries of the European Community

	Federal Republic of Germany	France	Italy	Netherlands	Belgium	Luxembourg	United Kingdom	Ireland	Denmark
Annual space and water heating requirements, kWh	20 000	17 200	14 000	16 000	16 000	16 500	15 800	15 400	19 600
Solar energy share, %	45	48	75	46	46	45	45	62	49
Annual fuel saving in the auxiliary heating system									
oil, liters	1385	1270	1615	—	1132	1142	—	1469	1478
gas, m^3	—	—	—	1253	—	—	1215	—	—

Source: Jäger, 1981, 133.

Increasing pressure on traditional, non-commercial energy sources in developing countries can have the end result of deforestation. Energy demand and supply balances therefore must be planned within the framework of integrated land management, which involves taking the climate into account. The question of fuelwood supplies will be addressed in more detail below.

A few studies have been made of the potential of renewable energy sources in developing countries. For example, Shakow *et al*. (1981) have discussed the potential in Kenya and conclude that renewable energy technologies do not offer much hope in the short term. They point out that the total hydropower potential in Kenya will not meet the country's needs (500–550 MW) by 1990 and that although extensive exploration for mini-hydropower is underway, much of this potential (750 MW) is distant from the integrated grid system where demand is growing rapidly. The authors also conclude that alcohol production competes directly with food production for land, critically important in countries like Kenya where there is a severe food deficit. They also suggest that although the wind energy resource is poorly evaluated, the potential for economic wind energy utilization is small in both the commercial and rural sectors. Thus, the authors conclude that the renewable technologies have little to offer the developing countries in the near future, although such alternatives are a necessity for the longer term.

Hutchinson (1974) has evaluated the potential of renewable energy sources in Zambia. He concludes that electrical power generation from the wind has little potential because of the difficulties in storage of electrical energy. Most of the energy would be produced in midmorning when the wind reaches its diurnal peak, whereas the heaviest consumption time is in the early evening. The price of required battery storage was considered prohibitive, and mechanical and hydraulic storage also were discounted. Hutchinson points out, however, that windmills are suitable for the pumping of water from wells. The author further suggests that the use of solar energy for domestic water heating appears to hold some promise in Zambia, since the energy available from solar radiation is, in theory, quite considerable.

Revelle (1980) has examined the economic costs and benefits of four renewable energy sources in rural areas of developing countries. He indicates that irrigation demands can, within limits, accommodate the intermittent nature and unpredictability of solar radiation and wind energy. An alternative to the use of windmills for irrigation is direct photovoltaic conversion. In hilly areas where running water is available throughout the year and cultivated areas are closely spaced, locally generated hydroelectric power can be used. Lastly, internal combustion engines for pumping irrigation water can be fueled with biogas.

Revelle (1980) also examines in detail the role of biomass supplies in the future world economy. He estimates that the sustainable yield from the world's forests, if the entire forested area were subjected to intensive silviculture, would be 11.1 billion metric tons of wood per year. If all the harvested wood were devoted to energy production, the total energy available would be near 6 terawatts per year. Revelle estimates that one net terawatt of energy in liquid hydrocarbons could be produced from sugar cane grown on about 10 percent of the presently cultivated arable land. This method of converting biomass energy would be most useful in countries with a warm climate and abundant water and sunshine.

The use of biomass in rural Asia is illustrated by Revelle with examples from Nepal and Bangladesh. In Nepal, for example, a virtually closed farming system has developed over many centuries. Until recent decades, forests, common pasture lands, cultivated fields, domestic animals and human beings existed in stable equilibrium. With rapid growth of the human population, the system has become destabilized. Now the people are destroying the resources upon which their future survival depends. Revelle suggests that a radical alteration of the farming system is needed and would be possible if a relatively small amount of non-biomass energy could be made available. This could be accomplished, according to Revelle, in many villages by the construction of small (15 kW) run-of-the-river hydroelectric plants. The establishment of fast-growing tree plantations is also recommended.

The problems associated with the use of fuelwood and charcoal in Kenya have been discussed by Shakow *et al.* (1981) and O'Keefe *et al.* (1981). Shakow *et al.* describe the problem as follows: woodfuel is the most important energy source in Kenya (75 percent of all requirements). The wood comes mainly from natural woodlands and forests. These are being decimated to make more land available for agricultural production. Kenya cannot continue using woodfuel in the present manner unless adequate arrangements are made to renew this resource. Current annual consumption of woodfuel is estimated to be about 30 million cubic meters per year. The demand in the year 2000 is estimated to be at least 55 million cubic meters. An area of 2 million hectares will have to be planted before the end of the century. Shakow *et al.* suggest that emphasis should be given to urban greenbelts, small-holder high-potential farms, small-holder dry land farms, and settlement schemes. Such a focus would place emphasis on sustained production in the areas of greatest demand.

These examples show that studies of the interaction of climate and energy supply and demand in the developing countries require different emphases than those for the developed countries. In particular, it is necessary to consider the supply of and demand for non-commercial energy. The question of the supply of and demand for fuelwood requires integrated studies considering the multiple uses of land for energy and agriculture, both of which are influenced by climate.

9.5 CONCLUSION

The impacts of climate and weather on energy supply and demand have received increasing attention in recent years. This chapter has examined some of the studies that have been made to determine the impacts of climate on energy supply and demand. Virtually all of the studies reported deal with impacts in the industrialized countries of the northern hemisphere. This is probably a reflection of the fact that most of the global energy consumption is concentrated in these areas, which are both developed and climatologically cold.

There are three basic methods which have been used so far to study the impacts of weather or climate on energy demand: case studies, physical models, and empirical or statistical models.

The case study approach is useful in illustrating the response of energy supply and demand to climatic variability. This approach, however, cannot be used directly for the prediction of future impacts. An extension of the case study method, referred to as scenario development, is more applicable in the policy-making realm. Quirk (1981b) has suggested that energy systems planners could use scenarios of the geographical pattern, amplitude and timing of typical climatic anomalies to determine the optimum way to make energy systems resilient to climatic variations. There is a need, as Quirk has pointed out, to develop scenarios for climatic anomalies not only on a national but also on an international scale, since there is evidence that climate anomalies can occur simultaneously in various regions. For instance, Quirk quotes studies showing that 40 percent of all winters have a pattern in which most of eastern North America and Europe have similar departures of temperature from normal. Lastly, with regard to case studies, Quirk has found that the most important actions that can be taken to prepare for climate anomalies are those that require years of effort, such as improvement of fuel storage and distribution, and weatherization of homes. Thus, Quirk concludes that the improved knowledge of statistics of past climatic variations is far more important than improved seasonal forecasts.

Physical models consider the actual heat losses from one building or a number of buildings and compute the changes of these heat losses, and therefore of energy demand for space heating, as a function of changes in climatic variables such as outside temperature and windspeed. A number of quite detailed physical models have been developed and used to compute the energy demand for the space heating of individual buildings. For studies of energy demand for space heating on a regional or larger scale, statistical sampling and data aggregation are required, although the introduction of statistical techniques is usually at the expense of model accuracy and is compensated by the reduction of the amount of input data.

The main argument usually used against purely statistical models of energy demand is that the models are accurate only if physical structures, use patterns and comfort levels remain constant. The advantage of statistical models is that they can be used for areas where there are only a small amount of data. Two empirical models (Crocker, 1976; Nelson, 1976) have been developed that circumvent the argument against the use of historical data by using cross-sectional data. In each case, the data upon which the empirical model is based are derived, for example, from each of the conterminous states of the USA. It can be assumed that factors such as physical structures, use levels and comfort patterns vary from state to state and that the relationships derived using the cross-sectional data are valid for the long-term analysis of the impact of climate on energy demand.

To date, there has been little or no application of national energy demand models to the study of the impact of climatic variations. Since these models usually take climate into account by means of a degree day variable, it would be relatively easy to make impact assessments on a national and perhaps on a world regional level. An extension of this would be to introduce the climate variable into national or regional models of energy supply. The impact of climatic variations on the balance of energy supply and demand could then be studied.

The impact of climate on energy supply is mainly due to the role that climate plays in renewable energy supply. Renewable energy sources do not yet make a large contribution to the total global energy supply, thus no large studies of the impact of climatic variations have been made. Since it is likely, however, that the developing countries could profit from the further use of renewable energy sources, the study of the impacts of climate is important. On the first level, assessments must be made of the resource potential for such renewable energy sources as the sun, wind and hydropower. Assessments of the average magnitude of the supply and of its variations in time and space are necessary. Computer models are also useful for simulating the performance of renewable energy conversion systems and the dependence of this performance on climate. Since, however, renewable energy sources are diverse, it is likely that large-scale systems could provide more resilience against climatic variability. On the other hand, some climatic anomalies could affect more than one renewable energy source. For instance, Quirk (1981b) points out that India gets 40 percent of its electricity from hydropower and half of its total energy from biomass. Thus, a failure of the monsoon could increase the demand for food and energy imports. It can be concluded that nations using renewable energy systems could reduce their vulnerability somewhat by making use of a variety of geographically dispersed and diversified sources. Nations that depend on renewable energy resources need an awareness of the potential impacts of climatic fluctuations and climatic information for the design of systems that are both resilient and economic.

REFERENCES

Ahti, K. (1975). *Application of Meteorology to Problems of Transmission and Consumption of Energy*. WMO Rapporteur Report. World Meteorological Organization, Geneva.

Aspliden, C. I. (1981). *Hybrid Solar-Wind Energy Conversion Systems, Meteorological Aspects*. Energy and Special Applications Programme, Report No. 2. World Meteorological Organization, Geneva.

Balling, R. C., and Vojtesak, M. J. (1983). Solar climates of the United States based on long-term monthly averaged daily insolation values. *Solar Energy*, **31**, 283–292.

Beltzner, K. (Ed.) (1976). *Living with Climatic Change. Proceedings of Toronto Conference*. Science Council of Canada, Ottawa, Canada.

Bhumralkar, C. M., Mancuso, R. L., Ludwig, F. L., and Renné, D. S. (1980). A practical and economic method for estimating wind characteristics at potential wind energy conversion sites. *Solar Energy*, **25**, 55–66.

Burroughs, W. (1978). Cold winter and the economy. *New Scientist*, **77** (19 January), 146–148.

Cohen, S. J. (1981). *Climatic Influences on Residential Energy Consumption*. Dissertation. Department of Geography, University of Illionois, Urbana-Champaign, Illinois.

Commission of the European Communities (1979). *Atlas of the Solar Radiation in Europe*. Vol. 1: *Global Radiation on Horizontal Surfaces*. W. Grösschen Verlag, Dortmund, Federal Republic of Germany (in German).

Crocker, T. D. (1976). Electricity demand in all-electric commercial buildings: The effect of climate. In Ferrar, T. A. (Ed.) *The Urban Costs of Climate Modification*. John Wiley & Sons, New York.

Diaz, H. F., and Fulbright, D. C. (1981). Eigenvector analysis of seasonal temperature, precipitation and synoptic scale system frequency over the contiguous United States, Part I: Winter. *Monthly Weather Review*, **109**, 1267–1284.

Diaz, H. F., and Quayle, R. G. (1980). An analysis of the recent extreme winters in the contiguous United States. *Monthly Weather Review*, **108**, 687–699.

Hardy, D. M. (1977a). Numerical and measurement methods of wind energy assessment. In Vol. 2, *Third Energy Workshop: Proceedings of the Third Biennial Conference and Workshop on Wind Energy Conversion Systems, September 19–21, 1977, Washington, DC* (CONF 770921/2), pp. 664–676. US Government Printing Office, Washington, DC.

Hardy, D. M. (1977b). Wind studies in complex terrain. In *Proceedings of the American Wind Energy Association Conference and Exposition, May 11–14, 1977, Boulder, Colorado* (UCRL-79430). Lawrence Livermore National Laboratory, Livermore, California.

Hardy, D. M., and Walton, J. J. (1977). Wind energy assessment. In Veziroglu, T. N. (Ed.) *Proceedings of the Miami International Conference on Alternative Energy Sources, December 5–7, 1977, Miami Beach, Florida*. University of Miami, Coral Gables, Florida.

Hörster, H. (1980). *Paths to the Energy-Saving Home*. Philips GmbH, Hamburg, Federal Republic of Germany (in German).

Hutchinson, P. (1974). *The Climate of Zambia*. Occasional Study No. 7, Zambia Geographical Association, Lusaka.

Jäger, F. (1981). *Solar Energy Applications in Houses: Performance and Economics in Europe*. Pergamon Press, Oxford.

Kahn, E. (1979). The compatibility of wind and solar technology with conventional energy systems. *Annual Review of Energy*, **4**, 313–352.

Lawford, R. G. (1981). Impacts of the climate of 1980 on the energy demand/supply cycle. In Phillips, D. W., and McKay, G. A. (Eds.) *Canadian Climate in Review, 1980*. Environment Canada, Atmospheric Environment Service, Ottawa.

McKay, G. A., and Allsopp, T. (1980). The role of climate in affecting energy demand/supply. In Bach, W., Pankrath, J., and Williams, J. (Eds.) *Interactions of*

Energy and Climate, pp. 53–72. D. Reidel Publishing Company, Dordrecht, Holland.

Mitchell, J. M., Felch, R. E., Gilman, D. L., Quinlan, F. T., and Rotty, R. M. (1973). *Variability of Seasonal Total Heating Fuel Demand in the United States*. Report to the Energy Policy Office, Washington, DC.

National Oceanic and Atmospheric Administration (NOAA) (ongoing). *Climate Impact Assessment United States* (a monthly bulletin). Assessment and Information Services Center, NOAA, US Dept. of Commerce, Washington, DC.

Nelson, J. P. (1976). Climate and energy demand: Fossil fuels. In Ferrar, T. (Ed.) *The Urban Costs of Climate Modification*. John Wiley & Sons, New York.

O'Keefe, P., Weiner, D., and Wisner, B. (1981). The tail that wagged the dog: A cautionary story of forestry planning in Kenya. In Buck, L. (Ed.) *Proceedings of the Kenya National Seminar on Agroforestry, November 12–22, 1980*. International Council for Research in Agroforestry, Nairobi.

Palz, W., and Chartier, P. (1980). *Energy from Biomass in Europe*. Applied Science Publishers, London.

Pimentel, D. (1981). Food, energy and climate change. In Bach, W., Pankrath, J., and Schneider, S. H. (Eds.) *Food–Climate Interactions*. D. Reidel Publishing Company, Dordrecht, Holland.

Quirk, W. J. (1981a). Climate and energy emergencies. *Bulletin of the American Meteorological Society*, **62**, 623–631.

Quirk, W. J. (1981b). *Energy Supply Interruptions and Climate*. UCRL-86254, Rev. 1. Lawrence Livermore National Laboratory, Livermore, California.

Quirk, W. J., and Moriarty, J. E. (1980). Prospects for using improved climate information to better manage energy systems. In Bach, W., Pankrath, J., and Williams, J. (Eds.) *Interactions of Energy and Climate*, pp. 89–99. D. Reidel Publishing Company, Dordrecht, Holland.

Reiter, E. R., and colleagues (1976, 1978, 1979, 1980, 1981). *The Effects of Atmospheric Variability on Energy Utilization and Conservation*. Environmental Research Papers Nos. 5, 14, 18, 24, 31. Colorado State University, Fort Collins, Colorado.

Revelle, R. (1980). Energy sources for rural development. In Bach, W., Manshard, W., Matthews, W. H., and Brown, H. (Eds.) *Renewable Energy Prospects*. Pergamon Press, Oxford.

Riches, M. R., and Koomanoff, F. A. (1978). The national insolation resource assessment program: A status report. Paper presented at the *American Meteorological Society Conference on Climate and Energy: Climatological Aspects and Industrial Operations, Asheville, North Carolina*.

Robinson, P. (1974). Evaluation of air conditioning energy costs. *The Building Services Engineer*, **42**, 195–198.

Shakow, D., Weiner, D., and O'Keefe, P. (1981). Energy and development: The case of Kenya. *Ambio*, **10**, 206–210.

Warren, H. E., and LeDuc, S. K. (1980). *Impact of climate on energy sector in economic analysis*. Paper presented at the Conference on Climatic Impacts and Societal Response, Milwaukee, Wisconsin.

Won, T. K. (1980). Climate and energy. In *Socioeconomic Impacts of Climate*. Northern Forest Research Center, Alberta, Canada.

Part III
Social and Economic Impacts and Adjustments

As the impact analyst moves from first-order biophysical impacts to higher-order consequences, the possibilities, outcomes and human choices attached to each link increase, and the causal chain becomes less distinct. Part III reflects that complexity. Three of the chapters, written from disciplinary perspectives, deal with second-order consequences on human health (Chapter 10), economy (Chapter 12), and society (Chapter 13). Three of the chapters deal with case study methods to assess so-called 'natural experiments', focusing on historical study (Chapter 11), climatically and economically marginal places (Chapter 14), and extreme weather and climate events (Chapter 15). Finally, two chapters deal with adjustment responses and mechanisms for their perception and choice (Chapters 16 and 17).

When studying self-provisioning societies it is convenient to think of human nutrition and related health effects as second-order consequences—climate-related yields of food leading to various health and demographic impacts. But there are few fully self-provisioning societies, therefore biological increases or decreases in the availability of foodstuffs must be traced through a network of existing social and economic relationships. Escudero's proposal for a case comparison of both shores of the Windward Passage is a case in point.

Economic and social relationships are the substance of the social sciences, particularly economics and sociology. Within the framework of market economies, but with some applicability to all economies, there are a set of well-defined quantitative approaches designed to answer two fundamental questions of how economies interact with climate. Given a change in either the mean or distribution of climate events, how will the allocation of resources change and which persons or places will lose or gain from such resource changes? As Lovell and Smith point out in Chapter 12, robust methods exist to answer these questions, but rarely have been applied. Nor have they been adapted to the special qualities of climatic variability and change: stochastic nature, large-area impact, and long time-horizon over which consequences take place.

Ironically, the more diffuse, less-defined social impacts appear to be better illustrated, with interesting and recent case examples from studies of weather modification and extreme events. With a broader view than economic analysis, social impact analysis, as Pilgrim notes in Chapter 13, is a class of policy analysis that arose from a concern with the 'hidden costs' of societal undertakings. Thus there is a strong emphasis on identifying the many different stakeholders affected by climatic variability or change and assessing the differential impact upon them.

Historical analysis, de Vries informs us in Chapter 11, employs the full array of social science methods and is limited only by the availability of data from the past. Climate adds to the rationale of historical explanation of human events, and historical events allow us to expand the stock of relevant climates and societies to examine. (See also the extensive discussion in Chapter 21.) In the latter case, historical analysis is often employed for the two types of natural experiments that are emphasized in this volume—a focus on vulnerable margins or groups and a focus on extreme climatic events, usually of interannual or decadal length.

Natural hazard research, described by Heathcote in Chapter 15, provides a rich body of relevant methodology to study the extreme events of the past as well as those of current experience. To illustrate, Heathcote reviews such methods within the ordered sequence of impacts and the interactive model of human response. The immediate dramatic impacts of violent storm or persistent drought are most easily identified and measured; the long-term impacts, however, are much in doubt. The margins of climatically sensitive activities that Parry documents in Chapter 14 seem to be more sensitive barometers of the impacts of longer-term changes in climate.

Woven throughout Part III are specific issues of adjustment and adaptation. Although some studies of biophysical impact attempt to ignore or to constrain societal interaction, studies of social and economic impacts are always interactive, analyzing the differential societal impacts of climate change and variability in the light of the differential ability to cope with or take advantage of such change. Indeed, several authors point out that the differential between societies and their resources is much greater than the differential between climatic regions or epochs in their impact on human beings.

At the level of individual and small-group decision-makers, studies of perception, alluded to in Chapters 11 and 15 and given full treatment by Whyte in Chapter 16, have served as a major link to studies of adjustment of the type described by Jodha and Mascarenhas in Chapter 17. Reports of the nature of adjustment to climate change and variability are scattered throughout the text. Lists of adjustments are given in most of the sectoral chapters. Given the bias towards industrialized nation experience, however, Jodha and Mascarenhas examine specifically developing country adjustment, drawing on their rich experience in South Asia and Africa.

When this set of disciplinary methods and case study opportunities for identifying human social and economic impacts is compared, a strong negative bias emerges. Most methods and case studies focus on climate as hazard; only scattered efforts have been made to study climate as a resource. Within the focus on climate as hazard, the balance of effort has been to identify the residual damages and losses caused by the impact of climate events on vulnerable groups or regions; less effort has been expanded on identifying and measuring the social cost of adaptation and adjustment. Assessing the social cost of climate adjustment and the opportunities of climate as a resource are important items for a research agenda.

Climate Impact Assessment
Edited by R. W. Kates, J. H. Ausubel and M. Berberian

CHAPTER 10
Health, Nutrition, and Human Development

JOSÉ CARLOS ESCUDERO

Department of Social Medicine
Universidad Autónoma Metropolitana
Xochimilco-México
Apartado Postal 22-444
México 22, D.F.

10.1 INTRODUCTION

The climate that we know, which has slowly developed over millions of years, is in general terms optimal for human life. It ought to be, as life (one of whose end products is human life) has developed, again over millions of years, within the constraints set by climate upon it.

Dialectically, life has also changed climate, as exemplified by the appearance of free oxygen in the biosphere through the action of photosynthesis. The subsequent appearance of that peculiar life form, man, has provided many more examples: changes in albedo, in vegetation and in water distribution, and industrial and other pollution—all of which modify climate.

Humans are an efficient and complex life form, but these very qualities produce in them great vulnerability to factors which alter their habitat beyond a very narrow range within which their demands are optimally satisfied. The fact that the human body is kept within a very narrow temperature range, for example, greatly increases its energy demands, but provides the background for a highly developed physiology, and offers a much more efficient use of energy through an oxidization process of high complexity. These high energy demands are a handicap for individual and group survival, and so is the outside temperature range within which humans must live, although this latter factor has probably been overestimated, as Darwin reflected when he observed the Fuegian Indians in the Beagle Channel.

When it comes to a discussion of climatic impacts, few, if any, societal effects can surpass in value, measured in whatever form, the chance to live and be healthy. Death is generally acknowledged to be the worst fate that an individual can suffer, and that one can inflict upon others; and the sensations of illness are among the most salient that an individual can feel.

This paper seeks to explore the most profound ways in which climate affects human life—by ending it in individual cases and by reducing it in aggregate below what can be expected from genetic potential—and human health, by increasing illness episodes beyond an acceptable minimum, and by making the outcome of some illnesses death, rather than recovery.

There are three general, related categories in which climatic factors manifest themselves in human health and wellbeing: through direct effects on individual health, through population movements and behavior, and through effects on diet and nutritional level.

With respect to the first category, consequences of climate can be such phenomena as sunburn, heat stroke, heat cramps, changes in sleep patterns, frostbite, drowning, and a host of other conditions, all of which can produce incapacitation, and some of which can eventually result in death. The literature on health impacts of climate (biometeorology) has usually focused on these immediate effects. Weihe (1979) has published a thorough review (see also Landsberg, 1984). It is clear from these reviews that climatic factors, defined this way, produce relatively few deaths and illnesses. For example, 1181 and 618 deaths were attributed to the severe summer heat waves of 1966 in New York and St Louis, respectively (Schuman, 1972). The method used to make calculations probably overestimates the role of climatic factors, as it assigns to climate all the deaths that exceeded a historical baseline. The populations at risk were about 8 million in New York and about 700,000 in St Louis, and most of the victims were elderly.

Climatic impact, of course, can be measured through human effects other than death or illness, for example, by indicators of migration or social disruption. Political upheaval, riots, and increases or decreases in crime rates may be associated with climatic extremes, although we have to leave aside the extremely value-laden question of whether these consequences are 'good' or 'bad'. Substantial movements of population may also be triggered as people abandon their homes and seek new locations, value judgments again being suspended here. These climate-induced political, behavioral and migratory outcomes have been explored as part of research on CO_2-induced climatic change (Meade, 1981) and under the heading of climate and human history (Rotberg and Raab, 1981).

This paper will not deal with these two categories of climatic impacts on human health. This is partly because the questions have been heavily examined, at least in the first case. More so, it is based on the judgment of relative importance of impacts. The climate-related mechanism which affects mortality and morbidity most is the drought-induced shortage of foodstuffs, which leads to malnutrition and a host of malnutrition-related conditions. Storms, floods, snow, lightning, and frost can be spectacular, but the human loss and illness that they cause is typically small* compared with that caused by chronic malnutrition and acute starvation. The critical links to explore are not the direct ones between climate and health, but the chain extending from lack of water for growth of food to inadequate diet, which produces in humans a vulnerability that makes them prey to infections and parasitic diseases of various sorts and stunts body growth. Compare the losses from a heat wave with the certainly more than 100,000 deaths—we will never know the exact figure—mostly of children, associated with the drought during 1973 in Chad, Mali, Mauritania, Niger, Senegal and Upper Volta, where the total population at risk was only about 15 million. Demographers can easily calculate the number of 'life years' lost when an infantile or an elderly population is struck, and the mechanism that we are stressing is selective in striking the former group.

Drought can be seen as a long-term phenomenon or a seasonal one. The repetitive pattern of seasonality would assign to it a label of 'normality', which makes its effects appear less salient than 'abnormal' droughts. Seasonality has been studied as a generator of death, illness and human suffering in various places (Chen *et al.*, 1979; Chambers, 1981; Wisner, 1980–81) and it also produces hazards other than those due to drought—those of the 'wet season' being caused by increased energy demands for agricultural production, less care of children and increased morbidity (Chambers, 1982). In any case, as we

* The two great twentieth-century exceptions to this statement are Indian and Bangladeshi experiences with typhoons, and Chinese experiences with floods. A 1970 typhoon caused perhaps as many as a quarter of a million deaths in Bangladesh and India (Wisner, 1978); one in 1977 in India caused 100,000 deaths (Whittow, 1980). The 1931 Hwang Ho floods in China may have caused as many as 3,500,000 deaths (Whittow, 1980).

shall see later, the development of productive forces would enable humans to do away with the consequences of either seasonality or long-term drought.

Ultimately, the underlying question that has to be answered in looking at the link between climate and human health is 'Is it climate, or is it society, or what combination of both that we are seeing?' The way to explore this question is through historical and cross-cultural comparison. Specific climatic perturbations of identifiable magnitude can be studied in their impacts on societies at different stages of development of their productive forces, and on societies which use different criteria in the allocation of their collective surpluses.

The general argument or hypothesis is that climatic damages to life and health decrease as the world's productive forces increase, as man's mastery over his environment improves. At the broadest level, the evidence for this line of argument is the increase in the world's human population, from 2 to 10 million around 10,000 BC, to about 750 million in 1750, about 1200 million in 1850, about 2500 million in 1950, and about 4500 million in 1980. Corresponding to this general increase in population is an increase in life span, to 60 years in the most advanced countries of the world by the turn of the twentieth century, and over 75 years today in the best circumstances.

The spatial diffusion of the human population is further evidence. Our ability to colonize almost the whole of the planet and to visit other celestial bodies is due to the ability to carry with us that particular range of environment in which we can survive and develop capabilities. Compared with most other mammals, humans have a low fertility, their offspring are helpless for a long period, and they reach sexual maturity relatively late. Thus, reproduction is a weak link in the process of maintenance of the human species. Yet, humans, through their mastery of nature, have derived an enormous strength beyond their biological capabilities. Control over the temperature affecting the species through the use of clothing, shelter, heating, cooling, ventilation, and a host of other measures allows humans to experience thermal variation under conditions which are usually quite favorable. Certainly air temperature, once an important determinant of our habitat and probably second only to the need for finding nutritional energy, now plays a minor role in constraining human activity.

Thus, the overall picture for human society appears to be one of strengthened human hold on the planet and decreased vulnerability to environmental factors, other than those we ourselves create or intensify. Why then is there the continuing spectacle of massive mortality and morbidity 'due' to droughts and other climatic anomalies? Why is the development of productive forces an uneven bulwark against climatic variability and change?

This paper seeks to establish tentative answers to these questions and to lay the basis for more careful argument in the future. The presentation is divided into four parts. The first section addresses the question of what exactly is the number of people who die or are taken ill 'due' to drought or other climatic phenomena. The second section sketches a case study, of comparative impacts of climate on health in Cuba and Haiti, which could shed light on the larger issues. The third section evaluates the capacity of the world food system to counter climatic aggression. The fourth section offers proposals for improvements in the measurement of climatic impact on life and health.

10.2 MEASURING MORTALITY, MORBIDITY, BODY GROWTH, AND PSYCHOMOTOR DEVELOPMENT

To estimate the impact of climatic phenomena, it is necessary to have a certain 'baseline' of data against which specific climatic impacts can be compared. With respect to health, the fundamental indicators are mortality and morbidity, along with measurement of body growth (a fairly specific and quite simple indicator of what must be the most widespread health consequence of climatic anomalies: drought-induced malnutrition) and psychomotor development. Reliable historical trends for these phenomena are very scarce, cover a very small percentage of the world's population, and should in any case be treated with much caution.

The phenomenon of mortality, which is the most 'exact' of the four indicators (death being a one-time, easily definable and highly salient circumstance), has fairly reliable quantitative series for some countries of northern and western Europe since about 1800. The description of cause of death, which is very important, as it should help to separate climate-related deaths from the rest, depends heavily on the state of medical knowledge and the bias of medical culture, both subject to variation over time. With advances in the structuring of modern states, reliable series have appeared for an increasing number of countries. For example, the US series are reliable since about 1900. The following review of the current status of registration of deaths worldwide thus also gives an idea of what was happening even in the most advanced countries of the world some generations ago.

Morbidity is much more difficult to assess, due partly to the potentially repetitive and much more ambiguous and culturally defined nature of the phenomenon. A good registration would imply a high medical coverage of the population, which does not exist now for most of the countries of the world. As an example, one-third of the population of Brazil today has no access whatever to medical care. In the area of morbidity, medical bias will also play a role.

Body growth is quite problematic, and psychomotor development even more so. It is not complicated to measure body parameters such as height, weight, and head, arm, leg circumference, and various psychomotor indices. Some physiological indicators of nutrition (onset of menstruation in women, for example) are easily measured or inquired about, and the results plotted against the chronological age of the subjects. Small size and degrees of psychological retardation can be taken as indicators of malnutrition. Unfortunately, measurements have to be undertaken on some sort of probabilistic population sample, for which techniques of evaluation are fairly recent, or in conjunction with a population census which, if it exists, is rarely up-to-date in a drought-stricken country, and which does not usually gather this type of information. There is also a theoretical question with regard to psychomotor development which compounds the problem: what constitutes a 'normal' development, against which deviance can be plotted?

10.2.1 Current Worldwide Measurement

The current worldwide measurement of the parameters of mortality, morbidity, body growth, and psychomotor development is not encouraging. There are standard demographic and epidemiological reference books published by various international agencies in which mortality and morbidity data are collected,* but these bulky and impressive tomes, with table upon table filled with rows and columns of numbers, must not be taken at face value. Much of the information presented in them is inaccurate, and it is more inaccurate in those parts of the world where climate would seem to impact most upon human populations. This relationship, which is a causal one, will be developed later on.

This sorry situation is not the fault of the international agencies which present the data. They are at the mercy of national governments which collect the data and transmit them to the agencies. For mortality the sources of these data are usually the country's records of vital statistics, and for morbidity, the country's health care systems. Little systematic data for body growth is currently being published.† The existing data are, furthermore, flawed by a selective underregistration of deaths (births are much more accurately registered) and by a massive underregistration of prevalent illnesses on the part of health care systems which have a very low coverage of population. Prompt improvement in this situation seems unlikely, and indeed the opposite phenomenon seems to be taking place: the current and deepening crisis of the world's economy is 'underdeveloping' many countries, with direct consequences in their data collection systems (Escudero, 1980).

Quality of statistics is related to a country's overall social fabric, which cannot change rapidly even in the best of circumstances. There are also sometimes political implications from an improvement in the quality of health

* These are: the *Demographic Yearbook*, United Nations, New York; and three World Health Organization (Geneva) publications: the *World Health Statistics Annual*, the *World Health Statistics Quarterly* (until 1976 the *World Health Statistics Report*), and the *Weekly Epidemiological Record*.

† Cuba appears to be the only Third World country that systematically publishes body growth data with nationwide coverage (see Jordán, 1979).

Table 10.1 Comparison of infant mortality rates in various Latin American countries provided by official statistics and by alternative methods

Country	Period	Official infant mortality rate (per thousand live births)	Infant mortality rate by other methods* (per thousand live births)
Argentina	1966	54	54
Bolivia	1971–72	—	161
Chile	1970	79	79
Colombia	1968–69	70	100
Costa Rica	1968–69	65	70
Cuba	1970	36	38
Dominican Republic	1969	62	110
Ecuador	1969–70	81	104
El Salvador	1966–67	63	118
Guatemala	1968–69	92	110
Honduras	1970–71	36	115
Nicaragua	1966–67	43	126
Paraguay	1967–68	44	64
Peru	1965	74	153

* From various 'indirect methods'. See Behm *et al.* (1976, 1977).
Source: Escudero, 1980, 423.

care figures; for example, the infant mortality rate in Nicaragua in 1966–67 was not 43 per thousand but approximately 126 per thousand; El Salvador's infant mortality rate at the same time was not 63 per thousand but approximately 118 per thousand. The former figures were gathered from Vital Statistics, reported by the countries' governments and transcribed in the relevant United Nations Demographic Yearbooks; the latter were calculated using alternative demographic methods for mortality data collection and analysis (Behm *et al.*, 1976, 1977)—the 'indirect methods'.*

Fortunately, the last decade has produced great advances in the study of mortality and fertility. The already mentioned 'indirect methods' have produced estimates of mortality, especially infant mortality, which for many countries revise upward drastically the figures published in the international reference books. Examples of this revision for Latin American countries, which are by no means those with the worst statistics, are shown in Table 10.1. For the countries whose data are published in the reference books, one must take into account a substantial underregistration of impacts, climatic or otherwise. In the case of malnutrition the situation is especially serious, and those population groups which suffer most (infants, young children, the rural

* For examples of the use of these, see Brass *et al.*, 1968; Sullivan, 1972; and Brass and Hill, 1973.

population) are those for whom the quality of data is worst. Overall, a significant proviso has to be made: if the structural factors which cause the poor quality of the data are more or less stable, the data can still show trends and can indicate, qualitatively if not quantitatively, the character of a given impact.

In terms of knowledge of mortality, Latin America is generally better off than Asia or Africa, where climatic impacts may be presumed to be more severe and where data usually range from very poor to nonexistent. An attempt to appraise the impact of the Sahel-Ethiopian drought of 1972 in terms of morbidity and mortality showed that the area of uncertainty was enormous. The dictum that the bigger the problem the worse the data, and that the worst problem might not produce any data whatever, was exemplified by the way the Ministries of Health of the countries concerned viewed the famine in their midst. The 1972 'Rapport sur l'activité des services de santé' of the Ministère de la Santé Publique et des Affaires Sociales of the Republic of Niger made no reference whatever to malnutrition. The Republic of Mali's health report for 1974 did not mention malnutrition (Imperato, 1976). The morbidity statistics of the Republic of Chad in 1974 quoted malnutrition as being 1 percent of the pathology observed, as reported in the *Annuaire du Tchad*, Vol. 1, Chapter 4, 'La Santé'. For a further discussion of this, see García and Escudero (1982).

The worst case with regard to mortality data would be no data at all, although it might be argued that complete absence of data at least does not mislead the investigator. The latest *World Health Statistics Annual* (1980) has no data on causes of death and no rates for such countries as China, India, Pakistan, Indonesia, the Philippines, Brazil, Bolivia, Haiti, Nigeria, Egypt, Sudan; in total for about two-thirds of mankind. In a review of most Asian countries, it was estimated that only five have 'reliable' vital registration statistics, including determination of cause of death: Hong Kong, Japan, Singapore, the Philippines and Thailand (Arriaga, 1979). Even in these five, the Philippines and Thailand have a high degree of underregistration of deaths, and a low quality of determination of cause, as shown by the fact that 50 percent of Thai deaths (themselves underregistered) are coded of 'ill defined' cause.

Records of morbidity are similarly troubling. The vital and health statistics of a country with good records by Third World standards, Mexico, showed that from 1967 to 1972, 14,725 deaths were reported from tetanus, while morbidity records showed only 5522 notifications of the same disease (Briceño *et al.*, 1979). The omission percentage of notification of measles, an easily diagnosable disease, ranged from 64 percent in 1971, through 63 percent and 50 percent in 1972 and 1973, to 64 percent in 1974 (Crevenna, 1978). The situation of most countries whose data are published in the reference books is even worse than this. As is the case for mortality, most of the countries of the world do not publish and basically do not know their prevalent morbidity.

Body growth, which is a simple and fairly reliable indicator of both chronic malnutrition and acute starvation, especially in children, is the least complete of the three indicators for which some bulk data are available. No data of this type are systematically published.* Nutritional assessments can serve as a proxy in certain cases, but these are carried out on an extremely unsystematic basis, with methodological discrepancies.

In summary, we know a great deal about the two thousand-odd lives lost in the Netherlands floods in 1953, or the few dozen deaths in the US heat wave of 1979; we know much less about the several tens of thousand lives, give or take a few thousand, lost every decade in typhoons in Bengal. Of the 1976 drought in Europe we can say with certainty that no lives were lost; we have a hazy notion that a few hundred thousand deaths took place during the Sahel and Ethiopian drought of the early 1970s.

10.2.2 The Problem of Recording Causes of Death

For countries that publish records including cause of death, a significant bias in registration appears. Some climatic phenomena, such as floods and heat waves, produce straightforward causes of death—drownings and heat strokes, respectively. Drought, in contrast, is mediated by malnutrition, which leads to subtle forms of bias. These pertain to the routine for selection of cause of death and to the view of malnutrition which is dominant among the medical profession. It is important to keep in mind that the medical profession is charged with the responsibility for recording death and has a large hand in designing the methodology regarding it.

Deaths worldwide are assigned a 'basic cause' through the use of the 'Rules for Selection of Basic Cause', which are stated in the Ninth Revision of the *International Classification of Diseases* (ICD). These selection rules are applied when more than one cause of death appears on the death certificate, a fairly common circumstance. Malnutrition hardly ever acts alone; it usually coexists with an infectious process or with a parasitic disease, and in any case is not easily perceived by physicians. When malnutrition and an infectious-parasitic disease both appear on a certificate, the relevant rule for 'selection of basic cause' of the Ninth Revision of the ICD (currently in use, with a preliminary revision scheduled for 1984 and no implementation of changes scheduled) states that only the infectious-parasitic disease and not malnutrition should be coded. Thus, malnutrition enters the limbo of non-events. The causes for this curious bias of the ICD, which goes against

* Cuba would appear to be an exception to this. For a compilation of body growth data, see Eveleth and Tanner, 1976.

current knowledge of the malnutrition–infection interaction, lie partly in the convenience of simple explanation and partly in the training of medical professionals, who prefer to explain the world in terms of anatomy and microbes than in terms of food or of sociopolitical systems.

Another subtle form of bias affects both physicians and laymen and makes for further underestimation of both mortality and morbidity caused by malnutrition. Perception of illness implies some sort of deviance from the normal. For example, we would not be so complacent with depression among the elderly if this condition were not so prevalent. Similarly, the spectacle of children with stunted growth and bloated bellies, or of emaciated adults, is 'normal', i.e., usual, with the massive chronic malnutrition which usually accompanies prolonged droughts, and is therefore not remarkable. So, it is rarely noted on death certificates. For an expansion of the above arguments, see Kotliar and Escudero (1974), Escudero and Kotliar (1975), Escudero (1978), Escudero (1980), Sabelli (1981), García and Escudero (1982), and Escudero (1984).

The magnitude of this underestimation of malnutrition as a cause of death by 'official' health statistics systems can be considerable. For Mexico, circa 1975, it was estimated that malnutrition caused approximately 70,000 deaths annually, whereas that country's vital statistics (as reported in the *World Health Statistics Annual 1973–1976*) gave a figure of 6498 deaths for 1973 (Escudero, 1984).

10.3 BOTH SHORES OF THE WINDWARD PASSAGE: A CASE COMPARISON

The Windward Passage is a narrow strip of ocean which separates Cuba from Haiti. At its narrowest point it is 90 km wide. Climate on the two sides of the Passage is fairly similar; droughts, floods, hurricanes, and other climatic disturbances occur on both shores of the Passage with more or less the same frequency and intensity. Yet the Windward Passage separates the greatest mortality and morbidity differential in the Western Hemisphere. Proving this point, nonetheless, involves certain difficulties. As mentioned above, where health is bad indeed, there are few or no statistics to prove the fact.

Cuban health statistics were poor prior to the 1959 revolution. For the period 1955–60 it has been estimated that underregistration of deaths was approximately 45 percent (Alvarez Leiva, 1964). Coverage of health services was very spotty and was centered on the cities of Havana and Santiago de Cuba. Beginning about 1963, great improvements started to appear in the vital and health statistics of Cuba, illustrating again that demographic and health statistics are rarely healthier than the social fabric that produces them. A very significant reduction of illiteracy and unemployment, and a great expansion in the country's health services, produced as a byproduct great improvement in

Cuba's vital and health data. A health care system was developed with total coverage of the population: 98.3 percent of births took place in hospitals in 1979 and the number of prenatal consultations per birth was 11; there were 4.5 annual medical and 0.9 dental consultations per inhabitant in 1979 (Ministerio de Salud Pública, República de Cuba, 1974, 1980; Puffer, 1974; Rojas Ochoa and Sánchez Teixidó, 1977; Ríos Massabot, 1981). These standards are comparable with those in the richest countries of the world.

What the Cuban data show now is a demographic and morbidity profile that is no different from that of European countries, Canada, Australia, or the United States: for 1980 a crude birth rate of 14.1 per thousand, an infant mortality rate of 19.6 per thousand (Ministerio de Salud Pública, República de Cuba, 1981); a life expectancy at birth of 73.5 years (*World Health Statistics Annual, 1980*); the virtual eradication of mortality due to diphtheria, measles, whooping cough, tetanus, malnutrition, and meningitic tuberculosis; and the eradication of malaria, polio, neonatal tetanus and human rabies. The so-called diseases of civilization—cardiovascular, cancer and suicide—have come to the fore. Cuban mortality is the lowest in the Western Hemisphere excepting the United States and Canada.

Haitian vital and health statistics are virtually nonexistent, but from various disparate sources a picture can be built of a 'constant catastrophe'. The Haitian birth rate is estimated at 37 per thousand. Haitian mortality appears to be the highest in the Western Hemisphere. The infant mortality rate has variously been estimated at 98–300 per thousand; life expectancy is estimated to be 50 or 51 years (King, 1978). As to nutrition, a survey undertaken by the US Agency for International Development with the Bureau de Nutrition d'Haiti in June–September 1978 showed that 73 percent of children were malnourished, of which Grade 3 (a very serious condition) accounted for 3 percent (Graitcer *et al.*, 1980). It has been estimated that 400,000 children under age 5 may be in the second or third degrees of malnutrition (Titus, n.d.).

Compare the situation in the Nord region of Haiti with that of Guantanamo Province, Cuba, 90 km away. The Haitian region comprises the provinces of Nord-Ouest and Cap-Haitien. No separate mortality or mobidity statistics for this region exist, but different malnutrition surveys have shown a high prevalence of malnutrition among children: Grades 2 and 3 malnutrition accounted for 17.5–35.2 percent and 5.6–16.3 percent, respectively (Titus, n.d.).

Taking Haiti as a whole, 60 percent of all deaths appear to occur among children of less than 5 years of age (King, 1978). A comparable percentage for the province of Guatanamo, Cuba, is 14 percent, where the infant mortality rate (well recorded) is of 24.1 per thousand (Ministerio de Salud Pública, República de Cuba, 1981).

These two settings, geographically close and socially distinct, provide a laboratory for the study of climate's relation to human health. How would climatic phenomena of the same magnitude affect life and health in Cuba and Haiti? It would be interesting to compare, for example, the impact of an 'average' hurricane on both sides of the Passage, to examine the relative quality of the emergency response apparatus, the possibility of evacuating population and livestock, the preparedness of the health care system in an emergency situation, durability of housing, capacity for delivery of relief to the affected population, provision of food and shelter, and so forth. Such a study, which would constitute a particular instance of methodology to study climatic impact, would be likely to show damages in Haiti much higher than those in Cuba, and societal response much weaker. It might remove from 'climate' much of the blame that it usually gets in naive approaches to climatic damage assessment. A study of comparative malnutrition associated with drought would likely lead in a similar direction.

Just as cross-cultural studies can shed light on the mechanisms which connect climate to human health and wellbeing, so can historical studies within a culture. Warrick (1980) has examined a hypothesis of gradual 'lessening' of the impacts of drought in the Great Plains of the United States. Additional interesting cases might include Ireland, where the subsistence crisis of the 1840s led to massive starvation and migration (Willigan, 1977), and Russia, where the great famine of 1891–92 was set off by climatic conditions comparable to those experienced more recently (Robbins, 1975).

An outcome of these studies should be a better vocabulary for describing the connections between climate and health. Even tentative exploration provides a strong argument against saying that a drought or a hurricane in Haiti 'caused' so many deaths, so many illness episodes, and so much malnutrition. If climate is not the cause, or is just the eventual trigger of a series of linked causes, should not analysis be directed at other causes which, unlike the proverbial weather, can be modified? A rapid view of the world food system and its capacity to counter damages attributable to climate is thus in order.

10.4 THE WORLD FOOD SYSTEM—ITS CAPACITY TO ABSORB CLIMATIC PERTURBATION

If assessed globally, the productive forces that the world has developed today to feed its population are ample. Indeed, there is 'slack' of such magnitude as to minimize—in some cases out of existence—the result of any climatic anomaly that is likely to appear in the near future. In 1960, calculations were made that by the year 2000 a world population of 6.5 billion (a 50 percent increase over current population) could easily be fed. The premises behind this calculation were that the area cultivated in 1950 could be doubled or even tripled, and that with 'know how' available in 1960, yields of 3–5 tons of grain per hectare could be achieved (Buringh, 1977). An advisory group to the US President calculated in 1967 that 19 percent of the planet's surface was under cultivation, but that the potentially cultivable areas were 24 percent (*The World Food Problem*, 1967).

More recent studies have estimated that a maximum use of photosynthesis for the production of grain would yield about 50 billion tons of grain per year (Buringh *et al.*, 1975; Buringh and Van Heemst, 1977). This is about 43 times what is currently produced, and would satisfy—using standards which have been shown to be adequate for the population of China—a human population in the planet which is 55 times larger than the current one (Escudero, 1983). It has further been calculated that global eradication of malnutrition—that is, the compensation of a daily deficit of about 350,000 kilocalories—would mean redistribution of 3.8 percent of the world's available cereals. In economic terms, this amount is equivalent to 2.4 percent of the gross national product (GNP) of 'developing' countries and 0.3 percent of the world's GNP, used to purchase grain (Reutlinger and Selowsky, 1976).

At the national and regional level, the question of potential food self-sufficiency is more difficult to assess. However, case studies of countries such as Bangladesh, whose name is synonymous with malnutrition, show that the country produces food in excess of basic nutritional needs, and that the prevalent malnutrition can be explained only in terms of its social and economic structure (Hartmann and Boyce, 1979).

Thus, for the globe and for most regions the leeway in food production to compensate for climatic damage is enormous. The facts that food production is not geared primarily to satisfy the needs of all the world's people and that abundant production coexists with massive malnutrition, due to climate or not, can be exemplified by the production of animal protein by the world's food industry, especially its most 'modern' segment.

10.4.1 The Animal Protein Diversion

The amount of food that is diverted into production of animal proteins with no nutritional justification is an example of how our productive forces are not geared to supplying human needs, among them the needs created by climatic perturbations.* US consumption of animal protein increased from 55 pounds per person in 1940 to 117 pounds in 1972 (García and Escudero, 1982). This hyperproteinization of diets may in itself be harmful to health; the fats which accompany animal proteins have a causal role in the production of atherosclerosis. More important is the fact that from five to ten units of vegetable energy are needed to produce one unit of animal protein energy. A study of diets in the People's Republic of China suggests that the Chinese population is adequately fed with 450 pounds of grain per person annually, of

* For criticisms of the organization of the world food market and its irrelevance to human needs, see George, 1977; Lappé and Collins, 1977; Groupe d'Information Tiers Monde, 1978; Tudge, 1979; Escudero, 1981.

which 350 are consumed directly and 100 are given to animals as fodder (Mayer, 1976). Currently, an enormous quantity of grain, about 490 million metric tons annually or about 43 percent of the world's production, is given to animals as fodder (FAO, 1977). These animals, or their produce, are going to be eaten by those who can pay for such foods as animal proteins and fats—the rich countries of the world, or the rich minorities in poor countries. The amount of grain used as fodder, using Chinese standards of human consumption, could provide food for 2000 million people, or 40 percent of the world's current population (García and Escudero, 1982).

Two additional points should be considered. The first is that by far the prevalent human malnutrition in the world is not due to a specific deficiency of proteins or animal proteins, but to a general deficiency of energy, that is, calories. To illustrate, a nutritional survey in 7000 Indian homes showed that 50 percent of those who had energy deficiencies also had protein deficiencies, whereas only 5 percent of those who had no calorie deficit showed a protein deficit (Sukhatme, 1970). Thus, while there have been frequent recommendations to cover protein needs, it seems unlikely that a food intake that is sufficient to cover energy needs might be insufficient to cover protein needs (UN World Food Conference, 1974). Moreover, if protein above a minimum is given to a malnourished person, it is not utilized as such, and is instead burnt as energy. This relationship between protein and calories leads to a re-evaluation of the nutritional advantages of traditional diets, such as the Mexican maize and bean combination, which provides at much lower cost an overall balance of essential amino acids from vegetable sources.*

The second point is that the emphasis on proteins, especially animal ones, stems from outdated sources. The historical origins of the emphasis on protein were some thorough epidemiological studies on human malnutrition, undertaken in West Africa in the 1930s. From a dietary point of view, West Africa is an exceptional region: the staple foods are roots, cassava, plantain, and breadfruit, with an exceptionally low protein content. The malnourished population exhibited a specific protein deficiency which the researchers baptized 'kwashiorkor'. This deficiency is rarely seen elsewhere in the world to such a degree. Indeed, it can now be said categorically that specific protein deficiencies are a rarity in the epidemiology of malnutrition. Nevertheless, the Protein Advisory Group of the United Nations took up the emphasis on protein as the nutrition problem of the world, and it has taken many years to reverse this trend (McLaren, 1974).

Dietary requirements for protein have been revised downward in recent decades. The US Food and Nutrition Bureau suggested 70 grams of protein daily per adult in 1942. By 1968 these requirements had been reduced to 65 grams, and by 1974 to 56. British standards asked from 66 to 146 grams in 1950

* For Mexico, see Chavez, 1980; and Ramírez *et al.*, 1971. For a general re-evaluation of traditional foods, see Béhar, 1976 (reprinted in García and Escudero, 1982).

and from 68 to 90 grams in 1969. By 1975, Canadian standards asked for 56 grams and West German ones for 63 grams.* A joint FAO/WHO ad hoc committee recommended 44 grams in 1973 (FAO/WHO, 1973).

10.5 PROPOSALS FOR IMPROVEMENT OF MEASUREMENTS OF CLIMATIC IMPACT ON LIFE AND HEALTH

This section presents proposals for improving the measurement of climatic impacts with respect to mortality, morbidity and body growth. A previously cited work (García and Escudero, 1982) recommends grading such proposals according to feasibility, that is, the human, material, temporal and political inputs needed to implement them. This approach is repeated here to some extent. We start with approaches which demand a low input of resources.

10.5.1 Mortality

For acute climatic impacts, causing a sudden loss of life in countries with a weak information infrastructure, a 'journalistic' approach, combined with longitudinal and retrospective case studies, would initially be useful. The margin of uncertainty surrounding estimations of loss of life, illness, migration and human suffering is going to be large.

For 'chronic' or long-term climatic phenomena, the initial effort must be to arrive at a more or less reliable baseline of data on mortality and morbidity. It is useless to suggest for the many countries with imperfect to nonexistent vital and health statistics the immediate development of a statistical network of the type which developed countries have evolved after many years and much expenditure of money. Indeed, spending large amounts of money on white elephants such as a statistical bureaucracy or a 'modern' computer data processing system may be counterproductive in this context, as the resources employed are likely to be diverted from more pressing societal needs, including direct action against climatic aggression. For the former elephant the excuse may be made that it at least provides employment; the latter elephant has no excuse whatever.

For a more or less correct measurement of mortality and fertility, the new 'indirect' methods referred to earlier are both very cheap and fairly reliable. These methods consist of a few questions put to respondents with regard to the fertility and mortality experienced over a fixed recollection period by the respondents themselves and their families, together with questions on family composition, birth order of siblings, and so forth. Analysis of the responses generates a Life Table for the population under study, and this in turn generates statistics on age-specific mortality risks, together with a fertility

* For US, British, Canadian and West German standards, see Munro, 1977.

table. These analyses can create a baseline, which was lacking, for example, when efforts were made to measure the impact of the Sahel-Ethiopian drought. If an infrastructure for the probabilistic collection of data exists in the countries (and the methodological problems in using this tool have been solved, by and large), the operational ones that remain need not be insurmountable. The method can also be employed in conjunction with a population census with fairly complete coverage. Given greater resources, it is also possible to perform in-depth mortality studies through sampling, like the one undertaken by Puffer and Serrano (1973), which threw so much light on the underregistration of deaths and the prevalence of malnutrition in Latin America. In general, for countries in very unfavorable data situations, it is advantageous to carry out population censuses at more frequent intervals than the 10-year one suggested by the United Nations. The census is a fundamental and comparatively cheap tool for the quantification of a society.

10.5.2 Morbidity

Morbidity is a more difficult proposition than mortality. Those countries that have good population coverage for their social services and an institutional health care system—for example the European ones, Canada, or Cuba—obtain good morbidity statistics as a byproduct. Those with good population coverage and with no institutionalized health sector, like the United States, know their morbidity basically from probabilistic sample surveys, which are expensive and, for exactness, must rely on complicated measurement methods. For the great majority of the poor countries, perhaps a simple monitoring system can be tried by using interrogation of population exclusively, and accepting an inevitable margin of error.

10.5.3 Malnutrition

Malnutrition, that most common biological manifestation of climatic aggression, can be assessed by using such simple measures as weight and height for age;* several body circumferences, such as that of the arm (Jelliffe and Jelliffe, 1969), or rapidity of growth (Martell *et al.*, 1979); age of onset of menstruation in girls; and prevalence of anemia or edema.† Probabilistic methods for these indicators can, again, be applied to groups at risk, to population samples, or to censuses. The depth of use must depend on resources available and time constraints facing investigators, who need to bear in mind that prompt diagnosis of malnutrition can produce the decision to give food to

* Several methodologies exist for this, and also several tables of 'normality' against which results can be plotted. For a discussion of what 'normal' constitutes, see Martorell *et al.*, 1975.

† There is a good brief review of these methods in Bourbour, 1981. For a methodological discussion of body measurements, see Jordán, 1979; and Eveleth and Tanner, 1976.

malnutrition victims. Food is a cheap therapeutic product with hardly any side effects, especially when compared to the products of the drug industry, or to the subsequent alternatives.

With respect to data on malnutrition, compromises can be put forward. Death certificates, if they exist, can be coded separately where mention is made of malnutrition (Escudero and Kotliar, 1975), disregarding the rules put forth by the International Classification of Diseases. Deaths 'due to' measles, diarrhea, bronchopneumonia, and parasites can be monitored, as they give an idea of the underlying malnutrition which plays a role in their occurrence. Paramedical personnel can be quickly trained to recognize malnutrition. All of this would also provide, with approximations, the baseline that is needed to know the prevalence and trends of mortality and morbidity phenomena.

For studies involving malnutrition, the selective blindness of the medical profession toward it must be overcome. During the 1972 Sahel famine, as noted, it was horrifying to observe that the publications of the Ministries of Health of the countries concerned did not mention, or hardly mentioned, the malnutrition that was striking the population. The retraining of the medical profession to recognize malnutrition is more difficult than it seems, as many of the countries where malnutrition is most prevalent use curricula copied from malnutrition-free countries (to which many of the physicians emigrate).* Change in curriculum is a good example of an apparently simple measure to improve defence against climatic aggression, which in reality implies the use of a great input of political power.

There are other indicators of human stress and suffering which should be mentioned, although they fall outside the scope of death, illness and modification of growth. These indicators of climatically induced stress are very real, even if they are not easily or commonly quantified. All of the 'adjustment' techniques that humans can devise to subsist under climatic (mostly drought-induced) stress can give rise to indicators of increased wretchedness: emigration; curtailment of the purchase of items that are not essential to survival; forced sales of household utensils, tools, animals, clothes; up to the number of unclaimed corpses found in the streets of cities (Currey, 1978; Rahaman, 1978). All of these can be measured and analyzed as timely pieces of evidence of the magnitude of human suffering involved.

Some measures are not advocated here, and it is important to explain why. Monitoring of food production is cumbersome and usually has little bearing on the nutritional status of the people who are supposed to eat the food. The 'animal protein diversion', already described, can produce the contradictory (or logical) situation of massive malnutrition in countries with agricultural production which, in 'per capita' terms, is well in excess of the nutritional needs

* A publication on medical migration is Mejía *et al.*, 1979; a description of the Bangladesh situation can be found in Hartman and Boyce, 1979. Latin American medical emigration to the United States is discussed in Escudero, 1977.

of the population. Situations of this type in Argentina and Brazil have been discussed elsewhere (Escudero, 1978; García and Escudero, 1982).

Nutritional surveys, which try to measure food intake quantitatively, either through the weighing of food through recall, are consuming of time and resources and much subject to measurement errors. It is usually better to measure nutrition by its end result rather than by the intermediary measurement of food intake at individual or family levels. An exception would be programs in which remedial measures are directly taken to correct food deficiencies suggested by the survey.

A promising new line of measurements involves intakes and losses of energy on the part of individuals. So far this has been undertaken at case study levels only, due to the laboriousness of the method (Spurr *et al.*, 1975; Brun *et al.*, 1979; Blaiberg *et al.*, 1980). A tentative finding is that the energy loss from work on the part of some agricultural workers is so large that their salaries are insufficient to compensate for it. As these workers have to be well fed in order to keep earning wages, the bulk of their families' food goes to them, thus creating malnutrition in their children who are, in effect, a captive population to any climatic misfortune that might weaken their already precarious hold on life (Gross and Underwood, 1971).

10.6 CONCLUSION

The problem of damages to human health from climatic perturbations is substantially political, that is, it pertains to the fashion in which societies allocate their collective surplus, as much as to the magnitude of the surplus itself. Ultimately, societies can be grouped by the ways in which they utilize their collective surplus to improve the quality of life: by minimizing mortality and morbidity; maximizing educational achievement, nutrition, lodging, community and political participation; or in other ways. Thus, the 'climate *vs* society' dichotomy is a fundamental one, and for its resolution we must look across cultures and backward in history, to see societal suffering, which we may assign in a much too facile way to climatic causes, in a different light. Were the Honan drought-famine of 1942–43 (2 million dead), the Bengal drought-famine of 1942 (over a million dead), or the Ethiopian drought-famine of 1972 (some hundreds of thousands dead) caused by a climatic agent or by a faulty societal response—a symptom of a fundamental failure of the societies and governments themselves?

It can be expected that societies which more or less fulfill the kinds of aims mentioned above will also be relatively impervious—as impervious as the development of productive forces allows—to climatic damages to health. For evidence, look to the negligible human consequences produced by the drought in Western Europe in 1976 or the unusually hot summer season in the United States in 1979. Or, look to the road conditions at the Dutch-Belgian border after a snowstorm in 1979: the Dutch roads were snow-free, the Belgian ones were not.

The Dutch roads evidently reflected a different perception of the costs and benefits of road clearance (de Vries, 1980). In the case of most of the poorer countries, this example can be magnified a thousand-fold, and it is not a hampering of transportation or the loss of a few lives from hazardous driving conditions that is at stake in a circumstance of climatic aggression, but the sheer fact of survival for millions of humans. The Belgians could purchase and run snow clearance equipment but did not; dozens of countries of the world can act so that some climate hazards result in virtually no deaths, but do not.

In parallel, and in a development that can be seen as paradoxical or as consistent, the societies whose health appears to be little affected by climate also have good monitoring systems to measure the unfavorable impacts of climatic aggression. A recommendation of an ultimate type would be for all societies to become like these fortunate ones, but no consensus exists on the way to reach their position, and the road in any case is certain to be difficult and even violent.

REFERENCES

Alvarez Leiva, L. (1964). *Some Attempts at Evaluation of the Degree of Integrity of Vital Statistics in Latin American Countries*. UN Document ST/ECLA/CONF 19/L/16 (November 16) (in Spanish).

Arriaga, E. E. (1979). Infant and child mortality in selected Asian countries. In UN Secretariat and World Health Organization, *Proceedings of the Meeting on Socioeconomic Determinants and Consequences of Mortality, June 19–25, 1975, Mexico City*, pp. 98–117. UN/WHO, New York/Geneva.

Béhar, M. (1976). European diets and traditional foods. *Food Policy*, (November), 432–436. Reprinted in García, R., and Escudero, J. C. (1982).

Behm, H., *et al.* (1976, 1977). *Mortality in the First Years of Life in the Latin American Countries*. CELADE, Series 'a', Nos. 1024–1032 (in Spanish). San José de Costa Rica.

Blaiberg, F. M., *et al.* (1980). Duration of activities and energy expenditure of female farmers in dry and rainy seasons in Upper Volta. *British Journal of Nutrition*, **43**, 71–82.

Bourbour, F. (1981). Practical techniques for front line health workers for the detection, treatment and prevention of nutritional disorders. *Journal of Tropical Pediatrics*, **27** (April), 114–120.

Brass, W., Coale, A., *et al.* (1968). *The Demography of Tropical Africa*. Princeton University Press, Princeton, New Jersey.

Brass, W., and Hill, K. H. (1973). Estimating adult mortality from orphanhood. In *International Population Conference 1973*, Vol. 3, pp. 111–123. International Union for the Scientific Study of Population, Liège.

Briceño, A., López, M. A., and Rozanes, M. (1979). *Tetanus in Mexico*. Mimeo (in Spanish). Maestría de Medicina Social, Universidad Autónoma Metropolitana, Mexico.

Brun, T. A., *et al.* (1979). The energy expenditure of Iranian agricultural workers. *American Journal of Clinical Nutrition*, **32** (October), 2154–2161.
Buringh, P. (1977). Food production potential of the world. *World Development*, Nos. 5, 6, and 7 (May to July).
Buringh, P., and Van Heemst, H. D. (1977). *An Estimation of World Food Production, Based on Labour-oriented Agriculture*. Center for World Food Market Research, Wageningen.
Buringh, P., Van Heemst, H. D., and Staring, G. J. (1975). *Computation of the Absolute Maximum Food Production of the World*. Dept. of Tropical Soil Science, Agricultural University, Wageningen.
Chambers, R. (Ed.) (1981). *Seasonal Dimensions to Rural Poverty*. Frances Pinter, London.
Chambers, R. (1982). Health, agriculture and rural poverty: Why seasons matter. *The Journal of Development Studies,* **18** (January), 217–238.
Chavez, A. (1980). *Food Needs in the Country and Suggestions for Solutions*. Mimeo (in Spanish). Mexico.
Chen, L. C., Chowdhury, A. K. M. A., and Huffman, S. (1979). Seasonal dimensions of energy protein malnutrition in rural Bangladesh: The role of agriculture, dietary practices and infection. *Ecology of Food and Nutrition,* **8**, 175–187.
Crevenna, P. (1978). *An Evaluation of Our Health Information Systems: Mortality*. Unpublished report (in Spanish). Mexico.
Currey, B. (1978). The famine syndrome: Its definition for relief and rehabilitation in Bangladesh. *Ecology of Food and Nutrition,* **7**, 87–98.
de Vries, J. (1980). Measuring the impact of climate on history: The search for appropriate methodologies. *Journal of Interdisciplinary History,* **10** (4, Spring), 599–630.
Escudero, J. C. (1977). The migration of Latin American physicians to the U.S. *Cuadernos del Tercer Mundo*, No. 15, Mexico (in Spanish).
Escudero, J. C. (1978). The magnitude of malnutrition in Latin America. *International Journal of Health Services,* **8** (3), 465–490.
Escudero, J. C. (1980). On lies and health statistics: Some Latin American examples. *International Journal of Health Services,* **10** (3), 421–434.
Escudero, J. C. (1981). Malnutrition in Latin America. In *Medicina Para Quien, Ediciones Nueva Sociología* (in Spanish). Mexico.
Escudero, J. C. (1983). Social damages caused by malnutrition. *Cuadernos Médico Sociales*, No. 25 (September, 1983), 5–16 (in Spanish). Rosario, Argentina.
Escudero, J. C. (1984). Deaths from malnutrition in Mexico. *Foro Universitario*, No. 40 (March) (in Spanish). Mexico.
Escudero, J. C., and Kotliar, M. (1975). Morbidity from malnutrition: A study of multiple causes of death. *Cuadernos de Salud Pública*, No. 10 (in Spanish). Buenos Aires.
Eveleth, P., and Tanner, J. M. (1976). *World-wide Variation in Human Growth, International Biological Programme 8.* Cambridge University Press, Cambridge, England.
FAO (Food and Agriculture Organization) (1977). *The Fourth World Food Survey* (FAO Statistics Series No. 11, FAO Food and Nutrition Survey No. 10). Rome.
FAO/WHO (Food and Agriculture Organization/World Health Organization) (1973). *Energy and Protein Requirements*. Report of a joint FAO/WHO expert committee. WHO Technical Report Series, No. 522; FAO Nutrition Report Series, No. 52. Geneva.
García, R., and Escudero, J. C. (1982). *Drought and Man: The 1972 Case History*. Vol. 2: *The Constant Catastrophe: Malnutrition, Famines and Drought.* Pergamon, New York.
George, S. (1977). *How the Other Half Dies: The Real Reasons for World Hunger*. Allanheld, Osmun & Co., Totowa, New Jersey.

Graitcer, P. L., Allman, T., Gedeon, M. A., and Duchett, E. (1980). *Current Breast Feeding and Weaning Practices in Haiti*. Mimeo (October 28).

Gross, D., and Underwood, B. A. (1971). Technological change and caloric cost: Sisal agriculture in Northeastern Brazil. *American Anthropologist,* **73**, 725–740.

Groupe d'Information Tiers Monde (1978). *Nestlé Kills Babies*. Maspero, Paris (in French).

Hartmann, B., and Boyce, J. (1979). *Needless Hunger: Vocies from a Bangladesh Village*. Institute for Food and Development Policy, San Francisco.

Imperato, P. J. (1976). Health care systems in the Sahel: Before and after the drought. In Glantz, M. H. (Ed.) *Politics of a Natural Disaster: The Case of the Sahel Drought*, pp. 282–302. Praeger, New York.

Jelliffe, E. F. P., and Jelliffe, D. B. (1969). The arm circumference as a public health index of protein-calorie malnutrition of early childhood. *Journal of Tropical Pediatrics,* **15** (4), 177-260.

Jordán, J. (1979). *Human Development in Cuba*. Editorial Científico Técnica, Havana (in Spanish).

King, J. (1978). *Analyses and Compilations of Nutrition Data and Studies,* prepared for the Bureau of Nutrition, Republic of Haiti. American Public Health Association, Washington, DC.

Kotliar, H., and Escudero, J. C. (1974). Malnutrition in Argentina. *Ciencia Nueva*, No. 31 (in Spanish). Buenos Aires.

Landsberg, H. E. (1984). Climate and health. In Biswas, A. K. (Ed.) *Climate and Development*. Tycooly International Publishing, Dublin.

Lappé, F. M., and Collins, J. (1977). *Food First: Beyond the Myth of Scarcity*. Houghton Mifflin, Boston.

Martell, M., Gaviria, J., and Belitzky, R. (1979). New method of evaluating postnatal growth until the age of two. *Boletín de la Oficina Sanitaria Panamericana,* **86** (2), 95–104 (in Spanish).

Martorell, R., Lechtig, A., Habitch, J. P., Yarbrough, C., and Klein, R. (1975). Anthropometric norms for physical growth in developing countries: National and international. *Boletín de la Organización Panamericana de la Salud,* **79** (December 6) (in Spanish).

Mayer, J. (1976). The dimensions of human hunger. *Scientific American,* **235** (September), 40–49.

McLaren, D. S. (1974). The great protein fiasco. *The Lancet,* **2** (872, July 13), 93–96.

Meade, M. (1981). *Effects of Climate Change on Human and Animal Health*. Report of Panel II.6, Climate Project, American Association for the Advancement of Science. AAAS, Washington, DC.

Mejía, A., Pizurki, H., and Royston, E. (1979). *Physician and Nurse Migration*. World Health Organization, Geneva.

Minesterio de Salud Pública, República de Cuba (1974). *Organization of Services and Level of Health* (in Spanish). Havana.

Ministerio de Salud Pública, República de Cuba (1980). *Annual Report 1979* (in Spanish). Havana.

Ministerio de Salud Pública, República de Cuba (1981). *Annual Report 1980* (in Spanish). Havana.

Munro, H. N. (1977). How well recommended are the recommended dietary allowances. *Journal of the American Dietetic Association,* **71** (November), 490–494.

Puffer, R. R. (1974). *Report on the Quality and Coverage of Vital Statistics, and on Research on Perinatal and Infant Mortality in Cuba.* Documento AMRO 3513. Pan American Health Organization, Washington, DC (in Spanish).

Puffer, R. R., and Serrano, C. (1973). *Patterns of Mortality in Childhood*. Scientific Publication No. 262, Pan American Health Organization, Washington, DC.

Rahaman, M. M. (1978). The causes and effects of famine in the rural population: A report from Bangladesh. *Ecology of Food and Nutrition,* **7**, 99–102.

Ramírez, H. J., Arroyo, P., and Chavez, A. (1971). Socioeconomic aspects of foods and nutrition in Mexico. *Rev. Comercio Exterior,* **21** (8) (in Spanish). Mexico.

Reutlinger, S., and Selowsky, M. S. (1976). *Malnutrition and Poverty: Magnitude and Policy Options*. World Bank Staff Occasional Papers No. 23. Johns Hopkins University Press, Baltimore, Maryland.

Revelle, R. (1974). Food and Population. *Scientific American,* **231** (3), 161–170.

Ríos Massabot, E. (1981). *Current Status on Information Systems on Morbidity in Cuba.* Instituto de Desarrollo de la Salud-MINSAP, Havana (in Spanish).

Robbins, R. G., Jr. (1975). *Famine in Russia 1891–1892: The Imperial Government Responds to a Crisis*. New York: Colombia University Press.

Rojas Ochoa, F., and Sánchez Teixidó, C. (1977). *Coverage and Quality of Statistical Information on Perinatal Mortality in Cuba.* Boletín de la Oficina Sanitaria Panamericana, Washington, DC (in Spanish).

Rotberg, R. I., and Raab, T. K. (Eds.) (1981). *Climate and History: Studies in Interdisciplinary History*. Princeton University Press, Princeton, New Jersey.

Sabelli, M. (1981) (pseudonym of J. C. Escudero). Epidemiology of malnutrition. *Cuadernos Médico Sociales*, No. 15 (in Spanish). Rosario, Argentina.

Schuman, S. (1972). Patterns of urban heat wave deaths and implications for prevention: Data from New York and St. Louis during July 1966. *Environmental Research,* **5** (March), 59–75.

Spurr, G. B., Barec, M. N., and Maksud, M. G. (1975). Clinical and subclinical malnutrition: Their influence on the capacity to do work. *First Progress Report to the Agency for International Development, C.S.D. 2943, January 1971–November 1975*. Washington, DC.

Sukhatme, P. V. (1970). Incidence of protein deficiency in relation to different diets in India. *British Journal of Nutrition,* **24** (2, June), 477–487.

Sullivan, J. M. (1972). Models for the estimation of the probability of dying between birth and exact ages of early childhood. *Population Studies,* **26** (1, March), 79–97.

Titus, R. (n.d.) *Epidemiology of Malnutrition in Haiti*. Mimeo (in French). Port au Prince.

Tudge, C. (1979). *The Famine Business*. Penguin Books, New York.

UN World Food Conference (1974). FAO, Rome.

Warrick, R. A. (1980). Drought in the Great Plains: A case study of research on climate and society in the U.S.A. In Ausubel, J., and Biswas, A. K. (Eds.) *Climatic Constraints and Human Activities*, pp. 93–123. Pergamon, Oxford.

Weihe, W. H. (1979). Climate, health and disease. In WMO, *World Climate Conference (February)*. World Meteorological Organization, Geneva.

Whittow, J. (1980). *Disasters: The Anatomy of Environmental Hazards*. Pelican Books, Harmondsworth, England.

Willigan, J. D. (1977). Mortality patterns during the great Irish Famine. *XVIII General Conference of IUSSP, August 8–13, 1977, Mexico City*.

Wisner, B. (1978). Letter to editor. *Disasters,* **2** (1), 80–82.

Wisner, B. (1980–81). Nutritional consequences of the articulation of capitalist and non-capitalist modes of production in Eastern Kenya. *Rural Africana,* **8–9**, 99–132.

World Food Problem, The (1967). A report of the President's Science Advisory Committee, Vol. 2. The White House, Washington, DC.

World Health Statistics Annual (1980). World Health Organization, Geneva.

Climate Impact Assessment
Edited by R. W. Kates, J. H. Ausubel and M. Berberian

CHAPTER 11
Analysis of Historical Climate–Society Interaction

JAN DE VRIES

Departments of History and Economics
University of California at Berkeley
Berkeley, California 94720 USA

11.1 INTRODUCTION

In this discussion of the techniques of analysis appropriate to the study of climate–society interactions in the past, it is desirable to begin by considering briefly those developments of modern historical research that have brought many historians to regard climate and environmental phenomena in general as appropriate objects of historical study. In so doing the reader will better understand

1. the position of historical research *vis-à-vis* the physical, biological and social sciences,
2. the type of role that the climate factor can hope to find in the explanation of historical change, and
3. the justification for the methodological recommendations to be made later in this chapter.

11.2 HISTORY AND SOCIAL SCIENCE

Historical research has always been rich in its variety, for the historian immodestly asserts his terrain to embrace all of past human experience. And yet, it cannot be denied that historians have long tended to restrict their interests and limit themselves methodologically by adhering, often implicitly, to the view that history consists of an accumulation of discrete human events. The historian's task appeared to be, first and foremost, to catalogue chronologically and describe as fully as possible these events, and, second, if possible, to establish connections among them that could hope to 'explain' how and why one thing led to another.

Since the historian felt compelled to work through the events step-by-step, ever aware that each was of potentially great significance, it is not hard to understand that the bewildering variety he encountered encouraged modesty about what the historian could hope to explain in a 'scientific' way. The historian was more inclined to see his task as the infusion of meaning into the past, much as a novelist infuses meaning into modern life. History could evoke and exhort, but it could not really do to the past what social scientists profess to be able to do to contemporary society.

This standpoint, reduced to a caricature by the brevity of my treatment here, has by no means disappeared. But it is now challenged by historians who argue for the systematic application of social scientific methodology, and by those (not always the same) who argue that the rigorous focus on [human] events is not only overly restrictive but positively misleading. The first stream strips away the historical profession's false modesty at the same time as it abandons literary pretensions by making history nothing more nor less than social science applied to past events. The second stream, often referred to as the 'Annales school' after the French historical journal that acted as the mouthpiece of its exponents, has more profound and interesting implications. In place of history as a profusion of 'mere' events, this school argues for the use of a tripartite concept of time, or duration: the event, the conjuncture and the structure. It is important to note that these progressively longer durations are not simply combinations of events into ever greater aggregations. On the contrary, it is the structures—very slow-to-change features of life such as institutions, ideologies, psychological dispositions, geographical features and ecology—that are basic to these 'structuralist' historians. Structures that channel and confine history until they give way, whether gradually or suddenly, are more important than the mere, often fleeting, event, which captures the newspaper headlines and briefly captures our attention, but is generally of no significance in actually changing things of importance. Between these two durations, one measured in days or a year, the other in centuries, is the conjuncture. Conjunctures are combinations of processes that join to bring about change in the medium term, usually in some patterned way. Demography, technological development, and price-forming markets are examples of forces that give rise to these cyclical processes.

With the acceptance of this framework of durations it is a logical next step to expand the range of phenomena studied by the historian to those that are not, strictly speaking, human events. History could become less anthropocentric with the recognition that 'non-human histories' play a role in the maintenance and demolition of structures, and join in the generation of conjunctural processes. An example will help make my point. 'Traditional' historians have often noted the outbreak of epidemic diseases—think of that celebrated, infamous event, the Black Death. These diseases could be explained by the knowledge of how rodents, bacilla, fleas, and the like carry and spread them. But beyond this, such events are exogenous to history. They are 'Acts of God' in the sense that insurance contracts use that term. Historians have also, of course, been concerned with how change in human society—in social organization, scientific progress, income inequalites—could render these medical facts of life more or less potent. Here the epidemiological variable is endogenized. What is new to the structural approach is the recognition that there could be an epidemiological history that is autonomous from human history but is capable of impinging on it at times.

In the same way, we can distinguish among three modes of historical treatment of the climate variable. The historian of the event has long recognized the role of climate, or perhaps more correctly of weather, as a random variable—an exogenous fact of life—that has influenced harvests and human health, and occasionally has wrought great destruction via floods, hailstorms, and other natural disasters.

Changes in human society that enable the forces of nature to be tamed (through technological changes reducing societal vulnerability to specific weather events) and human activities that themselves bring about climate change (for example, through deforestation or the release of CO_2) represent climate as an endogenized variable.

Finally, we come to a climate history that is exogenous and non-random (in the sense of exhibiting climate change as Hare defined it in Chapter 2). This non-human history, with its various and still imperfectly understood effects on human society, is this volume's chief object of interest, and it has come to attract the attention of historians who work in the structural tradition described above. The French historian Fernand Braudel expressed over 10 years ago his fascination with 'the possibility of a certain physical and biological history common to all mankind [which] would give the globe its first unity well before the great discoveries, the industrial revolution, or the interpenetration of economies' (Braudel, 1973, 19). What he had in mind was the contemporaneous occurrence of comparable events such as the collapse of empires, demographic change, and massive civil unrest in zones of the world not in intense contact with each other. These events, he speculated, might be explained by a world-embracing non-human history of climate and epidemiology.

The point deserving emphasis here is that through the deployment of various notions of duration, history is in a position to embrace new phenomena, including 'non-human histories' that expand the role of historical research. If social scientific history, strictly defined, seeks to merge the study of the past with the social sciences, the practitioners of the tradition described above have sought something different—to make history a meeting ground for various disciplines in the physical, biological and social sciences.

Historical research is sometimes said to be characterized by a careful scrutiny of sources (source criticism) and a devotion to chronology. Both are, of course, important to research in *any* discipline. It is more useful to emphasize history as a discipline of context. It is the historian's aim to explain events and processes by situating them in their full context. The contribution of recent historical research has been to enlarge our understanding of context to embrace differing durations and what I have called non-human histories.

For the benefit of the non-historian it is important to emphasize that the inclusion of the climate factor in the study of history must not be regarded as a search for an alternative, and deterministic, explanation of the past, but as an expansion of the context in which the workings of past societies are to be understood. It follows that we must not nurture extravagant expectations about the single-handed influence that climate change has had on society, nor can we expect the isolation and measurement of these effects to be easy or straightforward.

The research of historians and climatologists plus the example of recent events have prepared the ground for the acceptance of climate change as a vehicle of long-term historical explanation (for historians' views, see Davis, 1973; Parker and Smith, 1978). At the same time, scholars interested in contemporary policy feel the need for a better grasp of the types of historical experience with climate change. It is now recognized that the study of climate and society organized on the basis of impact models (see Kates, Figures 1.1 and 1.2 in this volume) are, by themselves, of limited value. They should serve as building blocks in the development of interaction models, ideally those embracing feedback processes (see Kates, Figure 1.5). Impact models are essentially ahistorical to the extent that they focus on an event and its direct consequences. Interaction models, on the other hand, merge with historical studies the more they seek to integrate climate events into the larger social and natural complex in which those events occur.

The great challenge for historical research is to find methodologies that do not ignore the requirement of treating society as an interacting complex—that is, to honor history as a discipline of *context*—while at the same time simplifying the task sufficiently to render it manageable. Historical research cannot make use of laboratory experiments that isolate the effect of variables one at a time; likewise, the time duration of historical research is almost always too great to permit the use of the *ceteris paribus* assumptions favored by the social sciences. On the other hand, the inclusion of all potentially relevant factors in an historical model is clearly beyond human capacities.

The dilemma as it applies to historical climate studies was presented by Anderson (1981) as an identification problem in which climate change is one of numerous changes working upon a society. These together produce numerous effects. No simple assessment of causation is possible in this situation, and it is particularly suspect to *infer* a climatic cause from an observed effect. The illegitimacy of this practice is sufficiently obvious to obviate the need for elaboration. Also objectionable is the common practice of treating contemporaneity of events as tantamount to causation. Parry (1981) offers a clear-headed discussion of this problem, and notes that:

> [Climate historians], despairing of the viability of a more rigorous approach ... have sometimes resorted merely to an investigation of the synchroneity of climatic and economic events ... Yet all that can be established by this argument is a space–time coincidence. (p. 321)

This practice is by no means confined to climatologists. Indeed, historians often encourage such inferences via their tendency to regard the establishment of a chronological narrative of events as, by itself, tantamount to explanation.

Whether only two discrete events are linked or two time-series are found to correlate well with each other, all that has been demonstrated in either case is association. To go further requires that a model of causation be established and that the hypotheses flowing from it be tested.

Here we come to a point where the objection will be raised, particularly by historians, that this—model building and hypothesis testing—is precisely what cannot or should not be done in historical research. After all, does it not do violence to the need to comprehend the multitude of factors and the unfathomable complexity of historical change emphasized above? Is it not better to rely on the inspired intuition and the insights of the historian, based on his laboriously acquired general knowledge of a particular time and place, than to seek a specious scientific accuracy?

The answer to these questions must be that the acknowledged difficulties of reconstructing past climate–society interaction offer no reason to adopt other scholarly standards than those for contemporary research in the social and physical sciences. There is no philosophical reason to distinguish historical explanation from [social] scientific explanation. Indeed, the nature of the phenomenon being discussed in this volume is such that historical analysis is unavoidable. Most of the climate changes and fluctuations that interest us require the inclusion of duration, and hence of historical research.

The specific techniques that may be useful in the study of climate–society interaction cannot be listed briefly. However, a survey of the historical studies now available (see Wigley *et al.*, Chapter 21) makes clear that it is important to begin with a sufficient base of knowledge about both the relevant physical and/or social processes *and* the historical context (institutional, economic and cultural) to specify the elements of a testable causal model. The further exploration of the model can proceed in different ways, depending on the character of the model and of the available data. The predicted consequences of a specific climatic phenomenon (in Parry's parlance, 'postdicted', since the predictions apply to events that have already occurred) can be specified through simulation (Parry, 1978), or can be confronted with the historical record via statistical techniques such as regression (de Vries, 1980; Lee, 1981), contingency tables (Ogilvie, 1981), or cross-spectral analysis (Lee, 1981).

The quantitative societal data that are most commonly used for the historical study of climate impact are, in order of general availability, commodity prices, demographic events (births, deaths, marriages), tax revenues, trade volumes, and production measurements.* Quantitative data for western countries after about 1650 are by no means scarce, but indicators appropriate for the testing of hypotheses about the impact of climate on the society often are. These hypotheses frequently require knowledge of production levels, which is among the least abundant types of quantitative information. Investigators often substitute for the unavailable production data the much more abundant price data. A great deal of confusion and erroneous inference has resulted from the uncritical use of this substitute. In order to accept price fluctuations as a proxy for fluctuations in the volume of output, very strong assumptions must be made (about the extent of markets, the volume of trade, the existence of substitutes, the elasticities of supply and demand), assumptions which are rarely sustainable in practice (see de Vries, 1980).

Where quantitative techniques are not applicable, or practicable, as is often the case, or when not enough is known even to specify plausible hypotheses, it is best to recognize frankly the limitations of current knowledge and build on it with careful observation leading to narrative reconstruction of the events at issue. Anthropological field research has popularized the term 'thick description' for this systematic assembly of detail uncontaminated by *a priori* interpretive biases. By developing a better understanding of sequential events and by expanding the context in which the events can be situated, 'thick description' can give rise to fruitful hypotheses at the same time that it helps supply the data with which they can be tested.

* The sources for these types of quantitative data are numerous, and no effort can here be made to offer a comprehensive bibliography. For the study of Europe, however, references to many of the relevant document collections can be found in Braudel and Spooner (1967), Mitchell (1975), Wilson and Parker (1977), Abel (1980), Flinn (1981), Wrigley and Schofield (1981), Le Roy Ladurie and Goy (1982).

In general, there is little to distinguish the range of techniques available to historical research from contemporary studies in the social sciences. A more specific historical contribution is located in the selection of the issues chosen for study and the time-scales selected for the examination of climatic effects. It is to these issues that we shall now turn.

11.3 HISTORICAL ANALYSIS

11.3.1 Climate Chronology

This discussion of historical interaction between climate and society must pass over in silence the role of historical research in constructing the record of climate variability and change. For this issue see Hare, Chapter 2, as well as Ingram *et al.* (1981), and Pfister (1980). It must be recognized that the selection of time-scale and the choice of research methodology is sensitively affected by the available climate record. It is important to continue the work of extending back in time climate data that are, in the words of the French historian Le Roy Ladurie, 'continuous, quantitative, and homogeneous' (1959, 138). Evidence of a less refined sort, evidence that indicates the direction of change between decades or 50-year periods, for example, or that identifies only extreme events, is not without its uses but limits severely the types of issues that can be addressed, as will become apparent below.

11.3.2 Perception

Fundamental to any modeling or formulation of hypotheses about societal adaptations to climate change is knowledge about how people perceived climate change before the establishment of meteorological offices and the broad dissemination of their data-collecting activities. This is important because 'adaptation' as opposed to 'impact' implies human choice, and this requires individual perception. Even when the object of study is very long-term climate change, the unit of perception must remain the individual and his interpretation of discrete events of relatively short duration.

Traditional historical research techniques would appear to be well suited to the study of perception. The source materials are generally descriptive (letters, diaries, log books, treatises, agricultural almanacs) and their interpretation depends very much on one's knowledge of contemporary customs, habits and beliefs.

Thus far, studies have focused primarily on the perception of climate in newly settled territories. The interest of historians in the American West has long been directed to how explorers and settlers interpreted the climate of the Great Plains. The climate itself exhibited cycles of drought, and it appears that settlers alternately overestimated and underestimated the region's agricultural possibilities. Eventually the accumulated knowledge and experience of the inhabitants narrowed the range of uncertainty, and institutions were established to deal with that. The settlement history of the Great Plains extends over less than 150 years. The settlement of Greenland and Iceland extends over a much longer period, although the documentation, particularly for the former, is much less complete. Here the studies of Ogilvie (1981) and McGovern (1981) use narrative description to identify the signals that reached the inhabitants and to determine how they were interpreted.

An interesting study that emphasizes the role of ideology (or presumed scientific knowledge) in limiting the ability of explorers and settlers to interpret correctly the climate with which they came in contact is Kupperman's (1982) study of early English experience in Massachusetts. Her study is of particular interest because of the comparisons made by the early settlers between their new habitat and the old England in which they had acquired their attitudes and expectations.

In non-marginal, long-settled areas with well-established material cultures the perception of climatic change will almost certainly be found to consist of incremental, subtle alterations in customary procedures and rules-of-thumb. It is possible that careful analysis of literary evidence can expose such decisions, but it appears likely that the historian will generally have to infer such perceptions from behavior and rely on a model of individual decision-making rather than on direct evidence.

The last word has by no means been spoken about this subject, but recent work points in the direction of seeing individual responses involving decision-making under conditions of uncertainty where the decision-maker tries to estimate the probability of certain events (usually, exposure to extremes of climate). The relevant information to the decision-maker—and hence, the information that the historical researcher needs—is not long-term averages of temperature or rainfall, but rather evidence of variability. One might say that it is the variability of the variability that is likely to be the bearer of signals to the individual decision-maker, the medium through which climate change works on society.

It was just such a problem of decision-making under uncertainty that the eighteenth-century Virginia tobacco planter Landon Carter pondered when, in 1770, he wrote in his diary:

> I cannot help observing as I have before done that this climate is so changing [that] unless it return to its former state Virginia will be no Tobacco Colony soon. (Quoted in T. H. Breen, 1983, 276–277.)

Carter was not yet ready to accept that the experience of the recent past (cold growing seasons) should be used to establish a new norm, governing future expectations. And we now know that the Virginia tidewater did not, in fact, cease to be a center of tobacco production. Carter remained committed to the crop but many of his fellow planters took steps to diversify and to reduce their dependence on tobacco in the 1760s and 1770s. In those years George Washington stopped the cultivation of tobacco and converted Mount Vernon to a grain-growing enterprise.

In view of the many other factors that could have influenced the production decisions made by Washington and his fellow planters, it would be rash to conclude on the basis of the contemporaneity of events alone that climatic variability induced the observed behavior. But any study of this episode must establish a model featuring decision-making under conditions of uncertainty.

At this point a word must be said in defence of the concept of 'decision-making' in preindustrial society. Eighteenth-century planters may have been rational businessmen, but did peasants process data, estimate probabilities, and perform cost/benefit calculations? The social scientist's custom of labeling all preindustrial societies as 'traditional' implies that peasants did none of these things. But this traditional view of the social scientist is certainly in error. By modern standards, the decision-maker in earlier centuries worked with insufficient information (for example, there were no weather bureaus), he operated under peculiar constraints (in particular, those imposed by the absence or imperfection of markets for such things as seed and insurance), and he may have been influenced by some considerations dismissed today as irrational. Indeed, as the 'Annales school' historians emphasize, significant changes in behavior may have occurred only when new dangers or new opportunities persisted for a long time. In other words, it took a particularly strong signal to evoke a response. But so long as these adaptive responses existed, and the preponderance of historical inquiry into the economic and demographic life of ordinary people deep into the European past insists that they did, investigations of the long-term relationship of climate to society cannot realistically take the form of impact models, i.e., where the human agent is assumed to be passive, or imprisoned in routine.

A useful analogy can be made to demographic history, where the ruling model linking society to demographic behavior is that bearing the name of T. R. Malthus. Malthus argued that the growth of a population could be stopped in two ways. Either the 'preventive check'—human decisions to restrain the 'passion between the sexes' and voluntarily restrain fertility—would operate to slow the growth of numbers, or, in the absence of decision-making, the 'positive check', the inexorable, inevitable rise of mortality as population outstripped food supply, would do so. The 'positive check', an impact model, is imposed by nature, and most historians used to believe that it was the dominant mechanism regulating population before the demographic transition of the nineteenth century. Peasants, it was argued, did not make fine calculations about their fertility behavior.

The impressive advances in historical demography of the past two decades (Flinn, 1981; Wrigley and Schofield, 1981) have amply demonstrated that the preventive check *did* function, that a decision-making model is necessary to account for the historical interaction of population and economy. Fertility control was, by modern standards, imperfect, and it was achieved differently than today (primarily through controls on marriage). But no one can now make the easy assumption that birth rates in the past were 'natural', that is, uncontrolled.

The model of perception and decision-making that I here urge upon the reader has important implications for the further design of historical studies of climate–society interactions. It asks us to identify the specific vulnerabilities of a society, given its institutional and technological characteristics, and to inquire into how the society accommodated to the risks to which it became exposed.

Further, for historical analysis it is obviously relevant to inquire how a society's vulnerability and its capacity to distribute risk changes over time. That is, in the interaction of climate and society over long periods of time both climate and society can change. Figure 11.1 offers a schematic presentation of some of the possibilities.

In these examples the narrowing or widening of the 'band of tolerance' characteristic of a society's technological and institutional endowments can be regarded as an autonomous structural change. That is, it may occur not in response to climatic change, but independently of it, in response to social phenomena. Scenario D can place a society under climatic stress just as much as alterations in climate (Scenario B) or climatic variability (Scenario C).

11.3.3 Climate Impact

Most studies of climate history have focused attention on the impact of climatic events—the events usually extreme and the impacts usually some form of damage, whether to crop production, human health, or life itself. This simplest form of climate–society study needs to be transcended, as I have argued above, both for the sake of policy relevance and of historical accuracy. But impact studies can nonetheless serve as building blocks for the more complex sorts of analysis. This is especially true of time-series studies of annual events as opposed to the study of isolated events.

Careful statistical studies, such as Lee's regression and cross-spectral analyses of the effects of rainfall and temperature on variation in mortality and fertility in preindustrial England (Lee, 1981) can help place the year-to-year influence of climate in a broader context in which its relative importance (compared, for example, to the incidence of warfare, price changes, or real wage changes) can be assessed. I tried to do the same with regression and difference-of-means techniques in identifying the impact of climatic variance

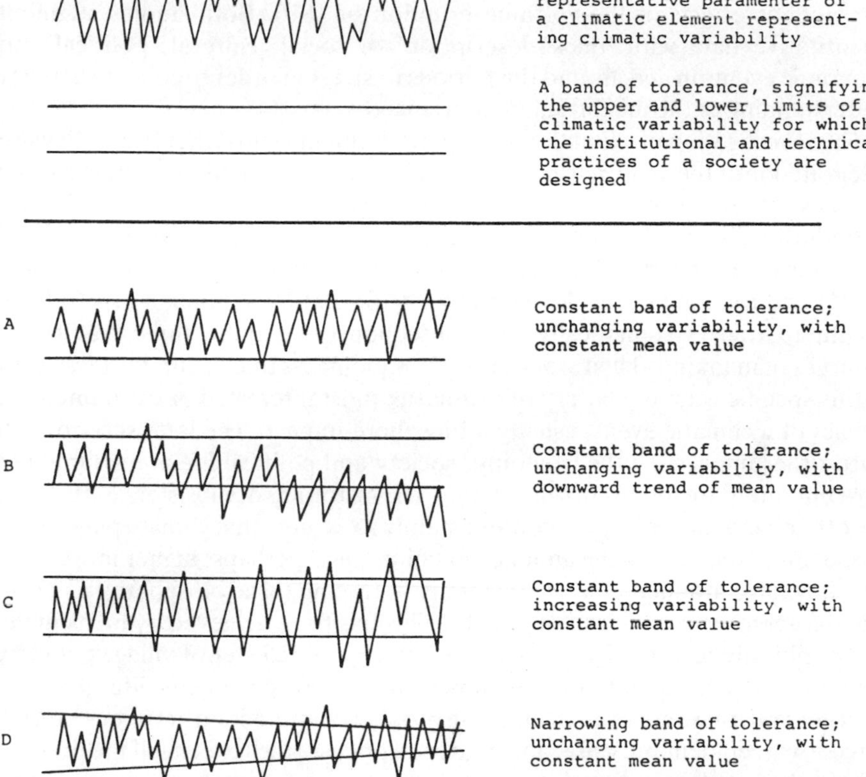

Figure 11.1 Idealized time-series of climate variability and societal 'band of tolerance' designed to portray scenarios of climate change and societal change that can give rise to climatic stress

on several types of agricultural production in preindustrial Holland (de Vries, 1980).

The results to date of such studies have been to assign to annual climatic variation a modest, but tangible, role in accounting for annual variation in demographic and economic phenomena. The preindustrial European societies that have been the object of study were certainly not perpetually held in thrall by the tyranny of mother nature. Yet there were undeniably periods of particular stress, usually produced by a succession of years of abnormal weather. The two most thorough monographic studies of climate history yet produced each focuses on such multiyear periods of climatic stress: Pfister's (1975) study of the Bern region in 1755–97 (with particular emphasis placed on the stress years, the 'katastrophenketten' 1768–71), and Post's (1977) study of the climatic effects of the volcanic eruption of 1816. Both studies augment quantitative data with 'thick description' of social, cultural, political and economic events in and around the periods of stress in order to come to a better understanding of the misfortunes experienced.

The investigation of 'climatic stress' periods might yield richer fruits if it were integrated into the *crise de subsistence* literature developed over the past two decades by historians of preindustrial Europe. *Fames, pestis et bellum* periodically disrupted economic and demographic life. A long list of historians, including the Frenchmen J. Meuvret (1971), P. Goubert (1960), and J. Jacquart (1974) and the Americans A. Appleby (1978) and M. Gutmann (1980) have sought to trace systematically the consequences of these three traditional scourges of mankind. Their work nicely complements that of climate historians in this specific way: the latter are ordinarily most interested in the immediate impact of a climatic event, usually a biosphere impact. The later second- and third-order impacts on the economy, society and political life are assumed to exist but rarely are followed in any detail. The historians of subsistence crises, on the other hand, ordinarily are content simply to assume that climate plays some unspecified role in bringing about crop failure, and perhaps, higher morbidity, but they have attained greater specificity in tracing these biosphere factors in demographic, economic and social life. This has been achieved by focusing on geographically restricted areas, such as a village, or cluster of villages, and by compiling detailed quantitative evidence on local socioeconomic life.

In general, these crises are seen as the product of interaction between external forces (war, epidemic disease, and high food prices—the latter itself the product of weather and/or external market forces) and internal structures, most notably the local economy, the demographic structure and the social organization. The nature of crop specialization and the transportation system, the function of church and local government in regulating distribution, the inheritance customs, and the compensatory demographic behavior of the population are among the factors woven into the historian's accounts of societal responses to periods of stress.

The consensus of these studies of subsistence crises is to reject the notion that catastrophe follows unavoidably upon the appearance of one of the external stress factors. Goubert (1973) states:

> The true demographic crisis, as studied principally in northern, eastern, and central France, where the population is densest and grain is the standard crop, stems from a series of climatic accidents (usually high summer rainfall) *in a given socio-economic context.* (p. 38) [Emphasis added]

Gutmann observed that the great catastrophes of early modern Europe took place when warfare, harvest failure, and epidemic disease occurred in combination. For the rural villages around the Belgian city of Liège, which he studied in detail, Gutmann (1980) concluded that:

> Bad weather [by itself] might produce a year or two of difficulties, and a few years of recovery, but [it] could not turn around the region's generally favorable economic and demographic climate. The dynamic mixed economy of the region ... probably explains this phenomenon. A bad harvest caused by bad weather might tip the balance in a locale with a poor or single-product economy; it could not do so in the Basse-Meuse [the region of Gutmann's study]. (p. 199)

These historical studies of stress periods agree on two things. First, the really lethal crises combined bad weather with war or epidemics; second, the severity of the crisis depended to a large extent on the socioeconomic characteristics of the locale. Goubert found the severe weather conditions of the 1690s to hasten the destruction of the small farmers in the monocultural villages of northern France; Gutmann found no lasting negative effects in the mixed-farming villages near Liège.

11.3.4 Cycles

Even when periods of climatic stress, studied in isolation, are found to impose problems that are well within the capacity of the social and economic institutions to absorb, it remains possible that their recurrence in a more-or-less patterned way can impose a rhythm, or cyclical process, on a society. Interest in such climate impact goes back at least to the late nineteenth century when the British economist W. Stanley Jevons (1884) sought to link business cycles to the 11-year sunspot cycle. More recently both Pfister and Post have maintained that the climatic fluctuations they have studied in such detail possessed an added importance as generators of economic and demographic cycles (Pfister, 1975, 190; Post, 1977, 267). Goubert, too, was stuck by the recurrence of subsistence crises about once every generation in seventeenth- and eighteenth-century France. But he attributed this more to the rhythm of demographic recovery and renewed population pressure than to any pattern of climatic fluctuation (that is, to scenario D rather than B or C in Figure 11.1). This reminds us of the need to place climatic variables in their historical context, preferably in well-specified models. It is also worth bearing in mind that in a society where 'normal' climate variability produces an average of one failure in every four harvests, chance alone will generate *katastrophenketten* of two harvest failures in succession on average once every 16 years, and of three successive failures every 64 years.

In our current state of knowledge, it seems prudent to recommend that investigations of the part played by climatic variation in feeding conjunctional patterns should emphasize their random and exogenous character. Indeed, a widely held view among economists is that cyclical processes are essentially statistical artifacts generated by the cumulative effects of random shocks to the economy (de Vries, 1980).

11.3.5 Climate Change

The study of the short- and medium-term consequences of extreme climate events is by no means a new interest of the historian. Rather, it is a traditional theme that is now being pursued with greater analytical sophistication. When we turn to the issue of the historical consequences of long-term climate change, we enter an area that was long *terra incognita* and is now just beginning to be explored. Elsewhere (de Vries, 1980) I surveyed the historical research and speculation in this field, concluding that:

> Historians are psychologically ready, even eager, for the rise of climate change as a vehicle of long-term historical explanation, but do not possess the means of distinguishing its impact from among the many other variables at work on human society. (p. 624)

This statement suggests that the key obstacle to progress is methodological. Indeed, the challenge of isolating the climate factor from the many at work on a society is daunting. But the prospects for progress in this difficult area depend at least as much on asking the right questions. Once this is achieved, a rigorous and ingenious use of existing social theory and statistical techniques can serve to advance our understanding of climate–society interaction.

Here, once again, we must begin with the question of perception. The response, if any, of a society to climate change must be predicated on some system of gathering, storing and assessing information about the environment. Only when such activities are entirely absent can one realistically utilize climate impact models. In all other cases long-term climate change involves a population in the process of decision-making under conditions of uncertainty. Specifically, decision-makers must calculate changes in the underlying probability of an adverse event (such as crop failure) on the basis of incomplete evidence.

The historian's *ex post* knowledge of changes in the frequency of extreme events may be superior to that of the decision-maker, but he must go further to relate that knowledge to the society with which the climate comes in contact. Questions need to be asked about the vulnerability, or sensitivity, of social and economic arrangements to climatic variation. The analyst of climate–society interaction cannot assume 'society' to be a constant any more than he can assume constancy in climate. In the long run, technological and institutional changes can alter the society's sensitivity to climate events even in the absence of real climate change. Among the factors that play a major role in altering vulnerability are transportation, storage and distribution systems, the crop-mix of agriculture, the degree of market integration, and the availability of insurance or cooperative practices. Both of the last two features influence the way the risk of loss due to climate is distributed among social classes and over larger or smaller populations of producers.

Examples of technical changes in the past that broadened or narrowed the band of tolerance are numerous: the replacement of oxen by horses as draught animals improved the chances for a good harvest by reducing the time required for plowing and planting (White, 1962). On the other hand, the replacement of bread grains by potatoes in the European diet increased the caloric output per unit of land but exposed society to crisis in the event of even a single year of crop failure because potatoes could not be stored from one harvest year to another (Mokyr, 1983). A systematic assembly of such information can lay the foundation for a long-term 'history of risk' that could shed light on many aspects of economic and political, as well as climate, history.

This brings us to the third type of question that can shed light on the process of long-term climate change. Besides *perception* of change, and *vulnerability* to climatic variation, we need to ask questions about societal *adaptation* to change. Here the main issue is likely to be how exposure to risk is minimized or diffused. The social unit that bears risk can vary from the individual producer to society at large. Similarly, the individual's response to an increased risk exposure can vary along three dimensions: *production changes* that alter the catastrophe threshold (for example, safety-first farming strategies), *political initiatives* that shift or spread risk to others, and *abandonment* of the risky activity, perhaps accompanied by migration. On the first type of response, see Jodha and Mascarenhas, Chapter 17 of this volume, and de Vries (1980, 626–629); on the second, see Warrick and Bowden (1979) and Bowden *et al.* (1981, 494–508); on the third, see Parry, Chapter 14 of this volume, and Parry (1978). Here it should be noted that the analysis of marginal areas, where the adaptation choices of the decision-maker are restricted to a 'yes or no' decision, is but the exposed tip of the iceberg of adaptations to climate change. Just as the pursuit of physical events through their biological, economic and political consequences involves the researcher in a 'cascade of uncertainty' (see Kates, Chapter 1), so the pursuit of adaptations to climate change from marginal, to recently settled, to non-marginal, to long-settled areas involves the researcher in a 'cascade of subtlety'. This difficulty can be avoided only by focusing on cases where a limiting factor enforces a yes–no decision in place of the more general situation where marginal adjustments are possible. Besides the geographical limitations exploited in the study of arable farming at high elevations or at the northern margin of cultivation, other such limiting factors can be identified in sectoral studies of specific economic activities such as fishing and rainfall-sensitive crop production.

In the analysis of all three types of questions discussed above it is important to bear in mind the dynamic character of long-run adjustment processes. Changes in perception and vulnerability and efforts at adaptation almost always occur incrementally. Their cumulative result in the long run can be unexpected and even unintended. The drilling of deeper wells to alleviate drought can intensify aridity in the long run; the assumption by a state of the risk of producing in a marginal area can expose the entire society to loss if the chance of crop failure rises permanently to a higher level. The narrowing of the genetic base of crop seed can reduce vulnerability to *present* climate while increasing the vulnerability to climate *change*.

Historical studies can advise contemporary policy-making in a tangible way by calling attention to the typically dynamic process of climate–society interaction. Society adapts to long-term natural processes via a succession of short-term responses, some of whose ultimate consequences are either unforeseen or incorrectly assessed. Part of the resulting policy failure is simply the product of incomplete information; another part flows from the discrepancy in the durations required for perception of change, for implementing various types of adaptation, and for feeling the effects of the changed practices.

In view of the insight of 'Annales school' historians that society is always waist-deep in routine and semi-immobilized by a combination of external, 'objective' constraints and internal, self-imposed limitations (Le Roy Ladurie, 1977; Braudel, 1977), we can expect the duration required to perceive a need for adaptation to be substantial, giving rise to discontinuous, 'jerky' responses. These might be triggered by multiyear stress periods that are themselves only incidental to a long-term climate change. In other words, societal responses might be to something other than a long-term natural process, giving rise to a chain of discrete, possibly contradictory, adjustments.

11.4 CONCLUSION

The techniques of analyzing historical climate–society interaction are numerous, drawn as they are from existing social theory and statistics. Their fruitful use depends on the formulation of strategic questions that capture the dynamic quality of historical change. I argue here that these questions should focus primarily on the process of decision-making in a probabilistic framework, in particular on decisions concerning the control and distribution of risk. It is not the spectacle of catastrophe that should be the primary object of interest.

The success of historical investigation also requires the acquisition of sufficient knowledge of the time and place being studied to enable specification of causal models of climate–society interaction. In the absence of such initial groundwork, statistical measures of association and narrative identifications of space/time contemporaneity cannot safely be interpreted. It follows that the larger the scope of the investigation, the more difficult it is to achieve a sufficient level of specificity in the causal models as well as in the record of climate variation itself. The prospects for success in historical research are far greater when the scope is restricted to small areas and/or individual sectors of socioeconomic life, such as Lee's analysis of fertility and mortality (1981), Parry's study of the Lammermuir Hills of Scotland (1978), or my investigation of dairy production in Holland (de Vries, 1977). Hope exists for an advance in understanding via an aggregation of such studies; this cannot be said for a frontal attack on the question 'what was the impact of the Little Ice Age in Europe?'

Finally, it bears repetition that the pursuit of the 'climate element' should be integrated into the larger historical context. Climate change should not be treated as an *alternative* explanation of history, but as an *additional* explanatory factor. Indeed, the numerous ways that societies have dealt with—or failed to deal with—climatic stress requires that the climatic factor be placed in the fullest possible historical context.

REFERENCES

Abel, W. (1980). *Agricultural Fluctuations in Europe from the Thirteenth to the Twentieth Century*. Methuen and Co., London.

Anderson, J. L. (1981). Climatic change in European economic history. *Research in Economic History*, **6**, 1–34.

Appleby, A. B. (1978). *Famine in Tudor and Stuart England.* Stanford University Press, Stanford, California.

Bowden, M. J., *et al.* (1981). The effect of climate fluctuations on human populations: Two hypotheses. In Wigley, T. M. L., Ingram, M. J., and Farmer, G. (Eds.) *Climate and History*, pp. 479–513. Cambridge University Press, Cambridge.

Braudel, F. (1973). *Capitalism and Material Life.* Translation by S. Reynolds. Harper and Row, New York.

Braudel, F. (1977). *Afterthoughts on Capitalism and Material Civilization.* The Johns Hopkins University Press, Baltimore, Maryland.

Braudel, F., and Spooner, F. (1967). Prices in Europe from 1450 to 1750. In Rich, E. E., and Wilson, C. H. (Eds.) *Cambridge Economic History of Europe*, Vol. 4, pp. 374–486. Cambridge University Press, Cambridge.

Breen, T. H. (1983). The culture of agriculture: The symbolic world of the tidewater planter, 1760–1790. In Tate, T. W., *et al.* (Eds.) *Saints and Revolutionaries: Essays in Early American History.* W. W. Norton and Co., New York.

Davis, R. (1973). *The Rise of the Atlantic Economies.* Cornell University Press, Ithaca, New York.

de Vries, J. (1977). Histoire du climat et économie: Des faits nouveaux, une interpretation differente. *Annales: Économies, Sociétés, Civilisations*, **32**, 198–226.

de Vries, J. (1980). Measuring the impact of climate on history: The search for appropriate methodologies. *Journal of Interdisciplinary History*, **10**, 599–630.

Flinn, M. W. (1981). *The European Demographic System, 1500–1820.* The Johns Hopkins University Press, Baltimore, Maryland.

Goubert, P. (1960). *Beauvais et le Beauvaisis de 1600 à 1730*. SEVPEN, Paris.

Goubert, P. (1973). *The Ancien Régime: French Society 1600–1750*. Harper and Row, New York.

Gutmann, M. P. (1980). *War and Rural Life in the Early Modern Low Countries*. Princeton University Press, Princeton, New Jersey.

Ingram, M. J., Underhill, D. J., and Farmer, G. (1981). The use of documentary sources for the study of past climates. In Wigley, T. M. L., Ingram, M. J., and Farmer, G. (Eds.) *Climate and History*, pp. 180–213. Cambridge University Press, Cambridge.

Jacquart, J. (1974). *La Crise Rural en Ile-de-France, 1550–1670*. Armand Colin, Paris.

Jevons, W. S. (1884). *Investigations in Currency and Finance*. Macmillan, London.

Kupperman, K. O. (1982). The puzzle of the American climate in the early colonial period. *American Historical Review*, **87**, 1262–1289.

Lee, R. D. (1981). Short-term variation: Vital rate, prices, and weather. In Wrigley, E. A., and Schofield, R. S., *The Population History of England 1541–1871: A Reconstruction*. Harvard University Press, Cambridge, Massachusetts.

Le Roy Ladurie, E. (1959). Histoire et climat. *Annales: Économies, Sociétés, Civilisations*, **14**, 3–34. Translation: *History and climate*. In Burke, P. (Ed.) (1972). *Economy and Society in Early Modern Europe*. Harper and Row, New York.

Le Roy Ladurie, E. (1977). Motionless history. *Social Science History*, **1**, 115–136.

Le Roy Ladurie, E., and Goy, J. (1982). *Tithe and Agrarian History from the Fourteenth to the Nineteenth Century*. Cambridge University Press, Cambridge.

McGovern, T. H. (1981). The economics of extinction in Norse Greenland. In Wigley, T. M. L., Ingram, M. J., and Farmer, G. (Eds.) *Climate and History*, pp. 404–433. Cambridge University Press, Cambridge.

Meuvret, J. (1971). Les crises de subsistances et la démographie de la France d'ancien régime. In Meuvret, J., *Études d'Histoire Économique*, pp. 271–278. Armand Colin, Paris.

Mitchell, B. R. (1975). *European Historical Statistics, 1750–1970*. Macmillan, London.

Mokyr, J. (1983). *Why Ireland Starved: A Quantitative and Analytical Study of Irish Poverty, 1800–1850*. George Allen and Unwin, London.

Ogilvie, A. E. J. (1981). Climate and settlement in eighteenth century Iceland. In Parry, M. L., and Smith, C. D. (Eds.) *Consequences of Climatic Change*. University of Nottingham, Nottingham, UK.

Parker, G., and Smith, L. M. (1978). *The General Crisis of the Seventeenth Century*. Routledge and Kegan Paul, London.

Parry, M. L. (1978). *Climatic Change, Agriculture and Settlement*. William Dawson & Sons, Folkestone, UK.

Parry, M. L. (1981). Climatic change and the agricultural frontier: A research strategy. In Wigley, T. M. L., Ingram, M. J., and Farmer, G. (Eds.) *Climate and History*, pp. 319–336. Cambridge University Press, Cambridge.

Pfister, C. (1975). *Agrarkonjunktur und Witterungsverlauf im Westlichen Schweizer Mittelland 1755–1797*. Geographisches Institut der Universität Bern, Bern.

Pfister, C. (1980). The Little Ice Age: Thermal and wetness indices for Central Europe. *Journal of Interdisciplinary History*, **10**, 665–696.

Post, J. D. (1977). *The Last Great Subsistence Crisis in the Western World*. The Johns Hopkins University Press, Baltimore, Maryland.

Warrick, R. A., and Bowden, M. J. (1979). The changing impacts of droughts in the Great Plains. In Lawson, M. P., and Baker, M. E. (Eds.) *The Great Plains: Perspectives and Prospects*. Center for Great Plains Study, University of Nebraska, Lincoln, Nebraska.

White, L. (1962). *Medieval Technology and Social Change*. Oxford University Press, Oxford.

Wilson, C. H., and Parker, G. (Eds.) (1977). *An Introduction to the Sources of European Economic History 1500–1800*. Cornell University Press, Ithaca, New York.

Wrigley, E. A., and Schofield, R. S. (1981). *The Population History of England 1541–1871: A Reconstruction*. Harvard University Press, Cambridge, Massachusetts.

Climate Impact Assessment
Edited by R. W. Kates, J. H. Ausubel and M. Berberian

CHAPTER 12
Microeconomic Analysis

C. A. KNOX LOVELL* AND V. KERRY SMITH†

* *Department of Economics*
University of North Carolina
Chapel Hill, North Carolina 27514 USA

† *Department of Economics*
Vanderbilt University
Nashville, Tennessee 37235 USA

12.1 INTRODUCTION

The purpose of this paper is to review the methods available for economic impact analysis of changes in exogenous conditions that affect economic activities. We specifically focus on the use of these methods for the evaluation of the economic impact of natural or human-induced climate variability and change. For the most part, we are deliberately vague concerning the specification of exposure units—the type of economic activity, the particular human population and the geographic area being impacted by the climate variation. We are also deliberately vague concerning the source and the nature of the climate variation whose impact is to be analyzed. The techniques we discuss, however, are flexible enough to be suitable for a wide variety of climatic impacts and a broad range of exposure units.

It is generally acknowledged that climate is an important natural resource that supports both production and consumption activities around the globe. An early recognition of this role of climate appears in Landsberg, 1946; more recent discussions can be found in Taylor, 1974; d'Arge, 1979; and World Meterological Organization, 1980. It is also generally acknowledged that climate is a hazard that can impose severe economic costs, globally as well as locally. For example, Burton *et al.* (1978) studied the economic effects of three climate-induced natural hazards (drought, flood, tropical cyclone) on pairs of industrialized and developing countries. The economic impact of natural disasters on human resources in developing countries and on economic resources in industrialized countries is reported in World Meteorological Organization, 1980. The economic impact of drought in the Sahel region of Africa has been studied by many scholars, including Kates (1981), García (1981), and García and Escudero (1982).

Despite this widespread acknowledgement of the dual role played by climate, however, there has been insufficient recognition of the fact that climate is fundamentally different from other exogenous shocks affecting economic activity. These special features affect how the analyst should represent the role of climate and its variations in evaluating its impacts on economic activities. Consequently, we begin our review in Section 12.2 with a discussion of the role of climate in economic activities. Before we can consider the implications of climate for the economic behavior of households and firms, it is essential to define climate and to understand the ways in which climate affects the economy at this disaggregated level. After these relationships are described, it is then possible to describe the specific objectives of economic impact analyses and the models used to undertake them.

Economic impact analysis means different things to different analysts and policy-makers. In the broadest terms it is a set of procedures for gauging the implications of changes taking place outside an economic system for the economic activities that take place within it. A clear evaluation of the prospects for using economic impact analysis to understand the prospective social and economic consequences of our climate resources requires that the conceptual foundation of economic impact analysis be described in some detail. In Section 12.3 we briefly discuss four types of models that have been used to conduct economic impact analyses—input–output, macroeconometric, microsimulation, and systems-dynamic models. Input–output models have been extensively used at regional and national levels in a wide variety of countries, and at a global level by Leontief *et al.* (1977). Macroeconometric models are also widely used at regional and national levels, although typically not for the purpose of examining the implications of climatic impact. Examples of macroeconometric models abound in the literature. Models have been constructed for the economies of most countries. Among the most well known of these models for the United States are the Wharton EFA and Data Resources Institute models. The Korean Agricultural System Simulation Model (KASM) illustrates the use of microsimulation models, while the SAHEL model of the human-ecological dynamics of the Sudano-Sahel region of West Africa provides an example of the systems-dynamic models (see Glantz *et al.*, Chapter 22). Each of these types of models is examined in Meadows and Robinson (forthcoming). In our discussion of these models we avoid referring to specific examples, but rather focus on the way in which each type of model establishes a linkage between the exogenous shock to the economic system and the response of that system.

One approach to the analysis of economic impacts is to measure the benefits and costs of an impact, or of alternative responses to an impact, or of alternative strategies to avoid or mitigate an impact. For example, d'Arge (1979) reports the results of a variety of benefit/cost analyses of human-induced climatic changes, many of which originally appeared in d'Arge *et al.* (1975). Other applications and critical evaluations of benefit/cost analysis to climatic impact can be found in d'Arge and Smith (1982), and d'Arge *et al.* (1980).

In some areas of economic policy-making (for example, regulatory impact analyses prepared as a part of the standard-setting process within the United States Environmental Protection Agency), economic impact analyses and benefit/cost analyses have served quite different roles. The same distinction has not been used in evaluating climate impacts. In this case, benefit/cost analyses have generally been treated as a type of economic impact analysis. These differences in vocabulary need not confuse our objectives. The reasons for these distinctions in other policy uses of economic impact and benefit/cost analyses follow from differences in the ways their results can be interpreted.

Economic impact analyses attempt to measure the effects of an exogenous change on the allocation of resources. If climate patterns change in the midwestern United States, will agricultural activities be maintained at the same levels? Will more or less irrigation, labor, fertilizer, etc., be used? The answers to these questions are descriptions of the changes in resource usage that would result from the change in climatic conditions. They do not necessarily imply that the individuals affected by these changes will have either enhanced or reduced levels of economic well-being as a result. By contrast, benefit/cost analyses seek to organize the information associated with a change (whether a conscious policy change or an external change outside man's control) in order to make judgments as to whether it improves the resource allocation (that is, increases the levels of economic well-being experienced by individuals and realized through the use of given resources). Therefore, in Section 12.4 we discuss the main features of benefit/cost analysis.

Finally, we discuss briefly in Section 12.5 the special features of climate that affect the use and interpretation of impact methodologies. The three most prominent features of climate are its stochastic nature, stressed by d'Arge and Smith (1982), Heal (1984), and McFadden (1984); the potentially broad geographic scope of its impact, which makes interjurisdictional cooperation desirable but difficult if different jurisdictions are affected differently; and the temporal nature of its impact (see also d'Arge *et al.*, 1980). Each of these features of climate requires that traditional impact methodologies be modified.

12.2 CLIMATE IN ECONOMIC ANALYSIS

12.2.1 A Definition of Climate

Climate describes the probable weather patterns that can be expected, on the basis of past experience, to prevail at a specific location during a given period of time. It can comprise a detailed array of features of these weather conditions, including temperature, rainfall, humidity, cloudiness and the like. In formal terms we can consider climate as a set of random variables used to describe the outcomes of a stochastic process—the interaction of atmospheric circulation with solar radiation and the oceans and land masses—to determine a pattern of weather events. Our description is based on the supposition that there is sufficient regularity in the processes determining climate that we can assume that these random variables have probability distributions with finite means and variances. Naturally these parameters will be specific to both the geographic location under consideration and the time of year. These definitions follow directly from Hare's description (Chapter 2 in this volume).

Within this framework it is possible to distinguish a variety of types of climate variation. To begin, it may be desirable to assume that the parameters associated with the probability distributions for the random variables describing the climate are themselves subject to change. Cycles of wet or dry conditions that occur with sufficient regularity may be considered (in this framework) to be the result of a particular functional relationship between the mean precipitation and time. Weather patterns may nonetheless exhibit variation about this cycle. The important point to note in this distinction is the ability to discern over time a pattern for the average weather conditions that describes how they are changing.

In order to define climate variation it is necessary to specify an appropriate time span. Although distinctions such as short-term (interannual), medium-term (decadal), and long-term (century) are useful, the current meteorological convention is to use 30-year 'normals', updated each decade. In practice, the use of 30-year normals gives fairly stable values of central tendency and variability for most features of climate. As we shall explain below, however, the use of such long time-frameworks is not always convenient for economic impact analysis, and it is of little relevance for the analysis of the economic impact of climatic hazards.

Within the context of any normal, we can identify three characteristics of climate—noise, variability, and change. Climate noise refers to the differences that occur in climate parameters between periods, as a sole result of the positioning of the start and end of the averaging periods. It has no real significance, and is henceforth ignored. Climate variability refers to fluctuations of the climate parameters within the averaging period, and is described by the 'spread' parameters of the probability distribution of the climatic random variables. These parameters describe the nonsystematic fluctuation in weather conditions and whether certain descriptions of the features of climate exhibit variations that move together. Finally, climate change refers to movements between successive averaging periods in the climatic parameters in excess of what noise can account for. Climate change is measured by the trend over averaging periods in various climatic parameters, and represents a systematic pattern of weather conditions (see Hare, Chapter 2 for further details).

For the purpose of our analysis we also distinguish natural sources from human-induced sources of climate variability and change. A human-induced variation occurs when one or more of the parameters of the probability distributions is affected directly or indirectly by human activities. These activities induce a change in the functioning of the atmospheric system. While such variations can occur over short or long periods of time, they are not the result of an evolution of the atmospheric interactions determining climate. This distinction may seem arbitrary and unnecessary for any physical description of climate and its effects on economic activities. It is, however, important to our analysis of the interaction between climate and economic activities, since it implies that there is the potential for a two-way relationship between the climate system and human economic activities.

12.2.2 Economic Implications of the Definition of Climate

The definition of climate leads naturally, and usefully for our purposes, to the interpretation of climate as a regional public good. That is, climate is a resource whose features and variation provide services to each economic agent (such as firms and households) in a given location without reducing the total amount of these services available for other agents in that location. (See Haurin, 1980, for a similar treatment of climate.) Obviously, the services provided by climate can be either beneficial or deleterious. Once an economic agent has selected a geographic location, it receives the climate services relevant to that region. Climate services are not the result of conscious consumption and production decisions by households and firms; they are the natural endowments to each region that influence these consumption and production decisions. Thus, in contrast to other public goods, such as police and fire protection or national parks, we are not concerned with determining the socially optimal level of provision of the services to meet the needs of consumption and production. Consequently, one might ask whether there is an economic problem associated with climate resources. That is, since we do not consider that the services of climate are provided by human actions and therefore subject to control, it may not be sensible to consider the economic implications of alternative climate regimes. Indeed, to the extent that there is full information on climate conditions, we would expect that firms and households would take these conditions into account in their decisions as to where to locate, what to produce and consume, and so forth. We do not expect to find farmers planting citrus trees in locations with unsuitable climate, or consumption activities sensitive to cold conditions to locate where low temperatures can routinely be expected.

To answer our own question, there is of course an economic problem associated with climate resources. The economic problem is not one of optimal *provision*, since climate is a regional public good, but rather one of *adjustment* to climate endowment, variability and change. It is, nonetheless, important to recognize the public attributes of climate because they preclude private markets from providing direct information on the value of climate services. Rather, this information must be derived by indirect means, through a recognition of the adjustment mechanism available to households and firms by means of migration and other responses. These adjustments are fundamental to the methods used for analyzing the economic impacts of climate and its variation.

It is reasonable to expect households and firms to react to climatic shocks in the same basic manner that they are supposed to react to other exogenous shocks to their economic environment. The cases where these conditions are most easily recognized involve agricultural production. For example, if production activities require controlled temperature or humidity conditions, they will be located (other things being equal) where these conditions can be maintained at least cost. Several authors have provided empirical evidence that individuals do consider climate among other environmental amenities in considering the real wages they are willing to accept (see Hoch, 1974; Rosen, 1979; Cropper and Arriaga-Salinas, 1980; Smith, 1983). If these empirical associations do reflect workers' attitudes toward climate, they will also influence labor supply conditions facing employers in different locations and thereby indirectly influence locational decisions.

The opportunities available to economic agents for behavioral response are governed by a number of factors. One factor is the length of time over which the response is observed. Long-term responses are termed 'adaptations' by geographers and 'full adjustments' by economists, while short-term responses are termed 'adjustments' by geographers and 'partial adjustments' by economists. The essence of the distinction in either parlance is that the planned or desired long-term response (through disinvestment, migration and reinvestment) to climatic shock is only partially completed in the short term. For examples of this temporal distinction see Wigley *et al.*, Chapter 21. The process of response is also influenced by the nature of the economic society being impacted by climatic shock—that is, the nature of the exposure unit. Thus, a technologically advanced, industrial society typically has more options available to it than does a pastoral or a self-provisioning society. For examples of this distinction see Burton *et al.* (1978); see also Chapter 21, this volume. A detailed analysis of response to climatic shock in self-provisioning societies is available in Jodha and Mascarenhas (Chapter 17, this volume). A third factor affecting the opportunities for behavioral response to climatic shock is the policies adopted by governments to deal with it. Such policies can influence the nature of the climatic shock (for example, the timing, magnitude and/or location) if it is human-induced, as in the cases of carbon dioxide, chlorofluoromethanes and other pollutants. They can also influence the nature of the response to either natural or human-induced climatic shock. Various policies for controlling, mitigating and adapting to the carbon dioxide problem are explored in Nordhaus (1980), Lave (1981) and Noll (n.d.).

How can economic models be structured to reflect these influences on the behavior of economic agents? The answer to this question depends rather importantly on how we define the climatic influences themselves. In our discussion of economic behavior we have implicitly accepted the view that economic agents make their decisions as if they were acting rationally in the pursuit of their own self-interest. Economists generally assume that households adjust their commodity demands and their labor supplies in an effort to maximize utility, and that business firms adjust their input demands and output supplies in an effort to maximize profit. Since household commodity demands and labor supplies can be expected to depend on climate, so too can the maximized utility of households. Similarly, since firm input demands and output supplies can be expected to vary with climate, so too can the maximized profit of firms. For most economic systems these behavioral models offer viable descriptions of the factors motivating economic responses. In some economic systems, where overt markets are not sanctioned, it may be more difficult to determine household preferences and firm objectives based on the goals pursued by households and businesses. Nonetheless, if households and businesses pursue any goals at all, then their optimizing behavior as reflected in their demands, supplies, and optimized objectives can be expected to depend on climate. Although goals and methods of pursuit may vary across economic systems, the essential dependence of purposeful, goal-seeking behavior on climate does not.

In such a framework the distinction between climate as a fixed flow of services (described by variables such as temperature, rainfall, etc.) versus a random flow of services (described by frequency distributions of these variables) delivered to firms and households is important. In the former case optimal behavior can be defined as conditional on a specific set of values for the variables defining climatic conditions. In principle, any change in one of these variables would imply a corresponding change in the optimal behavior patterns of the economic agents involved.

This conclusion is easily demonstrated with a formal example. Consider a general statement of the objectives of a profit-maximizing firm given in equation 12.1:

$$\max \quad \{P \cdot Q - \sum_{i=1}^{n} r_i X_i : Q \leqslant f(X_1, \ldots, X_n, \bar{C})\} \tag{12.1}$$

In this statement it is assumed that the firm employs inputs $X_i \geqslant 0, i = 1, \ldots, n$, available at fixed prices $r_i > 0, i = 1, \ldots, n$, to produce a single output $Q \geqslant 0$ for sale at fixed price $P > 0$. The activities involved in transforming inputs into output are assumed capable of being described by a production function $f(\cdot)$, a function that describes the maximum output levels that can be achieved for any combination of inputs. Finally, we have included in our statement of the problem the influence of climate, by assuming that the maximum output obtainable from various input combinations is conditioned on the realized vector $\bar{C}$ of climate variables. Using conventional methods of optimization, we can describe the maximum profit π^* that can be realized with this production constraint for alternative output and input prices and for alternative realizations of the vector of climate variables as

$$\pi^* = \pi^*(P, r_1, r_2, \ldots, r_n, \bar{C}). \tag{12.2}$$

Moreover, since maximized profit depends on input and output prices and on the realization of climate, so too do the input demands and output supply that maximize profit, and we have

$$\begin{aligned} Q^* &= Q^*(P, r_1, \ldots, r_n, \bar{C}). \\ X_i^* &= X_i^*(P, r_1, \ldots, r_n, \bar{C}), i = 1, \ldots, n. \end{aligned} \tag{12.3}$$

To the extent that climate variables influence production activities in a manner suggested by equation 12.1, they must also affect maximum profit, as well as the profit-maximizing output supply and input demand given by equations 12.2 and 12.3. Consequently, a change in any one climate variable, say the kth feature of climate, $\bar{C}_k$, which might be average temperature, may imply by equation 12.3 a reorganization in the patterns of input usage and an adjustment in the amount of output supplied. The partial derivatives of Q^* and X_i^*, $i = 1, \ldots, n$, with respect to $\bar{C}_k$ describe these adjustments, while the partial derivative of π^* with respect to $\bar{C}_k$ describes the consequences of such a change for the firm's profit. This approach to modeling the effect of climate on production activities by deriving the effect of climate on observable supply and demand equations is described in more general terms and in greater detail by Diewert (1982). A similar construct can be developed to model the effect of climate on utility-maximizing households.

Once we generalize the description of climate, by recognizing that it is described by a set of random variables, then we must inquire as to whether we can expect economic agents to deal differently with climate changes that are not certain as opposed to those that are. The answer is that we can. Analysis of climate in stochastic terms requires more complex models to incorporate some conception of how economic agents respond to climate-induced uncertainties.

Some insight into what such models might imply as compared with frameworks that treat climate variables as realized, and hence stochastic, can be derived from an analogy. An early paper by Stigler (1939) on the significance of uncertainty for business behavior considered the implications of uncertainty in the output level to be produced at a given plant for the way in which the plant might be designed to produce output. Conventional micro models unconcerned with uncertainty assume that firms know exactly what output levels they are required to produce at each plant. Thus, a plant can be designed so as to minimize cost for *the output level* it is producing. If the output level is not certain, however, the situation changes completely. Stigler argued that it is entirely reasonable to expect that firms seek to design plants that minimize cost over the most likely range of output rather than for a single output level. This alternative view provides a plausible conception of business behavior in the presence of demand uncertainty. It recognizes the need to accommodate, with little change in average cost, a range of production levels. Indeed, it may well be that the mean average cost over these output levels exceeds that realized with a single output level. Yet the best strategy remains the one associated with designing the plant to accommodate a range of outputs. Stigler referred to such modifications in plant design as a means of promoting *plant flexibility*.

The difference in unit cost between the two types of plant designs is the 'price' of the flexibility in the firm's operations. McFadden (1984) used the same type of framework to illustrate how climate considerations can be incorporated in a neoclassical cost function when climate is treated as a random variable. Indeed, using a fairly simple three-activity model, where demand for output varies with the random climate variable, he has clearly demonstrated the impact of variability in climate on both the technological design of the plant and the resulting cost. Climate variability in this framework provides an incentive to diversify activities and, with it, increases cost. However, these cost increases are *not* as great as they would be if only one activity were used to produce output over the range implied by the climate variability.

With a stochastic conception of climate, economic models describe the responses of economic agents to climate uncertainty as consisting of a restructuring of their activities to permit them to accommodate a range of values of the climate variables. The specific details of each type of economic agent's responses are likely to be related to a variety of factors, including:

1. each agent's perception of the nature of the likelihood for variation in climate attributes (that is, the individual's perception of the multivariate probability distribution describing the climate random variables);
2. each agent's attitude toward the risk of economic losses as a result of climate uncertainty; and
3. the costs of incorporating flexibility (that is, the ability to accommodate a range of climatic conditions) in the particular activities involved.

Thus, information, attitude, and the cost of resiliency are important determinants of the pattern of flexibility present in any sector and, in turn, the nature of the response to a change in climate. This conclusion must, however, be carefully interpreted. First, since this framework conceives of climate as a vector of random variables, any change in climate may not necessarily lead to an immediate change in the weather patterns which the economic activities have been designed to accommodate. The arguments we have described earlier, as well as McFadden's (1984) formal models, assume that all economic agents *know* the exact nature of the probability distributions for the climate random variables from the observed weather patterns.

Consider an example. Suppose there is a change in the atmospheric system that leads to a higher average temperature in a given region. Such an alteration implies a lower likelihood of the cooler temperatures that a given economic activity may have been designed to accommodate and a higher likelihood of warmer temperatures. Depending on the magnitude of the change in the mean, it may not be detected immediately by economic agents, since the resulting weather conditions may still be capable of being accommodated by the system. Of course, this conclusion depends on the extent of flexibility in the activity which, in turn, depends on each of the three factors discussed above. The important aspect of the comparison for our purposes is that the economic impact of a climate change may not be as immediate as would be implied by the nonstochastic conception of climate. Furthermore, the magnitude of the effect depends upon what the economic agents in the region perceive to be the likely variation in temperature and on their willingness to assume the risks of any economic losses associated with variations in temperature.

The complications to economic modeling that arise from treating climate services as random variables are important because they affect our ability to judge the economic impact of climate variability and change in response to human activities.

12.3 ECONOMIC IMPACT ANALYSIS

12.3.1 General Background

In most economies, policy-making relies on economic impact assessments to judge the effects of proposed policy initiatives or of factors outside a government's control that might nonetheless impinge on the economic activities taking place within its national boundaries. Because these studies have been undertaken for a wide variety of problems as well as to serve a diverse array of objectives, definition of what comprises an economic impact analysis is difficult. We stated, at the outset, that economic impact analyses attempt to gauge the magnitude and sectoral composition of the resource allocation changes that accompany an external change to an economy. To the extent that these changes require greater resources used to accomplish the same objective, then the external change has required adjustments that divert resources for one use to another where they were not previously needed. When the changes under study are the result of direct actions, it is often assumed that they represent attempts to improve an existing source of inefficiency in the economy. Consequently, for these cases, it is conventional practice to compare the benefits associated with the resource allocation changes with the costs of those changes.

Thus economic impact analysis attempts to measure the extent and types of adjustment that accompany an exogenous change, and benefit/cost analyses provide the basis for appraising the desirability of the change *as if* it were discretionary.

An example of an impact analysis would be an assessment of the influence of foreign steel manufacturers' export pricing behavior on the domestic steel industry of a nation. A second example would be an analysis of the consequences of a particular domestic regulatory program on specific industries.

These types of economic impact analysis seek to predict the nature of the changes in economic activities that accompany the external action. Often the particular action under study has not taken place before, so that analysis must relate the action to a parallel change that has occurred. In other words, the analysis must provide a mechanism for 'second-guessing' the response of economic agents. Of course, the process of second-guessing is easier if one can describe formally the behavior of these agents, as in equations 12.1–12.3 above, for example.

Applying such formal descriptions can involve detailed empirical analyses of the past responses of economic agents, as well as judgments about the limitations of these models as descriptions of economic behavior under alternative circumstances. For example, most economic theories distinguish the responses economic agents make in the short run from those possible in the long run. The distinction rests primarily with the cost of adjustment to certain resource allocations. Businesses do not easily modify their capital stocks. Movement from one location to another, by businesses or by households, is costly and difficult. Institutional constraints on economic behavior are not relaxed quickly. Therefore in gauging the economic consequences of an action one must recognize that the complete adjustment described by long-run economic models, such as that described in equations 12.1–12.3 above, does not provide an adequate characterization of short-run economic responses.

Economic impact analyses are typically organized to highlight the economic groups (or sectors) that gain resources and those that lose them. Attention is also focused on the degree of adjustment that such actions impose on particular types of economic agents. A good example of the use of such analyses to gauge the 'strain' imposed on segments of the economic infrastructure can be found in the analysis of boomtowns from the development of energy resources in the western United States (see Cummings and Schulze, 1978). Presumably, the objective of economic impact analysis in this case is to judge whether the local, private economic agents are capable of *efficiently* responding to the increased resource demands arising from the development.

To undertake this analysis, it is preferable to have a formal description of the economic activities affected by the action. Such descriptions must be consistent with the institutional restrictions governing resource allocations. Thus, in a market economy the model can be a characterization of the markets involved, while in a planned economy the model characterizes the planning mechanisms.

12.3.2 Describing Economic Activities

There are two essential ingredients to an economic impact analysis. The first is a formal description, usually a mathematical model, of the economic activities that are assumed to be affected by the action under evaluation. The second is a mechanism for linking the action to be evaluated to the model of economic activities. While the decisions made on these components to the analysis are clearly intertwined, for ease of exposition we deal with each separately, deferring the second to Section 12.3.3.

Four classes of models have been used to provide empirical descriptions of economic activities: input–output, econometric, microsimulation, and systems-dynamic models. We describe below, in simple terms, the main features of each of these classes of models. Since the models within each class are heterogeneous, our descriptions are intended only to highlight some of the most important features of the models for economic impact analysis of climate variability and change.

12.3.2.1 Input–Output Analysis

Input–output analysis is based on a recognition that production activities are interdependent. The outputs of some industries are inputs to others and vice versa. Thus, judgments as to the input quantities required to produce a given level of output for any particular industry depend on the input requirements for all industries. Of course, the specific degree of interindustry dependence is an empirical question. The objective of input–output analysis is to provide a consistent modeling framework for describing these interdependencies.

Input–output analysis incorporates the interconnections in production by enumerating the requirements for each potential input and output in the economy which is to be described. This economy may, in principle, be any size. The quality of description in each case depends on the accuracy of its characterization of the production activities taking place at that level. Input–output analysis generally assumes that the relationships of each unit of input to each type of output are constant (over the range of applications in the model).

The structure of input–output models imposes consistency between the defined production levels for each good or service and the internal (that is, interindustry) and external demands for them. If $a_{ij} \geqslant 0$ designates the requirements for input i to produce one unit of commodity j; X_j represents the amount of commodity j that is produced; and $d_j \geqslant 0$ corresponds to the external (that is, by households, government, and the foreign sector) demand for commodity j, we can describe input–output analysis with balancing conditions. These conditions are illustrated for an economic system consisting of k industries and an external sector by

$$\begin{aligned} X_1 &= a_{11} X_1 + a_{12} X_2 + \ldots + a_{1k} X_k + d_1 \\ &\;\vdots \\ X_k &= a_{k1} X_1 + a_{k2} X_2 + \ldots + a_{kk} X_k + d_k \end{aligned} \tag{12.4}$$

The model clearly recognizes that we cannot use more than we produce and that production activities often require some of their own outputs. For example, electricity generation requires some of the generated electricity for its production activities, but there remains a positive net output.

Input–output analysis was designed to facilitate planning the levels of the Xs that are required to meet some predetermined demands—the ds, given the constraints of production technology embodied in the a_{ij}s. It is not a full description of all the interactions involved in the equilibrium determination of all quantities and prices in an economy. There are no feedbacks from the description of production activities (and their implied costs) to final economic demands. Moreover, the approach treats interindustry interaction as purely a function of technical relationships. In a more general characterization of production activities we might recognize that the a_{ij}s are not constants, but depend upon how firms organize their production activities. Jones (1965) offered one of the first analytical descriptions of the general equilibrium interaction of economic activities in these terms. The empirical models of the role of energy in economic activities developed by Jorgenson and his associates (i.e., Hudson–Jorgenson, 1974; and Jorgenson–Fraumeni, 1981) reflect such considerations. These extensions are to be distinguished from input–output models that postulate that some or all of the a_{ij}s can change. The empirical models of Jorgenson provide a behavioral framework for describing why they change and, therefore, for predicting how they can be expected to change under specified conditions.

Input–output analysis, without these refinements, considers whether it is possible to solve the equations for the X_js, given the a_{ij}s and the d_js. That is, is there a potential mathematically consistent description?

The use of input–output models for economic impact analysis maintains that the changes induced by the action are determined exclusively by technical production considerations. The mutual interaction described by markets is ignored and the associated feedback effects are implicitly assumed to be unimportant. These are serious shortcomings, which must be balanced against the ease of application and detail of most input–output models.

12.3.2.2 Econometric Models

Econometric models, considered as a class of modeling structures for economic impact analysis, comprise the most heterogeneous category. For our purposes it is probably best to organize our discussion of them according to each model's relationship to a behavioral framework. More specifically, microeconomic theory provides analytical descriptions of the behavioral responses of economic agents to exogenous changes in conditions affecting them. These descriptions are provided by the demand and supply functions we derived earlier in equations 12.1–12.3, for example.

The correspondence between the econometric model and economic behavior is closest when the analysis relates to the most disaggregate level—describing the behavior of representative businesses or households. While analysts may legitimately question in some cases the processes through which such models have been estimated (that is, using micro data on individual economic agents or 'averages' for broad classes of agents), the frameworks themselves can be evaluated by their consistency with the economic behavior they are designed to depict.

As the level of aggregation increases and the correspondence between the specified components of an econometric model and the behavioral responses of economic agents diminishes, it is more difficult to interpret the models, evaluate their plausibility, and, especially important, use them in all forms of economic impact analysis. This last issue will be particularly important to benefit/cost analysis because without a direct association between the empirical models used to predict effects of external actions and a behavioral model, it is not possible to translate those predicted effects unambiguously into a consistent measure of change in economic well-being.

Regional econometric models (see Harris and Hopkins, 1972, as an example) are the most popular econometric structures used in economic impact analysis. These structures typically seek to describe economic activities for an arbitrarily defined regional unit, such as a county or district. They do so without clear behavioral foundations for the models. Regional models are based on hypothesized associations between economic aggregates that are assumed to be realized as 'approximations' to the responses taking place at the micro level. While there are good reasons to organize the description of economic activities according to the nation in which they are undertaken, the same economic rationale for small regional units, such as counties or districts, defined because these units are the basis for the data available, is more suspect. (For a discussion of this issue in terms of regional energy models see Freedman, 1981.)

An important distinction which can be used to classify econometric models is whether they describe the processes through which a set of economic variables is jointly determined or focus instead on the impacts of exogenous variables on the measures of economic activity. A member of the first class of models is generally described as a structural model, while a member of the second is a reduced form model. Most structural models are used to describe aggregated economic activities, that is, a region, nation, or, along another dimension of aggregation, an industry. It is, of course, possible in principle to solve structural models for their implied reduced form models.

To illustrate the distinction between structural and reduced form models, consider a very simple macroeconomic model introduced by Lawrence Klein (1950) as one of the first econometric applications of Keynesian economics. It includes:

1. a consumption function describing aggregate consumption (C_t) as a function of aggregate profit (P_t) and the aggregate wage bill (the sum of private wages W_{pt} and government wages W_{Gt});
2. an aggregate investment function relating investment expenditures (I_t) to current and past levels of profit and the past capital stock (K_t);
3. a wage equation relating private wages to current and lagged levels of private output (that is, disposable income Y_t plus taxes T_t less the government wage bill W_{Gt}) and
4. definitions of investment and total output in terms of components of income (W_{pt}, W_{Gt}, P_t) and in terms of the types of goods produced (C_t, I_t, G_t = government spending, T_t). Equations 12.5a through 12.5f detail the model.

$$C_t = a_1 + a_2 P_t + a_3 (W_{pt} + W_{Gt}) \tag{12.5a}$$

$$I_t = b_1 + b_2 P_t + b_3 P_{t-1} + b_4 K_{t-1} \tag{12.5b}$$

$$W_{pt} = c_1 + c_2 (Y_t + T_t - W_{Gt}) + c_2 (Y_{t-1} + T_{t-1} - W_{Gt-1}) \tag{12.5c}$$

$$I_t = K_t - K_{t-1} \tag{12.5d}$$

$$Y_t = W_{pt} + W_{Gt} + P_t \tag{12.5e}$$

$$Y_t = C_t + I_t + G_t - T_t \tag{12.5f}$$

The subscript t refers to the tth year. This structural model includes six linear simultaneous equations for the determination of consumption, investment, the private wage bill, capital, profit and disposable income. Time-series information on the relevant variables would be used to estimate the model's parameters (i.e., the *a*s, *b*s and *c*s).

We might also solve the model relating current values of each endogenous variable (i.e., C_t, I_t, K_t, P_t, W_{pt} and Y_t) to the values of these endogenous variables determined earlier in time (i.e., at $t-1$, $t-2$, etc.) and to exogenous variables determined outside the model (i.e., G_t, T_t, W_{Gt}). This description of the same endogenous variables is the reduced form model.

One of the most important difficulties in applying the econometric models associated with aggregated relationships for economic impact analysis arises in establishing the linkage between the action under study and its implications for the model. Unless it can be expected to change the features of one or more component equations in these models, because of the effects on the microeconomic responses underlying these equations, there is no sound means of making the change. For example, in the simple model described above, where would climate most logically enter?

12.3.2.3 Microsimulation Models

This aggregation problem provides much of the motivation for interest in microsimulation models. These models attempt to mimic economic activities at the micro level with a manageable number of economic agents. That is, they describe commodity demands and labor supplies of a limited number of households in an optimizing framework, with distinctions among households in the model made on the basis of characteristics thought to be important to the analysis (such as income or location). Similarly, a limited number of firms are described as the suppliers of goods and services consumed by the households and the employers of the labor services supplied by the households. The models generally assume ideally functioning markets.

What distinguishes many of the microsimulation models is how they specify the two dimensions of the economic interactions of a market-oriented economy and how they define an equilibrium condition. One of the most appealing approaches for structuring models uses a numerical general equilibrium (NGE) framework (see Scarf, 1973; and Shoven and Whalley (1984), for an overview and review). In these structures, the equilibrium condition is defined by the non-negative price vector (for goods and services) that assures all excess demands are nonpositive. At these prices the market demands can be satisfied with the available supplies.

In principle this approach offers detailed micro descriptions of economic agents' behavior *and* permits an assessment of the full impact of proposed actions with a consistent accounting of the economic interactions. This also implies that benefit/cost analysis can be based on these models. These gains, however, are not realized without cost. The NGE models rely on the analyst's ability to characterize a limited number of representative economic agents so that the sources of economic interaction are manageable within the NGE framework. Given the rapid acceptance of these models for policy analysis and their flexibility in dealing with policies that impact at the level of individual economic agents, this limitation does not appear to have posed problems in most policy-related uses (e.g., Whalley 1977, 1980).

12.3.2.4 Systems-dynamic Models

A final category of models—systems-dynamic frameworks—has gained considerable attention since the controversial Meadows *et al.* work, *The Limits to Growth*, was published in 1972. They are based on the notion that it is easier to characterize the dynamic relationships within large complex economic systems than it is to describe the behavioral and institutional motives that give rise to these relationships. In general, these models have not survived the test of a careful analysis of their implicit behavioral assumptions when they have been used to describe economic processes (see, for example, Nordhaus, 1973). Therefore, we do not regard these structures as a sound basis for consistent economic impact analyses. A discussion of applications is available in Robinson, Chapter 18, this volume.

12.3.3 Linking the Action to the Model

In order to use any of the modeling structures just described for evaluating the economic impacts of some action (or change in the external conditions affecting economic activities), the models must be altered to reflect the changes implied by the action. For example, if a change in climate is expected to change precipitation so that agricultural production will be altered, the exact nature of these changes must be known. In the context of input–output models, this implies that we must know which input requirements coefficients change and by how much. In an econometric model based on a demand and supply framework, we require some mechanism for linking the climate variation to the demands for or the supplies of the affected commodities. With numerical general equilibrium models, the linkage that must be known is the association with, for example, the agricultural production function; the model then translates the changes in production conditions into the implied supply changes.

The linkage in each case is broadly similar, but the form it takes is model-specific. How does one obtain the information for determining the appropriate linkages to be specified? The answer to this question depends on the action to be evaluated. We can specify three broad, interlinked sources of information:

1. observed patterns of past economic activities and responses to comparable changes (in the action);
2. theoretical models of the process linking the action to activities that are associated with economic behavior;
3. informed judgment.

12.4 BENEFIT/COST ANALYSIS—A TYPE OF ECONOMIC IMPACT ANALYSIS

12.4.1 General Background

Two objectives have provided consistent motivation for economic impact analysis as a part of the policy-making process. The first is an efficiency objective—to appraise whether the external action impinges upon the ability of the economic system to allocate resources to their highest valued uses. The second objective arises from equity concerns—both among income groups and, perhaps more importantly, among regions and/or sectors comprising the economic system.

Economic impact analysis as we have discussed it in Section 12.3 primarily serves this second objective. That is, a description of the reallocation of resources associated with an external action does permit one to assess the regions, sectors and (if sufficiently detailed) the income groups losing resources, as well as those remaining unaffected or gaining from the action.

These descriptive analyses do not, however, permit one to evaluate the efficiency effects of the action. Their objective is to measure changes in the levels of activities without necessarily attempting to measure either individuals' valuation of the changes or increments to firms' costs as a result. A judgment on the efficiency of an action usually implies that it can be controlled (directly or indirectly). The action is subject to choice. There are circumstances in which one would wish to use this reasoning for climate changes. For example, once it is recognized that man's activities can alter climate, then one can consider evaluating the merits of restrictions to those activities.

Benefit/cost analysis does permit judgments to be made that are consistent with evaluating the efficiency effects of certain actions. It does so in a rather special sense which deserves elaboration. Benefit/cost analysis is the practical implementation of welfare economics and therefore maintains that consumer values provide the basis for implementing the efficiency maxim—resources must be allocated to their highest valued uses. Theoretical statements of the problem of welfare economics (that is, the Pareto efficiency conditions) imply that practical implementation of them for efficient resource allocation decisions requires a comparison of the marginal benefits with the marginal costs of those actions under the control of the policy-maker making the judgment. (For further discussion see Bohm, 1973 and Pearce and Nash, 1981; Krutilla, 1981, has also recently offered a historical perspective on the evolution of the analysis.) Unfortunately, in practice the marginal benefit and marginal cost functions typically are not known; judgments are made on the basis of the net benefits (total benefits less total costs) of an action. Nonetheless, Bradford (1970) has convincingly argued that benefit/cost analysis using net benefits can be considered a check on the efficiency of any existing resource allocation. If there are positive net benefits from the change under evaluation, the existing position cannot be efficient in the Paretian sense. Such judgments do not assure that the new resource allocation is the welfare-maximizing one. They do imply, for a reasonably wide class of functions for describing benefits and costs, that the change under consideration provides an improvement.

To illustrate the relationship between the analysis implied by a theoretically ideal approach versus a conventional benefit/cost analysis, consider Figure 12.1. On the vertical axis we have plotted dollars and on the horizontal axis some measure of the activity, A (i.e., controlling the emission of a pollutant thought to influence regional climate) under policy control. $B(A)$ describes the aggregate benefits from the activity and $C(A)$ the aggregate costs. The maximum net benefits $[B(A) - C(A)]$ arising from A are realized for this illustration at A^* where marginal benefits equal marginal costs. Bradford's (1970) interpretation of benefit/cost analysis is that it establishes whether a change represents a movement toward efficiency. This is easily illustrated with the diagram. Suppose the level of activity is OA_1 and we are considering actions that will lead to OA_2. The aggregate net benefits associated with the change from position A_1 are given by the difference between the aggregate net benefits at A_2 (FH) and those at A_1 (EG), or (GH − EF). If positive the movement is toward the efficient level OA^*. We can consider movements either from below or from above A^* with these benefit and cost functions and realize that they are positive only when the movement is in the 'right' direction. Thus, economic impact analysis directed toward efficiency judgment must be conducted in benefit/cost terms, and can be conducted using the net benefits approach.

12.4.2 Measuring Net Economic Benefits

This rationale for benefit/cost analysis seems fairly clearcut, but as a practical matter it relies on the analyst's ability to gauge economic agents' true valuations of the actions and their full costs, that is, the $B(A)$ and $C(A)$

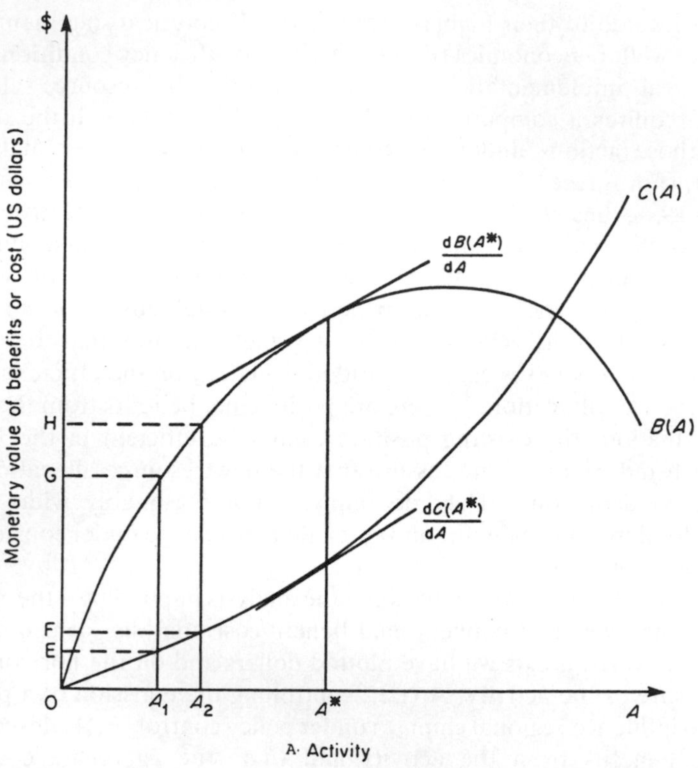

Figure 12.1 A typical distribution of benefits, $B(A)$, and costs $C(A)$, of an activity under policy control, with rising costs and diminishing benefits. A^* is the amount of activity with the largest net benefit, the point where marginal benefits equal marginal costs. An analysis of benefits increase by changing a policy of A_1 to A_2 demonstrates a movement towards efficient use of resources

functions. To do so, benefit/cost analysis has developed a set of measures that have achieved some degree of professional acceptance. It should, however, be acknowledged at the outset that these measures rest on theoretical foundations which are based in comparative analysis of static situations. Among the most important assumptions of this mode of analysis are that the changes under evaluation are small and that adjustments to them are instantaneous and complete.

The measurement of these net economic benefits is based on the behavioral relationships that economists use to characterize the responses of economic agents in markets—demand and supply functions. The demand function describes the maximum amount an individual (who is treated as synonymous with a household for our purposes) is willing to pay for each amount of a good or service. In perfectly competitive markets the supply function describes the

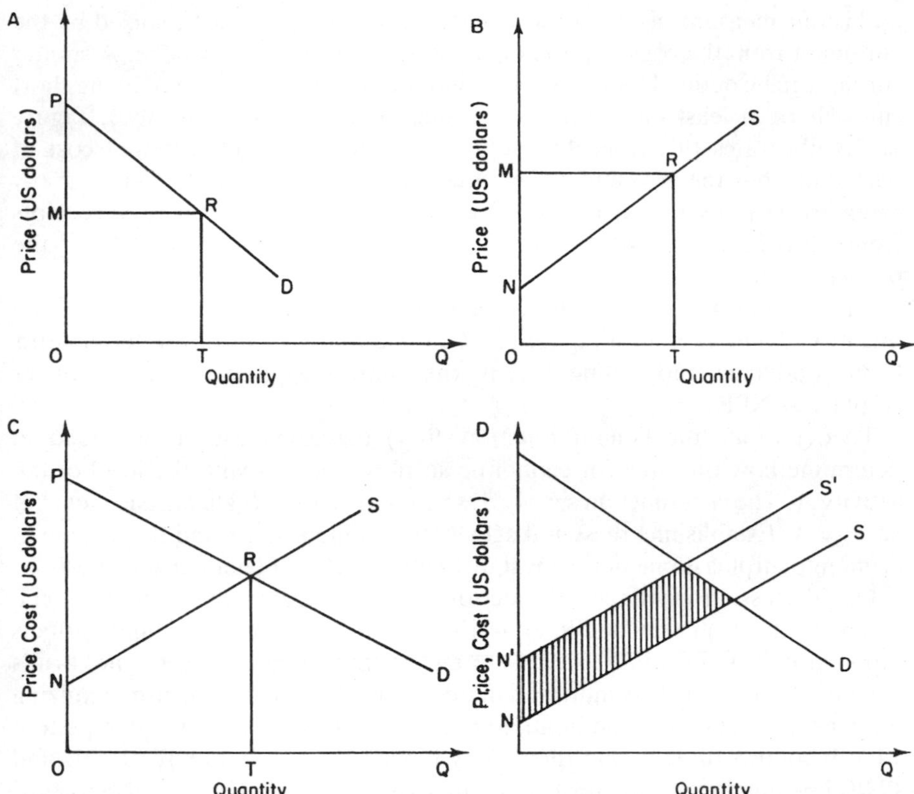

Figure 12.2 A traditional demand function (2A), supply function (2B), equilibrium point (2C) and effect of changes in supply for economic surplus (2D). 2A: A traditional demand function. For OT units, OPRTO is the maximum amount an individual is willing to pay; OMRT, the actual expenditure; and MPR, the consumer surplus. 2B: A traditional supply function. For OT units, OMRTO is the maximum receipts, ONRT the actual cost, and NMR the producer surplus. 2C: Market equilibrium at price of RT, and quantity, OT, with economic surplus, NPR. 2D: Illustration of a negative effect of a climate change, reducing supply from NS to N′S′. Shaded area designates loss in economic surplus

incremental cost of providing each additional unit of the good or service. Considering demand first, Figure 12.2A plots a demand function in the traditional mode (that is, with price on the vertical axis; the function describing demand in this form would actually be referred to as an inverse demand function). OPRTO is the maximum amount an individual is willing to pay for OT units of the good. Given a fixed price per unit for all goods consumed, say OM, total expenditure is OMRT, leaving an excess of the total willingness to pay over expenditure of MPR. This is generally referred to as *consumer surplus* and is our measure of the excess benefits (over what is paid) realized by the consumer from the consumption of OT units of the good or service. A similar surplus can be defined for firms (provided the analysis is conducted in the short run, where at least one input to production is not easily adjusted). Figure 12.2B illustrates this case. The supply schedule is the incremental cost of each unit, thus the area under the schedule is the total cost of producing any given level of output, say OT. With a fixed price, say OM, the total receipts from sales of OT are OMRT less costs of ONRT yields a surplus of NMR—the *producer surplus*.

Putting the two functions together to describe the equilibrium price and quantity in a market as in Figure 12.2C, the *economic surplus* or net benefits from producing and selling OT is the sum of consumer and producer surpluses—NPR.

To determine the benefit function $B(A)$ described earlier, we need to determine how this area (or economic surplus) changes with the level of the activity A. The cost function simply describes the cost of realizing each level of activity A. Establishing these linkages between the activity and the change in economic surplus is one of the most difficult aspects of benefit/cost analysis.

Our discussion of climate in Section 12.2 anticipated this outcome. The behavior of a firm sketched in equation 12.1 maintained that climate affects production activities. As a consequence, climate also affects the firm's maximum profit level, as indicated in equation 12.2, and its profit-maximizing output supply equation and input demand equations, as indicated in equation 12.3. It follows directly that the supply function drawn in Figures 12.2B and 12.2C is a function of climate. Thus changes in $\bar{C}$ will shift S, as from NS to N′S′ in Figure 12.2D. The shaded area represents the [negative] change in economic surplus.

The assumptions underlying this transition are important. If our characterization of behavior is incomplete, as the considerations in Section 12.2 might lead one to conclude, or if we question the assumptions of small changes and localized (partial equilibrium) analysis, then we must also reconsider the conventional approaches to evaluating economic benefits and costs.

12.4.3 Integrating the Components of Economic Impact Analysis

It is important to recognize that to the extent that both dimensions of economic impact analysis—projecting the resource reallocations associated with the action (or change) and evaluating the net economic benefits of the action—are needed, they are not necessarily within the same model. Indeed, for the case of impact analyses undertaken as a part of the policy-making process, it is unlikely that they will have a common model. Different analysts will be involved in each component. However, the failure to use a single model in these two components of impact analysis is not solely a matter of the organization of the analysis staff. The cause is probably best tied to the fact that the preferred models for the first component of impact analysis—input–output and aggregated econometric models—*do not* permit consistent benefit measurement in terms of the concepts discussed in Section 12.4.2.

12.4.4 Valuing Nonmarketed Outputs

Climate has been described as a regional public good. It does not exchange on organized markets, and thus we cannot rely on these markets to provide direct information on consumers' valuation of climate services. There are, however, a set of methods that can be used to infer these valuations. They have often been described as indirect market approaches to determining the valuation of goods or services that do not exchange on markets. Since services of climate as a resource are an ideal example, we will summarize in what follows some of the key features of these approaches.

All of the approaches to indirect market valuation require some type of assumptions that must be made to restrict the relationship between economic behavior as represented in a model and that observed in the real world. In some cases, these restrictions derive from specific assumptions on the nature of consumer preferences. In others they are technical associations between some marketed good or service and the nonmarketed good or service under study. Table 12.1 provides a simple taxonomy for the three approaches for determining the valuation of the nonmarketed good.

The first of these is usually associated with the work of Mäler (1974) and Bradford and Hildebrandt (1977). It requires an assumption that the utility function used to represent individual preferences has certain specific properties (such as weak complementarity). In the Mäler example, the property imposes a type of jointness in the consumption of the nonmarketed good and some marketed good or service. With such a restriction, it is possible to infer from information on the consumption pattern for the marketed good, the nonmarketed good's marginal valuation and, with it, the demand for the nonmarketed good. With this demand function it is possible to measure the total valuation as the aggregate willingness to pay for the good or service at any level of provision.

The second type of assumption is used where the delivery of the nonmarketed good or service is technically associated with some private good. In purchasing the private good, one is assured some level of the nonmarketed good. This case is perhaps most relevant to climate. That is, the exchange process for land (residential and industrial) reflects the attributes of the sites, including climatic conditions, air pollution and so on. Economists usually assume that the individuals and firms buying (or renting) land are aware of its attributes and understand their implications for all possible uses of the land. As

Table 12.1 A classification for benefit measurement methods

<table>
<tr><th colspan="2">Types of linkage between regulatory action and observed effects</th><th>Types of assumptions required</th><th>Measurement methods</th></tr>
<tr><td colspan="2">Physical linkages (no role for behavioral responses)</td><td>Responses are determined by engineering or technological relationships</td><td>Damage function</td></tr>
<tr><td rowspan="2">Behavioral linkages (behavioral responses are essential)</td><td>Indirect links</td><td>Restrictions on the nature of individual preferences or technically observed associations in the delivery of goods or services</td><td>Hedonic property or wage model</td></tr>
<tr><td>Direct links</td><td>Institutional</td><td>Contingent valuation
Contingent ranking</td></tr>
</table>

Source: Desvousges *et al.*, 1983.

a consequence, it is reasonable to expect that this information will affect demand and supply functions. Thus, the equilibrium prices for sites comparable in all relevant physical dimensions except climate will, under ideal conditions, reflect the equilibrium 'value' of the differing climatic conditions.

The last class of approaches may well be the most heterogeneous. It involves the use of mechanisms to alter directly the constraints facing an individual or a business so as to induce each to reveal their respective valuation for the nonmarketed good or service. This alteration can be either 'real' or 'hypothetical'. For example, one might consider the demand-revealing incentive schemes to be 'real' changes in the constraints imposed on individual choice by taxing the individual (or the firm) in response to the revelations. The contingent valuation (or bidding game) methods for soliciting individuals' valuation of nonmarketed goods rely on responses to hypothetical information (see Brookshire *et al.*, 1976, 1981). In these cases, the individual is confronted with a series of questions concerning proposed or hypothetical changes to the character of a nonmarket good or service. They may be changes in quality or quantity. The questions seek to elicit the individual's willingness to pay. As a rule, they also incorporate a series of checks for the biases that can be present in attempts to solicit directly an individual's valuations of nonmarket goods.

Each of these approaches reflects recognition that changes in nonmarketed goods and services do affect the behavior of economic agents. The patterns of production and consumption of marketed goods and services will change in response to alterations in their availability. However, this information can be effectively used only if the analysis is conducted at a level (that is, extent of disaggregation) that permits one to observe the behavior of 'representative' members of each class of economic agents.

12.5 APPLYING THE METHODS FOR ECONOMIC IMPACT ANALYSIS TO CLIMATE VARIABILITY

Any application of the methods for economic impact analysis to evaluate changes in the services provided by climate resources requires that these methods be modified to conform to the special features climate poses for the analysis. More specifically, we can identify at least three attributes of climate that are relevant to the analysis. First, as we noted in Section 12.2.2, climate is best regarded as a vector of random variables. The available microeconomic tools for analyzing the behavior of economic agents in the presence of uncertainty suggests that their behavior depends on how they perceive the uncertainty and on their attitudes toward risk (see Hey, 1979, for a good summary of the theory of microeconomic behavior under uncertainty). These behavioral responses to a stochastic change in the external conditions facing risk-averse households or firms do not generally coincide with the reactions arising from comparable certain changes (that is, of a magnitude equal to the expected value of the uncertain changes). Nearly all applied welfare theory deals with analysis of the efficiency of resource allocation decisions without the complications of uncertainty. Thus, there is a limited number of theoretical analyses of the appropriate implementation of benefit/cost procedures for these cases.

If the action or change to be evaluated cannot be anticipated, but still conforms to the other assumptions underlying economic impact analysis, then it is reasonable to expect that conventional approaches can be used with little alteration. By contrast, when the change is in the parameters describing the multivariate probability distribution for climate features, and we know how economic agents perceive the change, we must expect that the methods would need to incorporate explicitly those behavioral responses that arise for changes in the stochastic environment that affect their economic activities.

The difficulties posed for economic impact analysis by the stochastic nature of climate are partially offset by the fact that more impact-related questions can be asked when the action is random than when it is fixed. One such question that cannot be asked unless climate is treated as a vector of random variables involves the effect on economic agents of a reduction in the uncertainty surrounding any attribute of climate resulting, perhaps, from continued scientific research.

A second aspect of climate that affects the application of economic impact analysis is the scope of the change. Much conventional economic impact analysis is of a partial equilibrium nature, and is reasonably narrow in scope. However, climate variability and change are much broader in scope, their effects cannot be confined to single industries or regions, and their analysis thus calls for a general equilibrium approach. This is especially limiting to the application of benefit/cost methods, since these methods largely rely on partial equilibrium analyses of welfare changes. Moreover, it may also serve to accentuate the inconsistencies if different models are used to appraise the resource reallocation effects of the change versus the benefit/cost evaluation.

Finally, the timing also is important, because conventional methods are typically directed at comparative statics or involve fairly *ad hoc* specifications of dynamic adjustment. In either case, they are not likely to provide a satisfactory description of economic behavior in response to changes over a long period in the parameters describing a given climate regime, since such changes are unlikely to be either sudden or small.

12.6 CONCLUSION

At this time we can do little more than point up the problems with existing methods. To date no program of research has been established to evaluate the practical significance of any of these problems. It is nonetheless reasonable to suggest that models offering a consistent description of both the resource reallocation and the welfare changes (net benefits) associated with any particular action are more likely to be robust for expansion in scope. Equally important, a resolution of the practical significance of uncertainty and of the timing of adjustment patterns to economic impact analysis will be forthcoming only as a result of micro analyses of individual economic agents' behavior, not aggregated models.

For the present, economic impact analyses of climate variation have been forced to rely on conventional methods (see d'Arge, 1979, for example) and to ignore the problems climate poses for their estimates. Further research into the importance of these limitations is needed to judge the significance of these pragmatic responses to economic impact analyses of climate-related changes.

ACKNOWLEDGEMENT

Partial support for this research was provided by the National Science Foundation through the Climate Dynamics Program under grant No. ATM-8107990.

Thanks are due to Robert Kates and several anonymous reviewers for most helpful comments.

REFERENCES

Bohm, P. (1973). *Social Efficiency: A Concise Introduction to Welfare Economics*. John Wiley & Sons, New York.

Bradford, D. F. (1970). Benefit-cost analysis and demand curves for public goods. *Kyklos*, **23**, 775–791.

Bradford, D. F., and Hildebrandt, G. C. (1977). Observable preferences for public goods. *Journal of Public Economics*, **8** (October 1977), 111–132.

Brookshire, D. S., Ives, B. C., and Schulze, W. D. (1976). The valuation of aesthetic preferences. *Journal of Environmental Economics and Management*, **3** (December 1976), 325–346.

Brookshire, D. S., d'Arge, R. C., Schulze, W. D., and Thayer, M. A. (1981). Experiments in valuing public goods. In Smith, V. K. (Ed.) *Advances in Applied Microeconomics*, Vol. 1. JAI Press, Greenwich, Connecticut.

Burton, I., Kates, R. W., and White, G. F. (1978). *The Environment as Hazard*. Oxford University Press, New York.

Cropper, M. L., and Arriaga-Salinas, A. S. (1980). Inter-city wage differentials and the value of air quality. *Journal of Urban Economics*, **8**, 236–254.

Cummings, R. G., and Schulze, W. D. (1978). Optimal investment strategies for boomtowns. *American Economic Review*, **68**, 374–385.

d'Arge, R. C. (1979). Climate and economic activity. In *Proceedings of the World Climate Conference*, pp. 652–681. World Meteorological Organization, Geneva.

d'Arge, R. C., *et al.* (Eds.) (1975). *Economic and Social Measures of Biologic and Climatic Change*, CIAP Monograph No. 6. US Department of Transportation, Washington DC.

d'Arge, R. C., Schulze, W., and Brookshire, D. (1980). Benefit-cost valuation of long-term future effects: The case of CO_2. Paper prepared for the *Resources for the Future-National Climate Program Office Workshop on the Methodology for Economic Impact Analysis of Climate Change, April 24–25, Fort Lauderdale, Florida*.

d'Arge, R. C., and Smith, V. K. (1982). Uncertainty, information and benefit-cost evaluation of CFC management. In Cumberland, J., Hibbs, J., and Hoch, I. (Eds.) *The Economics of Managing Clorofluorocarbons: Stratospheric Ozone and Climate Issues*. Resources for the Future, Washington, DC.

Desvousges, W. H., Smith, V. K., and McGivney, M. P. (1983). *A Comparison of Alternative Approaches for Estimating Recreation and Related Benefits of Water Quality Improvements*. US Environmental Protection Agency, Environmental Benefits Analysis Series, Washington, DC.

Diewert, W. E. (1982). Duality approaches to microeconomic theory. In Arrow, K. J., and Intriligator, M. D. (Eds.) *Handbook of Mathematical Economics*, Vol. 2. North Holland, Amsterdam.

Freedman, D. (1981). Some pitfalls in large econometric models: A case study. *Journal of Business*, **54**, 479–500.

García, R. V. (1981). *Drought and Man: The 1972 Case History. Vol. 1: Nature Pleads Not Guilty*. Pergamon Press, Oxford.

García, R. V., and Escudero, J. C. (1982). *Drought and Man: The 1972 Case History. Vol. 2: The Constant Catastrophe: Malnutrition, Famines and Drought*. Pergamon Press, Oxford.

Harris, C. C., and Hopkins, F. (1982). *Locational Analysis*. D. C. Heath, Lexington, Massachusetts.

Haurin, D. B. (1980). The regional distribution of population, migration, and climate. *Quarterly Journal of Economics*, **95**, 293–308.

Heal, G. (1984). Interactions between economy and climate: A framework for policy design under uncertainty. In Smith, V. K., and Witte, A. D. (Eds.) *Advances in Applied Micro-economics*, Vol. 3. JAI Press, Greenwich, Connecticut.

Hey, J. D. (1979). *Uncertainty in Microeconomics*. New York University Press, New York.

Hoch, I. (1974). Wages, climate and the quality of life. *Journal of Environmental Economics and Management*, **1**, 268–295.

Hudson, E. A., and Jorgenson, D. W. (1974). US energy policy and economic growth, 1975–2000. *Bell Journal of Economics*, **5**, 461–514.

Jones, R. W. (1965). The structure of simple general equilibrium models. *Journal of Political Economy*, **73**, 557–572.

Jorgenson, D. W., and Fraumeni, B. (1981). Relative prices and technical change. In Berndt, E. R., and Field, B. C. (Eds.) *Modeling and Measuring Natural Resource Substitution*. MIT Press, Cambridge, Massachusetts.

Kates, R. W. (1981). *Drought Impact in the Sahelian-Sudanic Zone of West Africa: A Comparative Analysis of 1910–15 and 1968–74*. Background Paper No. 2, Center for Technology, Environment, and Development, Clark University, Worcester, Massachusetts.

Klein, L. R. (1950). *Economic Fluctuations in the United States*. John Wiley & Sons, New York.

Krutilla, J. V. (1981). Reflections of an applied welfare economist. *Journal of Environmental Economics and Management*, **8** (March 1981), 1–11.

Landsberg, H. (1946). Climate as a natural resource. *The Scientific Monthly*, **63**, 293–298.

Lave, L. (1981). *Mitigating Strategies for CO_2 Problems*. CP-81-14. International Institute for Applied Systems Analysis. Laxenburg, Austria.

Leontief, W. W., Carter, A. P., and Petri, P. (1977). *Future of the World Economy: A United Nations Study*. Oxford University Press, New York.

Mäler, K. G. (1974). *Environmental Economics: A Theoretical Inquiry*. Johns Hopkins University Press, Baltimore, Maryland.

McFadden, D. (1984). Welfare analysis of incomplete adjustment to climatic change. In Smith, V. K., and Witte, A. D. (Eds.) *Advances in Applied Micro-economics*, Vol. 3. JAI Press, Greenwich, Connecticut.

Meadows, D. H., Meadows, D. L., Randers, J., and Behrens, W. W., III (1972). *The Limits to Growth*. Universe Books, New York.

Meadows, D. H., and Robinson, J. M. (forthcoming). *The Electronic Oracle: Computer Models and Social Decisions*. John Wiley & Sons, New York.

Noll, R. G. (n.d.). *Adaptive Approaches to the CO_2 Problem*. Paper prepared for the American Association for the advancement of Science (AAAS) Climate Project. AAAS, Washington, DC (mimeo).

Nordhaus, W. (1973). World dynamics: Measurement without data. *Economic Journal*, **83** (December 1973), 1156–1183.

Nordhaus, W. (1980). *Thinking about Carbon Dioxide: Theoretical and Empirical Aspects of Optimal Control*. Department of Energy-Cowles Foundation Discussion Paper No. 565. Department of Economics, Yale University, New Haven, Connecticut.

Pearce, W., and Nash, C. A. (1981). *The Social Appraisal of Projects*. John Wiley & Sons, New York.

Rosen, S. (1979). Wage-based indexes of urban quality of life. In Mieszkowski, P., and Straszheim, M. (Eds.) *Current Issues in Urban Economics*. Johns Hopkins University Press, Baltimore, Maryland.

Scarf, H. (1973). *The Computation of Economic Equilibria*. Yale University Press, New Haven, Connecticut.

Shoven, J. B. and Whalley, J. (1984). Applied general equilibrium models of taxation and international trade. *Journal of Economic Literature*, **22** (September 1984), 1007–1051.

Smith, V. K. (1983). The role of site and job characteristics in hedonic wage models. *Journal of Urban Economics*, **13** (May), 296–321.

Smith, V. K., and Krutilla, J. V. (1982). Toward a restructuring of the treatment of natural resources in economic models. In Smith, V. K., and Krutilla, J. V. (Eds.) *Explorations in Natural Resource Economics*. Johns Hopkins University Press, Baltimore.

Stigler, G. J. (1939). Production and distribution in the short run. *Journal of Political Economy*, **47** (June), 305–327.

Taylor, A. S. (Ed.) (1974). *Climatic Resources and Economic Activity*. John Wiley & Sons, New York.

Whalley, J. (1977). The U.K. tax system 1968–70: Some fixed point indications of its economic impact. *Econometrica*, **45**, 1837–1858.

Whalley, J. (1980). Discriminatory features of domestic factor tax systems in a goods mobile-factors immobile trade model: An empirical general equilibrium approach. *Journal of Political Economy*, **88**, 1177–1202.

World Meteorological Organization (1980). *Outline Plan and Basis for the World Climate Program, 1980–1983*. WMO No. 540. Secretariat of the World Meteorological Organization, Geneva.

Climate Impact Assessment
Edited by R. W. Kates, J. H. Ausubel and M. Berberian

CHAPTER 13
Social Analysis

BARBARA FARHAR-PILGRIM

4600 Greenbriar Court
Boulder, Colorado 80303, USA

13.1 INTRODUCTION

From the evolution of custom and culture in different climatic zones and the seasonal rhythms of all societies, we have strong evidence that climate has social impact. However, the *systematic* assessment of the impact on society of climate variability and change, which is the subject of this chapter, remains a lightly developed area. By social impact assessment (SIA) we refer to a body of research that examines more strictly the behavioral or 'social' aspects of impacts on communities and individuals, arising from related demographic and economic factors. We draw from recent efforts to analyze the social impacts of technologies and environmental changes, with primary interest in the methods for and experiences in social impact assessment for such climatic phenomena as frost, hail, or sustained changes in snowpack, rainfall and temperature.

As orientation to the nature of the concerns under consideration, the chapter first offers two illustrations of social impact assessment, one of climate change in a developed country context and one of climate variability in a developing country context. Subsequent sections discuss the origins and methods of social impact assessment, present more examples of climate-related projects employing SIA, and identify research concerns and opportunities.

13.2 CASE STUDIES

13.2.1 Metromex

One major effort to assess the impacts of persistent climate change is METROMEX (for Metropolitan Meteorological Experiment)—the study of the St Louis urban weather anomaly. In brief, METROMEX was an unplanned natural experiment. Urban-induced, unintended changes in the local climate of up to 30 years in duration were established (Changnon, 1981c). For summer weather patterns, these included a 10 percent increase in local cloudiness, a 30 percent increase in rainfall (primarily from late afternoon and early evening thunderstorms), and increased severe storm activity (up to 100 percent). The changes occurred over the city and eastward within a 4000 km^2 area.

By 1950, concomitant changes were detectable in streamflow, flooding and crop yields. Measured impacts included more runoff (11 percent), more local flooding (up to 100 percent), and more stream and groundwater pollution (up to 200 percent). A net local-area average increase in grain crop yields of 3–4 percent was found, although there was a 100 percent increase in crop-hail losses. While the net effect of the anomaly on agriculture appears to have been beneficial, local, state and federal agencies have borne increased costs of water management and flood control. Figure 13.1 presents a chart summarizing first-, second-, and third-order impacts of the urban weather anomaly; not all of the impacts shown have been empirically established. As the chart suggests, the weather changes resulted in both economic winners and losers.

The social impacts of the St Louis urban weather anomaly were specifically studied after the anomaly's characteristics had been clearly identified by atmospheric scientists (Farhar *et al.*, 1979). A comparative case study design was

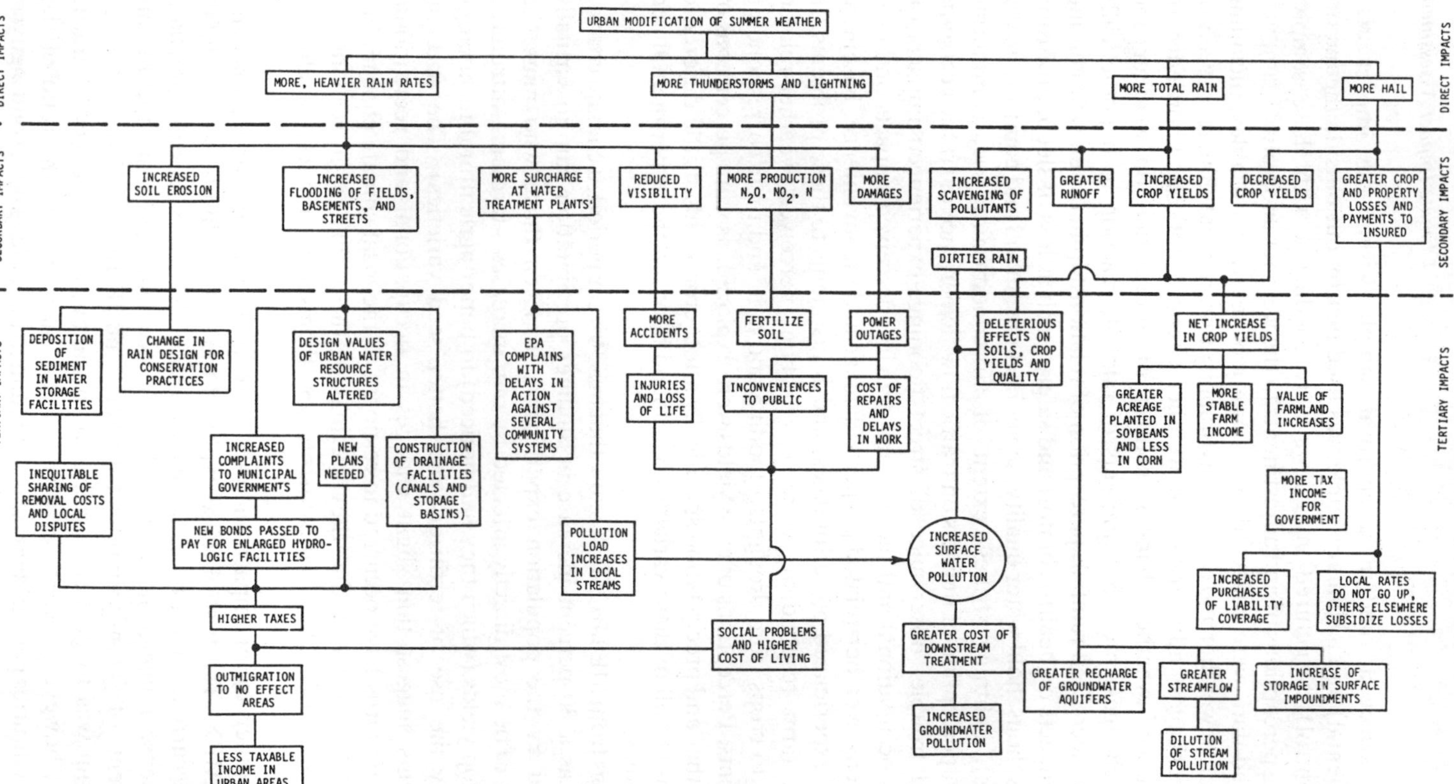

Figure 13.1 Interrelated impacts of certain urban-induced precipitation anomalies at St Louis, Missouri. Reproduced by permission of the American Meteorological Society from Changnon, *Meteorological Monographs*, Vol. 18, No. 40 (1981b), 156

used to select a random urban/rural sample within the impact area (experimental) and a similar sample was used nearby but outside the impact area (control). Structured interviews were conducted with these samples regarding perceptions of and adjustments to the weather anomaly.

Detailed data were collected on a variety of relevant variables, including perception of weather and climate, favorability toward local climate, perceived weather changes, adjustments to weather changes noticed, knowledge of urban-induced weather changes, sources of information on weather and weather changes, preferred amounts of rainfall, reaction to hypothetically increased/decreased rainfall, perception of relationship between weather and health, impacts on health, comfort and safety (such as loss of sleep, profusion of insects, household water quality, sewer backups, mood and comfort, health problems), lightning strikes, extent of air conditioning use, commuting behavior, perception of impacts on traffic flow, experience of traffic accidents and road trouble, observance of street flooding, experienced impacts on everyday occupational and recreational activity, perceived impacts on lawns and gardens, weather-related property damage and inconvenience, basement flooding, experienced power outages, and adjustments to these problems. In addition, farm respondents were asked about perceived weather-related damages to crops from droughts, flooding and hail, and their corresponding adjustments. Perceptions of crop yields were explored, as were perceptions of how weather and precipitation specifically affect crop yields. Little difference in response to all of these variables was found between the experimental and control areas.

Findings from METROMEX show that in a 30-year period, a climate change as great as a 30 percent increase in summer precipitation can go virtually unnoticed by the population experiencing it. Even those who might be expected to be most directly affected by rain increases—farmers—attributed higher crop yields (which they *had* noticed) to better agricultural technology (primarily the use of fertilizers and better seed varieties). The data on adjustments suggest that slight changes in occupational and recreational behavior patterns have occurred in response to the local climate change, but people remain unaware that they are making these changes in response to anything unusual. These adjustments are themselves impacts of climate change.

Three other social impact analyses were performed in connection with METROMEX. The first of these explored the relationship between rainfall and the occurrence of traffic accidents, using METROMEX precipitation data from a dense raingage network and traffic accident data from the Illinois Department of Transportation (Sherretz and Farhar, 1978). A linear relationship was found between increased rainfall and frequency of traffic accidents; however, severity of accidents (as measured by the ratio of the number of injuries per accident) was not found to increase. The second analysis examined the relationship between rainfall and selected criminal activity; no relationship was detected, possibly due to crime data problems (Farhar *et al.*, 1979). The third analysis explored responses to urban weather anomaly on the part of organizations that reasonably could be expected to be concerned (Farhar *et al.*, 1979). These included air traffic control, sanitation districts, water quality control, regional government, air pollution control, floodfighting, utility companies, engineering design firms, urban planning agencies, private industry, and the National Weather Service Forecast Center. The interests and responsibilities of these organizations were potentially directly affected by the urban climate anomaly. For example, sewage treatment facilities are not designed to handle 100 percent of historical precipitation, but are designed to handle a tolerably low number of surcharges per season. At the time of the study, thunderstorms and consequent system overloads caused untreated sewage to be passed directly from a sewage treatment facility into the Mississippi River three to six times each summer. One of the municipal water intakes on the Mississippi is downstream from this facility. The impact on water quality is not known, but one respondent said this was a low priority as the Mississippi is a multi-use river, and usage is not considered degraded by the problem.

Most organizations were relatively uninterested and felt that the magnitude of change measured was in the 'noise' of weather variability. Organizational response and adjustment to the anomaly were minimal to nonexistent. Organizations unanimously indicated that all systems are designed and operated at less than 100 percent of potential capacity. Designs assume periodic power outages, floods and surcharges, for example. Storm drainage systems are designed for a 5- or 10-year event, but not for a 100-year event. Increased precipitation meant that systems based on historical data were exceeding design more frequently. But at 30 percent increase, organizations responsible for existing systems were not very much interested in information about the anomaly. Organizations designing new systems were somewhat more interested. However, Changnon (1981c) reports that urban anomaly data were used in a federal court decision regarding water use in the Chicago area.

The METROMEX social impact assessments were done in piecemeal fashion, as limited resources would permit. Nevertheless, by accomplishing focused empirical investigations on elements of the problem (such as water quality, crop yields, occurrence of surcharges, perceived impacts, and traffic safety), the foundations for a comprehensive assessment of social system impacts of climate change were gradually laid. As Changnon (1981b) points out, the impact analysis remains incomplete, both because more research is needed and because the story is still unfolding.

13.2.2 Post-frost Adjustment in Brazil

Margolis (1980) has reviewed the post-frost strategies of coffee cultivators in southern Brazil. Frosts have received little attention from environmental impact researchers, probably because they present no direct threat to human life and, as such, are said to have low catastrophe potential (White and Haas, 1975). Then, too, as Margolis (1980) points out, the full impact of frost damage in agricultural communities may not be felt for many months, that is, until the normal time of harvest of the affected crops.

In early August 1975, as Margolis reports, a mass of cold Antarctic air swept up from the south and blanketed Brazil's major coffee region with the most severe frost within living memory. Temperature fell below 0°C in parts of Mato Grosso and Minais Gerais—areas which had never frozen before—but the greatest devastation was wrought in Paraná state, the most southerly in Brazil's coffee zone. Close to 100 percent of thc 915 million coffee trees in Paraná were affected. As the temperatures inched below freezing, the leaves on the trees blackened and fell to the ground, eliminating the two subsequent harvests. In the most severe cases, the cells of the tree trunks were ruptured by the expansion of the ice, killing the trees when the trunk split. As a result of this event, Paraná in 1980 had an estimated 600 million coffee trees, or nearly one-third fewer than before the frost.

The most immediate effect of the frost was large-scale unemployment. In the coffee zone as a whole, it is estimated that of the one million workers involved in the cultivation and processing of coffee, 600,000 were left jobless as a consequence of the frost. The response to widespread joblessness frequently was emigration, with people often moving to the large urban centers of the coast or newly opened frontier areas.

The aftermath of the unusual weather was by no means uniform in the coffee-growing communities where the impact of the frost was greatest, in the northern portion of the state of Paraná. By 1978, three years after the event, much of northern Paraná had been transformed, and three distinct responses to the frost were evident. In one area, thousands of small farms had been replaced by large mechanized estates devoted to soybean and wheat agriculture. In a second region, hundreds of medium-sized holdings had given way to extensive cattle ranches. In a third area, where cattle ranching had made important inroads before the 1975 frost, the conversion to pasture was halted after the frost and coffee trees started to be replanted.

Margolis concludes that the 1975 frost was a catalyst for rather rapid social and economic change in Paraná's coffee zone. But the specific responses to the extreme weather condition in each of the zone's three subregions were shaped largely by conditions existing prior to it. Land divisions, soil quality, water resources, labor legislation, and credit availability all influenced the frost's impact and the subsequent evaluation of the various areas. In conclusion, Margolis emphasizes the need for analyses of predisaster conditions for full understanding of post-disaster responses.

13.3 BACKGROUND AND METHODS

13.3.1 History

Although study of the social impact of environment and technology has a long history, explicit social impact assessment (SIA) methodology is a recent development, stimulated largely by legislation like the US National Environmental Policy Act (NEPA) of 1969 (Enk and Hart, 1978). A US Ad Hoc Interagency Working Group on SIA was formed in July, 1974 (Connor, 1977). In the United States, SIA has been used to assess the social impacts of such projects as proposed oil shale development, nuclear power plants, coal projects, pipelines, dams and offshore drilling. In developing countries, impacts on local populations have been assessed for rural roads, electrification, water supply, irrigation, family planning, and health care projects and programs (Albertson, 1983; Cano, 1983; Mohanty, 1983). The recently formed International Association for Impact Assessment and the Social Impact Assessment Center (along with its publication *Social Impact Assessment*) serve as clearinghouses in the United States for information from all countries on impact assessments in fields as diverse as health and climate. New designs and handbooks on SIA are being developed (Rossini and Porter, 1983; Flynn and Flynn, 1982; Branch *et al.*, 1982). Recent reviews and manuals include Chalmers and Anderson (1977), Bowles (1981), Finsterbusch and Wolf (1981), and Leistritz and Murdock (1981). The most highly developed area of study is probably in social impacts of energy resource development, where Denver Research Institute (1979), Murdock and Leistritz (1979) and Conn (1983) are indicative of the state of the art.

13.3.2 Public Participation

As a methodological approach, SIA can be viewed as a component of both environmental impact statements (EISs) and technology assessments (TAs). These approaches are a class of policy analysis that came into being to address public demands to assess the 'hidden costs' commonly associated with proposed projects and programs (Peterson and Gemmell, 1977). The approach has more recently been used to assess the consequences of proposed regulations. SIA is often anticipatory research that aims to predict and evaluate the impacts of policies, projects, or programs on society (Wolf, 1981).

When an SIA is undertaken on behalf of a government, it typically incorporates a broad mandate for public involvement, including affected publics, organizations, and communities. Indeed, techniques for public participation and reviews of findings are generally included as integral parts of SIA, EIS and TA projects. The many parties-at-interest, or 'stakeholders', that can be affected by the change or project in question are involved not merely as objects of study.

13.3.3 Patterns of Assessment

The incidence and distribution of social costs and benefits is a key matter for assessment. In fact, the central question for SIA is frequently social equity: who wins and who loses in a given situation. What to do about it once the assessment is completed is a matter for public policy.

SIA draws freely on all social science disciplines. Since every impact situation has unique features, methods must be adjusted to suit each assessment problem. No one 'best way' of conducting assessment is accepted. However, Wolf (1981, 18–19) presents a 'main pattern' that assessment projects tend to follow. The assessment steps comprising the pattern are:

- scoping the problem,
- identifying the problem,
- formulating alternatives,
- profiling the system,
- projecting effects,
- assessing impacts,
- evaluating outcomes,
- mitigating adverse impacts,
- verifying results,
- specifying who wins and who loses, and
- designing institutional arrangements and management.

With regard to style of work, it is generally desirable for assessment projects to be

- interdisciplinary and interactive
- synthetic, aggregative, integrative
- inclusive (involve users of the assessment information in the process), and
- policy relevant.

13.3.4 Types of Data

Assessment projects usually rely on already-existing data. Four basic types of social data that can be used in assessment are:

1. *statistical social data* (for example, censuses of population/housing, traffic counts, vehicle registrations, mortality, farm size, employment, hospital beds, police cars);
2. *written social data* (for example, letters to editors, novels,* prepared testimony, historical documents, reports, newspaper articles);
3. *observational social data* (for example, systematic observation of relevant

* For example, *The Grapes of Wrath* may be viewed as a potent statement on the social impacts of the 1930s Dust Bowl in the United States.

events, unobtrusive measures, land modifications, measured responses to experimental situations); and

4. *respondent contact social data* (for example, polls and surveys, interviewing, ballots, citizen/expert comment on impacts, registration applications and other forms (Miller, 1970; Gale 1977b).

These types of social data result from accepted techniques used routinely in behavioral science research.

How data are analyzed and synthesized to complete the basic function of SIA research will always be subject to revision and improvement. Connor (1977) lists 14 different data collection and analytical techniques that have been used in various combinations in SIAs:

demographic analysis	evaluative research
community studies	institutional analysis
causal modeling	value analysis
social indicators	multivariate analysis
ethnomethodology	social network analysis
archival research	social forecasting
survey research	matrix methodologies

13.3.4.1 Impact Categories

Drawing on studies of the forest sector, Gale (1977a) presents a comprehensive list of social impact categories that can be used to guide assessment projects (see Table 13.1). The value of such an heuristic device is that it provides a framework for the systematic assessment of impacts on society's sectors and processes, helping to ensure that major interests are not overlooked. It also serves to organize frameworks for the multiplicity of data that are inherent in any assessment project.

13.3.4.2 Data-poor Countries

Given SIA's reliance on existing data, what can be done in data-poor countries, such as the developing countries? Some preliminary attempts have been made to deal with this problem in the context of impact assessment efforts (Castro *et al.*, 1981). A number of evaluation studies conducted or sponsored by the US Agency for International Development have sought to include social characteristics such as extent of village social cohesion, degree of village isolation, demographic characteristics, leadership, and access to major urban areas. A 'social soundness analysis' produced by the US Agency for International Development (1978) recommends a form of qualitative stakeholder analysis, specifying who wins and who loses as a result of proposed developmental projects. Using the household as a unit of analysis, Castro *et al.*

Table 13.1 Forest-use social impact categories, variables and components

Social impact categories	Social impact variables	Social impact components
SOCIAL INSTITUTIONS	Community culture change (subculture, trait, or theme)	
	Leisure and 'cultural' opportunities	Carrying capacity Available land and facilities
WAYS OF LIFE	Recreational opportunities	Recreational demand 'Optimal recreationist'
	Special group access (elderly, handicapped, poor, transit-dependent) Security (anxiety, unpredictability, and the 'unknown') Open space	
		Minority group impacts Civil rights
SPECIAL CONCERNS	Minority and civil rights Historical and archeological sites	
COHESION AND CONFLICT	Physical cohesion (barriers) Demographic cohesion (class characteristics) Attitude and value cohesion Proposed action activities cohesion and conflict Community activities	
LAND TENURE AND LAND USE	Land allocation and use	Actual use compatibility Suitability (environmental carrying capacity) Esthetic effects (viewer access)
	Land use regulation	Conditional use and building permits Comprehensive planning and zoning

POPULATION DYNAMICS	Population size (growth, stability, decline)	Population size perspectives Population size change
	Population density Displacement of people	Physical displacement Use displacement
	Population distribution Population of people	Geographical mobility Social mobility
	Population structure (age and sex)	
COMMUNITY CONTEXT	Community identity 'Sense of place'	
SYMBOLIC MEANING	Places Practices 'Things'	
BASIC VALUES	Value orientations Value dimensions Value rankings	

Source: Gale, 1977a, 2.

(1981) developed indicators of economic stratification for rural villages in developing countries (see Table 13.2). Mohanty (1983) examined the sociopolitical conflict arising from the Hirakud Dam Project in the state of Orissa, India. The environmental and sociopolitical impacts of the Sahel drought have been explored (Glantz, 1976, 1977). Albertson (1983) claims that the world hydropower potential is vast and urges worldwide study of hydropower's effects on agriculture, fisheries, forestry, mining, health, employment, income distribution, political factors, standard of living, ecology and environment, and self-reliance of local people.

13.3.5 Types of Methods

The use of scenarios, or 'social forecasting', can be a useful method where empirical data are relatively scarce or the phenomenon to be studied, such as climate change impacts, will occur in the distant future. Social forecasting involves the analysis of probable social consequences of current trends and events. The social future cannot be forecast without ambiguity; present knowledge and methods are not sufficiently powerful to permit accurate predictions of social behavior over long periods of time. Nevertheless, as

Vlachos (1977) argues, use of scenarios is one class of methods that relies on informed disciplined imagination; this approach is worthwhile in outlining the alternative futures that society might face. Scenarios are future histories describing how the world might look. (For a view on scenario uses as a more formal modeling device, see Lave and Epple, Chapter 20, this volume.)

The scenario is but one method that can be used in social impact assessment. Vlachos (1977) has defined four additional classes of methods. These are:

1. consensus—opinions by themselves; agreement among experts (delphi); conjecture; brainstorming; heuristic programming; moot courts.
2. historical extrapolative—historical determinism based on how the past was; historical surveys; social trends analysis; monitoring; correlation and regression; probabilistic analysis; growth metaphors; historical analogies.
3. problem structuring —
   (a) modeling: identify a set of elements, concepts, etc., explore their relationships within a situation and generate a graphical representation of structural relations. Models are partial representations that organize elements in time and space.
   (b) cross-impact matrix: relates elements to each other based on expert judgment.
   (c) other: simulation, iterative system projection, systems analysis, input–output analysis.
4. decision methods — determine a goal, then 'backcast' to analyze what should be done now to get there; includes forecasting and casting around, using morphological analysis, contextual mapping, objective trees, relevance trees, judgment theory.

Each of these classes of methods is appropriate for different parts of the SIA. For example, scenarios and extrapolative methods are useful in establishing the context of the problem, consensus is useful in defining the range of alternatives and potential impacts, and decision methods are useful in policy option analysis. Any assessment project could employ one or more of these kinds of methods in combination.

Data-poor countries can also begin to accumulate 'bits and pieces' of social data bases, using resources which are limited but available for a variety of projects. Eventually, as happened in the METROMEX project, a more comprehensive picture which is based on accumulating empirical evidence will emerge. Employing available frameworks will guide the effective selection of variables on which to collect data.

13.4 EXAMPLES

Prior to 1960, little research on weather is reported in the social science literature (see Riebsame, Chapter 3, this volume). In the last two decades social science research in the climate area has focused on natural hazards

Table 13.2 Indicators of rural household economic status

Surveys of economic stratification. Household is unit of analysis.
Land ownership
Quality of land
Intensity of cultivation and types of crops
Land outside sample area
Tenants and landless
Ownership of capital equipment and consumer durables
Income
Agricultural income
Sale of crops
Rent
Hired labor
Marketing and processing
Non-agricultural income
Ownership of livestock
Ownership of non-productive property
Housing
Furnishings and consumer goods
Access to fuel
Ceremonial expenditures
Diet, nutrition and health
Education
Household size and composition

Source of indicators: Castro *et al.*, 1981.

(White and Haas, 1975; Drabek, 1983) and human modification of weather and climate.

13.4.1 Snowpack Augmentation

Some early work in assessing the impact of weather on human activity was spurred by cloud seeding projects proposed to increase snowfall in Colorado and California. Two studies are worthy of mention here:

1. the technology assessment conducted by Stanford Research Institute for the US National Science Foundation (NSF) on proposed widespread orographic showpack augmentation in the San Juan Mountains of Colorado (Weisbecker, 1974) and
2. the partial social assessment conducted by Human Ecology Research Services for the Bureau of Reclamation and NSF on proposed orographic snowpack augmentation in the Sierra Nevada, California (Farhar and Mewes, 1974; Farhar and Rinkle, 1976).

The Weisbecker study was an interdisciplinary technology assessment involving the following steps:

- defining a weather modification system,
- identifying effects on water quantity and quality
- assessing impacts of water supply on economic systems,
- assessing impacts of increased snow on environmental and ecological systems,
- assessing impacts of increased snow on social systems, and
- addressing public policy.

To assess the *social* impacts of more snow, Logothetti (1974) identified potentially involved sectors of society, or stakeholder groups (see Table 13.3). Many stakeholders were interviewed, using scenarios to collect data on their *perceptions* of impacts, but sampling was unsystematic. Demographic analysis describing mountain populations and growth were performed. Tables 13.4 and 13.5 list some advantages and disadvantages of snowpack augmentation to supporters and opponents.

The Sierra study consisted of a partial assessment of citizen and organizational perception of snowpack augmentation and its potential effects; it was conducted through systematic surveys in and around the proposed project area during 1974 and 1975. Citizen testimony at public hearings in the project area provided another data source.

A random sample of area citizens thought snowpack augmentation was a less desirable way to augment water supply than building more reservoirs and practicing water conservation. They were concerned about the risk and controllability of the proposed project and about potential adverse impacts.

The organizational perception study was based on a systematic sampling of stakeholders. A sampling matrix was used to identify the relevant groups located in, or with interests in, the study area (see Figure 13.2). The sampling frame compares sociopolitical levels of concern (federal, regional, state, county, city) to functional interests (such as agriculture, energy and transportation) with relevance to snow. Specific individuals and organizational representatives were identified for each cell of the matrix (representing domains of responsibility) and a structured instrument was used to interview each of them personally. Sampling frames such as this can be useful in identifying specific parties-at-interest.

The study found that stakeholders tend to adopt positions toward snowpack augmentation consistent with their assessment of how increased snow will affect their interests. If they feel their interests are unaffected, their position is neutral or indifferent. Heterogeneity of weather needs characterizes the area, so it is not surprising that some respondents perceived benefit from increased snowfall (for example, ski areas and utility companies) and some perceived harm (for example, county government because of highway snow removal costs).

A few generalizations can be drawn from the snowpack studies. Increases in snowpack in the order of 10–20 percent per winter season have not yet been found to have serious adverse impacts on society or the environment. Changes of

Table 13.3 Stakeholders, as represented by 29 groups, concerned with snowpack augmentation in the Colorado Rockies

Bureau of Reclamation	Irrigators
Cattlemen and ranchers	Loggers
Cloudseeding contractors	Mountain communities and community leaders
Colorado Mining Association (individuals, mining companies, mining communities)	Panel on Public Information in Weather Modification
Congressmen	People afraid of interfering with nature
Downwind residents	People with weather-related problems
Ecologists	Protectors of the public interest
Elected representatives of mountain communities	Public Health Service
Electric power companies	Senators
Farmers	Ski area owners and operators, skiers
Federal government	State and local governments
Federal weather-modification regulatory agency	Target area residents
Grand Mesa Water Users Association	Tourist industry, tourists
Industries	Water users (especially in Arizona and California)
	WOSA scientists, forecasters

After Logothetti (1974).

Table 13.4 Possible advantages to stakeholders

More water for irrigation
More water for electric power
More water for municipal uses
More water for industrial uses
More snow for ski area corporations
More water for livestock and grass
More snow for skiing and snowmobile use
More water for Los Angeles residents
More water for irrigation and agriculture in southwest Colorado

After Logothetti (1974).

Table 13.5 Possible disadvantages to stakeholders

More avalanches, causing revenue losses to mining companies and communities
More avalanches, causing losses of life and limb
More unintended snow on residents in downwind areas
More snow feeding avalanches that are threats to property
More snow increasing snow removal costs
More snow preventing employees from getting to work
More snow reducing geological exploration by mining companies
Heavier snow load on roofs
More water and spring flooding
More snow affecting logging operations
More snow shortening the summer tourist season
More floods keeping tourists away

After Logothetti (1974).

that magnitude, being within normal climate variability, might go unnoticed by the populations experiencing them for relatively long periods of time (if they are not from publicized weather modification programs). The heterogeneity of weather needs on the part of major sectors of society was made clear: some will benefit and some will lose as a result of any weather event or pattern. Systematic data on who the winners and losers are and how they are affected is sparse indeed, however. Additionally, interregional problems arise in that a weather pattern that possibly harms one geographic area (more mountain snowfall) may benefit another (downstream water users).

	Forest	Highway	Political	Land/water use administration	Environmental organizations	Economic interests*	Media	Other**
Federal								
State/ Regional								
County								
Community								

* For example: lumbering, recreation, agriculture, utilities, mining.

** For example: community 'opinion leaders'

Figure 13.2 Sampling matrix to identify weather stakeholders in a study area (for snow). (After Farhar and Rinkle, 1976)

13.4.2 Severe Winters

Work on the impacts of natural severe winters on society has been conducted in Illinois (Changnon, 1979; Changnon *et al.*, 1980). Urban and rural residents completed questionnaires on the impacts of the severe winter weather of 1977–78 on their households. Specific impacts and their frequency of occurrence were identified, ranging from heating costs to absences from school and work, morbidity and medical costs, and family arguments arising from being snowbound. The average added cost per household attributed to the severe winter was $93; extrapolating to the state from these findings, the authors estimated economic costs to state residents, as well as the inconvenience, anxiety, extra work and injuries. Decreased tax revenues, loss of work and other costs to government and industry were not calculated. The specificity and detail of the data represent an advance over the earlier snow impact studies, yielding information more useful in building systematic assessments of how climate and its variation can affect society.

13.4.3 Hail Suppression

Following the assessments of snowpack augmentation, a major national technology assessment for hail suppression was conducted in the United States—TASH, for Technology Assessment of the Suppression of Hail (Changnon *et al.*, 1977, 1978; Farhar *et al.*, 1977). The TASH project team were experts, each from a different discipline,* with prior experience in weather modification. Rather than gathering the experts under one roof, team

* Represented were atmospheric sciences, agricultural economics, law, sociology, political science, environmental science, assessment methods, and—unlike earlier efforts—representatives from businesses, such as crop-hail insurance and cloud seeding.

members remained at their organizational homes across the country, while the project was managed at the University of Illinois. Project integration and coordination were achieved through frequent team meetings interspersed with conference calls. The project was iterative; team members prepared working papers in their own areas of expertise. These were reviewed by other team members until a common vocabulary and understanding developed that superseded disciplinary lines. As work progressed, each element of the team found itself dependent on the input of others. For example, economists had to produce crop-yield data by crop-producing region before social impacts could be fully assessed, but sociologists projected the probable adoption (social acceptability) of the hail suppression technology across the nation before economists could assess macroeconomic impacts.

Assessment study users (stakeholders) representing national and local perspectives carefully reviewed the final draft of the project report at workshops held for the purpose. The TASH project, which employed scenarios, computer modeling, consensus, existing survey data, historical case study analysis, and a limited amount of new data, was fully interdisciplinary, iterative in process, and synthetic in product.

One of the key findings from TASH was that, while hail damage represents a serious problem to those experiencing it, on a national scale its impact is relatively minor. Much more significant social, economic, legal and political impacts arise from changes in precipitation—both increases and decreases. Figure 13.3 shows the interrelationship of impacts of less hail and more rain in one of the conceptual models employed.

13.4.4 Global Cooling, Global Warming

The examples described so far in this study illustrate methods and concerns which have arisen primarily in the assessment of local and regional climate variation. The question of the social impact of a global climate change has also been addressed on at least two occasions, once for the Climate Impact Assessment Program (CIAP) of the US Department of Transportation, which focused on the possible consequences of a global cooling (see Glantz and Robinson, Chapter 22, this volume), and more recently by the US Department of Energy's Carbon Dioxide Program, which has focused on global warming. For CIAP, social impacts were analyzed by Sassone (1975), who sought to estimate the costs for municipal governments and the like, and by Haas (1975), who explored the effects on family and community activity. Table 13.6 outlines changes which might be associated with cooler and wetter or cooler and drier conditions. The studies for the carbon dioxide case (see Chen *et al.*, 1983) stress the need for a variety of perspectives: cultural, psychological, historical, political and legal. They outline research needs and largely refrain from the assessment of possible outcomes.

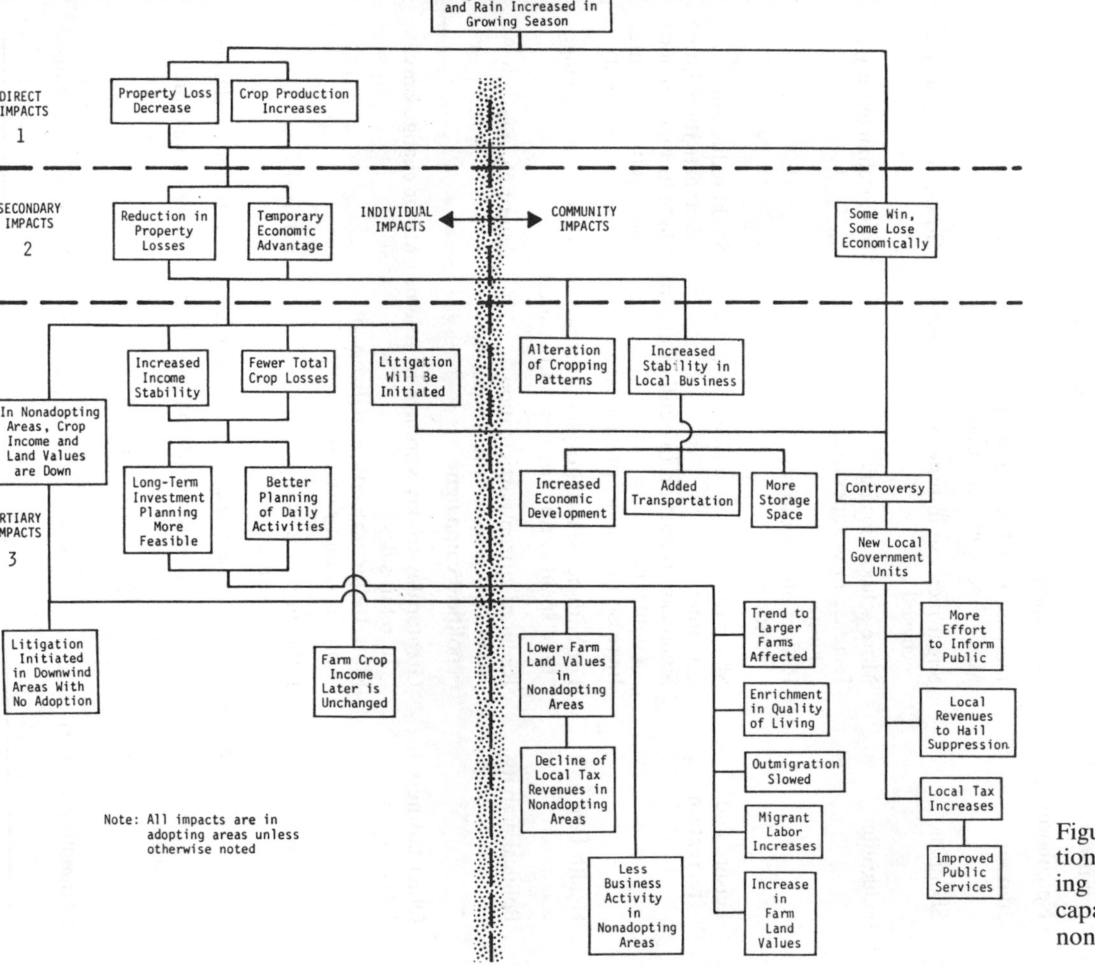

Damaging Hail Decreased and Rain Increased in Growing Season
DIRECT IMPACTS 1
Property Loss Decrease
Crop Production Increases
SECONDARY IMPACTS 2
Reduction in Property Losses
Temporary Economic Advantage
INDIVIDUAL IMPACTS
COMMUNITY IMPACTS
Some Win, Some Lose Economically
TERTIARY IMPACTS 3
Increased Income Stability
Fewer Total Crop Losses
Litigation Will Be Initiated
Alteration of Cropping Patterns
Increased Stability in Local Business
In Nonadopting Areas, Crop Income and Land Values are Down
Long-Term Investment Planning More Feasible
Better Planning of Daily Activities
Increased Economic Development
Added Transportation
More Storage Space
Controversy
New Local Government Units
Litigation Initiated in Downwind Areas With No Adoption
Farm Crop Income Later is Unchanged
Lower Farm Land Values in Nonadopting Areas
Decline of Local Tax Revenues in Nonadopting Areas
Less Business Activity in Nonadopting Areas
Trend to Larger Farms Affected
Enrichment in Quality of Living
Outmigration Slowed
Migrant Labor Increases
Increase in Farm Land Values
More Effort to Inform Public
Local Revenues to Hail Suppression
Local Tax Increases
Improved Public Services
Note: All impacts are in adopting areas unless otherwise noted

Figure 13.3 The interrelationship of impacts resulting from hail suppression capability. (Source: Changnon *et al.*, 1977, 356)

Table 13.6 Outline of anticipated changes in family and community activity due to two types of climatic change

Activity	Average weather conditions are: Colder and wetter	Colder and drier
(Local) Within the nuclear family	Increased adult–child interaction Increased sibling interaction	Less significant change in same variables
Neighboring	Reduction in frequency	Minor change only
Journey to work	Increased tardiness and absenteeism More time spent on road	No significant change
Shopping	Slight increase in time per unit purchased	No significant change
Education	More school closings Increase in unsupervised children at home Reduction in outdoor exercise for children	Minor change only
Leisure and recreation	Sharp reduction in outdoor activity (winter) Significant reduction in outdoor activity (summer) Increase in indoor, passive leisure	Slight reduction in outdoor activity (winter) Slight increase in outdoor activity (spring, summer and fall)
Health care	For both home and hospital, increase in time spent caring for sick and injured	No significant change
Political activity	Reduced citizen participation Additional reliance on TV during political campaign	Minor change only
Other business activity	Greater day-to-day variance in volume of retail sales More frequent delays in delivery of products and services	Minor change only
Religious activity	Reduction in participation in organized activity	No significant change
(Distant) Familial	More interrupted and cancelled trips	No significant change
Leisure and recreation	Increase in interrupted and cancelled trips (winter) Decline in trips to certain areas (all seasons)	Minor change only
Commuting to work	Decline in number of commuters using autos	No significant change

Source: Haas, 1975, 4–105.

13.5 RESEARCH CONCERNS AND OPPORTUNITIES

Some specific concerns arise in attempting to apply SIA methods to the study of climate.

13.5.1 Type of Problem

SIA is not appropriate in studying how social processes affect climatic processes; for example, how social behavior gives rise to the CO_2 accumulation in the atmosphere, public and decision-maker awareness of the CO_2 problem and its climatic consequences, willingness to change behavior and under what conditions, and institutional arrangements and normative structures that would avert CO_2 accumulation. However, SIA could be applied to study the social impacts of fossil fuel use, such as 'boomtown studies', which document the phenomena of rapid-growth communities.

SIA can appropriately be used to study adaptations to actual and projected climate change. Adaptations can be efforts to benefit from expected or actual changes or efforts to mitigate anticipated or actual adverse impacts. Adaptations to benefit from such climate change effects as precipitation increases could involve, for example, constructing irrigation systems or changing cropping patterns. Adaptations to mitigate adverse impacts from such climate change as precipitation decreases could involve construction of water storage facilities, planned slow migration, and food trade arrangements among nations.

13.5.2 Study Sequence

Behavioral science is concerned with the *indirect* effects of climate change. Indirect effects are best studied after something is learned—or projected—about the first- and second-order impacts of climate change. Regional changes in precipitation (and other weather variables), their direction and magnitude, and their subsequent effects on, for example, water supply, crop yields and energy use must be investigated or defined by actual data or in scenario form before social impacts can be assessed. SIAs cannot be conducted in isolation from other disciplines.

13.5.3 The Rate of Climate Change and the Rate of Social Change

An important question in assessing social impacts is whether social adjustments and adaptations to climate change will be slow and incremental, or will climate change be ignored until its effects (if adverse) reach catastrophic proportions? At what point and under what conditions does a quantitative change in climate become a qualitative event that affects the quality of life, increasing or decreasing social well-being? When does it become socially disruptive?

As climate (experienced, measured and anticipated) is the aggregation of discrete weather events statistically distributed over time, so social impacts are discrete events, multiplied across populations and aggregated over time. Since large climate changes are likely to be experienced as an unusual sample of historical weather and climate, behavioral scientists should be studying current situations involving repetitive droughts, flooding, precipitation increases, frost, and higher and lower temperatures and their impacts on society. How are populations adjusting to such phenomena? How great a change seems tolerable for existing infrastructures without undue strain? Studies of societal response to some kinds of catastrophic disasters (such as volcanic eruption, earthquakes, tornadoes and hailstorms) may be helpful in assessing the relevant social impacts of climate change. However, their onset is sudden and intensely disruptive, while climate change is more likely to involve a gradual accumulation of opportunities and hazards.

13.5.4 Perception

Whether citizens, stakeholder groups, and organizations perceive slow climate changes and how significant they feel these are is an important research question. While there is little doubt that everyone in an affected area is aware of an event like an earthquake, it is unclear whether most people sense slow, incremental climate changes and adjust to them appropriately. 'What people define as real is real in its consequences' has become a sociological truism that applies as well to climate change as to other phenomena. In the case of climate change, however, what people *ignore* can be just as real in its consequences. The difference between ordered, minimally disruptive and maximally beneficial adaptation to climate change and more chaotic, disruptive response to adverse impacts (or missed opportunities for beneficial ones) may lie in the perceptual arena (see Whyte, Chapter 16, this volume).

The *credibility* of climate forecasts is probably crucial in this regard. Adaptive response will depend on expectations for future climatic conditions. If historical methods of determining likely future weather/climate conditions are to be replaced by long-range scientific predictions, those predictions will have to be believed before important decisions based on them will be implemented. Such belief will be critical in determining responses. It is important to keep in mind that adaptive behaviors in turn produce *impacts*—beneficial or adverse—upon various sectors.

13.5.5 Equity Issues

Studies on society and weather/climate interactions have shown that some win and others lose. Analysis should be performed to identify stakeholders and assess the likely impacts on them. Distribution of impacts by age, gender and social class should be included in these analyses. Such analyses should be extended from the 'arena of effect' to include interregional trade-offs in costs/benefits from the changes expected. Studies of current weather-change situations can address this distribution-of-effects problem. Policy analysis can be used to design potential institutional mechanisms that could be employed to meet values of social equity—that is, to distribute beneficial and adverse effects more equitably throughout society and between societies. It remains a political issue to decide which mechanisms, if any, are ultimately employed.

13.5.6 Infrastructure

Existing agricultural and water resource systems are designed and operated within known levels of probability or frequency of adverse or beneficial weather conditions. Farmers and insurance companies, for example, expect certain levels of crop risk from weather events; urban water supply and drainage systems are designed to handle certain flow levels that are known to be exceeded some of the time. Climate change is likely to change the probability of occurrence of events to which the infrastructure's systems are designed. As change occurs, affected systems may become increasingly inadequate to handle increases in frequency and/or intensity of, say, precipitation events, or to make use of them.

Organizational adaptive strategies to such situations should be studied. Determining domains of responsibility, interest in protecting existing systems, denial that change is occurring, concerns about organizational maintenance—these are potential organizational responses to changing environments and are areas for study in current relevant weather situations. Determining the allocation of political responsibility for climate change impacts is a central problem. By studying how these responsibilities are being handled in current weather situations, insights useful for future situations could result.

13.5.7 Research Agenda

An array of methodological and topical research recommendations has been made concerning how to approach the problem of assessing the impacts of climate variation on society. These include:

- study indirect effects (such as multipliers, social disruption) (Warrick and Riebsame, 1981).
- focus on process (identifying the pathways and linkages within social systems through which effects of climate change are transmitted, how they

are transmitted, decision processes) (US Department of Energy, 1980; Warrick and Riebsame, 1981).

- focus on adaptation (how peasants handle drought; how societies have responded to a variety of natural disaster situations through such mechanisms as insurance, contour plowing, etc.) (Wisner, 1977; Berry and Kates, 1980; Meyer-Abich, 1980; US Department of Energy, 1980).
- study 'natural disaster' stress situations (US Department of Energy, 1980).
- perform interdisciplinary integration (Ad Hoc Panel on Climate Impacts, 1980; Glantz *et al.*, 1982 and Chapter 22, this volume; Kates, Chapter 1, 1, this volume).
- distinguish between research and assessment (Glantz *et al.*, 1982 and Chapter 22, this volume).
- include stakeholder analysis (objective and perceptual analysis of impacts and adjustment of parties-at-interest) (Glantz *et al.*, 1982).

To these can be added the research recommendations arising from this review:

- study the existing situation of known weather/climate change in various climate regimes. These would involve drought, precipitation increases, flooding events, and high and low temperature events.
- monitor trends over time (longitudinal studies).
- study perceptions of weather/climate variation on the part of citizens, stakeholders and organizations. Awareness of changes and adaptations to them are important issues for these studies.
- study the allocation of political responsibilities and decision processes in affected and potentially affected areas.
- study interregional trade-offs involving impacts.
- examine the credibility of weather and climate forecasting.
- in data-poor countries, build incrementally the data on relevant SIA variables as funding is available.

13.6 CONCLUSION

Social impact assessment can fruitfully be employed in countries throughout the world to initiate and extend an understanding of what impacts localized and global changes will have on societies. In countries with existing data bases, longitudinal studies of current situations involving repetitive droughts, flooding, precipitation increases, frost, and higher and lower temperatures will extend their utility. In data-poor countries, collecting data on a piecemeal basis in connection with various projects will begin to build the data bases needed to approach with more certainty the question of climate impacts. The immediate use of SIA methods mentioned in this chapter that do not rely heavily on empirical data will also be beneficial. To the extent that the goal is to study social change, much can be gleaned from reviews of related literatures on natural hazards and natural resources development and management.

Consideration of the potential part social analysis might play in assessing the impacts of global climate change is best limited to multiyear persistent episodes and longer periods of gradual climate changes. What will these changes mean to the world's societies? Interannual and persistent, multiyear episodes can be studied by focusing on current situations of drought, urban weather anomolies, flooding, and so on, using the approaches mentioned above. Longer, century-scale 'little ages' can best be addressed using scenarios and decision methods. These methods permit imaginative synthesis of climate change problems, consideration of various outcomes emergent from current trends, and identification of action alternatives.

At all relevant time-scales, social impact analysis has a demonstrated capability to make an important contribution to studies of the interaction of climate and society. Just as applied climatologists strive to disentangle the complexities of the atmospheric system into the components important to human affairs, so social impact assessors systematically reduce the complexities of social interaction to entities amenable to survey and study. As with all such reduction, some essence of the whole is lost, but a considerable improvement in analytic power is gained.

REFERENCES

Ad Hoc Panel on Climate Impacts (1980). *Report of the Climate Board, Ad Hoc Panel on Climate Impacts, to the National Climate Program Office Regarding Social Science Impact Research*. C. Howe, chairman, R. Cummings, A. Eddy, J. Edwards, P. Porter, A. Schnaiberg. Available from the Board on Atmospheric Sciences and Climate, National Academy of Sciences, Washington, DC.

Albertson, M. L. (1983). The impact of hydropower on society. *Impact (UNESCO)*, No. 1, 69–81.

Albrecht, S. L. (1978). Socio-cultural factors and energy resource development in rural areas in the West. *Journal of Environmental Management*, **7**, 78–90.

Berry, L., and Kates, R. W. (Eds.) (1980). *Making the Most of the Least: Alternative Ways to Development*. Holmes and Meier Publishers, New York.

Bowles, R. T. (1981). *Social Impact Assessment in Small Communities: An Integrative Review of Selected Literature*. Butterworths, Toronto.

Branch, K. J. T., Creighton, J., and Hooper, D. (1982). *Guide to Social Assessment*. Mountain West Research, Inc., Billings, Montana.

Cano, G. J. (1983). The influence of science on water institutions. *Impact (UNESCO)*, No. 1, 17–25.

Castro, A. P., Hakansson, N. T., and Brokensha, D. (1981). Indicators of rural inequality. *World Development*, **9**, 401–427.

Chalmers, J. A., and Anderson, E. J. (1977). *Economic-Demographic Assessment Manual: Current Practices, Procedural Recommendations, and a Test Case*. US Bureau of Reclamation, Denver, Colorado.

Changnon, S. A., Jr. (1979). How a severe winter impacts on individuals. *Bulletin of the American Meteorological Society*, **60**, 110–114.
Changnon, S. A., Jr. (1981a). Hydrologic applications of weather and climate information. *Journal of the American Water Works Association* (October), 514–518.
Changnon, S. A., Jr. (1981b). Impacts of urban modified precipitation conditions. In Changnon, S. A., Jr. (Ed.) *Metromex: A Review and Summary*. Meteorological Monograph 18 (October), pp. 153–177. American Meteorological Society, Boston, Massachusetts.
Changnon, S. A., Jr. (Ed.) (1981c). *Metromex: A Review and Summary*. Meteorological Monograph 18 (October). American Meteorological Society, Boston, Massachusetts.
Changnon, S. A., Jr., Davis, R. J., Farhar, B. C., *et al.* (1977). *Hail Suppression: Impacts and Issues*. Final report, ERP75-09980, Office of Exploratory Research and Problem Assessment, Research Applied to National Needs Program, National Science Foundation. Illinois State Water Survey, Urbana.
Changnon, S. A., Jr., Farhar, B. C., and Swanson, E. R. (1978). Hail suppression and society. *Science*, **200**, 387–394.
Changnon, S. A., Jr., Changnon, D., and Stone, P. (1980). *Illinois Third Consecutive Severe Winter: 1978–79*. ISWS/RI-94/80. Illinois Institute of Natural Resources, Urbana.
Chen, R. C., Boulding, E., and Schneider, S. (1983). *Social Science and Climate Change*. D. Reidel, Dordrecht, Netherlands.
Conn, W. D. (Ed.) (1983). *Energy and Material Resources: Attitudes, Values, and Public Policy*. Westview Press, Boulder, Colorado.
Connor, D. M. (1977). Social impact assessment: The state of the art. *Social Impact Assessment*, **13/18** (June), 4–7.
Cortese, C., and Jones, B. (1977). The sociological analysis of boom towns. *Western Sociological Review*, **8**, 76–90.
Denver Research Institute (1979). *Socioeconomic Impact of Western Economy Resource Development*. Council on Environmental Quality, Washington, DC.
Drabek, T. (1983). Sociology research needs. In *A Plan for Research on Floods and Their Mitigation in the United States*, pp. 107–133. Illinois State Water Survey, Champaign, Illinois.
Enk, G. A., and Hart, S. L. (1978). *Beyond NEPA Revisited: Directions in Environmental Impact Review*. Institute on Man and Science, Renesselaerville, New York.
Farhar, B. C. (1979). *Discussion Summary: DOE/SERI Workshop on Social Costs/Benefits of Energy*. Draft paper. Solar Energy Research Institute, Golden, Colorado.
Farhar, B. C., Changnon, S. A., Jr., Swanson, E. R., *et al.* (1977). *Hail Suppression and Society: Summary of Technology Assessment of Hail Suppression*. Illinois State Water Survey, Urbana (June).
Farhar, B. C., Clark, J. A., Sherretz, L. A., *et al.* (1979). *Social Impacts of the St. Louis Urban Weather Anomaly*. Institute of Behavioral Science, University of Colorado, Boulder.
Farhar, B. C., and Mewes, J. (1974). Public response to proposed snowpack augmentation in the Sierra Nevada. In *A Summary of the Initial Public Involvement Meetings, Correspondence and Public Response to a Proposed Research Program for Snow Augmentation in the Sierra Nevada, California*. US Bureau of Reclamation, Denver, Colorado (December).
Farhar, B. C., and Rinkle, R. (1976). *A Societal Assessment of the Proposed Sierra Snowpack Augmentation Project*. US Bureau of Reclamation, Denver, Colorado (December).

Finsterbusch, K., and Wolf, C. P. (Eds.) (1981). *Methodology of Social Impact Assessment*, 2nd edn. Hutchinson Ross, Stroudsburg, Pennsylvania.

Fitzsimmons, S. J., Stuart, L. I, and Wolff, P. C. (1975). *Social Assessment Manual: A Guide to the Preparation of the Social Well-being Account*. Abt Associates, Cambridge, Massachusetts.

Flynn, C. B., and Flynn, J. H. (1982). The group ecology method: A new conceptual design for social impact assessment. *Impact Assessment Bulletin*, **1** (4, Summer), 11–19.

Gale, R. P. (1977a). *Social Impact Assessment: An Overview*. Forest Service, US Department of Agriculture, Washington, DC (June).

Gale, R. P. (1977b). *Social Impact Assessment Handbook*. Technical review draft. University of Oregon, Eugene (June).

Glantz, M. H. (Ed.) (1976). *The Politics of Natural Disaster: The Case of the Sahel Drought*. Praeger, New York.

Glantz, M. H. (Ed.) (1977). *Desertification: Environmental Degradation in and Around Arid Lands*. Westview Press, Boulder, Colorado.

Glantz, M. H., Robinson, J., and Krenz, M. (1982). Climate-related impact studies: A review of past experiences. In Clark, W. (Ed.) *Carbon Dioxide Review: 1982*. Oxford University Press, New York.

Haas, J. E. (1975). Social impact of induced climate change. In d'Arge, R., *et al.* (Eds.) Climate Impact Assessment Program (CIAP), Monograph 6: *Economic and Social Measures of Biologic and Climatic Change*. US Department of Transportation, Washington, DC.

Leistritz, F. L., and Chase, R. A. (1982). Socioeconomic impact monitoring systems: A review and evaluation. *Journal of Environmental Management*, **15**, 333–349.

Leistritz, F. L., and Murdock, S. H. (1981). *Socioeconomic Impact of Resource Development: Methods for Assessment*. Westview Press, Boulder, Colorado.

Logothetti, T. J. (1974). Social systems. In Weisbecker, L. W. (Ed.) *The Impacts of Snow Enhancement*, pp. 353–388. University of Oklahoma Press, Norman.

Margolis, M. (1980). Natural disaster and socioeconomic change: Post-frost adjustments in Parana, Brazil. *Disasters*, **4**(2), 231–235.

Meyer-Abich, K. M. (1980). Chalk on the white wall? On the transformation of climatological facts into political facts. In Ausubel, J., and Biswas, A. K. (Eds.) *Climatic Constraints and Human Activities*. Pergamon, New York.

Miller, D. C. (1970). *Handbook of Research Design and Social Measurement*, 2nd edn. David McKay Co., Inc., New York.

Mohanty, R. P. (1983). Multi-purpose reservoir system impact. *Impact (UNESCO)*, No. 1, 83–95.

Murdock, S. H., and Leistritz, F. L. (1979). *Energy Development in the Western United States: Impact on Rural Areas*. Praeger, New York.

Peterson, G. L., and Gemmell, R. S. (1977). Social impact assessment: Comments on the state of the art. In Finsterbusch, K., and Wolf, C. P. (Eds.) *Methodology of Social Impact Assessment*, pp. 374–387. Hutchinson Ross, Stroudsburg, Pennsylvania.

Rossini, F. A., and Porter, A. L. (1983). *Integrated Impact Assessment*. Westview Press, Boulder, Colorado.

Sassone, P. (1975). Public sector costs of climate change. In d'Arge, R., *et al.* (Eds.) Climate Impact Assessment Program (CIAP) Monograph 6: *Economic and Social Measures of Biologic and Climatic Change*. US Department of Transportation, Washington, DC.

Sherretz, L. A., and Farhar, B. C. (1978). An analysis of the relationship between rainfall and the occurrence of traffic accidents. *Journal of Applied Meteorology*, **17**, 711–715.

US Agency for International Development (1978). *Social Soundness Analysis*, Appendix 4A, dated 2/15. Washington, DC.

US Department of Energy (1980). *Environmental and Societal Consequences of a Possible CO_2-induced Climate Change: A Research Agenda*, Vol. 1. DOE/EV/10019-01, Washington, DC.

Vlachos, E. (1977). The use of scenarios for social impact assessment. In Finsterbusch, K., and Wolf, C. P. (Eds.) *Methodology of Social Impact Assessment*. Hutchinson Ross, Stroudsburg, Pennsylvania.

Warrick, R. A., and Riebsame, W. E. (1981). Societal response to CO_2-induced climate change: Opportunities for research. *Climate Change*, **3**, 387–428.

Weisbecker, L. W. (1974). *The Impacts of Snow Enhancement*. University of Oklahoma Press, Norman.

White, G. F., and Haas, J. E. (1975). *Assessment of Research on Natural Hazards*. MIT Press, Cambridge, Massachusetts.

Wisner, B. G., Jr. (1977). *The Human Ecology of Drought in Eastern Kenya*. PhD dissertation, Department of Geography, Clark University, Worcester, Massachusetts.

Wolf, C. P. (1981). Social impact assessment. *Impact Assessment Bulletin*, **1**, 9–19.

Climate Impact Assessment
Edited by R. W. Kates, J. H. Ausubel and M. Berberian

CHAPTER 14
The Impact of Climatic Variations on Agricultural Margins

Martin L. Parry

Department of Geography
University of Birmingham
PO Box 363
Birmingham B15 2TT, UK

14.1 INTRODUCTION

Human activities, within the constraints of accessibility and the competition of other more valued activities, are generally located in areas favourable for their practice. Over time the fit is usually improved upon by various adaptive actions. As the most favourable areas are taken up, activities expand to their margin, where they are barely worthwhile or barely compete with other activities. Here, at the margin, the impact of climatic variability can be more pronounced, and marginal areas may thus be appropriate 'laboratories' for studying the interaction between society and climate. That is the thesis of this chapter.

Over the past decade or so, several researchers have evolved relatively sophisticated methods for assessing climatic impacts in marginal areas. Their studies, some of which I shall describe below, give some assurance to the assessment of the nature, scale and location of impacts which might be expected for certain types of climatic variability and climatic change.

14.1.1 Definitions

We can identify three types of marginality—spatial, economic and social. The first relates to locations and areas where activities are at the edge of their ideal climatic region, for example, where warmth or moisture is barely sufficient to enable an adequate return from a particular type of farming. In a sense these areas are akin in their sensitivity to the ecotone (or transition zone in tension between two communities), which has frequently been employed by the biologist as a laboratory for the study of ecological change. In the case of farming, many such margins are determined economically as much as climatically—they mark the boundaries of comparative advantage between competing farming types (for example, between the corn belt and the wheat belt in the United States). But climatic change can bring about shifts in economic margins as well as in biophysical margins. The important point is that climatic impacts can be expressed as shifts of the margins (farmers changing their farming systems or abandoning farming entirely). For the climatic impact analyst, margins are lines or zones between arbitrarily defined classes which undergo a spatial shift for given climatic variations, and thus provide an indication of the type and location of impact.

In reality there are other types of marginality which can complicate this form of analysis: *economic marginality*, where returns to a given activity barely exceed costs, and *social marginality*, where groups are isolated from their indigenous resource base and are forced into economies which contain fewer adaptive mechanisms for survival (Baird *et al.*, 1975). These provide two further dimensions to the spatial one: while spatial marginality implies differentiated environmental resource complexes, different economic resources and different political power can allow different access to the environmental resources.

To start with we shall simplify this real-world complexity by considering only spatial marginality. We review methods available for identifying marginal areas at risk from short-term climatic variations or longer-term climatic changes, outline a general research strategy, and exemplify some variants of this strategy to illustrate how it can be employed to identify the location of areas of increased impact.

14.2 STRATEGIES FOR IDENTIFYING IMPACT AREAS

On the face of it, similar impacts may occur as a result of quite different factors. For example, adjustments in agriculture may occur from changes in farming objectives (such as levels of expectation or tolerance levels), changes in farming systems (in technology, labour, demand) and changes in the agricultural resource base (such as soil erosion, climatic change). These three groups of factors (and others not considered here) are sometimes closely interconnected. The problem is to disentangle them so that we can specify, with some confidence, the effects of climatic change and variability.

One way to tackle this problem is to attempt a prediction of the areas and types of impact on the basis of an understanding of the interactions between agriculture and weather, and then proceed to test these predictions against historical actuality. It will subsequently emerge that different studies in different regions of the world have, for obvious reasons, employed variants of this strategy to suit their local circumstances, but the overall approach has remained broadly the same. There are four steps in this approach:

1. to isolate the important climatic variables by modelling crop/climate relationships;
2. to establish critical levels of these variables by relating them to farming behavior (such as through changes in the probability of reward or loss);
3. to resolve climatic fluctuations into fluctuations of the critical levels (such as probabilities); and
4. to map these as a shift of isopleths to identify impact areas.

14.2.1 Modelling Crop/Climate Relationships

In order to have any confidence in an assessment of likely changes in agricultural output that would occur as a result of certain fluctuations in climate, we need to ascertain the weather variables which account for most variability of yield now (that is, under 'normal' or 'unperturbed' conditions).

14.2.1.1 Isolating the Weather Variables

Various means of analysing crop/climate processes have been discussed by Nix (Chapter 5, this volume). We can distinguish three types of models:

1. crop-growth simulation models (which attempt to represent the physical, chemical and physiological mechanisms underlying plant- and crop-growth processes);
2. crop/weather analysis models (which attempt to represent the functional relationship between certain plant responses—for example, yield—and variations in selected weather variables at particular stages of plant development); and
3. empirical-statistical models (which seek to identify those weather variables which show a strong association with crop yield by virtue of high correlation coefficients).

Each of these models has varying data requirements and disadvantages in assessing impacts over long rather than short time-scales, but they at least provide some scientific basis for impact assessment.

14.2.1.2 Selecting Critical Levels of Weather Variables

By specifying levels of the climatic variables which correspond to apparently critical biophysical margins of certain crop types, or critical margins of yield or profitability, it is possible subsequently to evaluate climatic impacts as shifts of such margins. Both the variables and their levels will, of course, vary according to different farming systems and crop types. For example, at high levels in southern Scotland the oat crop is especially sensitive to exposure, and the warmth and wetness of the growing season. By an empirical study of crop limits, critical levels were established as being 6.2 m/sec average windspeed, 60 mm potential water surplus, and 1050 degree days above a base of 4.4 °C (Parry, 1975).

On the Canadian Great Plains, the northern margin of cereal growth is controlled by temperature and photoperiod. A biophotothermal time-scale has been established for wheat (Robertson, 1968) and barley (Williams, 1974) to determine the region within which these crops would normally reach various phenological stages: at the northern boundaries of each region the crops would be expected to ripen by first freeze in 50 out of 100 years (Williams and Oakes, 1978). It has also been possible to compute where, on average, the crops would ripen 0–20, 20–40, and more than 40 days before first fall freeze (Williams *et al.*, 1980). In a similar fashion Uchijima (1978) has drawn isopleths of minimum effective temperature for rice cultivation in northern Japan.

14.2.2 Climate and Farming Decisions

The selection of supposedly critical levels of climate variables for agriculture assumes a knowledge both of what is critical to the crop (such as levels of warmth, moisture and sunshine) and what is critical to the farmer (such as levels of yield, profit and the like). Thus far we have considered only what is critical to the crop. In order to assess impacts on agriculture we need to ask: to what exogenous perturbations is the farming system most vulnerable? For example, to changes in average yield or changes in minimum yield? To the probability of profit or loss? And what levels of these? Once again, our assessment should consider the different levels of vulnerability and different responses which tend to characterize different farmers and farming systems.

14.2.2.1 The Probability of Crop Failure, Net Loss or Critical Shortfall

Previous studies have tended to focus on climatic impacts on average yields. There are several reasons, however, for believing that changes in the probability of success or failure due to climate variability are more important to many marginal farmers than changes in average yield. First, marginal farmers, by definition, operate towards the limits of profitability, have a slender buffer against hardship, and thus are more concerned with survival than with wealth. Their strategies emphasize risk avoidance rather than maximizing outputs. Second, even the profit-maximizing farmer (including those in nonmarginal areas) knows well that net returns are not simply a function of average yield, but also of the balance struck between gambling on 'good' years and ensuring against 'bad' ones (Edwards, 1978). Third, the boundary between profit and loss for particular farming activities over the medium or long term may depend on the relative frequencies of favorable and adverse weather; for example, a major constraint on profitable wheat production in Alberta is related to the probability of first autumn freeze occurring before the crop matures (Robertson, 1973).

The appropriate measure of this kind of impact is likely to vary from place to place. Among noncommercial farmers it may be failure to give a minimum yield. Among commercial farmers it may be the balance between inputs and outputs (for example, net loss) or some measure of shortfall below the expected yield (for example, 10 percent below average).

14.2.2.2 The Probability of Consecutive Impact

Marginal farmers may be especially vulnerable to losses in successive years. Consecutive harvest failures, by removing the reserve of seed corn, have been taken as a cause of famine in subsistence communities in the past, and today, at the international level, consecutive failures can lead to a sudden and marked drop in food reserves that is difficult to make up by increases in productivity or by the extension of the cultivated area, except over the longer term. The chance of two successive shortfalls (90 percent of average output) in US foodgrains production is certainly not insignificant. Sakamoto *et al.* (1980) estimate it to be 9 percent.

Moreover, the probability of the occurrence of two extremes in consecutive years is far more sensitive to a change in mean climatic value than is the probability of the single occurrence. To illustrate, consider a numerical example and suppose that extremely cold winters or dry summers occur with a probability of $P = 0.1$. The return period for the occurrence of a single extreme is, therefore, 10 years, while the return period for the occurrence of two consecutive extremes is 100 years (assuming a normal distribution of frequencies). Any change in climate will lead to a change in P, either through changing variability which will change P directly, and/or through a change in mean conditions that must also change P if extremes are judged relative to an absolute threshold. Alternatively, P may change through changes in some critical impact threshold as a result of land use changes, new crops or crop mixes, increasing population pressure, and so forth. If P becomes 0.2, then the return period for a single severe season is halved to 5 years. The return period for consecutive severe seasons, however, is reduced by a factor of four to only 25 years.

We may conclude that, in certain cases, climate impact on agriculture can appropriately be expressed as changes in the probability of some critical occurrence or, in other words, as a change in risk. This notion can be incorporated in the overall strategy so that changes in the frequency of critical levels of selected weather variables are expressed as a shift of isopleths of the probability of risk or reward. The strategy may be summarized as a flow diagram (Figure 14.1). In this example the weather, described by a set of meteorological data for a number of years, is expressed as a probability of crop failure. When calculated for a number of stations this probability level can be mapped geographically as an isopleth. Scenarios of changing climates can then be used as inputs to the model to identify geographical shifts of the probability isopleths. The area delimited by these shifts represents areas of specific climatic impact. The method will now be illustrated by reference to a case study in upland Europe, where the major limiting factor to cereal cropping is summer warmth.

14.2.2.3 Calculating the Frequency of Failure, Loss or Critical Shortfall

At many locations the climatic variables that influence rates of plant growth (such as temperature, precipitation, sunshine) decrease in a roughly linear fashion towards the margin of cultivation. For example, in areas where cereal cropping is limited largely by temperature (namely at high latitudes and high elevations), accumulated warmth decreases approximately linearly with increasing elevation and increasing latitude. While this is, of course, a generalization, the point is that, assuming annual levels of warmth or moisture to be normally distributed from year to year, the *probability* of a minimum level of warmth or moisture required to avoid failure, loss or critical shortfall would increase, not linearly towards the margin of cultivation, but in the S-shaped curve characteristic of the cumulative frequency of a normal distribution

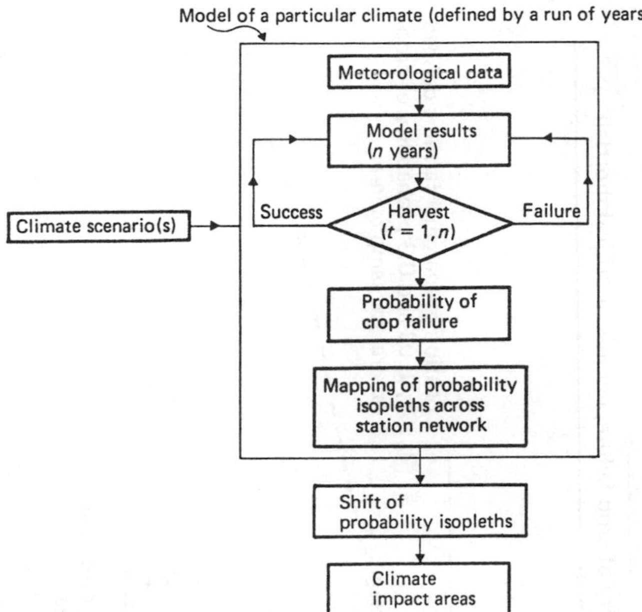

Figure 14.1 Steps in the identification of climate impact areas. (After Parry and Carter, 1983)

(Figure 14.2). At the lower end of this curve there is a marked, indeed quasi-exponential, increase in the probability of failure, and it will be shown that it is precisely at this part of the curve that marginal land is frequently located. It seems, therefore, that marginal areas are frequently characterized by a very steep 'risk surface'. A consequence is that any changes in average warmth or aridity, or in their variability, would have a marked effect on the level of risk. In reality, the interannual distribution of warmth or moisture is frequently non-normal, so the real levels of risk may be even higher.

To illustrate, in southern Scotland, where the main constraint on cereal cropping is the intensity of summer warmth, the oat crop (variety Blainslie) will not mature before mid-September in growing seasons of less than 970 growing degree days (Parry, 1975). The frequency of this type of crop 'failure' increases quasi-exponentially with elevation, in a fashion similar to that described in Figure 14.2. It follows that changes in temperature over time (due to a change in climate rather than a shift of location) have a magnified effect on levels of risk: at the margin of cultivation (about 300 metres) in Scotland a 0.5 °C reduction in mean monthly temperature would, *ceteris paribus*, lead to a 10 percent fall in accumulated warmth, a doubling in the probability of crop failure, and a *six-fold increase* in the probability of two successive failures (Parry, 1976). This result would hold for an equal decrease in each monthly mean across the whole growing season. An increase in variability would further magnify the probabilities.

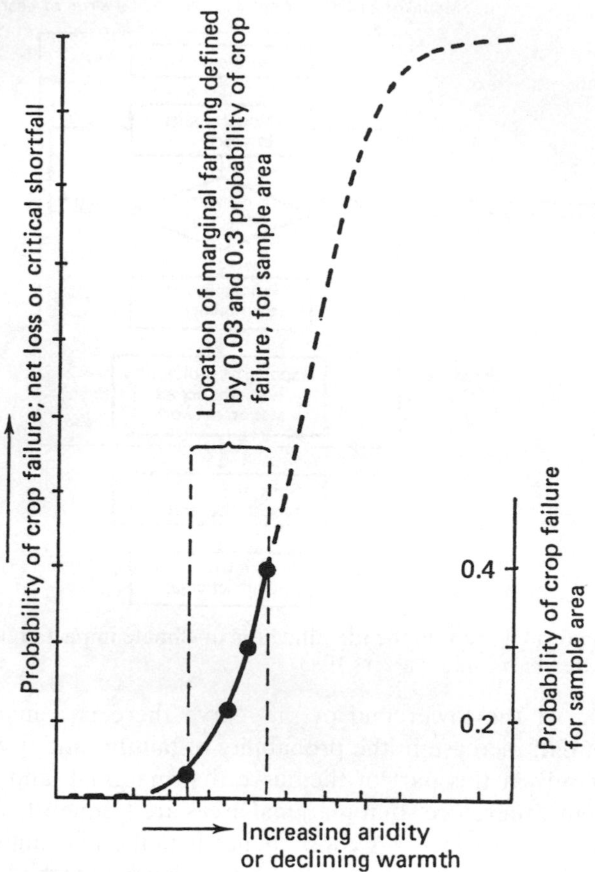

Figure 14.2 Probability of crop 'failure', net loss or critical shortfall, with linear (and normally distributed) gradient of aridity or warmth. Sample area is S. Scotland, probabilities of crop 'failure' are for oats (var. Blainslie). Probabilities of crop failure which define the marginal areas are derived from empirical data on farming strategies in S. Scotland. (Adapted from Parry, 1976)

From a study of the temperature lapse rate with elevation we can also gauge the gradient of the risk surface: at an elevation of about 200 metres in Scotland the frequency of crop failure increases 100 times with only a 150-metre increase in elevation. From meteorological data that enable us to estimate the different lapse rates in different regions, it is possible to map areas of climatic risk by drawing isopleths of tolerable levels of crop failure. The levels of tolerance need to be based upon some empirical assessment (such as interviews with farmers) that takes account of local social and economic factors. As an

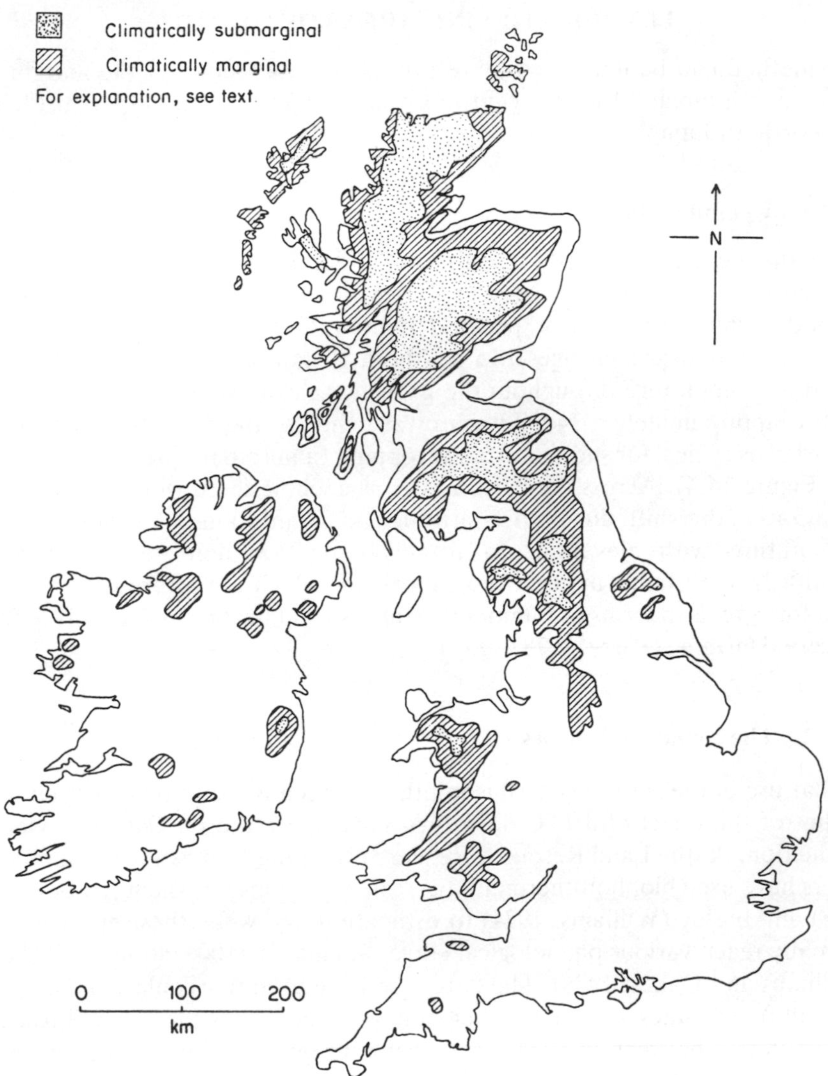

Figure 14.3 Areas that are marginal for oats cropping in the British Isles. Calculated from data relating to temperature requirements for the ripening of oats (var. Blainslie). (After Parry, 1978)

example, the levels chosen for oats cropping in Britain were frequencies of failure of 1 in 3.3 and 1 in 33—between these levels oats cropping as an enterprise was taken to be highly precarious (Parry, 1978). Isopleths of these risk levels (Figure 14.3) point to areas where this type of agriculture is especially sensitive to climate. The next step is to examine what shift of these isopleths occurs for specified changes of climate.

14.3 IDENTIFYING AREAS OF IMPACT

The method can be illustrated by reference to case studies at the margin of cultivation in upland Europe, central Canada, the southern US Great Plains and northern Japan.

14.3.1 Upland Britain

Using the method described for southern Scotland, and constructing isopleths of a crop failure frequency of 1 in 3.3, we can analyse the shifts of these isopleths which occur for a specified temperature change (or a variety of possible temperature changes). In northern Britain a 1 °C increase in mean monthly temperature throughout the growing season would, *ceterus paribus*, lead to approximately a 140-metre upward shift of the probability isopleths regarded as critical for successful oats cropping (a failure frequency of 1 in 3.3) (see Figure 14.4). Across the British Isles as a whole there would be regional variations in this shift, due both to latitude and to variations in the lapse rate of temperature with elevation. In total about 2 million hectares (about one-third) of Britain's unimproved moorland, which are at present submarginal for cereals in terms of summer warmth, would become marginally viable for cereal farming (Parry, 1978).

14.3.2 The Canadian Prairies

Similar use of the shift of critical isopleths has been made in two independent studies of the effect of a 1 °C downturn in temperature on Canadian wheat production. At the Land Resource Research Institute (Ottawa), Williams and others have used biophotothermal time-scale equations for wheat (Robertson, 1968) and barley (Williams, 1974) to estimate if and when these crops would normally reach various phenological stages at each of 1100 stations in Canada (Williams and Oakes, 1978). The data also were used to calculate the normal first fall freeze dates and the number of days from ripening to first fall freeze. Extrapolation between these stations enabled isopleths of the limits for three phenological stages (heading, soft dough stage and ripening) to be drawn for Olli barley and Marquis wheat. To compute the climatic resources for a cooler climatic regime, 1 °C was subtracted from the temperature normals for every month. This made the assumed planting date later, extended the time required to mature as computed by the biophotothermal time-scale equations, and brought forward the date of first fall freeze. Figure 14.5 illustrates the shift of isopleths bounding the wheat-maturing zone: the area suited to wheat production (not constrained by soil or terrain) would be reduced by one-third. The area suited for barley would contract by only one-seventh because it extends farther north and therefore is more limited by terrain than by

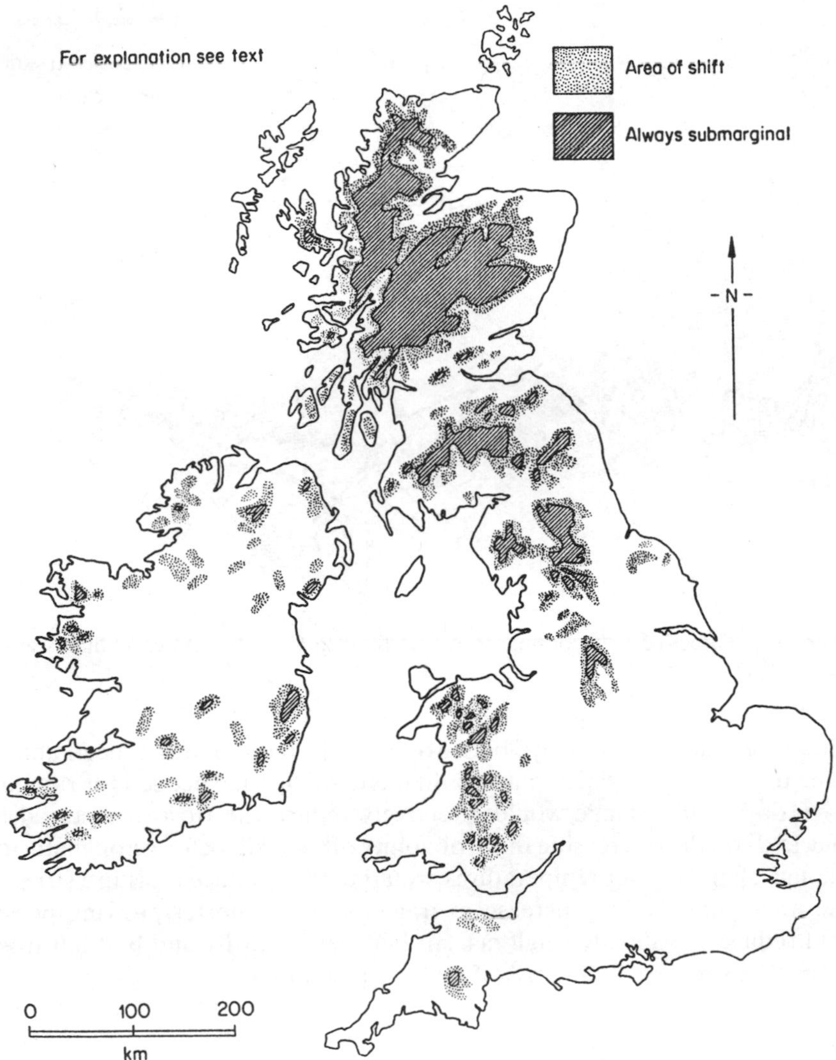

Figure 14.4 Shift of 1 in 3.3 failure frequency for oats in the British Isles for 1 °C increase in mean annual temperatures (normals 1856–95)

temperature. These are, of course, average estimates; no account has been taken of changes in the degree of risk.

A similar, though unpublished, study in the Environmental Systems Branch of Environment Canada gave more attention to changes in the *probability* of ripening (Winstanley, 1974, personal communication). The study concluded that a decrease in mean annual temperature of 1 °C would reduce the frost-free

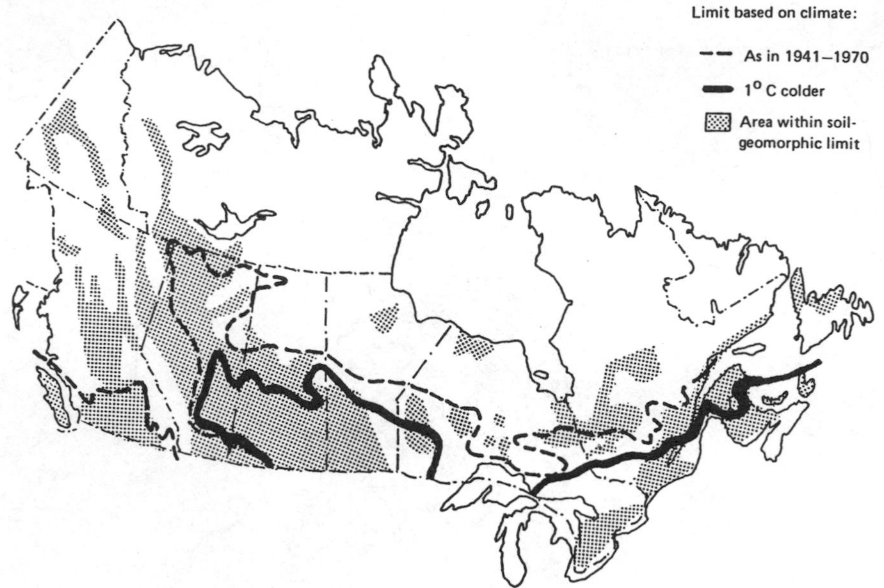

Figure 14.5 Effect of 1 °C cooling on wheat limit in Canada. (After Williams and Oakes, 1978)

period in southern Canada by about 10 days but, by decreasing mean annual degree day totals by 4–6 percent, would increase the time needed for ripening by 4–6 days. This change would effectively reduce the frost-free period by about 15 days, thus increasing the probability of frost kill before crop maturity. Although slightly lower temperatures would tend to reduce moisture stress in some areas and thereby increase average yields, a shorter growing period would reduce the already small margin between maturity and first fall frost, and thus greatly increase the risk of total crop failure.

14.3.3 The US Corn Belt

A third variant of the isopleth-shift approach can be illustrated by reference to the work of Newman (1980) on the US Corn Belt. Newman applied daily differences of ± 1 °C to growing degree days (GDD) for 18 stations in Indiana over a 10-year period and determined that the average north–south distance between isolines of 1600 GDD for the three different conditions (normal, $+1$ °C and -1 °C) was 144 km per degree C. He also calculated the effect of a temperature change on potential evapotranspiration (PET), concluding that a 1 °C change in mean annual temperature in Indiana produced a change of 5.9 cm in annual PET. These values translate into a west–east shift of approximately 100 km in annual PET values per degree C. The 100 km west–east shift

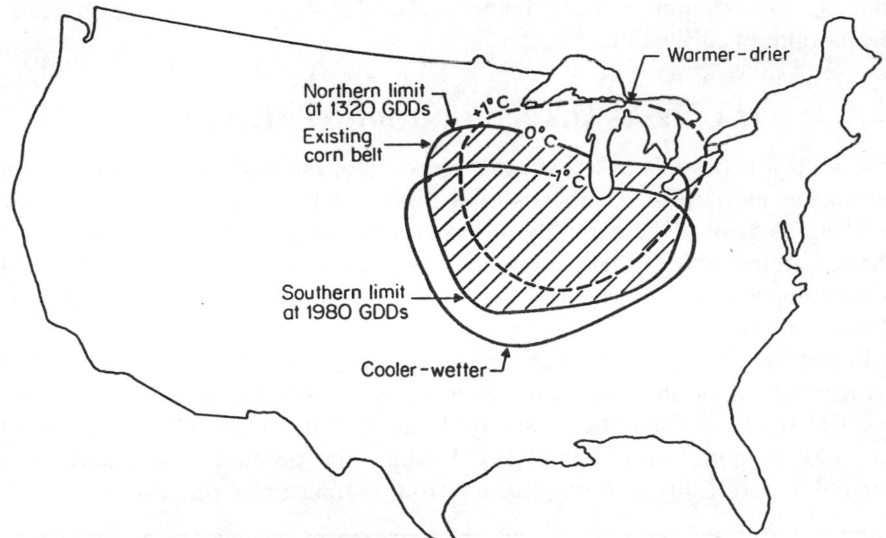

Figure 14.6 Simulated geographical shift in the US corn belt with temperature changes (based on frost-free growing season thermal units, GDDs). (After Newman, 1980)

in annual PET coupled with the 144 km north–south shift of the 1600 GDD line per degree C were used to estimate the geographical shift in the corn belt per degree C. Figure 14.6 illustrates the simulated shift for a 1 °C warmer and drier climate, which is a plausible scenario for the future, given continued increases in the CO_2 content of the atmosphere, and for a 1 °C cooler and wetter climate, which is a plausible simulation of conditions which probably occurred for some cool decades in the seventeenth century.

14.3.4 Northern Japan

It may be more realistic to predict the maximum and minimum thermal resources which can be expected in a given period by calculating a return period using ordered data on GDD and growing season. In Japan, Uchijima (1978) found that the magnitude of the possible maximum and minimum GDD and growing season increases linearly with the increment of the coefficients of variance for these fluctuations. The isopleths of GDD in the southern, middle and northern parts of Japan could be expected to shift southward by 150, 200 and 300 km, respectively, if an anomalistic decline in temperature expected every 30 years should occur. Such a southward shift in the isopleths of GDD would cause a reduction in rice production in the Hokkaido district by about 40 percent, but would have much smaller effects further south. The conclusion, once again, is that the risks stemming from climatic variability increase markedly towards the margins of cultivation.

14.4 TESTS AGAINST HISTORICAL ACTUALITY

Insofar as it is possible to use climatic episodes of the past as approximate (but by no means perfect) analogues of a future scenario, it is possible to test predictions against the actuality of impacts in the past. The problem is that changes in technology make the scenarios difficult to compare; the scale and type of impact could be very different in the future than they were in the past. But at least the *location* of the impacts is likely to be comparable.

In southern Scotland there is a close temporal and spatial fit between the distribution of permanently abandoned settlements and farmland and the predicted 'fall' of theoretical climatic limits to cultivation between AD 1600 and 1800 (Figure 14.7) (Parry, 1975). Since the predicted correlation was derived theoretically and empirically from estimates of process links, it is

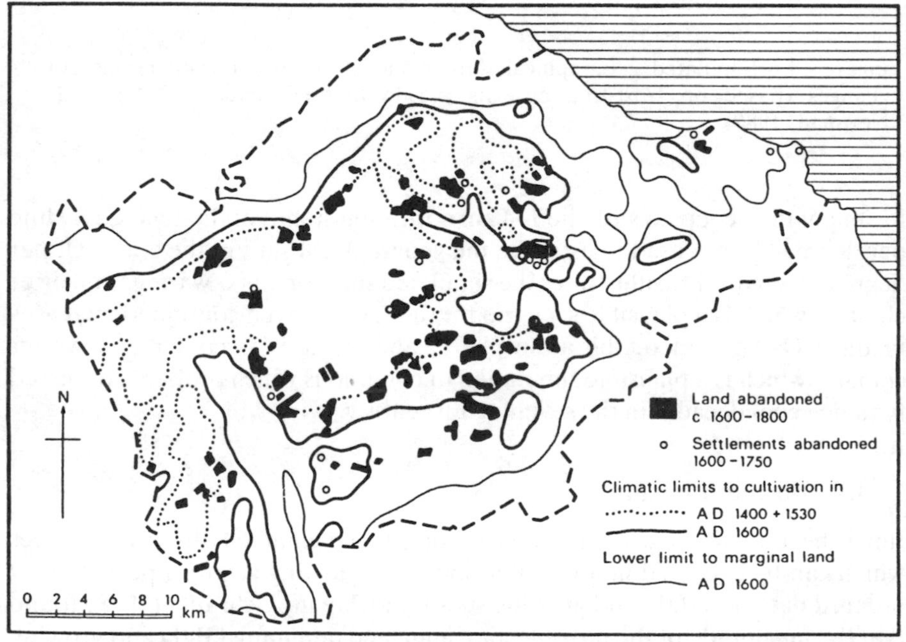

Figure 14.7 Abandoned farmland and lowered climatic limits to cultivation in southeast Scotland, AD 1600–1750. (After Parry, 1975)

logical to assume that the observed correlation reflects some kind of causal connection between climatic change and land abandonment.

One problem, however, in making historical tests of predicted impact is that the evidence for historical shifts in agriculture and settlement has

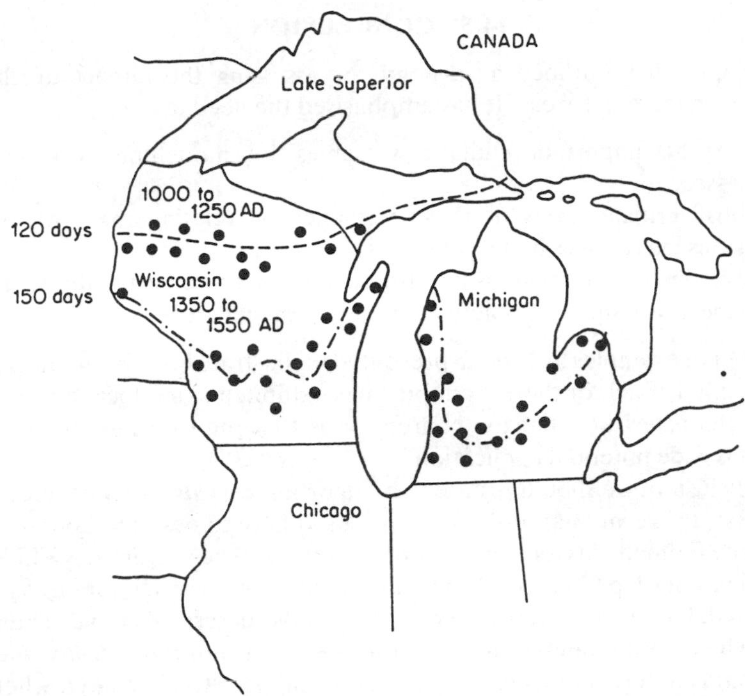

Figure 14.8 Distribution of prehistoric ridge-furrow maize gardens in relation to present-day frost-free seasons. Reproduced by permission of Swets Publishing Service from Newman (1980), *Biometeorology*, **7**, 128–142 (on p. 137), after Riley and Friemuth, *American Antiquity*, **44**, 271–285 (1979)

frequently been used to reconstruct past climates. To test the predicted impact of climate change against 'response' in agriculture and settlement would thus be to invite criticism of a circular, or at least elliptical, argument—although this has unfortunately not discouraged some from trying.

One region for which there may be sufficient and 'uncontaminated' evidence for impact is in the upper Great Lakes region of the United States. The northern limit of prehistoric maize fields appears to have retreated up to 320 km southward concurrently with cooling in the thirteenth and fourteenth centuries (Figure 14.8) (Riley and Friemuth, 1979; Newman, 1980). At present we must be content to class this as a 'space–time' coincidence, but it does seem to mirror Newman's simulated shifts in the Corn Belt which would occur with a 1 °C fall in mean summer temperatures.

The impact of cooling on early agriculture may, however, be more readily detectable than the impact of drought, probably because it is both temporally and regionally less variable. Studies of impact in cold marginal areas thus hold the greater prospect of reward. Early indications of success are now emerging from work on Greenland (McGovern, 1980) and Iceland (Ogilvie, 1981).

14.5 CONCLUSION

This chapter has outlined a strategy for assessing the impact of climatic exchange in marginal areas. It has emphasised the need to

1. isolate the important climatic variables by modelling crop weather processes,
2. establish critical levels of these variables by relating them to farming decisions via changes in the probability of reward or loss
3. resolve climatic fluctuations into fluctuations of these probabilities, and
4. map these as a shift of isopleths to identify impact areas.

A number of examples have been presented to illustrate how the method can be implemented. Each of these examples has exhibited some local and perhaps unique characteristics, but the contention is that the overall strategy is one which has wide potential application.

In conclusion, we should restate two important *caveats* and introduce a new one. First, the term 'margins' have been used here to describe boundaries of arbitrarily defined classes (for example, land-use types, average yield levels, frequencies of crop failure). Marginal areas are not always 'poor' areas. They are areas delimited by a set of criteria such as those described above. Thus, land at the wheat/corn boundary in the United States is not *intrinsically* marginal land (that is in terms of its fertility, etc.). It happens to be located where the concerted influence of present-day environmental, economic, social and other forces leads to a finely balanced competition between wheat and corn. The thesis of this paper has been that finely balanced margins of this kind can provide us with sensitive measures of climatic impact.

Second, we have not considered how different economic and social resources allow different access to environmental resources. Those in farming, as in other walks of life, may be rich or poor, good managers or poor managers, risk-averse or risk-taking. Such variety is bound to place many new dimensions on the issues we have discussed and, at local levels, factors like these would certainly need to be considered.

Finally, there is no suggestion here that it is possible to predict areas of climatic impact on the basis of global interpretations emerging from climatic history. One approach has been described, modified for certain case studies, and, in some instances, tested against the past. Wherever else it is applied, it would require a similar degree of modification to suit local problems and local data.

REFERENCES

Baird, A., O'Keefe, P., Westgate, K., and Wisner, B. (1975). *Towards an Explanation and Reduction of Disaster Proneness*. Occasional Paper 11. Disaster Research Unit, University of Bradford, Bradford, UK.

Edwards, C. (1978). Gambling, insuring and the production function. *Agricultural Economics Research*, **30**, 25–28.

McGovern, T. H. (1980). Cows, harp seals and churchbells: Adaptation and extinction in Norse Greenland. *Human Ecology*, **8** (3, September) 245–275.

Newman, J. E. (1980). Climate change impacts on the growing season of the North American Corn Belt. *Biometeorology*, **7** (2), 128–142. Supplement to *International Journal of Biometeorology*, **24** (December, 1980).

Ogilvie, A. E. J. (1981). Climate and economy in eighteenth century Iceland. In Delano Smith, C., and Parry, M. L. (Eds.) *Consequences of Climatic Change*, pp. 54–69. Department of Geography, University of Nottingham, Nottingham, UK.

Parry, M. L. (1975). Secular climatic change and marginal agriculture. *Transactions of the Institute of British Geographers*, **64**, 1–13.

Parry, M. L. (1976). The significance of the variability of summer warmth in upland Britain. *Weather*, **31**, 212–217.

Parry, M. L. (1978). *Climatic Change, Agriculture and Settlement.* William Dawson & Sons, Folkestone, UK.

Parry, M. L., and Carter, T. R. (1983). *Assessing Impacts of Climatic Change in Marginal Areas: The Search for an Appropriate Methodology*. IIASA Working Paper WP-83-77. Laxenburg, Austria.

Riley, T. J., and Friemuth, G. (1979). Field systems and frost drainage in the prehistoric agriculture of the Upper Great Lakes. *American Antiquity*, **44** (2), 271–285.

Robertson, G. W. (1968). A biometeorological time scale for a cereal crop involving day and night temperatures and photoperiods. *International Journal of Biometeorology*, **12** (3), 191–223.

Robertson, G. W. (1973). Development of simplified agroclimate procedures for assessing temperature effects on crop development. In Slatyer, R. O. (Ed.) *Plant Response to Climatic Factors*, pp. 327–341. *Proceedings of the Uppsala Symposium, 1970*. UNESCO, Paris.

Sakamoto, C., Leduc, S., Strommen, N., and Steyaert, L. (1980). Climate and global grain yield variability. *Climatic Change*, **2** (4), 349–361.

Uchijima, Z. (1978). Long-term change and variability of air temperature above 10 °C in relation to crop production. In Takahashi, K., and Yoshino, M. M. (Eds.) *Climatic Change and Food Production*, pp. 217–229. University of Tokyo Press, Tokyo.

Williams, G. D. V. (1974). Deriving a biophotothermal time scale for barley. *International Journal of Biometeorology*, **18** (1), 57–69.

Williams, G. D. V., and Oakes, W. T. (1978). Climatic resources for maturing barley and wheat in Canada. In Hage, K. D., and Reinelt, E. R. (Eds.) *Essays on Meteorology and Climatology: In Honour of Richard W. Longley*. Studies in Geography, Monograph 3, University of Alberta, Edmonton, Alberta, Canada.

Williams, G. D. V., Mackenzie, J. S., and Sheppard, M. I. (1980). Mesoscale agroclimatic resource mapping by computer, an example for the Peace River region of Canada. *Agricultural Meteorology*, **21**, 93–109.

Climate Impact Assessment
Edited by R. W. Kates, J. H. Ausubel and M. Berberian

CHAPTER 15
Extreme Event Analysis

R. L. HEATHCOTE

The Flinders University of South Australia
Bedford Park, South Australia 5042

15.1 INTRODUCTION

Most societies, in developing the physical resources of the environment in which they are located, have evolved strategies to cope with climatic events which have a limited range of magnitudes. Such a limited range, experienced over time, becomes accepted as the 'normal' spectrum of climatic events for that society. Extreme climatic events, that is, those events with magnitudes outside the normal spectrum, may pose severe and unexpected stresses upon the society.

It is the aim of this chapter to provide an overview of the methods used in the analysis of such stresses and to raise some questions to be considered in such future analyses. To do this we shall examine first the nature of the extreme events; second the nature of their impacts; and third, the methods used in the analysis of those impacts and societal adjustments and adaptation.

It is hoped that this review will show that the methods of extreme event analysis have a wider relevance, indeed may be pertinent to the analysis of the societal impacts of fluctuations within the normal spectrum of climatic events and to the analysis of the impact of climatic change itself.

15.2 THE NATURE OF EXTREME CLIMATIC EVENTS

As energy flows in the global ecosystem, variations in winds, temperatures and precipitation can be measured for specific locations and conventional periods of time (Chapter 2). The extreme climatic events, as defined in this chapter, are those short-term perturbations of the energy flows which provide magnitudes outside the normal spectrum or range of the typical averaging period. Such events at any one location, as we shall see, may be measured in minutes or in years' duration. Their frequency, however, is likely to be limited to return periods of at least 10 years.

15.2.1 Definitions

The general definition above glosses over the considerable difficulties facing any attempt at broad definitions of extreme climatic events. To define such events purely in terms of their physical parameters, for example, ignores the question of their impact upon the society. To society the significance of the events is not only that they are of a certain magnitude, but that this magnitude creates an unexpected impact upon society. Whether such an impact occurs depends as much upon the vulnerability of the society to stress as the magnitude of the event itself. Thus on Figure 15.1 the variations in the physical variables of climatic events are considered 'normal' as long as they do not exceed the range of values within which they provide the basis for effective resource management. Magnitudes in excess of that range become stressful to the society because they cannot be accommodated in the normal resource management system.

The practical difficulties of providing a specific global definition of extreme climatic events, given this changing interface between physical parameters and societal vulnerability based in part upon expectations of future events, is obvious. Even if it were possible to forecast the 50- or 100-year return periods for certain magnitudes, and the most recent research suggests that this is as yet impossible within any acceptable levels of accuracy (see Kishihara and Gregory, 1982), the future vulnerability of the society to stress would itself probably have changed.

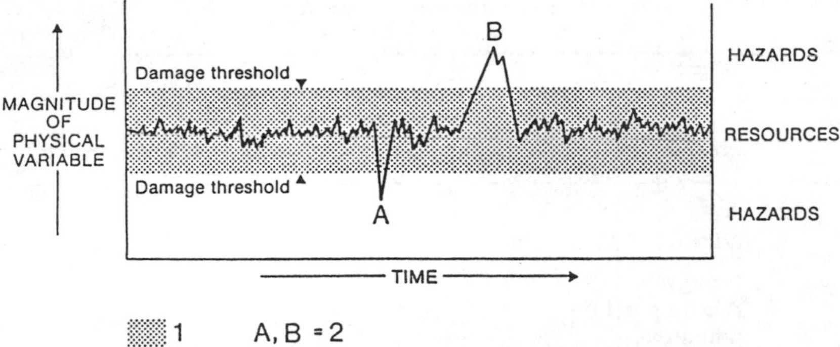

Figure 15.1 Identifying the extreme events. 1, zone of insignificant damage from variations in magnitude of the physical variable with upper and lower limits. Within this zone human activity regards the variations as resources. 2, lower (A) and upper (B) magnitudes of the physical variable which constitute hazards for human activity and which form the extreme events beyond the lower or upper damage thresholds of the resources zone. (Adapted from Burton and Hewitt, 1974)

In practice we are faced by multiple definitions of extreme events resulting from the analysis of specific combinations of certain physical magnitudes of phenomena and the associated societal reactions. Thus, for drought in Australia the official physical indicator is rainfall in the first decile of records, but this is justified by evidence of the coincidence of such occurrences over time with widespread reporting of drought impacts from other sources (Gibbs and Maher, 1967).

Using such composite definitions in the same spirit as the definitions of natural hazards in White (1974) and Burton *et al.* (1978), we can identify a variety of extreme events in the past which have been the subject of specific analysis. Such events would include the cyclone and storm surge in Bangladesh in 1970, the Sahel drought of 1968–73, the blizzard in the United States of February 1978, and the tropical cyclone 'Isaac' which hit Tonga 2–3 March 1982. Listings of some of these events will be found in Dworkin (n.d.) and Thompson (1982), and a recent spatial overview of 'natural hazards' is provided in MRG (1978).

15.2.2 Classification

A useful classification of extreme climatic events, using energy concentration, duration and spatial extent and in which particular events may reflect physical characteristics along a scale from 'pervasive' to 'intensive', has been suggested by Kates (1979, Table 15.1). The contrasts in spatial and temporal contexts implied by this scale for such events require that monitoring systems and

Table 15.1 Characteristics of extreme climatic events

Criteria		Spectrum of characteristics: Pervasive events	Intensive events
Spatial	Area affected	Extensive	Localized
	Energy release per unit area	Low	High
Temporal	Frequency	Frequent (?)	Rare (?)
	Onset	Slow	Rapid
	Duration	Long	Short
Event Type		Drought	Tornado
		Windstorm	
		Hailstorm	
			Avalanche
		←Cyclone→	
		←Flood→	

Source: Modified after Kates, 1979, 517.

societal responses be able to cope with a variety of phenomena. On the one hand such phenomena might occur over a track of some few meters wide, but perhaps several kilometers in length, occurring over minutes or at the most hours (as with tornadoes and windstorms). On the other hand they may occur over half a continent for 2–3 years or more (as with some droughts). Such variations have obvious relevance also in any assessment of the impacts of such events—a variation simplistically stated as between the potential for rapid total local destruction or the slow but steady attrition of regional resources.

15.2.3 Monitoring the Events

The varying spatial and temporal characteristics of extreme events pose specific monitoring problems. The basic technical problems are of equipment failing under extreme conditions. The anemometer at Darwin airport, Australia, for example, failed after recording a wind gust of 217 km/h during

Cyclone Tracy's impact in 1974, when maximum gusts were *estimated* to be up to 250 km/h. In addition, however, the network of weather stations may be too sparse to enable adequate regional monitoring of even basic data such as precipitation and temperatures—this is a particular problem in developing countries where data have to be interpolated over large rural areas from a few, often urban, weather stations (Rijks, 1968). Further, it is generally easier to monitor the intensive extreme events because of their energy magnitudes and their relatively short duration, both of which show up easily in time-series analysis and enable the onset and retreat of the event to be fairly easily identified. In contrast, the onset of events such as drought and sea level fluctuations (associated with climatic changes) may be very difficult to recognize because of the long lead time and gradual, almost imperceptible, transformation of the environment.

The cross-cultural natural hazards research studies show that most societies are aware of critical thresholds of climatic stresses and have identified them in terms, for example, of the required timings of the 'opening rains' for crop plantings (White, 1974). Independent studies have confirmed this kind of predictive awareness. In northeastern Brazil's semi-arid *sertão*:

> tradition holds that if it does not rain by 19 March, St. Joseph's Day, a *sêca* [drought] is bound to follow (Hall, 1978, 18).

In Nigeria there is a considerable body of peasant farmer knowledge of both 'normal' and abnormal climatic stresses upon crop potentials, which is encapsulated in folklore, oral histories and local vocabularies for climatic events (Oguntoyinbo and Richards, 1978). Analysis of such sources reveals those events which the society had, from past experience, identified as crucial to its future well-being.

15.3 THE NATURE OF EXTREME CLIMATIC EVENTS IMPACTS

The occurrence of an extreme event does not automatically imply an impact upon a society. Apart from the obvious situation where an event occurs in an uninhabited area, impacts will vary depending not only on place of occurrence but also with spatial and temporal dimensions and the population and material wealth at risk. The relationship between the magnitude of the event and the nature of its impact can be very complex, as in the qualitative descriptions measuring the height and significance of Nile floods (Figure 15.2). A reduction of annual flood volumes resulting from low rainfalls in Ethiopia could lead to increasing societal distress as the flood levels decrease and fewer areas can be irrigated. However, heavier than normal rainfalls in the river catchment could produce an extremely high level of flooding which could destroy the society so precariously perched on the river banks.

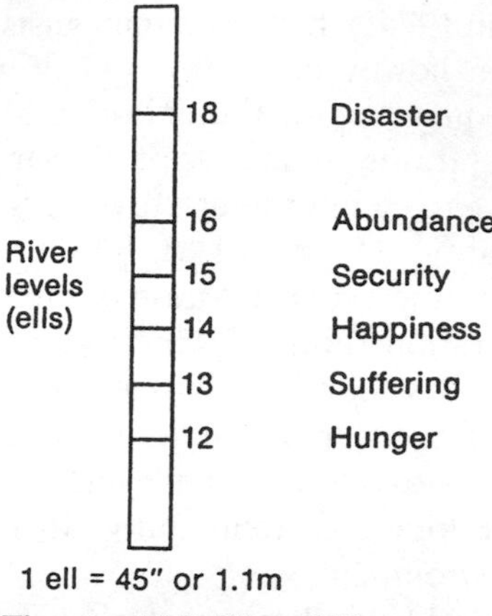

Figure 15.2 The Nilometer. (Reproduced by permission of Oxford University Press from Drower, 1956 (after Pliny), in Suiger *et al.*, *A History of Technology*, Vol. 1)

15.3.1 Monitoring the Impacts

Given the variety of physical characteristics of extreme events and the variety of global societies, the monitoring of impacts is fraught with difficulties. Their reporting may be distorted and the analyses themselves may be distorted by biases in the researchers and their monitoring systems.

Because extreme events have impacts upon societies, and because societies have a long history of regarding these impacts as resulting from external forces and therefore of considering the victims as deserving of the support of the entire society, there can be situations where both some underestimation and some exaggeration of impacts occur.

On the one hand, there may be attempts by government officials to play down the significance of the extreme event because the impact is seen to affect a minority group or segment of society currently in disfavor politically (Swift, 1977), or because the distress is seen to have resulted from unwise resource management or even reckless exposure to climatic risks from practices such as 'compensation farming' (Amiran and Ben Arieh, 1963). On the other hand, the 'victims' may be tempted to exaggerate their losses in order to gain public sympathy and official relief (Heathcote, 1969a), or politicians may exaggerate impacts in order to obtain external financial aid for their own regions in the nation (Glantz, 1976). In the former cases the full impact may be hidden and the plight of the victims worsened by being denied relief, while in the latter cases relief donors might recognize a 'cry wolf' syndrome and regard future disaster reporting with undue suspicion.

Whether the impacts are monitored at all seems to reflect the operations of

Table 15.2 'The many perspectives of hunger'—researchers' biases in famine studies

Research discipline or ideology	Diagnosis of causes of famine	Recommended solution for famine problem
Medicine	Nutritional disorders; environmental stress; disease	Vaccination; breastfeeding/ weaning food; environmental sanitation
Agriculture	Low food supply	Food production; food aid; post-harvest technology; marketing
Education	Ignorance; food habits	Nutrition education; mass communications
Population sciences	High population density and high growth rate	Population control; resettlement
Neoclassical economists	Maldistribution of food	Fiscal policies; income-generating projects; employment programs
Marxists	Capitalism	Revolution
Planners	Lack of planning and coordination	Food and nutrition councils; training

Source: Jonsson, 1981, 1.

the 'Issue-Attention Cycle' elaborated in Foster and Sewell (1980, 2). A sequence of popular concern is suggested for particular environmental issues in the developed countries. From the situation prior to recognition of any problem to alarm at recognition of a problem, opinions change. Enthusiasm for measures to mitigate the problem is followed by the realization that the problem may be a necessary consequence of 'progress', and public interest fades until the stage is set for the next cycle of public attention to a specific issue. This rapid discounting of past events (by planners as well as by the general public) has been noted by Linstone (1973) and is a serious constraint upon the effectiveness of monitoring systems.

Individual researchers bring a variety of biases, which affect the data collated, to their studies. A recent review of studies of the causes of, and remedies suggested for, famine provided a salutary comment on the distortions of such potential biases, not least on what is monitored (Table 15.2, see also Chapter 10, this volume). Only an overview which recognizes all such possible viewpoints could hope to have the basis for a balanced assessment of the impacts of famine, and the same is true for the analysis of extreme events.

Where and when attention is focused, however, various monitoring systems have been adopted. Different types of extreme events may need different impact assessment periods. The recently inaugurated climate impact assessments of the Center for Environmental Assessment Services (CEAS) in the United States distinguish two time periods for reporting upon climate impacts:

1. monthly, for events over land, and
2. quarterly, for events over the marine environment.

The first appears to be an arbitrary but convenient time period, but the second is dictated by 'the extended intervals that frequently exist between the climate-induced event and the observed impact' and the details relate to dominant seasonal characteristics of the marine environment (Center for Environmental Assessment Services, 1982a, i).

Monitoring systems may use maps of areas affected, tables of the sequences of events, time-series graphs or trend lines, or even computer-simulated sequences of actual past events. Specific indices calculated to show relative intensities of losses by area may be attempted. Thus the United States climate impact assessment monitoring system provides national maps of areas affected by 'major weather events' together with maps showing, for example, areas of above- or below-normal demand for energy (using calculations of daily temperature departures from means weighted by the population within the weather station area), as well as maps of transport disruptions (Center for Environmental Assessment Services, 1982a, 12–13). Australia's 'Drought Watch System' uses monthly rainfalls to predict the likelihood of a drought (defined as rainfall in the first decile) and provides national maps of rainfall deficiencies by deciles. These maps are used as one component in federal government drought relief policies. It has been proposed to adapt them for drought surveillance in Botswana (Lee, 1979).

A standard method of monitoring impacts has been to use time-series correlations of climatic and socioeconomic phenomena. Graphs of a 'Business Index' (weekly coal, raw steel, motor vehicle and lumber production against temperatures*) were used to suggest the socioeconomic impacts of the 1976–77 winter freeze conditions in the United States, for example (Center for Environmental Assessment Services, 1982b). Studies of the impacts of climatic change also make extensive use of time-series correlations.

The method, despite its plausibility, can be criticized on two grounds. First, although the correlation may be shown to exist, no causal relationship is usually established; and second, often a mechanistic relationship between

* The 'Business Index' was derived from a weekly index combining production of raw steel, automobiles, trucks, electric power, crude oil refinery runs, bituminous coal, paperboard, paper, lumber, machinery, defense and space equipment and rail and truck traffic. The temperature data were departures from the normal number of days during the month in which the temperature dropped to 0°F or lower at seven industrial cities.

climate and the state of the societal economy through agricultural yields is assumed, and again without necessary proof (Anderson and Jones, 1983).

15.3.2 Identifying the Victims

In terms of the spatial relationship between extreme events and their impacts, most evidence suggests that the area over which the event occurs tends to be smaller than the area over which the impact of that event is felt. In part this may result from the type of event. The blocking of a vital road or rail route by an avalanche or flood will result in societal costs far beyond the area of the avalanche or flood itself through delays in transport and rerouting of materials or passengers.

Research has suggested that most extreme events produce a 'cascade' of impacts—in bureaucratic societies by sharing some of the economic costs among all taxpayers (Figure 15.3), and in less bureaucratic societies by sharing losses among relatives and kinfolk (Cochrane, 1975; Waddell, 1983). In simplistic terms, therefore, in addition to the directly affected population devastated by storm, flood or drought, we must recognize those indirectly affected through increased voluntary charitable donations, involuntary taxation, or social responsibilities for kinsfolk.

In terms of time-scale a similar relationship between the duration of the event and its impact can be suggested. The destruction during an event, whether of life or property, usually occurs over a shorter period than the post-event impact. Injured victims die long after the tornado has struck, flood-damaged structures take time to be repaired, grazing herds may take up to ten years to reach predrought levels and, for some victims, the psychological traumas of the event may remain throughout their lifetimes.

15.3.3 Identifying the Impacts

Accepting the definition of first-, second-, and *n*th-order impacts as noted in Chapter 1, a wide spectrum of sources for analysis may be identified—from the personal testimony of eyewitnesses to satellite imagery. To accommodate this spectrum and some of the problems of biases in reporting, it may be convenient to identify at least three distinctive groups among the 'observers' of the impacts. For each group there are implicit constraints upon the accessible sources of information and the methods by which the data are analyzed.

For immediate victims who have survived the disaster, the impact is clear enough—deaths and injuries, property damaged and food shortages. Their experiences provide sources for three different types of interpretation of the disaster. First, the experience becomes part of the collective societal experience of the environment, expressed perhaps as folklore, poetry,

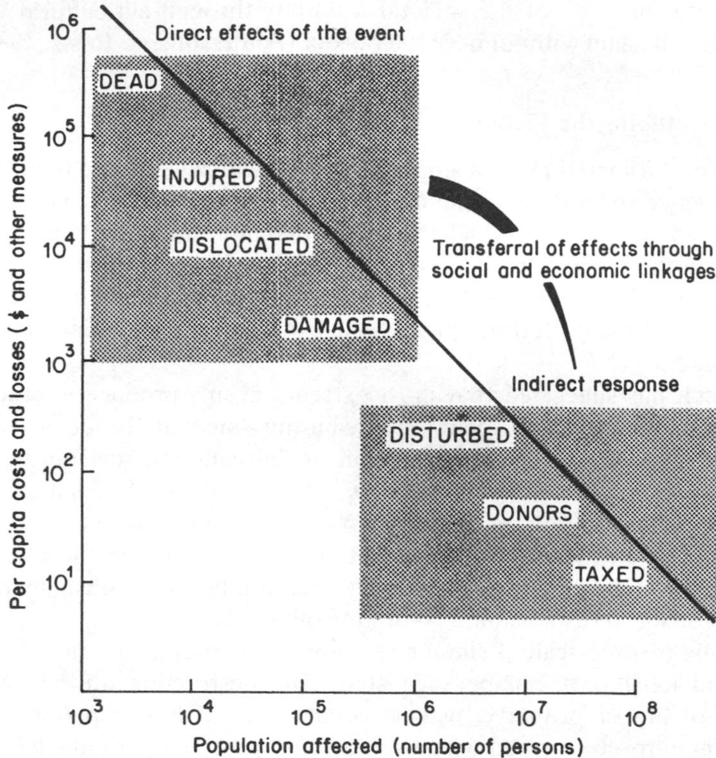

Figure 15.3 The cascade of disaster impacts. (Source: White and Haas, 1975, 76)

aphorisms or preferred sequences of activities such as the timing of crop plantings. Second, the experience becomes the source for research by academics or government officials enquiring into the disaster, and finally the experience, as translated through the media, becomes the basis for the general public's view of the event.

The technical specialists—the academics, the insurance loss assessors, the disaster relief agencies, the government officials—form the second group of observers attempting to identify the impacts. They use a combination of field observation of areal impacts after specific events; historical study of chronological associations of climatic events and other data on human activities through time–series analysis; historical study of media reporting of events and impacts; and simulation of actual or hypothetical interactions of the climatic events and human responses through modeling (for example, of ecological relationships) or developing scenarios of future alternative response strategies. (For a discussion of these methods, see Chapters 11, 18, 20 and 21.)

Such assessors face considerable problems in trying to identify the disaster impacts. There may be differences in the assessment of the significance of the impacts between the victims and the researchers, stemming perhaps from a fatalistic acceptance of the disasters by the impacted society (Burton *et al.*, 1978, 103), lack of awareness of the long-term effects of the extreme events (such as the effects of droughts upon *future* crop yields through accelerated soil erosion—see Tennakoon, 1980), or a failure to link climatic events and human activities with environmental impacts (such as Navajo attitudes to soil conservation—see Fonaroff, 1963). Indeed the requisite data for impact assessment may not be available; for example, a recent study noted that although a useful model for estimating the 'net severity of a disaster' existed, namely

$$\text{Impact Ratio} = \frac{\text{Losses from National Disaster}}{\text{Total Community Resources}} \tag{15.1}$$

the ratio could not always be used because 'information was not readily available' even in a developed society such as the United States (Rubin, 1981, 7). Apart from the lack of data, any attempt to quantify the concept of 'total community resources' would be difficult in itself.

A further general problem facing any attempt at assessment of the significance of the impacts relates to the distinction between what natural hazards research has identified as human 'adjustments' and 'adaptations' to stress situations (White, 1974; see also Chapter 1, this volume). The difference lies partly in the duration of the impact and partly in the characteristics of the human reactions. Adaptations imply long-term responses (over 100–100,000 years) to environmental stresses, responses which may be biological or cultural. Adjustments are those more immediate and short-term responses, both incidental and purposeful, which may last only one or up to 100 years after the stressful event (Figure 15.4). Such distinctions may be too fine to be recognized either by victims or technical specialists. Adaptations may not be recognized to be responses to particular extreme events because of their distance in time from the event, for example, and as a result no significant impact would be perceived (see Chapter 16).

This dilemma faces historians attempting to assess the significance of climatic change upon societies. In the short term the coincidental chronology of climatic variabilities with socioeconomic fluctuations (such as crop failure and human starvation) may exaggerate the climate impact assessments, whereas longer-term societal adaptations to climatic variations (such as changing agricultural practices or life styles) may be overlooked and the climate impacts underestimated (de Vries, 1980; Post, 1980). The impact may indeed relate less to the absolute fluctuations of the climate and more to whether those

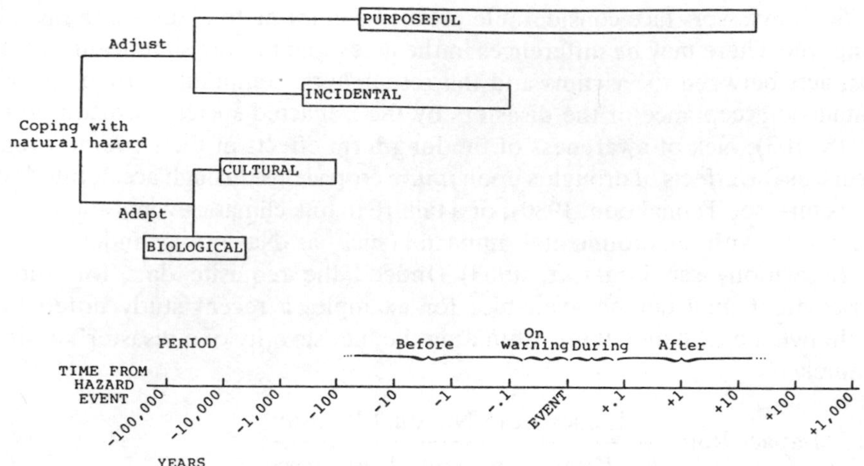

Figure 15.4 Adjustments and adaptations to natural hazards. From *The Environment as Hazard* by Ian Burton, Robert W. Kates and Gilbert F White. Copyright © 1978 by Oxford University Press Inc. Reprinted by permission

fluctuations came after a period of relatively stable or unstable socioeconomic conditions (Parry, 1975). In any event, impacts should be viewed not merely as the direct consequences of climatic events but analysis should attempt to include the effort and cost involved to adjust and to adapt to both past and future events.

For the general public, its view of the disaster impact is constrained by the media coverage of the event and to some extent by the official response to the potential 'cascade' of effects from the disaster (Figure 15.3). Apart from their varied ability to cover the disaster, the news media (newspapers, radio and television news services) have their own biased policies on the reporting and presentation of news items, stemming from their ownership and capacity to search out and relay information. Thus the immediate, short-term, spectacular impacts may be reported, but the longer-term post-event impacts may be overlooked.

15.3.4 Displaying the Impacts

For the technical specialist, the nature, location and duration of the impacts need to be displayed so that the significance can be assessed and, if necessary, a relief response coordinated. For the victim there is no need for display, he or she is fully occupied during and immediately after the event in trying to assess the losses and staying alive. For the general public the newspaper headlines and graphic reports on radio and television provide the brief display.

For the technical specialist concerned with disaster relief or mitigation, the impact of the disaster must be displayed within a relevant time period if an effective response is to be mounted. Since communication networks and power lines are often the first casualties in a disaster, obtaining relevant information promptly is difficult. Some success in locating lightning-induced forest fires, by means of strategically located sensors coupled to a computerized reporting network, has been claimed in Canada (Ellis, 1982) and presumably similar techniques could be adopted for river flood levels in remote locations. For the developed countries at least, microcomputer technology does offer the chance of impact displays within relevant time periods.

In cases where the specialist concern is to display the environmental effects of disasters, techniques developed to monitor land degradation and environmental pollution have been found useful. Perhaps the most useful are those for pervasive events such as droughts. Changes of vegetation patterns along climatic transition zones have been studied, using literary and field sources, in North America, Africa and Australia. In all cases a combination of human activities with stressful climatic conditions (that is, droughts) seems to have produced significant changes in vegetation patterns, leading to claims of desertification or deterioration in the resource potential of the vegetation (Kokot, 1955; Harris, 1966; Mabbutt, 1978). The difficulty with all such studies is, first, to distinguish the role of the extreme climatic events. The often unresolved question is: were the extreme climatic events the trigger or merely the final link in a long sequence of environmental stresses? The second difficulty is to convince the decision-makers that such long-term trends are relevant to short-term decision-making and that they should not be ignored as they have been in the past.

Some specialists have attempted to predict and display future disaster impacts, using computer-based simulations and scenarios. In nearly all cases, however, it is the intensive events, such as floods, which have been studied. Reviewing the United States experience, White and Haas commented that simulation allowed factors such as variations in the event, land use in the areas at risk, building codes affecting vulnerability, and the role of insurance to be incorporated in the models. Against these advantages were set the crudity of the assumptions which had to be made, the lack of information on some relevant variables, the complexity of the data required, and the geographical uniqueness of some of the characteristics of each event which could not be covered in the model (White and Haas, 1975, 134–135). Ericksen (1975) considered scenarios more effective in the sense that they can be used for analyses of interrelationships of events and impacts at one time or over a sequence of time, while a combination of scenarios and computer simulation offers an even better tool (see also Lave and Epple, Chapter 20 of this volume).

15.4 ANALYZING FIRST-ORDER IMPACTS

Extreme climatic events initially affect local and regional ecosystems, both natural and managed, in ways that may modify the character of those ecosystems as resource bases for human activity. Documentation of those impacts may involve methods ranging from those requiring complex technology, such as satellite imagery analysis for drought impacts in vegetation and crop patterns (Chakraborthy and Roy, 1979), to those drawing upon field surveys of the condition of the ecosystems and analysis of local peasant environmental monitoring systems (Oguntoyinbo and Richards, 1978).

The techniques for the documentation of these initial impacts have generally been successful, although they have required considerable technical ingenuity and there are some limitations. For example, there can be no doubt of the general value of satellite imagery in global environmental monitoring of both climatic events in the making (particularly the intensive types such as hurricanes and windstorms) and their impacts (such as flooding). However, local weather conditions, particularly cloud cover in the tropics, can considerably reduce the value of such imagery (Currey *et al.*, in press).

Human deaths and injuries are usually the most obvious evidence of first-order impacts from intensive events, although the accurate documentation of deaths may be difficult. Indeed, in the greatest disasters bodies may not be recovered—for example in the Bangladesh cyclone of 1970 total fatalities will never be known, as whole communities were wiped out and prior census data were probably wrong to begin with. The exact cause of death or injury also may not be clear in all cases. Sorting out 'abnormal' from 'normal' death rates may itself be difficult, particularly when the 'normal' is unknown (Caldwell, 1975; Kates, 1981) or there are regional variations in death rates which may themselves reflect regional climatic differences, as is suggested for southern versus northern United States city death rates in the eighteenth century (Fischer, 1980, 829).

Field evidence of past impacts can provide useful clues to potentially risky sites for human activity. Thus vegetation scars from avalanche tracks (Prowse *et al.*, 1981) and past flood levels (Figure 15.5) can furnish evidence of the physical dimensions of the event. Research in geomorphology, pedology and hydraulics can provide estimates of the likelihood of soil erosion from rainfalls of different intensities (Ormerod, 1978, 359) and of actual losses by 'soil loss equations' such as those developed for the United States (Beasley, 1972). The role of extreme events in landform modification, particularly the importance of the rapid erosion/deposition associated with events of 15- to 25-year return periods, has also been demonstrated (Wolman and Miller, 1960).

In terms of relevance to the society, however, such analysis of these first-order impacts must be followed up by examination of the ramifications of the second- to *n*th-order impacts.

15.5 ANALYZING SECOND-ORDER IMPACTS

Second-order disaster impacts are defined as those short-term effects upon a society of the ecosystem disruptions which form the first-order impacts. The

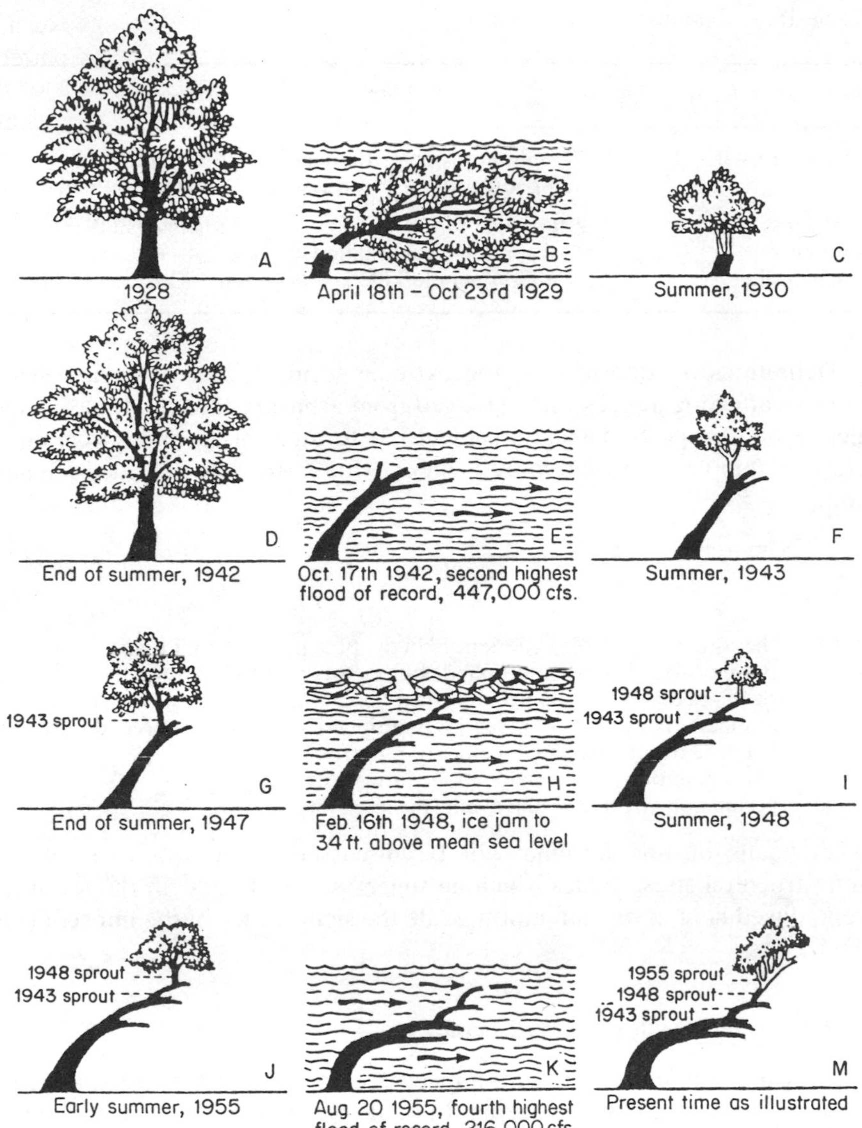

Figure 15.5 Vegetation indicators of flood levels. (Reproduced with permission from Foster, 1980, 57)

complexity of these second-order impacts is impressive: the resulting negative impacts or disasters have several components (post-event fatalities and injuries, financial losses and societal disruption), the precise documentation of which may be very difficult, while the positive impacts may be even more difficult to recognize.

Table 15.3 Definitions of disasters

Descriptive term	Definition	Source
Major disaster	> \$1 million property damage > 100 people killed or injured	Sheehan and Hewitt (1969)
Accidents	1–999 people killed	Michaelis (1973)
Disasters	1000–1 million people killed	
Catastrophes	> 1 million people killed	

Definitions of 'disasters' as the extreme form of general environmental stresses affecting a society have ranged from arbitrary thresholds based upon deaths and property damage (Table 15.3) to attempts to index community stress. Examples of the latter are Foster (1976, 1980), who provided a general formula:

$$CS = a + b + cd \tag{15.2}$$

where

CS = the community stress experienced in the time period under review
a = number of deaths [multiplied by factors scored differently between developed and developing countries]
b = number of injuries or illnesses [multiplied by factors scored differently between developed and developing countries]
c = stress resulting from damage to the infrastructure ...
d = population affected by the event (Foster, 1980, 38).

The details of this formula will be discussed below and the table of infrastructural stress values which he suggested for item 'c' in the formula is reproduced here as one attempt to scale the significance of the impact (Table 15.4).

15.5.1 Human Health and Well-being

Deaths and injuries as first-order impacts are only part of the stressful societal impacts of extreme climatic events. In their turn they, together with financial and material losses, generate further mental anguish on a wider section of the society, and may even generate population movements leading to the collapse of the entire society.

Foster (1976, 1980) has provided an attempt both to assess such mental anguish and to assess what he suggests are possible differences in the magnitude of those stresses between the developed and developing world. Adapting the 'Social Readjustment Scale' of Homes and Rahe (1967) and using deaths as a surrogate for stress, he makes assumptions about the size of families

Table 15.4 Infrastructural stress values

Event intensity	Designation	Characteristics	Stress value
I	Very minor	Instrumental.	0
II	Minor	Noticed only by sensitive people.	2
III	Significant	Noticed by most people including those indoors.	5
IV	Moderate	Everyone fully aware of event. Some inconvenience experienced, including transportation delays.	10
V	Rather pronounced	Widespread sorrow. Everyone greatly inconvenienced, normal routines disrupted. Minor damage to fittings and unstable objects. Some crop damage.	17
VI	Pronounced	Many people disturbed and some frightened. Minor damage to old or poorly constructed buildings. Transportation halted completely. Extensive crop damage.	25
VII	Very pronounced	Everyone disturbed, many frightened. Event remembered clearly for many years. Considerable damage to poorly built structures. Crops destroyed. High livestock losses. Most people suffer financial losses.	65
VIII	Destructive	Many injured. Some panic. Numerous normal buildings severely damaged. Heavy loss of livestock.	80
IX	Very destructive	Widespread initial disorganization. Area evacuated or left by refugees. Fatalities common. Routeways blocked. Agriculture adversely affected for many years.	100
X	Disastrous	Many fatalities. Masonry and frame structures collapse. Hazard-proofed buildings suffer considerable damage. Massive rebuilding necessary.	145
XI	Very disastrous	Major international media coverage. World-wide appeals for aid. Majority of population killed or injured. Wide range of buildings destroyed. Agriculture may *never* be re-established.	180
XII	Catastrophic	Future textbook example. All facilities completely destroyed, often little sign of wreckage, surface elevation may be altered. Site often abandoned. Rare survivors become lifelong curiosities.	200

Source: Foster, 1980, 38–39.

and circles of close friends and combines these with estimates of infrastructural stress values (Table 15.4) to offer two versions of his basic community stress formula (equation 15.2 above):

$$TS_{DD} = 445a + 280b + cd \quad (15.3)$$

or

$$TS_{DG} = 630a + 410b + cd \quad (15.4)$$

where a, b, c and d are as for equation 15.2 above, and TS_{DD} and TS_{DG} = total stress caused by a calamity occurring in the developed world (TS_{DD}) or developing world (TS_{DG}). On the basis of these formulae he assessed a variety of 'calamities' (from a motorist's parking ticket, through the Cyclone Tracy [1974] impact in Australia, to the stress generated by the Second World War) on a Calamity Magnitude Scale (Foster, 1976, 246). As a means of comparison between different types of disasters the formulae provide extremely crude, but nonetheless useful, preliminary measures.

Societal stress may further be shown in attempts to modify events by prayer. This seems to be a standard component of responses to drought-induced crop and livestock losses in both the developed and developing world (Burton *et al.*, 1978). One study even attempted to use the incidence of prayers conversely to indicate the severity of the event! namely drought (Fischer 1980, 821). Sociologists and psychologists have indicated that post-disaster stresses can be considerable and have demonstrated the extent to which stress can be buffered by social groups (Young, 1954; Barton, 1969) and can vary among the victims (Raphael, 1979). Such studies have identified the morbidity effects of disasters and the high-risk groups within societies, using interviews of survivors and control (unaffected) groups immediately after the impact and longitudinal follow-up of the survivors' success or failure in adjusting to their changed personal and family circumstances.

In terms of the demographic impact of the disasters, the scale may range from the proven loss of a generation and whole communities, as with the Bangladesh cyclone of 1970, to dubious overestimates of deaths, as with the '100,000' deaths attributed to the Sahel drought (Caldwell, 1975, 23–24). In the latter case Caldwell found little evidence of an increased death rate but convincing evidence of an increase in the traditional stress reaction, namely migration of the people affected. Kates (1981), assessing the evidence 5 years later, agrees with Caldwell's claims regarding the specific estimate of Sahelian deaths yet concludes that overall, deaths were probably underestimated by the '100,000 claim'. A basic problem for the Sahel is the absence of detailed longitudinal mortality data for so-called normal conditions.

15.5.2 Socioeconomic Gains and Losses

Probably because of the immediacy of the impacts and the possibility of reducing them relatively quickly to a common denominator—namely, money—research has been most prolific in this sector of the extreme event impacts. The bibliographies of natural hazards research (for example Cochran, 1972; Torres and Cochran, 1977; Morton, 1979) contain many references relating to the socioeconomic impacts of intensive events such as windstorms (tornadoes to hurricanes) and floods, and to a lesser extent of pervasive events such as droughts.

The immediate question facing any analysis of these impacts is 'what to measure?' The recent ongoing climate impact assessment in the United States identifies eight categories of societal activities which are to be monitored—from human resources to food and energy production and government spending (Center for Environmental Assessment Services, 1982a; see also Chapter 4, this volume). At the national level this would be an example of the technical specialist's view of the impacts in the same way that the checklist of items to be investigated in the 'Field Report Outline' used for comparative analysis of six United States disasters (Rubin, 1981, 26–28) would provide a local scale for the same view. In the latter case the checklist has four components: first, description of the event's magnitude and estimates of deaths, injuries and property damage set in the context of similar prior events and impacts; second, the recovery activities, differentiated by participants; third, the mitigation measures, divided according to their short- or long-term character; and fourth, the assessor's perception of the impacted community's commitment and capacity to carry out these mitigation strategies.

In the developing world, one of the few studies to address the question of what to measure proposed an evaluation of six major adjustments in terms of four disaster relief objectives for Bangladesh. The six adjustments (improvement of disaster warning systems, evacuation procedures, emergency relief, land use controls, coast protection measures and improvement of economic conditions) were to be evaluated according to whether they met the four objectives: reducing loss of life; reducing property and crop damage; generating employment; and generating higher income (Islam and Kunreuther, 1973). The study recognized, however, that not all objectives could be equally met and that compromise in disaster mitigation planning was inevitable.

15.5.3 Economic Losses

What are the economic costs of the disaster? Most disaster reports list the number of human dead and injured separately from the value of property damage. However, a human life can be given an economic value, usually by estimating the loss of expected earning capacity over a normal life span based upon actuarial tables. On this basis and on figures from insurance payouts and civil court actions, human life has been variously valued at from $US 9 thousand to $US 9 million (Foster 1980, 37) and from $A 12 thousand to

$A 35 million (Cordery and Pilgrim, 1979, 222–223). However, what would the premature death of Albert Einstein have cost the world?

Although there are formidable problems of data collection and interpretation, the assessment of property damage by insurance companies, government agencies and relief organizations is based upon standard procedures, even though there may be no standardization among the assessing agencies! For the intensive impacts—the crops flattened by hail, the vehicles swept away by floods and the buildings demolished—the job of assessment is relatively easy. For the pervasive impacts the task is more difficult.

One of the most difficult impacts to cost has been that from drought. This is partly because of its slow onset and broad spatial extent, which makes the impacts difficult to recognize; partly also because of the ramifications of the long-term effects upon other components of the ecosystem (for example, soils and vegetation composition); and partly because some members of the impacted society might actually benefit from the drought (Heathcote, 1969a). Estimates of economic losses based upon crop production shortfalls, price fluctuations and government and private drought relief payments do exist, however, as well as estimates of drought-related research costs (Heathcote, 1967, 1979; Warrick, 1975), although their crudity as technical specialists' views is acknowledged.

The problem of assessing drought impacts raises a further question, namely 'What is the long-term economic cost of the impacts?' This question impinges in part upon the problem of distinguishing adjustments from adaptations as mentioned in Section 15.3.3 above, but if the time-scale is reduced to less than a century to concentrate upon the impact responses as adjustments there are some longitudinal studies available.

Using local area statistics and interviews with local officials, an investigation of the impact of Hurricane Audrey upon a Louisiana parish found that, in comparison with the conditions before the event, 4 years later more people were in debt but the quality of the housing had improved, social institutions had become more formalized, and prior social trends had been accelerated (Bates *et al.*, 1963). Expanding the social survey techniques used in that study, Friesema *et al.* (1979) tried to establish trends of both economic and social measures of 'community performances' for four impacted communities in the United States for 10 years before and after an extreme event. Time-series analysis of over 200 sets of data suggested that because in each case the basic economic resource of the community was not seriously affected, serious economic impacts did not last more than a year.

This assessment of the long-term, community-level effects of natural disasters has been further confirmed by Wright *et al.* (1979) who claimed that the 'Impact Ratio' (Section 15.3.3, equation 15.1 above) could be calculated using time-series data for census tracts in the United States and concluded that 'natural disasters do not cause long-term, community-level effects'.

What of the spread of the economic costs of disasters through the impacted society? Cochrane (1975) analyzed the disaster losses, official disaster relief strategies, insurance and bankruptcy data and urban renewal programs for the United States. Apart from establishing the validity of the 'cascade' effect of disaster impacts noted in Section 15.3.3 above, he concluded that different sectors of the society suffered differently:

> the lower income groups consistently bear a disproportionate share of the losses: they receive, in most instances, the smallest proportion of disaster relief; they are the least likely to be insured (for either health, life or property); and they live in dwellings which are of the poorest construction and most subject to damage.
> (Cochrane, 1975, 110)

The middle- and upper-income groups offered the safest security for relief loans; with the greatest value of property at risk they generated the greatest flow of insurance claims, and were most aware of the possibilities of external disaster relief.

Such conclusions on the economic impacts of disasters relate mainly to the developed world. What, however of the situation in the developing world?

15.5.4 Social Losses

Natural hazards research, which has provided many of the studies of human adjustments to environmental stresses (White, 1974), has been criticized for its apparent bias towards developed world case studies and concentration upon monetary economic cost assessments. In the developing or Third World nations, so the argument runs, not only are the economic reserves of the society lower, but intrusive, often foreign-derived, policies of resource development—particularly associated with pressures for expanded cash cropping to develop a monetary economy—push the rural communities into resource management strategies and locations which are more vulnerable to extreme impacts. This process, which is claimed to illustrate a 'theory of marginalization' (Susman *et al.*, 1983), is exacerbated as reliance upon traditional socioeconomic adjustments to climatic stresses is reduced (Wisner *et al.*, 1976; Oguntoyinbo and Richards, 1978; Watts, 1983). The Third World studies, relying much more upon anthropological and sociological observations of village activies and on folklore, are more aware of the views of the potential victims. This is an important attribute because the studies are of societies where the members can expect much less disaster relief from outside the impacted community than similar victims in the developed world.

In fact, there seems to be some evidence that social impacts magnitude per event tend to be less in the developed than the developing world. Thus, the United States longitudinal studies noted above suggested that long-term social impacts were relatively small and 'the basic social structure survived their natural disasters virtually unscathed' (Friesema *et al.*, 1979, 107). In the developing world, however, severe social disruption may result, either because the traditional emergency management strategies are no longer feasible, as where the traditional emergency food stores have not been maintained because of the conversion to cash crops or a cash economy (Apeldoorn, 1981), or because official relief measures have temporarily reduced the need for these strategies without adequately encouraging local future self-reliance (Waddell, 1983). The resulting disruptions may take the form of family disintegration as well as community breakdown and increased levels of prostitution, banditry and other crimes of violence (Hall, 1978, 12).

15.5.5 Socioeconomic Gains

The disastrous impacts of extreme events bring benefits to some members of the impacted society and even to other nations. For example, drought-induced crop failures benefit farmers in grain-producing regions in Australia, Argentina, Canada and the United States, since massive purchases of grain on the world market force up grain prices. Given the global trade in plant and animal products, shortfalls in production from disasters in one nation are likely to prove beneficial to competing producers.

Within the nation also, there may be benefits from disasters. Reconstruction following extensive property damage creates employment and larger profits for the construction industry; emergency evacuation of people or livestock generates increased business for the transport industry; government relief brings money into local and regional economies which would otherwise not benefit; and changes in the regional infrastructure (for example), modernization of buildings and land-use zoning controls) may be easier after total destruction of the urban fabric than before. The methods for analyzing such gains incorporate all those used for estimates of losses plus, for example, estimates of the proportion of regional income from government relief payments as opposed to normal commercial activities. Borchert (1971) demonstrated the role of relief in maintaining the Great Plains economy in the United States through the droughts of the 1930s and 1950s. Another method used noncompliance with relief legislation requirements as an indicator of local beliefs that relief was in fact not needed, because the benefits from recapitalization following the disaster outweighed the losses (Friesema *et al.*, 1979, 13).

Although most of the studies that show gains from the impacts refer to the monetary economies of the developed world, there is some evidence of similar benefits in the developing world. Studies of the drought impacts in the Sahel in the early 1970s suggest that commercial cash cropping was still profitable at times when other subsistence sectors of the various national agricultures were showing considerable stress that was merging into famine (Meillassoux, 1974;

Kates, 1981). Hall's study (1978) of land tenure, crop production, population trends and rural incomes in drought-prone northeastern Brazil suggested that the drought stresses brought considerable losses for small holders and landless laborers, whereas for larger landholders the stresses brought opportunities to buy up bankrupt small holders and provided the government with surplus labor for relief works. Such benefactions were in addition to:

> the *industrias da sêca* [drought industrialists], a term used to describe anyone who exploited the drought situation for his own profit. These included simple thieves of relief shipments, corrupt officials in charge of public works and those in search of quick profits caused by food shortages. (Hall, 1978, 8)

There may be political benefits from extreme climate impacts. Regional political representatives will press for official relief from the national treasury for their own districts and the exaggeration of impacts may be part of their strategy. Victims may exaggerate losses to increase their share of the national budget or international relief, as we saw in Section 15.3.1. Resource mismanagement leading to increased loss from climatic events may be concealed by putting the sole blame on the climatic event, rather than on the socioeconomic system which has proved inadequate to cope with the vagaries of the regional ecosystem (Hall, 1978, 125).

15.5.6 Human Movements and Migrations

There is evidence of both the impact of climatic change and of extreme climatic events upon human population movements. A historian commented recently.:

> There is abundant evidence of the importance of climate as a determinant of population movements in Scandanavian history ... . The beginning and end of Iberian expansion may have been influenced in part by ... the desiccation of the Spanish peninsula. (Fischer, 1980, 827–828)

Flight is a standard human response to extreme events, and over time it might result in the permanent relocation of population. This seems to have been the case with the nomad population shifts associated with the Sahel droughts (Caldwell, 1977) and with rural labor migrations from northeastern Brazil (Hall, 1978). In both cases the researchers were able to distinguish abnormal from normal movements by comparing pre-event with post-event patterns or comparing movements from stressed as opposed to non-stressed areas. In the case of Brazil, for example, rural/urban migration (1960–70) in the worst drought-affected state involved 36 percent of the population compared with neighboring state figures of 14 and 22 percent (Hall, 1978, 124). As such figures suggest, however, the impact of the event is to accelerate an existing process, not necessarily to initiate a new one.

15.6 ANALYZING SUBSEQUENT IMPACTS: THE ADJUSTMENT/ ADAPTATION RESPONSES

> In the short or relatively short term (on an intradecennary, decennary, and in certain cases interdecennary scale), agricultural history is vulnerable to the caprices of meteorology which produce bad harvests and used to produce food crises. But in the long term the human consequences of climate seem to be slight, perhaps negligible, and certainly difficult to detect.
>
> (Le Roy Ladurie, 1971, 119).

The difficulties of identifying these long-term consequences—the ramifications of the first- and second-order impacts—require that studies look for relevant data far and wide. On a time-scale of centuries, evidence of a climatic change in the northern hemisphere *c.* 1150–1200 AD (drawn from tundra advances in North America, expansion of the Greenland–Iceland sea ice areas, decrease of summer temperatures from pollen analysis in Michigan, and desiccation of vegetation in western Iowa from palynological and paleozoological data) tempted Baerries and Bryson to reconstruct the hypothetical pattern of rainfall which would have resulted in the formation of the central plains of the United States and to compare it with actual human activity during the period. They found that a sporadic reduction of 30–50 percent in July rainfall from 1150 to 1200 AD over the northern plains was accompanied by *wetter* Julys on the southern plains. Coincidentally, archaeological evidence suggested that:

> On the northern high plains ... thousands of small villages characterized by rainfed maize agriculture before 1100 AD had completely disappeared by 1200 AD and many became covered with wind-drifted soil ... Farther east, in western Iowa ... maize farmers ... had occupied valley floors with forested terraces in a region of generally tall-grass prairie. After 1200 AD the forests were gone and short-grass prairie dominated ... After a little less than two centuries the culture succumbed ... About 1200 AD a number of villages practicing the rain-fed maize agriculture were established in the Panhandle regions of Texas and Oklahoma, in a region where that had not been possible before and is not at present.
>
> (Bryson and Padoch, 1980, 596–597)

Over similar time-scales in Europe, the relevant indicators of stress and human response may be the changes in the types of crops grown, as suggested for the spread of buckwheat (a hardy crop) in the Netherlands during the cooler, wetter, Little Ice Age of the seventeenth century and its decline in the milder winters of the eighteenth century (de Vries, 1980, 625).

If we reduce the time-scale to less than a century, other indicators may be relevant. Trends in attitudes to, and policies of, resource management may be identifiable from the analysis of changing legislation on land management—particularly that relating to the principles and administration of public disaster relief or subsidies for settlement in what are designated as stressful environments (Gates, 1968, especially Chapters 18 and 22). In this context the transformation of societal controls from Toennies' *Gemeinschaft* to

Gesellschaft systems indicates the codification of customs, possibly in response to extreme environmental stresses over time (Heathcote, 1969b).

At the other end of the time spectrum, individual extreme events may give rise to weather manipulators—individuals who claim to be able to influence weather events and whose activities, plotted over time, provide further evidence of a society's reaction to unfavorable weather conditions. Significantly, the most recent and comprehensive study for the United States, although restricted to the period up to the Second World War, identified and concentrated upon the 'age of "pluviculture", from roughly 1890 to 1930' when the advance of agriculture onto the subhumid to semi-arid western plains made the activities of rainmakers a relevant strategy to resource managers beset by droughts (Spence, 1980, 2).

Indeed, one of the most successful studies of the scope of regional adjustments to extreme events has been the study of drought impacts on the Great Plains of the United States (Warrick, 1975). Reviewing responses since the droughts of the 1930s a variety of adjustments were identified and trends over time suggested (see Chapter 1, this volume and Figure 15.6).

A further study used the same basic methodology—a careful time-series analysis of detailed population data, farm transfer rates, and regional

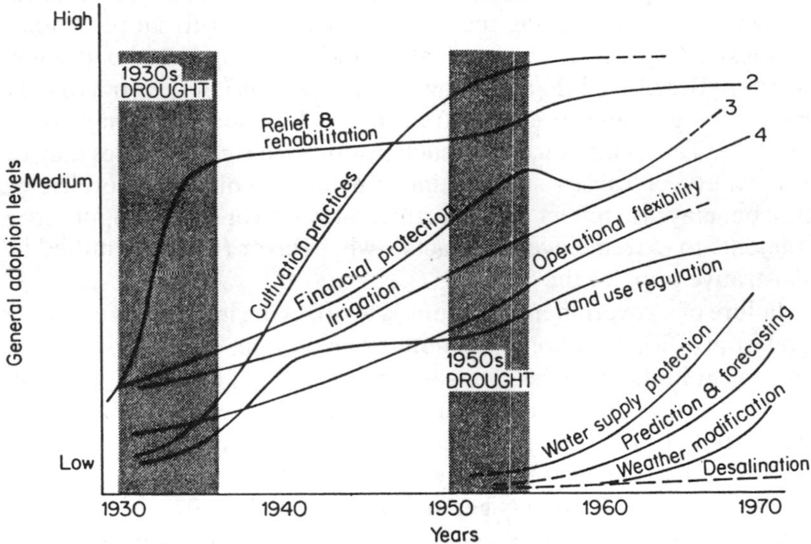

Figure 15.6 Generalized historical trends of drought adjustment on the Great Plains of the USA (1, Very rough approximation of relative levels of adoption). 2, institutional arrangements for R & R (not payments); 3, shape of curve generalized from number of acres insured and amounts of loans in the United States (dip in 1950s reflects lower adoption of insurance at that time); 4, based on total irrigated acres in the United States. (Reproduced with permission from Warrick, 1975, 102)

production data alongside the meteorological sequences for the study area—to compare drought impacts in the Great Plains (1880–1979), the Tigris-Euphrates Lowland (6000 BP to the present) and the Sahel (1910–74) (Bowden *et al.*, 1981). The findings were that in the Tigris-Euphrates area lack of sufficiently detailed information prevented any firm conclusions on the significance of the role of extreme climatic events on the societal collapse in the area, but for the Sahel and the Great Plains the impact of the extreme events had lessened over time. In the latter two cases, however, catastrophic potential remained, but rested less upon climatic extremes than on socioeconomic and demographic factors which would increase their vulnerability should disaster strike.

There is little doubt that many of the responses to extreme climatic events both reflect the sociopolitical character of the impacted society and, in turn, modify that character. Governments supported by a centralized bureaucracy can, in theory, coordinate responses at a regional or a national level. Where a special problem appears, perhaps in response to the issue–attention cycle noted in Section 15.3.1, a standard response is to create a special branch of the administration to cope with it. Indeed, one method by which the seriousness of the impact on a society is assessed uses the chronology of the growth of departmental responsibilities as indicators, the creation of a separate government department being seen as evidence of the official recognition of the seriousness of the problem. Hall's study of responses to drought in northeastern Brazil used changing government responsibilities both to indicate changing strategies and to demonstrate the lack of success of most of them (Table 15.5). The lack of success of such institutionalized strategies may result from a variety of causes such as incompetence, corruption or irrelevant policies, but may also reflect the departmentalization or sectoral approaches of governments to extreme event impacts—what Baker (1976) identified as the 'administrative trap' for the Sahel.

The failure of a government's responses to offset the impacts may result in its removal from office by national elections or revolution. The specific role of the extreme event in the political changes may be difficult to identify—was it the sole cause or the final straw after a sequence of other societal stresses inadequately met? The drought and associated famines in the Sahel in the early 1970s have been assumed to be partly responsible for the Ethiopian revolution in 1974 and the revolutions in Niger in 1974 and Chad in 1975.

Successful official disaster relief may solve the immediate problem but create many more. Greater vulnerability of the society to future disasters may result from an inreasing dependence on external sources of relief. In a colonial or politically dependent situation, the reinforcement of institutional controls which official relief programs provide may also increase vulnerability (Waddell, 1983).

Table 15.5 Government drought relief institutions in Brazil, 1877–1970s

Droughts	Institutions		
	Established	Title	Policy
1877–79	1877	National Commission of Inquiry [into drought]	↓
	1906	'Superintendencia dos Etudos e Obras Contra os Efectos da Seca' (SEOCES) [Superintendency for study and mitigation of drought effects]	↓
	1908	This became 'Inspectoria de Obras Contra as Sêcas' (IOCS) [Inspectorate for the mitigation of drought effects]	Water storage in man-made reservoirs
	1918	This became 'Inspectoria Federal de Obras Contra as Sêcas' (IFOCS) [Federal Inspectorate for mitigation of drought]	↓
1930–32			
1945	1945	This became 'Departmento Nacional de Obras Contra as Sêcas' (DNOCS) [National department for the mitigation of drought]	↓
1951			- - - - - - -
	1952	'Banco do Nordeste do Brasil' (BNB) [Bank of Northeastern Brazil] set up as commercial bank	↓
1958	1956	'Grupo de Trabalho para o Desenvolvimento do Nordeste' [Working group on development of the northeast—as part of BNB]	Irrigation works
	1959	'Superintendencia do Desenvolvimento do Nordeste' (SUDENE) [Superintendency for development of the northeast]	↓
1970			
	1974	'Programa de Desenvolvimento de Areas Integrados do Nordeste' (POLONORDESTE) [Integrated development program for the northeast]	↓

Source: Hall, 1978, 6–14.

One method by which such offical strategies to mitigate extreme societal stresses may be assessed has been put forward by Foster (1980). A series of criteria for evaluating the strategies together with 'strategy-related questions'

Table 15.6 Evaluating mitigation strategies

Context of assessment	Foster's criteria for evaluating mitigation strategies: 'Criteria'*	'Strategy-related questions'
Time	2. Timing	Will the beneficial results of this strategy be quickly realized?
	6. Continuity of effects	Will the effects of the application of this strategy be continuous or merely short term?
	7. Compatibility	How compatible is this strategy with others that may be developed?
	11. Hazard creation	Will this strategy itself introduce new risks?
	12. Hazard reduction potential	What proportion of the losses due to this hazard will this strategy prevent? Will it allow the safety goal to be reached?
Environment	10. Effects on the environment	What will be the environmental impacts of this strategy?
Economy	4. Cost to government	Is this strategy the most cost-effective or could the same results be achieved more cheaply by other means?
	9. Effects on the economy	What will be the economic impacts of this strategy?
Society	1. Equity	Do those responsible for creating the hazard pay for its reduction? Where there is no man-made cause, is the cost of response fairly distributed?
	3. Leverage	Will the application of this strategy lead to further risk-reducing actions by others?
	13. Public and pressure group reaction	Are there likely to be adverse reactions to implementation?
	14. Individual freedom	Does the strategy deny basic rights?
Institutions	5. Administrative efficiency	Can it be easily administered or will its application be neglected because of difficulty of administration or lack of expertise?
	8. Jurisdictional authority	Does this level of government have the legislated authority to apply this strategy? If not, can higher levels be encouraged to do so?

* Numbers refer to Foster's list.
After Foster, 1980, 28.

were suggested and may be grouped in terms of the context for their assessment (Tablc 15.6). He further suggests a matrix for the planning of general disaster mitigation activities spread over at least 4 years (Foster 1980, 36). Such activities include 'risk mapping' through 'disaster simulation and prediction' to 'planning for reconstruction'. The aim is to design a 'forgiving environment' where planners 'assume a relatively high incidence of destruction or misuse' and plan their defenses on the assumption that disasters *will* occur (Foster, 1980, 107 and 113). Whether such an analysis would identify the range of positive and negative aspects of the official relief measures remains to be seen.

15.7 CONCLUSION

Such a brief review cannot hope to do more than hint at the wealth of studies relevant to the assessment of extreme climatic event impacts. At the risk of some generalization we may conclude that a considerable number of methods exist for the study of disasters (natural or man-made) and many are relevant sources for the assessment of climatic impacts. Most studies have assessed such impacts in the developed world and in economic terms. Studies of social and psychological impacts are fewer and less well-documented (see Chapter 13). Studies of climatic event impacts for the developing world are sparse and stress more the social than the economic impacts. The most successful methods—successful in the sense of appearing to present a logical description of the complexities of human response to environmental stress—seem to be those which combine the detached viewpoint of the technical specialist with the intimate knowledge of the potential and actual victims. Such methods, in fact, draw upon a wide range of disciplinary expertise, are truly interdisciplinary in their approach, and promise the greatest future insights.

REFERENCES

Amiran, D. H. K., and Ben-Arieh, Y. (1963). Sedentarization of Bedouin in Israel. *Israel Exploration Journal*, **13**, 161–181.

Anderson, J. L., and Jones, E. L. (1983). *Natural Disasters and the Historical Response*. Economics Discussion Papers, 3/83. School of Economics, La Trobe University, Melbourne, Australia.

Apeldoorn, G. J. V. (1981). *Perspectives on Drought and Famine in Nigeria*. Allen and Unwin, London.

Baker, R. (1976). The administrative trap. *The Ecologist*, **6**, 247–251.

Barton, A. H. (1969). *Communities in Disaster: A Sociological Analysis of Collective Stress Situations*. Ward Lock, New York.

Bates, F. L., *et al.* (1963). *The Social and Psychological Consequences of a Natural Disaster: A Longitudinal Study of Hurricane Audrey*. National Academy of Sciences, Washington, DC.

Beasley, R. P. (1972). *Erosion and Sediment Pollution Control*. Iowa State University, Ames, Iowa.

Borchert, J. R. (1971). The Dust Bowl in the 1970's. *Annals of the Association of American Geographers*, **61**, 1–22.

Bowden, M. J., Kates, R. W., Kay, P. A., Riebsame, W. E., Warrick, R. A., Johnson, D. L., Gould, H. A., and Weiner, D. (1981). The effect of climate fluctuations on human populations: Two hypotheses. In Wigley, T. M. L., Ingram, M. J., and Farmer, G. (Eds.) *Climate and History*, pp. 479–513. Cambridge University Press, Cambridge, UK.

Bryson, R. A., and Padoch, C. (1980). On the climates of history. *Journal of Interdisciplinary History*, **10** (4), 583–597.

Burton, I., and Hewett, K. (1974). Ecological dimensions of environmental hazards. In Sargent, F. (Ed.) *Human Ecology*, pp. 253–283. North-Holland, Amsterdam.

Burton, I., Kates, R. W., and White, G. F. (1978). *The Environment as Hazard*. Oxford University Press, New York.

Caldwell, J. C. (1975). *The Sahelian Drought and its Demographic Implications*. Overseas Liaison Committee Paper No. 8. American Council on Education, Washington, DC.

Caldwell, J. C. (1977). Demographic aspects of drought: An examination of the African drought of 1920–74. In Dalby, D., Harrison Church, R. J., and Bezzaz, F. (Eds.) *Drought in Africa*, **2**, 93–102. International African Institute, London.

Center for Environmental Services (CEAS) (1982a). *Climate Impact Assessment United States*. National Oceanic and Atmospheric Administration, US Dept. of Commerce, Washington, DC.

Center for Environmental Services (CEAS) (1982b). *U.S. Economic and Social Impacts of the Record 1976–77 Winter Freeze and Drought*. National Oceanic and Atmospheric Administration, US Dept. of Commerce, Washington, DC.

Chakraborthy, A. K., and Roy, P. S. (1979). Influence of droughts in ecosystem as observed from Landsat. In *Hydrological Aspects of Droughts, International Symposium, 3–7 December 1979*, pp. 47–65. Indian National Committee for International Hydrological Programme, New Delhi.

Cochran, A. (1972). *A Selected Annotated Bibliography on Natural Hazards*. Natural Hazards Research Working Paper No. 22. Institute of Behavioral Science, University of Colorado, Boulder, Colorado.

Cochrane, H. C. (1975). *Natural Hazards and Their Distributive Effects: A Research Assessment*. Institute of Behavioral Science, University of Colorado, Boulder, Colorado.

Cordery, I., and Pilgrim, D. H. (1979). Acceptance of hazard in design of flood-prone structures. In Heathcote, R. L., and Thom, B. G., (Eds.) *Natural Hazards in Australia*, pp. 216–226. Australian Academy of Science, Canberra.

Currey, B. C. (1979). *Mapping Areas Liable to Famine in Bangladesh*. PhD dissertation, Department of Geography, University of Hawaii.

Currey, B., Bardsley, K., and Fraser, A. S. (in press). Landsat image availability, cloud cover and agricultural activities in the wet rice growing areas of Asia. *Nature*.

de Vries, J. (1980). Measuring the impact of climate on history: The search for appropriate methodologies. *Journal of Interdisciplinary History*, **10** (4), 599–630.

Drower, M. S. (1956). Water-supply, irrigation and agriculture. In Singer, C. S., Holmyard, E. J., and Hall, A. R. (Eds.) *A History of Technology, Volume 1: From Early Times to Fall of Ancient Empires*, pp. 520–557. Clarendon Press, Oxford, UK.

Dworkin, J. (n.d.) *Global Trends in Natural Disasters 1947–1973*. Natural Hazards Research Working Paper No. 26, Institute of Behavioral Science, University of Colorado, Boulder, Colorado.

Ellis, G. (1982). The computerized forest. *Forestalk*, Summer, 3–8.

Ericksen, N. J. (1975). *Scenario Methodology in Natural Hazards Research*. Institute of Behavioral Science, University of Colorado, Boulder, Colorado.

Fischer, D. H. (1980). Climate and history: Priorities for research. *Journal of Interdisciplinary History*, **10** (4), 820–830.

Fonaroff, L. S. (1963). Conservation and stock reduction in the Navajo tribal reserve. *Geographical Review*, **53**, 200–223.

Foster, H. D. (1976). Assessing disaster magnitude: A social science approach. *Professional Geographer*, **28** (3), 241–247.

Foster, H. D. (1980). *Disaster Planning: The Preservation of Life and Property*. Springer series on Environmental Management. Springer-Verlag, New York.

Foster, H. D., and Sewell, W. R. D. (1980). *Water: The Emerging Crisis in Canada*. Canadian Institute for Economic Policy Series, Ottawa.

Friesema, H. P., Caporaso, J., Goldstein, G., Lineberry, R., and McCleary, R. (1979). *Aftermath: Communities After Natural Disaster*. Sage Publications, Beverly Hills, California.

Gates, P. W. (1968). *History of Public Land Law Development*. US Government Printing Office, Washington, DC.

Gibbs, W. J., and Maher, J. V. (1967). *Rainfall Deciles as Drought Indicators*. Bulletin No. 48. Commonwealth Bureau of Meteorology, Melbourne, Australia.

Glantz, M. H. (1976). Nine fallacies of natural disaster. In Glantz, M. H. (Ed.) *The Politics of Natural Disaster: The Case of the Sahel Drought*. Praeger, New York.

Hall, A. L. (1978). *Drought and Irrigation in North-east Brazil*. Cambridge University Press, Cambridge, UK.

Harris, D. R. (1966). Recent plant invasions in the arid and semi-arid southwest of the United States. *Annals of the Association of American Geographers*, **56**, 408–422.

Heathcote, R. L. (1967). The effects of past droughts on the national economy. In *Report of the ANZAAS Symposium on Drought*, pp. 27–45. Commonwealth Bureau of Meteorology, Melbourne, Australia.

Heathcote, R. L. (1969a). Drought in Australia: A problem of perception. *Geographical Review*, **59**, 175–194.

Heathcote, R. L. (1969b). Land tenure systems: Past and present. In Slatyer, R. O., and Perry, R. A. (Eds.) *Arid Lands in Australia*. ANU Press, Canberra, Australia.

Heathcote, R. L. (1979). The threat from natural hazards in Australia. In Heathcote, R. L., and Thom, B. G. (Eds.) *Natural Hazards in Australia*, pp. 3–14. Australian Academy of Science, Canberra.

Holmes, T. H., and Rahe, R. H. (1967). The social readjustment rating scale. *Journal of Psychosomatic Research*, **11**, 213–218.

Islam, M. A., and Kunreuther, H. (1973). The challenge of long-term recovery from natural disasters: Implications for Bangladesh. *Oriental Geographer*, **17** (2), 51–63.

Jonsson, U. (1981). The basic causes of hunger. *United Nations University Supplement to the Newsletter*, **5** (2), 1.

Kates, R. W. (1979). The Australian experience: Summary and prospect. In Heathcote, R. L., and Thom, B. G. (Eds.) *Natural Hazards in Australia*, 511–520. Australian Academy of Science, Canberra.

Kates, R. W. (1981). Drought in the Sahel: Competing views as to what really happened in 1910–14 and 1968–74. *Mazingira*, **5** (2), 72–83.

Kishihara, N., and Gregory, S. (1982). Probable rainfall estimates and the problems of outliers. *Journal of Hydrology*, **58**, 341–356.

Kokot, D. F. (1955). Desert encroachment in South Africa. *African Soils*, **3** (3), 404–409.

Le Roy Ladurie, E. (1971). *Times of Feast, Times of Famine: A History of Climate Since the Year 1000*. (Translated by B. Bray.) Doubleday, Garden City, New York.

Lee, D. M. (1979). Australian drought watch system. In Hinchey, M. T. (Ed.) *Proceedings of the Symposium on Drought in Botswana*. Botswana Society, University Press of New England, Hanover, New Hampshire.

Linstone, H. A. (1973). On discounting the future. *Technological Forecasting and Social Change*, **4**, 335–338.

Mabbutt, J. A. (1978). *Desertification in Australia*. Water Research Foundation Report No. 54, Kingsford, Australia.

Meillassoux, C. (1974). Development or exploitation: Is the famine good business? *Review of African Political Economy*, **1**, 27–33.

Michaelis, A. R. (1973). Disasters past and future. *Emergency Measures Organization National Digest*, **13**, 4–14.

Morton, D. R. (1979). *A Selected, Partially Annotated Bibliography of Recent (1977–78) Natural Hazards Publications*. Natural Hazards Research and Applications Information Center, Institute of Behavioral Science, University of Colorado, Boulder, Colorado.

MRG (1978). *World Map of Natural Hazards*. Münchener Rückversuchungs-Gesellschaft, Munich.

Oguntoyinbo, J., and Richards, P. (1978). Drought and the Nigerian farmer. *Journal of Arid Environments*, **1**, 165–194.

Ormerod, W. E. (1978). The relationship between economic development and ecological degradation: How degradation has occurred in West Africa and how its progress might be halted. *Journal of Arid Environments*, **1**, 357–379.

Parry, M. L. (1975). Secular climatic change and marginal agriculture. *Transactions of the Institute of British Geographers*, **64**, 1–13.

Post, J. D. (1980). The impact of climate on political, social, and economic change. *Journal of Interdisciplinary History*, **10** (4), 719–723.

Prowse, T. D., Owens, I. F., and McGregor, G. R. (1981). Adjustment to avalanche hazard in New Zealand. *New Zealand Geographer*, **37** (1), 25–31.

Raphael, B. (1979). The preventive psychiatry of natural hazard. In Heathcote, R. L., and Thom, B. G. (Eds.) *Natural Hazards in Australia*, pp. 330–339. Australian Academy of Science, Canberra.

Rijks, D. A. (1968). Agrometeorology in Uganda—a review of methods. *Experimental Agriculture Review*, **4**, 263–274.

Rubin, C. B. (1981). *Long-term Recovery from Natural Disasters: A Comparative Analysis of Six Local Experiences*. Academy for Contemporary Problems, Columbus, Ohio.

Sheehan, L., and Hewitt, K. (1969). *A Pilot Survey of Global Natural Disasters of the Past Twenty Years*. University of Toronto, Toronto, Canada.

Spence, C. C. (1980). *The Rain Makers: American 'Pluviculture' to World War II*. University of Nebraska, Lincoln, Nebraska.

Susman, P., O'Keefe, P., and Wisner, B. (1983). Global disasters, a radical interpretation. In Hewett, K. (Ed.) *Interpretations of Calamity from the Viewpoint of Human Ecology*, pp. 263–283. Allen and Unwin, London.

Swift, J. (1977). Sahelian pastoralists: Underdevelopment, desertification and famine. *Annual Review of Anthropology*, **6**, 457–478.

Tennakoon, M. U. A. (1980). Desertification in the dry zone of Sri Lanka. In Heathcote, R. L. (Ed.) *Perception of Desertification*, pp. 4–33. United Nations University, Tokyo.

Thompson, S. A. (1982). *Trends and Developments in Global Natural Disasters, 1947 to 1981*. Natural Hazards Research Working Paper No. 45. Institute of Behavioral Science, University of Colorado, Boulder, Colorado.

Torres, K., and Cochran, A. (1977). *A Selected, Partially Annotated Bibliography of Recent (1975–1976) Natural Hazards Publications*. Natural Hazards Research and Applications Information Center, Institute of Behavioral Science, University of Colorado, Boulder, Colorado.

Waddell, E. (1983). Coping with frosts, governments and disaster experts: Some reflections based on a New Guinea experience and a perusal of the relevant literature. In Hewitt, K. (Ed.) *Interpretations of Calamity from the Viewpoint of Human Ecology*, pp. 33–43. Allen and Unwin, London.

Warrick, R. A. (1975). *Drought Hazard in the United States: A Research Assessment*. Institute of Behavioral Science, University of Colorado, Boulder, Colorado.

Watts, M. (1983). On the poverty of theory: Natural hazards research in context. In Hewitt, K. (Ed.) *Interpretations of Calamity from the Viewpoint of Human Ecology*, pp. 231–262. Allen and Unwin, London.

White, G. F. (Ed.) (1974). *Natural Hazards: Local, National, Global*. Oxford University Press, New York.

White, G. F., and Haas, J. E. (1975). *Assessment of Research on Natural Hazards*. MIT Press, Cambridge, Massachusetts.

Wisner, B., Westgate, K., and O'Keefe, P. (1976). Poverty and disaster. *New Society*, **37**, 546–548.

Wolman, M. G., and Miller, J. P. (1960). The magnitude and frequency of forces in geomorphological processes. *Journal of Geology*, **68**, 54–74.

Wright, J. D., Rossi, P. H., Wright, S. R., and Weber-Burdin, E. (1979). *After the Clean-Up: Long Range Effects of Natural Disasters*. Contemporary Evaluation Research Series 2. Sage Publications, Beverly Hills, California.

Young, M. (1954). The role of the extended family in a disaster. *Human Relations*, **7**, 383–391.

Climate Impact Assessment
Edited by R. W. Kates, J. H. Ausubel and M. Berberian

CHAPTER 16
Perception

ANNE V. T. WHYTE

Institute for Environmental Studies
University of Toronto
Toronto, Ontario M5S 1A1
Canada

16.1 INTRODUCTION

Environmental perception is what makes us take our umbrella with us, even when it doesn't rain, or complain about neighborhood air pollution while we are smoking. It helps the farmer to adjust his crop planting patterns to the forthcoming rains and the municipal engineer to get his snow ploughs out in time.

Environmental perception is the means by which we seek to understand environmental phenomena in order to arrive at a better use of environmental resources and a more effective response to environmental hazards. The processes by which we arrive at these decisions include direct experience of the environment (through the senses of taste, touch, sight, hearing and smell) and indirect information from other people, science, and the mass media. They are mediated by our own personalities, values, roles and attitudes. The study of environmental perception has to encompass all these means of processing environmental information and to place the individual psychological processes of prediction, evaluation and explanation into a relevant social and political framework.

The main objective of a perception approach to environmental management is to analyze decision-making and choice of adjustment from the inside-out, or from the perspective of the decision-frame (that is, the decision and its context) as it appears to the decision-maker, with all its imperfections. Indeed, it is often the limitations and inconsistencies in subjective decision-frames that become the focus of attention in perception research.

Subjective decision-frames can provide post-hoc explanations for human behavior, some of which may seem at first to be 'irrational', and they can indicate directions for education and improved public information and public policy. The learning process is not just one-way, from the policy-makers to the public. Some of the most useful perception research has revealed to policy-makers both the value of 'folk' environmental knowledge and the need to incorporate lay people's values into scientific and policy models.

Thus, research has shown what is intuitively obvious: that choices are made within the framework of *perceived* alternatives and *available* information. Alternatives and information are profoundly affected by people's attitudes and values and the roles they play in relation to the decision to be made.

There are many techniques available within the methodological resources of the social sciences to measure the psychological and social parameters of decision-frames and decision-making. Some of these techniques have been used successfully in developing countries as well as in the modern industrial societies for which they were originally developed.

The major problem in adopting a perception approach is not so much in finding appropriate techniques to measure specific variables, but in knowing what variables to measure. The issue of knowing where to bound the system to be studied is one familiar to all scientists. It is a particularly intractable problem for environmental perception research because human response to the natural environment is everywhere mediated by strong social forces.

For example, the eastern Caribbean is one of the most seismically active areas in the world. The small islands are subjected to earthquakes, major volcanic

Table 16.1 Public perception of main problems facing people living in five East Caribbean islands

Problems in island	Barbados $n = 682$ %	St Kitts $n = 103$ %	Nevis $n = 51$ %	St Lucia $n = 321$ %	St Vincent $n = 259$ %
Lack of jobs	52	45	90	62	65
Poor education	3	11	14	12	6
Inflation	49	53	69	57	38
Crime	27	7	—	11	2
Rastafarians	9	4	—	8	2
People's attitudes and behavior	9	12	14	17	5
Low wages	11	32	55	18	17
Crowding	4	3	—	1	2
Transportation	4	3	8	3	1
Tourists	1	—	—	—	—
Politics	4	10	—	25	4
Shortages of goods	1	1	4	2	4
Poverty	3	2	4	1	1
Land/Labor	1	1	—	1	—
Environmental hazards	1	—	—	1	—
Lacks facilities	1	12	28	4	4
Other problems	10	6	4	9	9
No problems	2	—	—	1	—
Total %*	192	202	290	233	160

* Percentages total more than 100 percent because more than one response from each person is included.
Source: Whyte, 1982.

eruptions and hurricanes as well as a number of less devastating natural hazards. Yet these events hardly impinge on the daily lives of the local people. Their concerns for their countries and neighborhoods are dominated by economic and social problems (see Table 16.1). Even individual and governmental decisions that directly affect the vulnerability of the population to natural hazards, such as house construction style and land use zoning, are determined primarily by other considerations, such as traditional land tenure systems and the desire for increased tourism (Whyte, 1982).

It is factors like these that have led critics of natural hazard research to argue against the unrealistic 'roping-off' of human response to hazards from the rest of man–environment relations (Hewitt, 1983) but the problem is a more general one for environmental perception.

16.2 FRAMEWORKS FOR STUDYING CLIMATE PERCEPTION

This discussion of perception focuses on the factors that lead to decision-making and the choice of alternative adjustments to climatic processes and events. Perception studies, however, are considerably broader than this in scope; for example, one could be designed to measure the contribution of weather to the value of a landscape, or the role of climate in people's nostalgia for their childhood environments.

A simple framework is to imagine choice as the outcome of interactions between climatic variables, decision-maker variables, and perception processes (Figure 16.1). This enables the researcher to define what aspects of climate, what kinds of decision-makers and what types of decisions represent the focus of study.

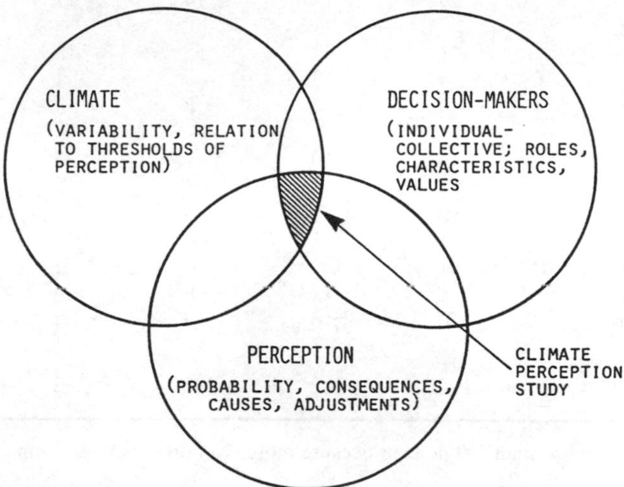

Figure 16.1 Main components in the design of a climate perception study for impact analysis

In the context of the models of Chapter 1, this framework, simply stated, represents a form of the interaction model with feedback and underlying process depicted in Figure 1.5C. It encompasses climate variability and change, individual and collective exposure units, impacts and consequences, and the choice of adjustments. It also incorporates some underlying processes related to the subjective perception of these elements, as well as the socially defined roles, characteristics and values of the decision-makers.

16.2.1 Climate as Perceived

A useful breakdown of climate in relation to human response is to consider it in terms of variability from normal, expected values and in relation to thresholds of

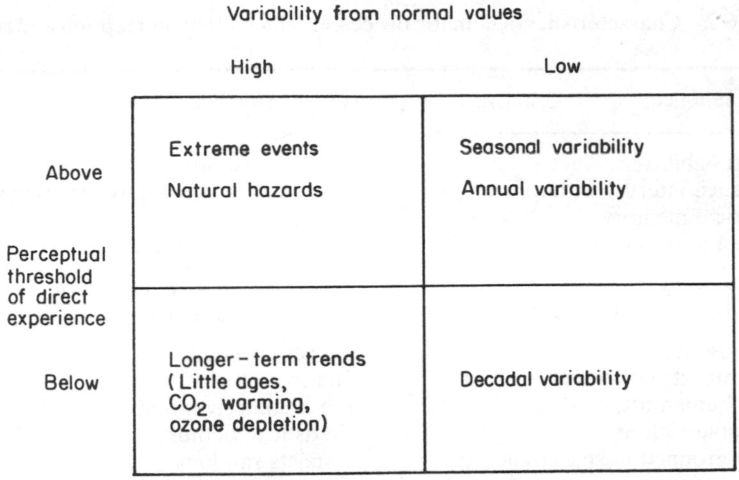

Perceptual threshold of direct experience	Variability from normal values: High	Variability from normal values: Low
Above	Extreme events Natural hazards	Seasonal variability Annual variability
Below	Longer-term trends (Little ages, CO_2 warming, ozone depletion)	Decadal variability

Figure 16.2 Climatic variability and direct experiential perception

direct human perception (Figure 16.2). In terms of the three classes of events addressed in this volume (see Chapters 1, 2) it can be hypothesized that extreme interannual events and natural hazards are likely to produce the greatest behavioral response because they are above the perceptual threshold of direct human experience and are easily recognizable as extreme events. Longer-term climatic events such as 'little ages' or CO_2 warming of the atmosphere cannot be directly perceived by individuals, and although short-term seasonal and annual variability can be felt, they are not usually regarded as extreme enough to be significant. Decadal variability appears to be below both the level of expected variability and direct perception. Pilgrim, in Chapter 13, reports on thc METROMEX study in which a 30 percent increase in precipitation over a 30-year period went virtually unnoticed.

Drawing from the diverse literature on risk, hazard and decision-making, it can be further hypothesized that varying characteristics of climatic processes affect the salience of climate for human perception and response. For example, the public and policy-makers alike tend to disregard future risks and put resources instead into responding to more immediate problems. They are prone to attach greater importance to events which are likely to occur and about which there is some experience, or at least agreement, about what will happen. Thus, scientific uncertainty and controversy become translated into public apathy in a world where problems compete for attention and resources.

Some of the characteristics of climate that affect perception relate to the magnitude and frequency distribution of the processes involved in relation to the time perspective of the decision-maker (see Table 16.2). Other characteristics relate to the size, nature and distribution of the impacts. These

Table 16.2 Characteristics of climatic processes which affect perception and response

Higher salience	Lower salience
High probability	Low probability
Recurrence interval within living or historical memory	Impacts not previously experienced/long time in past
Expected to occur soon	Longer time in future
Extreme event	Lower variability from norm
Imaginable/definable event	No clear beginning/end
High consequences	Lower consequences
Direct impact on people's welfare	Indirect effects
Loss of human life	No human lives lost
Identifiable victims	Statistical victims
Impacts grouped in space/time	Impacts random
Reasonably certain to occur	Uncertain/controversial
Effects/mechanisms understood	Effects/mechanisms not understood
Dramatic impacts	Less perceptible impacts

characteristics stem from the interaction of the physical nature of climate and the type of socioeconomic system being impacted. In turn, the nature of past or expected impacts affects the way in which they are perceived.

One way to understand this complex set of interactions between past events, future expectations, and perception processes within the decision-maker is to imagine it in terms of a decision-frame. The decision-frame is composed of all the information, values and attitudes which the decision-maker brings to bear on a particular choice. The inclusion or exclusion of a particular piece of information can profoundly affect the eventual choice. Sometimes information is excluded from consideration in a particular decision, not because it is unknown, but because it is deemed not to be relevant or to belong to a different category.

This subjective process is a recognized aspect of medical and legal judgments, where careful sifting of much evidence is required. Several experimental studies have examined the effects of decision-framing on choice (Kahneman and Tversky, 1979). One study of the use of automobile seat belts found that reluctant users tended to frame their decision in terms of the risk of death per *individual trip*. This is very low (1 in every 3.5 million person-trips in the USA). If, however, the users considered the risk over their whole lifetime (40,000 trips or a probability of being killed of 0.01), they were more likely to believe in the efficacy of wearing seat belts (Slovic *et al.*, 1978).

In the climate context, the decision-frame can be powerfully affected by the imaginability of the climatic processes or events (Slovic *et al.*, 1974). Thus, for the layperson, hurricanes are more imaginable than droughts because they are more clearly defined in space and time. They are more memorable because the physical processes are more dramatic and their impacts on human well-being are more direct and better understood. Hurricanes provide well defined 'events', each of which can be added to the mental category of 'hurricane'. The victims of a hurricane are sufficiently well grouped in space and time that they are attributed to a *named* event.

In contrast, drought regions and drought periods have more blurred edges. When the 'events' are less well defined, they are less likely to be remembered as falling in a sequence, or as units in a larger category of events. The proportion of indirect victims is greater for droughts than for hurricanes and many losses will never be included in drought impacts. Put simply, the decision-frame for drought is likely to be based on a subsense of past events and past impacts. For newly recognized climatic hazards like ozone depletion or CO_2 warming of the atmosphere, the framing of the decision is determined less by memory and subjective categories than by individual values and belief in authoritative predictions.

16.2.2 Categories of Decision-makers

The definition of the decision-makers who are the focus of a particular study (see the discussion of exposure units in Chapter 1, stakeholders in Chapter 13, and victims in Chapter 15) also influences the selection of perception variables to be measured. The level of description will depend primarily on whether individuals, a large group, or an anonymous collective of decision-makers are involved. For example, a perception study of a small group of farmers may be able to measure adequately their risk-seeking or risk-aversive personalities, whereas a study of public perceptions at the national level is more likely to use standard demographic measures such as age and education as explanatory variables.

Standard measures of individual and group characteristics include age, sex, occupation, education, income, ethnic origin, language and religion. These descriptors are relatively easy to obtain and use for the analysis of individual decision-making and public surveys. They cause some difficulty, however, when applied to other levels of decision-making, such as households, which may contain great variability and for which mean or median values are not helpful. They are least useful for the analysis of collective decision-making, as for example in different levels of government.

Less recognized, but equally important for perception studies, is that each individual simultaneously plays several roles, and that this multiplicity of roles affects the ways in which decision-frames are constructed. A government administrator is also a member of a household, has a professional specialization but is also a layperson, and is a member of the public and a member of diverse smaller social and other groups. Some measure of these roles, and particularly the conflict or inconsistency that they produce in making choices, is important in the delineation of decision-makers.

Another significant aspect of role is the self-identity of individuals and their sense of belonging to particular groups, communities or places. The ways in which a researcher classifies and codes the sample population may not include the significant ways in which the individual categorizes himself or herself. It has been shown, for example, that scientific information can be more effectively passed within self-reference groups (in this case, defined by professional training) than within the organizations where people work (Whyte, 1976). In studying individual or household responses to a disaster, a better predictor of evacuation behavior is whether the individuals are parents, especially with small children, than characteristics like their age or education (Burton *et al.*, 1981).

It is useful to think of decision-making as a nested hierarchy in which each level, from the individual to higher government levels, affects and is affected by other levels. This is part of the social context of perception and decision-making and it is important to include it as part of any perception study. The exposure units of population, places and activities discussed in Chapter 1 are another context that affects decision-making.

16.2.3 Perception Processes

Figure 16.3 illustrates one attempt to simplify, into a very general model, the many variables that have been measured in perception studies. The variables are arranged from right to left approximately in order of generality to specificity for a particular decision, for two levels of decision-makers (individual and collective).

Linking the individual and social variables are four independent processes which together act as the main organizing force in the system. There are the 'perception processes' which link all the components. In this model they are considered as four process elements on the pragmatic grounds of what are measurably different components of perception at the field level. Thus, *categorization and judgment* are grouped together in the model because they are usually observed together, although they are conceptually different parts of the perception process.

The other three major divisions of perception used here are: *sensory perception* (e.g. sight, smell); *attitudes*; and *communication and information flow*. In the field, these processes (either separately or together) can be investigated as links between any subset of variables relevant to the study.

The model was designed as an heuristic device to help the research planning task (Whyte, 1977). It does not provide any help with developing specific hypotheses, nor in delineating the direction of flows between the components.

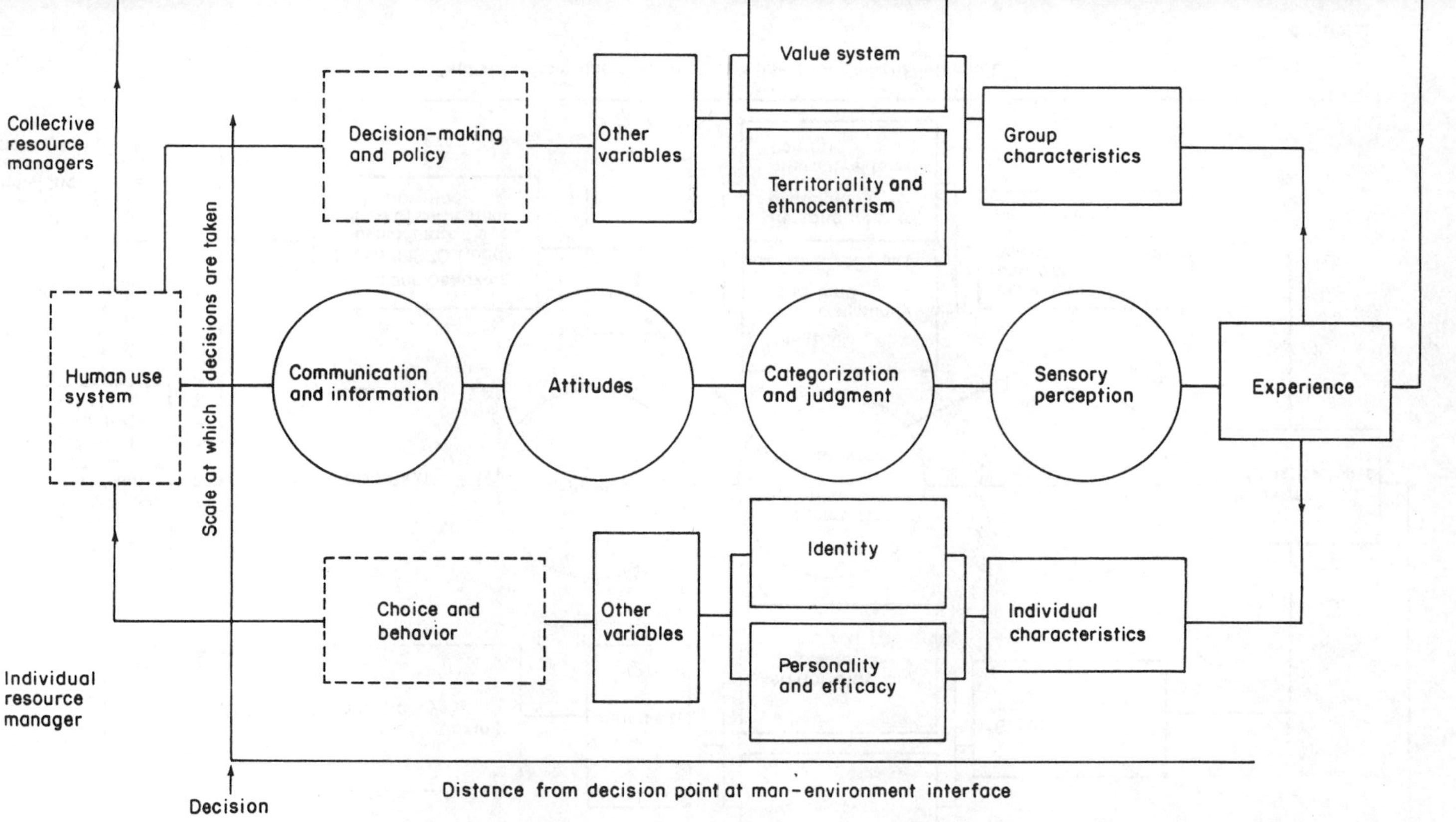

Figure 16.3 General model of variables and related perception processes commonly measured in environmental perception studies. The variables are arranged from right to left approximately in order of generality to specificity relevant to a specific decision. The upper stream relates to collective levels of decision-making and the lower to individual levels. From Whyte, A. V. T., *Guidelines for Field Studies in Environmental Perception*. Map Technical Note 5. © Unesco 1977. Reproduced by permission of Unesco

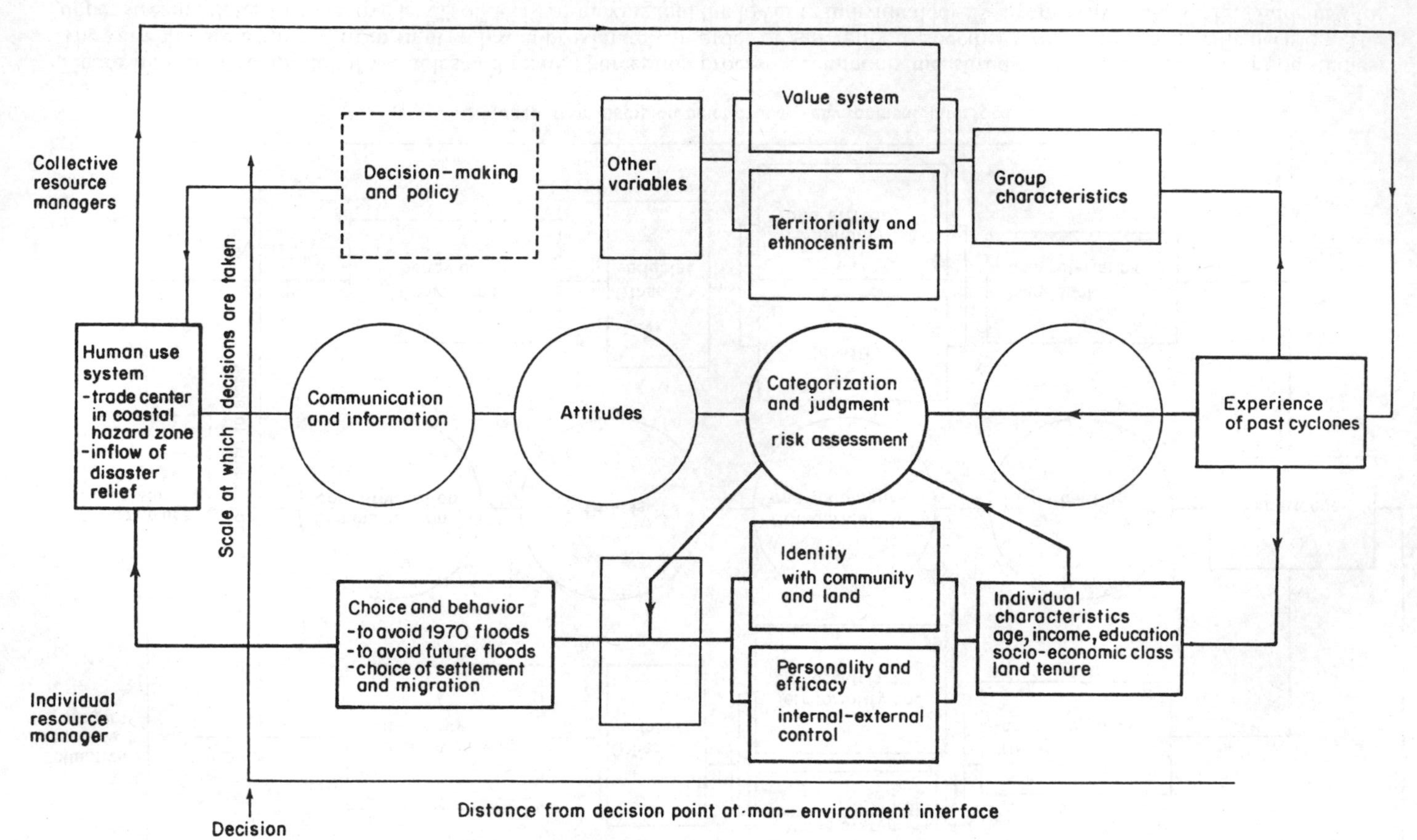

Collective resource managers
Individual resource manager
Human use system
-trade center in coastal hazard zone
-inflow of disaster relief
Scale at which decisions are taken
Decision-making and policy
Other variables
Value system
Territoriality and ethnocentrism
Group characteristics
Communication and information
Attitudes
Categorization and judgment
risk assessment
Experience of past cyclones
Choice and behavior
-to avoid 1970 floods
-to avoid future floods
-choice of settlement and migration
Identity with community and land
Personality and efficacy
internal-external control
Individual characteristics
age, income, education
socio-economic class
land tenure
Decision
Distance from decision point at·man—environment interface

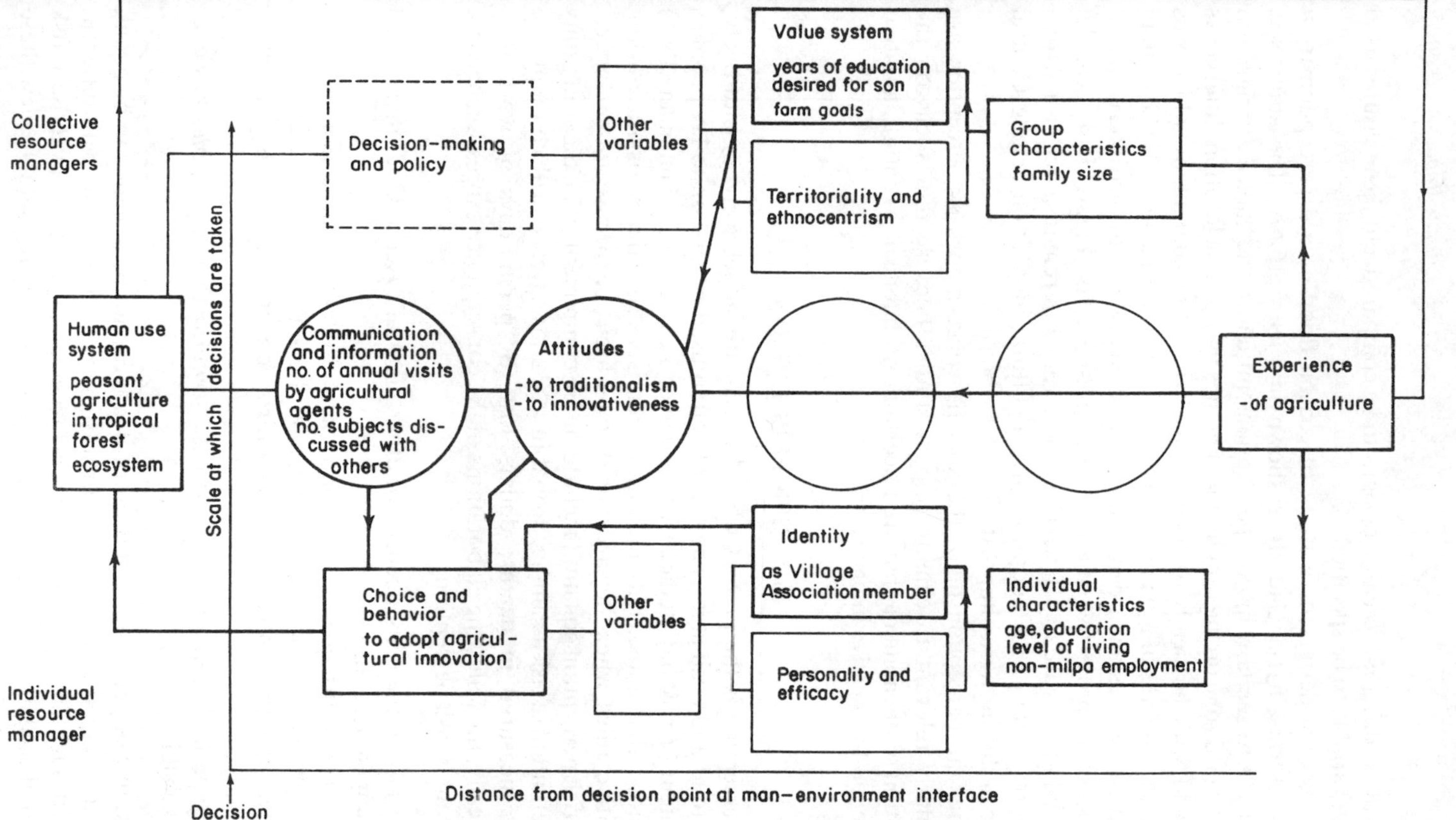

Figure 16.4 Two examples of perception research designs: A, for perception of tropical storm hazard in coastal Bangladesh; B, for adoption of agricultural innovation among Central American peasants. From Whyte, A. V. T., *Guidelines for Field Studies in Environmental Perception*. Map Technical Note 5. © Unesco 1977. Reproduced by permission of Unesco

It does allow different research designs to be compared and the measures used for each variable to be specified.

The research design for a study of coastal flood-hazard perception from tropical storms in Bangladesh is shown in Figure 16.4A. The study was undertaken to see how hazard perceptions influence the choice of migration from a hazard zone on the rebuilding of homes in the same location. Interviews were held with 66 heads of households living in a community which was inundated to depths of 3–9 meters in the 1970 disaster (Islam, 1974). The main method used was a standard questionnaire given in a face-to-face interview that included some projective (indirect) perception tests. Figure 16.4A can be compared with Figure 16.4B, which shows a very different research design for a perception study of the adoption of agricultural innovation among Central American shifting agriculturalists (Feaster, 1968).

Within the perception field, the specific concerns of *risk* perception are particularly relevant to climate impact studies because they focus on the perception of phenomena that are uncertain, probabilistic in nature, and that have generally negative impacts.

16.3 RISK PERCEPTION

Risk perception is the process whereby risks are subjectively, or intuitively, understood and evaluated. The term is often used in relation to lay people's assessment of risk but studies (for example, Tversky and Kahneman, 1974) have shown that statisticians and other scientists also estimate risks according to intuitive 'rules' when they are outside their area of expertise or familiarity.

The risk perception equation includes more components than the commonly used definition of risk as simply 'probability X' consequence, with consequence usually measured in deaths, dollars lost, or person-days of work lost. Specifically, the following elaborations of the consequence expression become significant in risk perception:

1. the *cause* of the consequence (whether 'natural', Act of God, managerial incompetence, etc.);
2. the *type* of consequence (e.g. long, painful death *vs* quick 'heart attack'; great physical incapacity producing a burden on the family *vs* death);
3. the *victim* of the consequence (old, infirm person *vs* breadwinner of family, *vs* child);
4. the type and scale of the *worst case scenario* that is *possible*, however improbable.

Thus, in risk perception, all ways of death cannot be assumed to be equal, nor all lives equal in value. Nor, at the same time, can events with different causes (though similar consequences) be expected to be viewed by those at risk with equal acceptance, resignation or outrage. It is more useful to view these elements in risk perception not as the indiosyncracies of the ignorant, but as challenges to the scientific community to improve its own measures and expression of risk.

16.3.1 Perception of Probability and Uncertainty

More attention, particularly by psychologists, has been given to the perception of probability than to the perception of consequences in risk perception research. This may be because perceived probabilities are more easily quantified and compared with mortality and morbidity statistics.

An example of this approach is shown in Figure 16.5, taken from Slovic *et al.* (1979). The relationship between laypeople's perceived probability of death from 41 causes is compared with statistical estimates for the United States. Each point in the figure represents the average value for the subjective data obtained from 111 university students and 77 members of the League of Women Voters in Eugene, Oregon. Bars for selected points show the 25th and 75th percentiles of the range of perceptions observed.

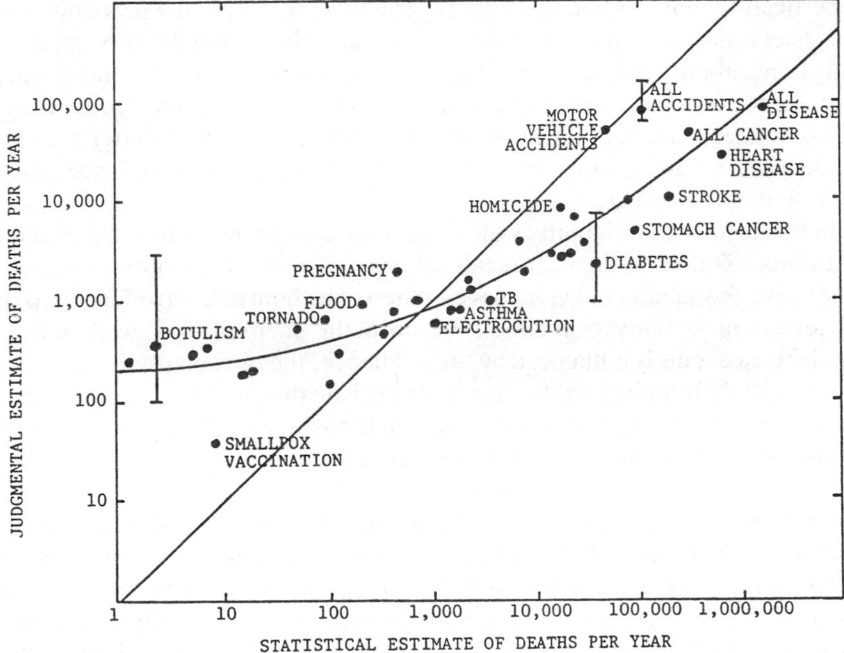

Figure 16.5 Relationship between frequency estimated (perceived) by 188 subjects and actual number of deaths per year for 41 causes of death in the USA. The straight line is the hypothetical perfect correlation between mean estimated and actual frequencies, the curved line a fit to the actual data. (Source: Slovic *et al.*, 1979)

Respondents were found to underestimate the differences in the probabilities of the most and least frequent causes of death. In other words, the subjective estimates narrowed the range of probabilities from six to three to four orders of magnitude. The hazards whose probabilities were exaggerated by laypeople tend to be characterized by high visibility, sensationalism and easy imaginability (for example, botulism, tornado) while the incidence of common diseases like cancers and heart disease was underestimated.

One important explanation for these and similar findings is framed within decision-making models of risky choices. The errors in subjective probabilities are ascribed to limitations in human information processing. Many well-designed experiments have shown that people (including scientists and statisticians away from familiar subject matter) rely on a relatively small number of shortcuts for estimating probabilities. Reliance on these intuitive rules leads to systematic biases in judgment (Nisbett and Ross, 1980; Kahneman *et al.*, 1982). Three important rules, or heuristics, are representativeness, availability and anchoring.

People tend to evaluate probabilities without taking adequate account of base-rate frequencies or relative sample sizes. They are overly influenced by superficial similarities and by stereotypes when estimating the probability that an observation X falls within class A or B. The heuristic also applies to judgments about chance events. The layperson's perception of randomness is that chance is a self-correcting process and that short random sequences will mirror longer sequences. The well known gambler's fallacy that after a long run of 'heads', a tail is now due, is also based on intuitive reliance on the representative heuristic.

In this context, availability is the dependence on memory to bring examples of events to mind. If one can recall other similar events or observations, the subjective probability of event X is elevated. The heuristic introduces bias into perceived probability estimations because the memory, or availability, of similar past events is influenced by their salience, the time elapsed since the last event, and their imaginability. It is this heuristic that seems to be operating to increase subjective probabilities of natural hazards after the recent occurrence of one. The availability bias could also account for the exaggeration of dramatic events like tornadoes as causes of death.

Anchoring is another decision heuristic that has been demonstrated in several experiments. When asked to estimate quantities or probabilities, people tend quickly to select an initial value and then to adjust it before reaching a final figure. Typically, the adjustments made (by mentally searching for more information and 'rationalizing') are insufficient. The initial, more intuitive, value acts as an anchor. It is this initial value that is most influenced by the availability and representativeness biases. Thus anchoring compounds any error in probability judgment that has been made (Tversky and Kahneman, 1974).

One of the ways in which the uncertainty associated with probabilistic phenomena is dealt with psychologically is to perceive a pattern of events. This can provide the individual with a sense of control over the environment or allow him to believe that he is in the hands of a higher authority who has both motivation and command (see also Section 16.3.4). A common reaction to news of an event is to seek to attribute a cause to that event (causal attribution or attribution theory). It is apparently difficult for us intuitively to accept that events can be 'uncaused' or, more accurately, be random. Indeed, experts in climatology struggle with tendencies to perceive trends or cycles which may or may not exist (Chapter 2).

The notion that there is motivation, and therefore pattern, to events is a very powerful one. For example, many studies of natural hazards, particularly hazards with a recurrence interval measured in years rather than in decades or longer, have found that laypeople perceive patterns where statistical records show none. Common patterns perceived for floods, hurricanes, tornadoes, and so forth is that they come once in every 3 or 7 years (Burton *et al.*, 1978). The biblical view of drought is that it comes once in every 40 years.

A study of Mexico of the perception of rainfall by peasant farmers in a semi-arid area found that commonly believed rainfall patterns included 3-, 4-, 5- or 7-year cycles as well as autocorrelation between years (that is, one wet year was followed by another). When these perceived patterns were compared with meteorological records and with memories of specific past wet years, the commonly perceived pattern of a 3-year cycle could be seen as an idealization of actual annual rainfall variability (Figure 16.6). The pattern of remembered past events, however, indicated that a more extreme recent event (higher rainfall year) effectively blotted out the memory of earlier, less extreme events and seemed to act as a reference point against which to calibrate later episodes. Thus, the most recent wet year was remembered, but memory of preceding wet years extended back only to the one which was even wetter, and so on (Kirkby, 1973).

16.3.2 Perception of Impacts

Limitations in human information processing partly explain how differences arise between statistical probabilities and perceived risks. Another explanation is that the finer points of subjective probability are not very important to risk perception because our attitudes to risk are more powerfully shaped by perceived links between causes and effects and by the perceived nature of the consequences. For example, the efforts of the nuclear industry in several countries to convince the public of the acceptibility of nuclear power, based on its low probability as a cause of death, have not been successful. Public attention is not focused on the question of probabilities, however low, but on consequences, which conceivably could be catastrophic (Whyte, 1983).

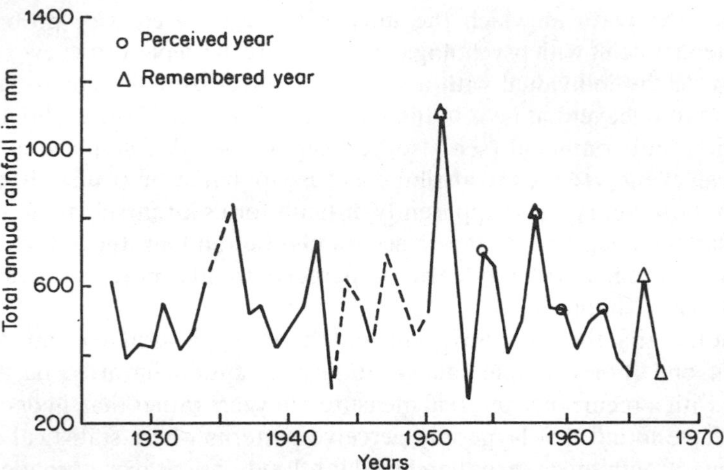

Figure 16.6 Patterns of rainfall memory and perceived wet years related to annual rainfall record (broken lines inferred) among peasant cultivators in Mexico. *Remembered* years are specifically named by respondents. Perceived years are inferred by the author based on patterns described by respondents. (Reproduced by permission of The University of Michigan, Museum of Anthropology, from Kirkby, *Memoirs of the Museum of Anthropology*, No. 5 (1973) Kirkby, 1973)

Long-term climatic change presents somewhat similar characteristics to the layperson. The probabilities of disastrous change are perceived as low, but the scale of the consequences can be imagined as global and catastrophic. Scientific controversy about the direction, magnitude and time-scale of such long-term climatic changes, however, significantly reduces the concern and attention that the public gives to the matter.

There is evidence, both for extreme weather events like hurricanes and for longer imperceptible climate changes, that the characteristics of the impacts affect perception and response more than do their probability distributions. Expected utility measures and simple risk equations based on standardized measures of impact (deaths or dollars lost) fail to capture people's sensitivities to different types of impacts and ways of dying. This is an important consideration in designing climatic risk perception studies.

It is often regarded as axiomatic that people will accept higher risks if they expect to be compensated directly or indirectly by higher benefits. Perception studies of climatic impacts therefore need to include perception of the distribution of the associated benefits.

Formal cost/benefit and risk/benefit analysis have their own parallels in intuitive decision-making. In both cases, the framing of the analysis is critical to the outcome. What costs and benefits are to be included? The public is aware (or can be made aware) that tradeoffs inevitably have to be made between production of energy and acid rain. The issue is not so much public inability to recognize the existence of tradeoffs as it is who pays and who benefits. An individual is more likely to accept risk if he also benefits, but not if he perceives the benefits to be going elsewhere.

Social surveys have probed the risk/benefit equation for the public on a number of environmental issues. Several US surveys since the 1970s have asked the public to choose between economic growth and environmental protection. In polls conducted in 1975 and 1978 it was found that about 60 percent of the US public preferred higher prices and a cleaner environment to lower prices and more air and water pollution (Harris and Associates, 1975, 1978). These findings have been replicated in several other surveys (Mitchell, 1980; US National Science Board, 1981). Another survey found that 52 percent of the US public favored 'the environment is more important than growth' scenario rather than the opposite (Harris and Associates, 1979).

In the environment versus energy equation, however, the US public generally has been found to favor somewhat higher environmental costs to ensure energy supplies and development. These differences in findings support the argument that the public understands cost/benefit analysis and can discriminate between alternative tradeoffs.

16.3.3 Attribution of Causality

One of the important findings of risk perception research is that the scientific distinction between 'natural' and 'manmade' events does not necessarily coincide with the causes of those events as perceived.

For natural hazards, the influential force sometimes may be viewed as supernatural, but increasingly in modern societies it is seen as a natural process compounded by human choice. Thus the public in the United States tends to regard floods as caused more by engineering and zoning decisions than by 'Acts of God'. Similarly, explanations of the causes of the Sahelian desertification of the 1960s and 1970s tend to include both natural climatic variability and human activity, such as the concentration of animal herds around technologically developed waterholes.

Experimental studies have shown that, in general, initial attributions about cause, effect and responsibility, once made, are remarkably resistant in the face of later, conflicting information (Ross and Anderson, 1982).

The significance of attribution of causality for behavior is that, when impacts are seen as not falling randomly, there is an urge to blame some other sector of society. The choice of adjustment becomes more dependent upon expectations of where responsibility lies and upon the perceived motivations, credibility and competence of other people and groups (see Section 16.4.3). In this case, the research model becomes more focused on social variables than on environmental perception.

16.3.4 Perception of Control

Starr (1972) pointed out that the risk of death in many voluntary activities is higher than that considered acceptable for involuntary activities. Private flying is more risky than commercial air transportation, and skiing or canoeing is more dangerous than travelling by car. He estimated that the risks accepted in voluntary activities were a thousand times greater than for involuntary ones. While there has been criticism of the reliability of Starr's data, the view that the voluntary–involuntary dimension is important to risk perception is accepted by many analysts.

In a climate context, we may infer that we will probably willingly tolerate higher levels of the risk of skin cancer from voluntarily sunning ourselves on the beach than we will from involuntary exposure to increased radiation because of our occupation or anthropogenic changes in the Earth's atmosphere.

16.4 THE ROLE OF INFORMATION

16.4.1 Direct versus Indirect Information

There are several components to direct information. The first is the perception of a hazard directly through one's own senses of sight, hearing, taste, touch or smell. But a hazard can exceed the thresholds relating to human sensory perception, and direct experience then becomes a powerful information source.

A second experiential source of information is that of an individual who is, or personally knows, the victim of an afflication such as cancer. In this case, the direct experience is of the consequences of a hazard that may be beyond one's own sensory perception, but observed in another.

A third aspect of direct information is the directness and certainty that can be attached to cause and effect. A flood that drowns, a fire that burns, or a plane that crashes are more direct in their effects than a carcinogen with latent effects that show some 40 years (and a lifetime of carcinogens) later.

Direct information usually influences individual decision-making more than indirect. For example, Adams (1971) examined the comparable effects of direct, personal weather observations and television weather forecasts on outdoor recreation decisions, and found that personal observation clearly was more significant.

In the case of indirect information, risk perceptions are more likely to reflect the quality of reporting and the credibility of the information source than actual levels of exposure or risk.

16.4.2 The Role of the Media

For many climatic changes, the magnitude of the hazard and the links between cause and effect can be obtained only from indirect sources. For those who learn about hazards and their consequences indirectly, the mass media (television, radio, newspapers and magazines) are the principal sources of information. In the last decade, the amount of information given to climate change in major newspapers of record has tripled (see Figure 16.7).

One of the functions of the mass media is to distribute information from official sources to a wider audience, but the information that finally appears is selective, is usually accompanied by evaluative comment, and is put into a particular context. All of this changes the impact and meaning of the message.

For example, Riebsame (1983) tracked the news media coverage of the US National Weather Service's 1982–83 winter forecasts by monitoring 1710 daily and 8200 weekly newspapers. He found that the forecasts were quite accurately conveyed but that prior, widely publicized disagreement between government and private forecasters was prominently reported on virtually every news story on the forecasts.

Scientific and quasi-scientific media reporting can also shift perception. Warrick and Bowden (1981) suggest that partly because of 18 major media articles reporting on an expected drought on the US Great Plains in the 1970s, laypeople, government and the media itself acted as though a major drought actually was occurring, an event not borne out by observation.

The biases that often are criticized in media reporting (focus on rare events, large-consequence events, dramatic stories, 'big bang' events with strong visual images) are but explicated versions of the layperson's perceptual biases. Media reporting thus exaggerates the already existing intuitive biases of risk perception. For example, an analysis of the reporting of causes of death over a 1-year period of two US newspapers (one in Massachusetts and one in Oregon) found that many of the statistically common causes of death (such as diabetes, emphysema, various cancers) were rarely reported. Relatively rare violent events such as tornadoes, fires and homicides, however, were more frequently reported. Diseases take 16 times as many lives as accidents in the United States, but two newspapers reported accident deaths three to seven times as often (Combs and Slovic, 1979).

The same study found that the correlation between people's judgments about the frequency of death from various causes and newspaper reports (holding statistical frequency constant) was 0.89 and 0.85 for the two newspapers. Although intuitive risk perceptions can be related to newspaper reports, the direction of the relationship is not known: do newspapers reflect intuitive biases of the public or help to create them?

16.4.3 Credibility and Expectation

An important aspect in the reception of information is how credible it is perceived to be. Although the public may recognize that journalistic accounts of events can be biased, it often feels that 'there is something in them'. The

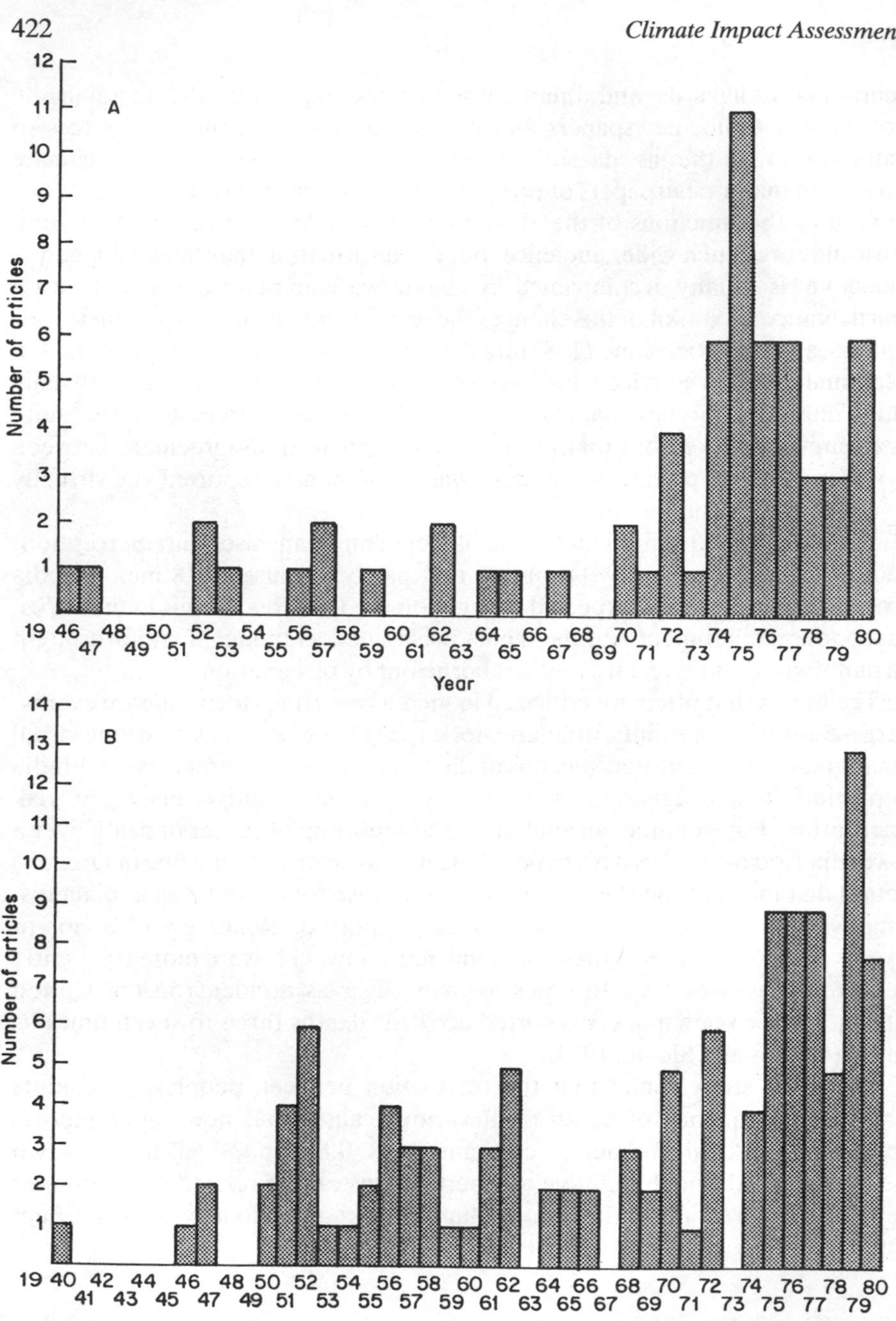

Figure 16.7 Numbers of articles on climatic change appearing in A, *Toronto Globe and Mail*, 1946–80; B, *New York Times*, 1940–80. Reproduced by permission of the American Meteorological Society from Harrison, *Bulletin of the American Meteorological Society*, **63**, 730–738 (1982)

issue of credibility is more critical to governments, official agencies and the private sector than it is to the mass media in determining how the information that is disseminated will be received.

A preeminent example of the credibility problem is nuclear power, where the mass media are more believed by the public than are the governments and government agencies. Declining trust in the national government is a major factor in public risk perception in some countries. There is evidence of a mismatch between public expectations of governmental responsibility in solving environmental problems and public ratings of past governmental performance. The credibility issue is also important in climate risk perception at the international level. International effort is seen as necessary for reducing problems, but usually little hope is held out for any effective action.

16.4.4 Historical Dimensions

As each wave of environmental crises and hazards to health successively hits the headlines, the public is often bombarded with information about industrial irresponsibility, government lapses, or incompetence. We are told of increased hazards in the home, on the road, at the workplace, and even in the very air that we breathe and water that we drink. Within the last 10–15 years the public has become sensitized to risks and has come to recognize 'risks' as a category of phenomena to which it is exposed in everyday living.

A survey in the United States indicates that the public is more risk conscious than in the past (Marsh and McLennan, 1980). Almost 80 percent of those asked believed that risks to society are increasing and that the US public is becoming more aware of risk (Table 16.3).

Table 16.3 Perception of increasing future risk and public awareness of risk as seen by the US public in 1980

	n = 1488
Risks to society will be	%
Greater	55
Less	18
About the same	22
Not sure	5
	n = 1488
US public is	%
Overly sensitive to risk	15
More aware of risk	78
Both	2
Not sure	5

Source: Marsh and McLennan, 1980.

The longer-term effect of crisis events is cumulative. Not only is there increasing public concern about risks within a category (for example, air pollution), the concern extends to other risks as well. A study of the effects of the 1979 Mississauga derailment (near Toronto) on risk perception found that, compared to a control group, evacuees had a higher expectation of a major nuclear accident happening in Ontario within the next 10 years. The accident experience had thus increased their sensitivity to other hazards (Burton *et al.*, 1981).

16.5 METHODS

16.5.1 Choice of Research Approach

Although there are a number of specialized techniques and instruments available for measuring environmental perceptions, the three basic methods of investigation are observing, listening and asking questions. Of the three, asking questions is the most commonly used method and is usually done through structured questionnaires or semi-structured interviews in face-to-face situations.

There is no single 'best' method for studying perceptions. Different techniques have been developed for answering different research questions and for different research situations. Methods of observation, listening and asking questions can provide mutually reinforcing and complementary data so that they should be used in combination wherever possible.

The range of techniques available to measure perceptions is illustrated in Figure 16.8 and is ordered according to the degree of control exercised by the researcher versus the flexibility of response open to the respondent. This is an important consideration in the selection of techniques because this spectrum also represents a tradeoff between more controlled, experimental designs and replicable observations and more idiosyncratic, less statistically reliable but often more relevant, findings of 'situation-defined' research.

These differences in research style are typified in the contrast between public opinion surveys and in-depth case studies. Each has merits and drawbacks as well as appropriate research techniques. The less structured approaches to perception studies, however, place greater demands on the individual capabilities and credibility of the researcher, and are also more dependent on content analysis of perception data. They are likely, therefore, to require more research time per unit of observation.

16.5.2 Limitations on Existing Studies

Data on risk perception have been obtained primarily through three methods:

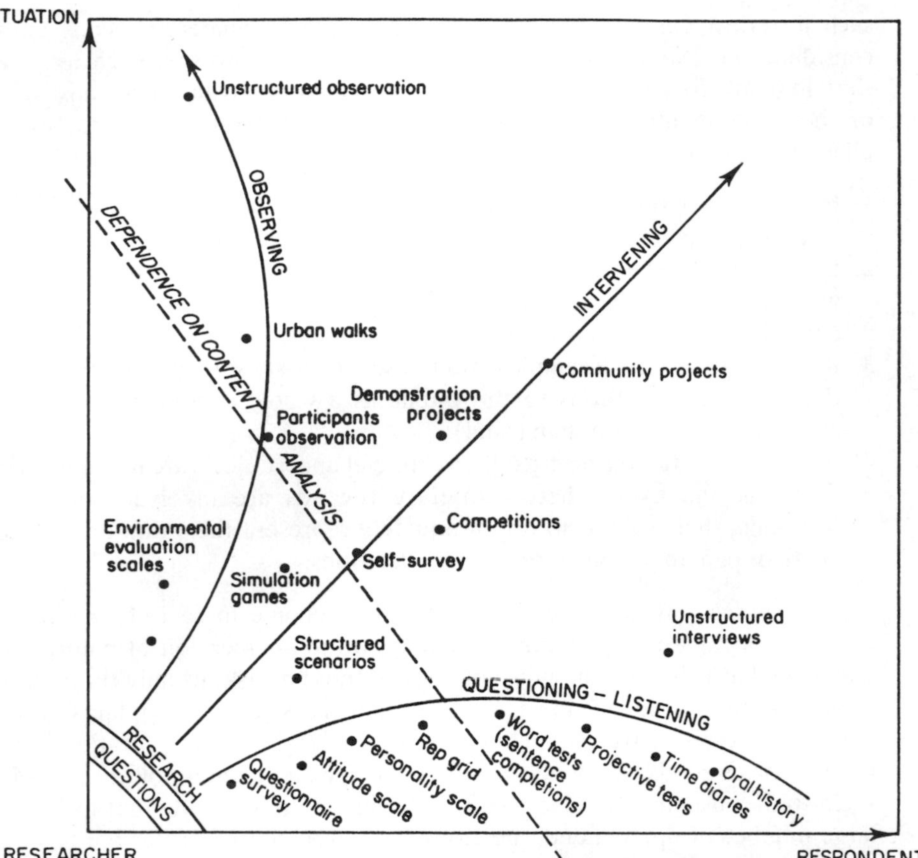

Figure 16.8 Techniques for measuring perception in relation to the degree of control exercised by the researcher. From Whyte, A. V. T., *Guidelines for Field Studies in Environmental Perception*. Map Technical Note 5. © Unesco 1977. Reproduced by permission of Unesco

1. *laboratory experiments*, such as those conducted by social psychologists on heuristic biases;
2. *field studies*, such as those carried out by geographers on perception of natural hazards in tornado zones, floodplains, etc.;
3. *social surveys*, such as those conducted by sociologists and public opinion pollsters on national samples of the population.

Each of these methods brings with it various strengths and weaknesses.

In a *laboratory setting*, the parameters of the risk decision can be simplified and manipulated to isolate and test individual dimensions of perception, such as the anchoring effect. When similar effects are found to be repeated under different experimental conditions, one can with some confidence accept that such an effect exists generally in risk perception, whatever risk is under consideration. (Some repeatedly observed effects, however, such as 'risky shift' in small group decision-making, are now believed to have been an artifact of the experimental situation rather than a reflection of any real-world phenomenon.) The main limitations of such experimental data are that:

1. by isolating parameters in risk perception, they fail to predict outcomes when all parameters are combined, as they are in real-world decision-making;
2. they have tended to focus on the probability component of risk perception, whereas other evidence would indicate that it is perceived differences in consequences that is more important to public perception;
3. they tend to use small samples and to use university students as subjects, so that any generalizations to the public as a whole should be made with caution (more caution than usual);
4. the experimental setting is itself an artificial one; subjects are not facing real decisions and even where monetary rewards are involved for 'right' decisions, their values do not adequately represent the value of avoiding death or pain to people in real-world decisions.

Field studies, on the other hand, do study people in real-life situations. Examples include many empirical projects on the perception of natural and manmade hazards from the point of view of those at risk, usually through the location of their residence in hazard zones (such as coastal floodplains subject to hurricanes; residents around nuclear reactors or along transportation routes for dangerous goods; residents of areas with high air or water pollution levels). Field studies usually use a face-to-face questionnaire schedule that includes a large number of open-ended questions, and even simple projective tests, to probe explanatory variables for the perceptions measured. Their focus on one hazard, while it may bias response in one direction, does provide a coherence and logic to the interview that can be grasped by the respondent.

The main limitations of many (not all) of these studies are:

1. their inadequate sampling frames and procedures, which are the result of cutting cost but which limit their validity for larger populations, although it is demographic measures that are the main independent variables;
2. their focus on one hazard, rather than 'all hazards at a place', and the definition of that hazard by the researcher rather than by the population. This approach has been known to measure perceptions of air pollution or earthquake risk for people who were unaware that the hazard existed before the interviewer came along;
3. their emphasis on the individual as the decision-maker and their relative neglect of the social dimensions of decisions, particularly in the household or family context, and (for developing countries) the community (tribe, village, etc.) context. These contexts certainly limit the choices that most

individuals make and in many cases such decisions are really better modeled as family or group decisions.

The main strength of *social surveys* is their sampling methodology and their generalization to large populations. Their limitations are similar to those of field studies except that they are even more severc, namely,

1. social surveys may include a wide range of items that bear no relation to one another. Respondents may be asked their opinion on the effects of acid rain, the use of forest sprays, whether they think the government is doing a good job in relation to unemployment, foreign policy, and so on, in rapid succession. There is no context to these questions and responses beyond basic independent demographic measures such as age, sex, education and income;
2. there is no independent measure of respondents' exposure to risk, as there is in localized field studies, and absolutely no social context in terms of the household or larger decision-making unit;
3. economic measures such as 'willingness to pay', which are sometimes included in social surveys, are as hypothetical to the respondent as are experimental situations. They clearly measure something but it is not 'willingness to pay' in terms of willingness (or ability) actually to hand over the money.

16.6 EXAMPLES OF CLIMATE PERCEPTION STUDIES

The literature on climate perception is very limited. Some work has been done on public perception of weather and climate information and forecasts (see, for example, Murphy and Brown, 1983), and on the use of manuscript records, such as travel logs and diaries, in the reconstruction of past climates (Lawson, 1974), but the literature lacks studies of climate perception *per se*.

One story that stands out is a survey of senior citizens' contemporary and recollected climate perceptions that was conducted by Oliver (1975) and his colleagues in Terre Haute, Indiana. They interviewed 93 people, ages 60–92, with an emphasis on their perceptions of climate change during their lives. Most (70 percent) felt that the climate had changed, but they expressed different views as to the types of changes. Comparing respondents' recollections of past climate to the instrumental record, the researchers concluded that their sample was able to compare quite accurately past and current climate. The respondents were also quite accurate in their recall of specific weather events and memorably severe seasons (such as particularly snowy winters).

16.6.1 Perception of Annual Rainfall Variability

A perception study being conducted in the Mapimi Biosphere Reserve in the states of Durango, Chihuahua and Coahuila in northern Mexico is concerned with the choice of adjustment to annual rainfall variability and the related carrying capacity of arid rangeland for cattle (Whyte, 1984). The average annual rainfall of the area is 200 mm, 80 percent of which falls in heavy thunderstorms between June and September.

Land tenure in the area is undergoing considerable change as new ejidos (agricultural community-based cooperatives) are established in land that was formerly in private ownership. The oldest settlements in the area were established around 1940; the most recent ejido was granted in 1981 and has not yet built houses in the area. Within a small area, therefore, the effects of private versus cooperative land tenure, experienced and inexperienced cattle herders, and different administrative jurisdictions (three state and four municipal governments) can be compared.

There are nine small settlements within the Reserve, ranging in size from one to five family units. Public services are minimal—there is no electricity, housing conditions are poor, cooking fuel is mostly wood, and there are no paved roads. The major economic activity of the area is raising cattle for meat export to the United States. Some horses, goats and a few sheep also are grazed. The animals are allowed to range freely and only the privately owned ranch is completely fenced. There are presently between 16 and 25 hectares per head of cattle.

The main components of this perception study are illustrated in Figure 16.9. The data on perceptions are obtained through unstructured interviews in Spanish with ejiditorios and ranchers, in their homes, without the use of a formal questionnaire or projective techniques, which would arouse distrust in this setting. Data on behavior are obtained through participant observation, and on range management through interviews and the examination of written livestock records. Some climate records and historical data on past resource use in the area are also available.

The focus of this study is to analyze the annual decisions made about stocking levels in terms of the decision-makers' expectations of the carrying capacity of the semi-arid pasture over the dry season (October–March). It is possible, on the basis of summer rainfall (April–September) to have a good estimate of the carrying capacity of the land in the critical dry period. This, then, is an interaction study of interannual variation and response to it.

The largest private owner in the Reserve does annually adjust his herd size in response to his expectations of rainfall-dependent variations in rangeland carrying capacity. The ejiditorios, especially those with least experience in the area, do not. A major objective of the study, therefore, was to discover the reasons for the differences in choice of adjustment.

The critical differences between the ranchers and ejiditorios are *not* to be found in their perceptions of the annual variability in carrying capacity, but in important differences in their aspirations, attitudes and the degree to which they have control over their own management decisions. The Mapimi case study, therefore, illustrates very well the need for perception studies to look

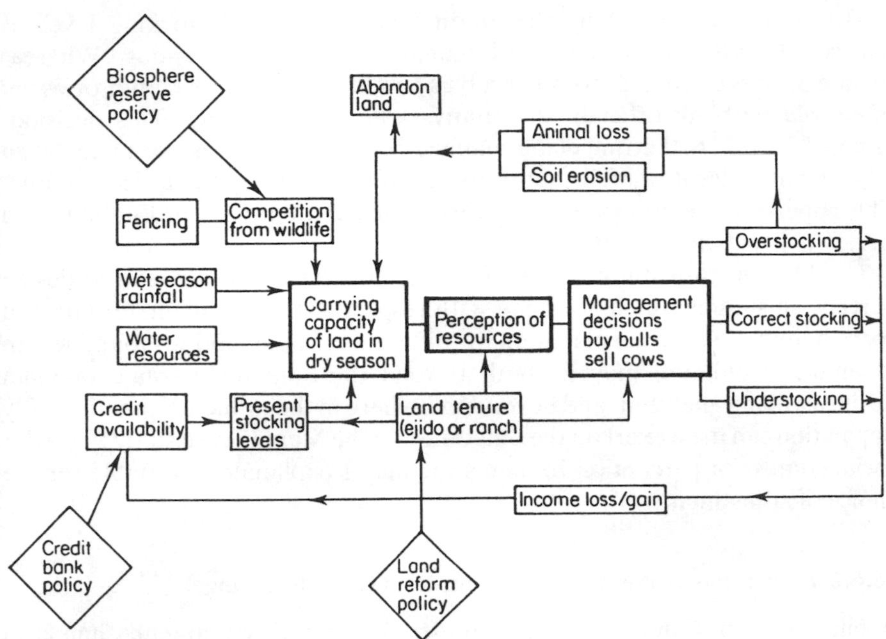

Figure 16.9 Model for annual range management decisions based on perception of seasonal carrying capacity, Mapimi, Mexico. From Whyte, A. V. T., (1984), in Di Castri *et al.*, *Ecology in Practice*, Part II. © Unesco 1984. Reproduced by permission of Unesco

beyond perceptions *per se*, and to consider the decision-frame and the social context of decision-making.

For example, the decision-frame for a private owner is to make annual adjustments in herd size in response to rangeland carrying capacity so as to maximize economic return and maintain rangeland productivity in the long term. He has owned the ranch since 1942, lives in town, and has other economic resources. He keeps careful accounts of annual productivity, has fenced the land so that he knows its productivity, and has adopted modern herd management policies.

The ejiditorio's decision-frame is dominated by aspirations to improve his economic situation and that of his children, which he sees as dependent on building up his herds as rapidly as possible. His perceptions of annual variability in rainfall become translated into adjustments in the rate of increase of stocking levels rather than adjustments of total herd size.

The second important lesson to be learned from the Mapimi perception study is the need to take into account the effect of other levels of decision-making on choices that are made at any one level. In the Mapimi case, three policy areas—credit banks, land reform and biosphere reserve policies—influence the range management decisions taken (see Figure 16.9).

The control exercised by the credit banks over herd stocking levels is particularly critical for rangeland management in some ejidos. Whereas private owners can decide for themselves how many cows to sell, ejiditorios are often *told* by bank officials how many cattle they will buy. This decision, although it affects the rangeland management in Mapimi, is not made as an adjustment to local rainfall, but in the context of the bank's decision-frame. This appears to be dominated by international cattle prices, profitability and cattle quality standards for export.

The Mapimi case study is, therefore, an illustration of the need to design perception studies which are open with respect to their immediate time and space boundaries. The explanation for differences in perceptions found within a sample population may lie both beyond the immediate frame of main decisions being analyzed, and beyond the sphere of the influence that the study population can itself exert on the outcomes. In the Mapimi case study the wider social context of perception becomes the major explanatory variable for the choice of adjustment.

16.6.2 Perception of Weather and Long-term Climate Change

A pilot study in Ontario, Canada, undertaken for Environment Canada in 1981, is one of the few studies known on the public perception of long-term climatic change (Whyte and Harrison, 1981). The study used a structured questionnaire administered in a telephone survey to samples of urban and rural resident and snow plough operators.

The study probed the relationship between memory of past winters, especially of the winter of 1980–81, and expectations of the weather of the next winter before asking questions about long-term climatic change. This approach was adopted in order to establish rapport and interest in the survey. Most respondents feel comfortable in giving opinions about the weather, and in Ontario, winter conditions are particularly salient to rural and urban dwellers alike.

The choice of a telephone survey to obtain data was made on the grounds of least cost, but such a method severely limits the quality of the perception data and the number of independent and dependent variables which can be obtained within a 10-minute interview. It is therefore appropriate only at the exploratory stage of a perception survey or when large population samples are needed. One of the objectives of the Canadian study was to explore if public awareness of a climatic change is sufficiently widespread to justify a future perception study.

The contrast between the two case studies could hardly be greater. The focus of attention in Mapimi was on an annual event, one which could be directly perceived, and was translated into decisions which directly affected the decision-makers' wellbeing. In Canada, not only is the socioeconomic situation of the respondents very different from that of a poor area of Mexico, but the object of perception is a low-probability, highly uncertain future event which may never affect the respondents directly, and about which no direct decisions are required of them. Further, direct perception of weather is formed from experiencing climatic noise (short-term variations) rather than from information about climatic change. Public awareness of climatic change comes largely from indirect sources, especially the mass media; part of the Canadian study therefore focused on an analysis of newspaper coverage of climatic change from 1946 to 1980.

When asked to predict the following winter's weather, only 13 percent of the respondents declined, and the majority made reasonably correct predictions of colder than normal temperature and heavy snowfall. One common explanation given for the predictions was the observation of nature and climate during the fall, which was particularly cold. Many people believe that a cold fall augurs a cold winter (37 percent of the respondents). The behavior of animals, insects, birds and plants, as well as the fuel stockpiling of local Indians, were cited as other signs of a hard winter to come (14 percent). Another group of respondents (18 percent) subscribed to the view that nature is a self-correcting patterning process, so that one mild winter 'will be paid for' by a cold one.

When asked if the climate in Canada is changing, 66 percent of the respondents said yes, 21 percent said no, and only 5 percent said that they did not know. Changes in winter, rather than summer, weather were the major indicator referred to, although the perceived direction of the changes shows a large variation from respondent to respondent (Table 16.4).

The two main groups of responses for the perceived *cause* of climatic change were specific references to *manmade* causes (38 percent) and general references to natural trends, variability and change (28 percent). In addition, 31 percent of the respondents could give no reasons for climatic change. Teleological explanations ('God's will') were offered by only two people. Thus, many people see climatic change in Canada as a manmade problem (Table 16.5).

Questions were included in the survey to probe whether people had heard about loss of ozone, the 'greenhouse' effect, or an increase in CO_2 (all of which have appeared frequently in news media reports), and whether they believed that they were linked to climatic change. The results show that in 1980 people were more aware of the ozone problem than the increase in CO_2, but that both issues were regarded as part of climatic change.

The inclusion of three sample populations in the study was designed to test whether the level of awareness of weather and climate was a function of life style and 'contact' with weather. Snow plough operators, rural dwellers and urbanites spend greater to lesser amounts of time dealing directly with winter weather. The results showed that perceptions of weather and long-term climatic change did not vary significantly among the three populations, nor

Table 16.4 Responses to the question: 'In what ways is the climate changing in Canada?' given by 422 respondents in Ontario, 1980

Respondent mentions (up to three responses)	Total	% of total responses*
Winters		
shorter	14	2.6
longer	15	2.8
colder	13	2.4
warmer	114	21.1
more snow	6	1.1
less snow	60	11.1
	222	41.1
Summers		
cooler	69	12.8
hotter	10	1.9
drier	6	1.1
wetter	6	1.1
longer	1	0.2
shorter	24	4.4
	116	21.5
Years		
colder	26	4.8
warmer	36	6.7
wetter	13	2.4
drier	6	1.1
	81	15.0
Trends		
less predictable	73	13.5
more temperate	11	2.0
equilibrium	17	3.1
seasons later	12	2.2
windier/cloudy	4	0.7
worse	2	0.4
	119	22.0
Don't know	2	3.7
Total number of responses	540	99.6

* Percentages do not add to 100% because of rounding of decimals.
Source: Whyte and Harrison, 1981.

were they a function of individual characteristics like age, sex or educational level. Perceptions of climatic change seem to be part of the Canadian public's general environmental awareness. Awareness of issues, in contrast to

Table 16.5 Public perception of the reasons for climatic change in Canada

Respondent mentions (up to three responses)	Total	% of total responses*
People's occupation and activities on earth and in space		
Space exploration	55	14.5
People/pollution/urbanization	46	12.1
Nuclear explosions/tests	38	10.0
Loss of ozone	5	1.3
		37.9
Natural long-term global reasons		
Global changes	36	9.5
Natural variability	34	8.9
Ice age coming	17	4.5
Long-term trends	13	3.4
Volcanoes	8	2.1
		28.4
Other		
Know from experience	5	1.3
God's will	2	0.5
People's paying attention	2	0.5
Don't know	119	31.3
		33.6
Total number of responses	380	99.9

* Percentages do not add to 100% because of rounding of decimals.
Source: Whyte and Harrison, 1981.

knowledge of issues and attitudes towards them, does not usually show strong demographic or life style differences (Whyte and Burton, 1982).

16.7 CONCLUSION

There are many aspects to the perception of any environmental phenomenon and a number of ways to study perception. As was suggested in Section 16.2, the first research task is to clarify what components of climate, what types of decision-makers, and what aspects of perception form the focus of the investigation. From these decisions flow the choice of method, sampling frames, and specific measurement techniques to be used.

The section of risk perception (16.3) provides an overview of some of the main perception parameters that are considered relevant for climate impact studies. Few of these aspects of risk perception, however, have been examined specifically with respect to climate impacts, except for natural hazards. The reader will note some discrepancy between the discussion of risk perception in Section 16.3 and the parameters measured in the two examples described in Section 16.5.

In the field, many of the more psychological aspects of risk perception, such as attribution of causality or perception of probability, became masked, or dominated by, contextual parameters. In Mapimi, these included the constraints of decisions made by others on the choice of adjustment of the study population. In Ontario, awareness of climatic change is more influenced by the mass media than by direct experience, so that a more sociological and less individual-oriented approach is relevant.

The bulk of this discussion has dealt with climate as hazard (which is discussed also in Chapter 3) and within this context, more attention has been given to the perception of probabilities of risk than to the perception of consequences or adjustments. Still less attention has been given to the perception benefits, and least attention to the perception of the benefits of climate itself or climate as a resource.

A well-designed and conducted perception study does not come cheaply, and research costs and time are typically underestimated. The value of perception studies lies in the insight that they can bring to explaining human behavior and the directions that they can provide for public policy in an increasingly uncertain and risk-conscious world.

REFERENCES

Adams, R. L. A. (1971). *Weather Information and Outdoor Recreation Decisions: A Case Study of the New England Beach Trip*. Department of Geography, Clark University, Worcester, Massachusetts.

Burton, I., Kates, R. W., and White, G. F. (1978). *The Environment as Hazard*. Oxford University Press, New York.

Burton, I., Victor, P., and Whyte, A. (1981). *The Mississauga Evacuation: Final Report to the Ontario Ministry of the Solicitor General*. Ontario Government Publications, Toronto, Canada: 426 pages.

Combs, B., and Slovic, P. (1979). Newspaper coverage of causes of death. *Journalism Quarterly*, **56** (4), 837–843, 849.

Feaster, J. G. (1968). Measurements and determinants of innovativeness among primitive agriculturalists. *Rural Sociology*, **33**, 339–348.

Harris, L., and Associates (1975). *A Survey of Public and Leadership Attitudes Toward Nuclear Power Development in the United States*. Conducted for Ebasco Services, Inc., New York.

Harris, L., and Associates (1978). *Environmental Problems Worry Public*. Press release of July 20, 1978.

Harris, L., and Associates (1979). *Survey of Public Opinion on Environmental Issues Conducted for the US Soil Conservation Service, October 19–November 21, 1979*. Press release of January 17, 1980.

Harrison, M. R. (1982). The media and public perceptions of climatic change. *Bulletin of the American Meteorological Society*, **63** (7), 730–738.

Hewitt, K. (Ed.) (1983). *Interpretations of Calamity: From the Viewpoint of Human Ecology*. Edward Arnold, London.

Islam, M. A. (1974). Tropical cyclones: Coastal Bangladesh. In White, G. F. (Ed.) *Natural Hazards: Local, National, Global*, pp. 19–25. Oxford University Press, New York.

Kahneman, D., Slovic, P., and Tversky, A. (Eds.) (1982). *Judgement Under Uncertainty: Heuristics and Biases*. Cambridge University Press, New York.

Kahneman, D., and Tversky, A. (1979). Prospect theory: An analysis of decision under risk. *Econometrica*, **47** (2), 263–292.

Kirkby, A. V. T. (1973). The use of land and water resources in the past and present valley of Oaxaca, Mexico. *Memoirs of the Museum of Anthropology*, No. 5. University of Michigan, Ann Arbor, Michigan: 168 pages.

Lawson, M. P. (1974). *The Climate of the Great American Desert*. University of Nebraska Press, Lincoln, Nebraska.

Marsh and McLennan Companies (1980). *Risk in a Complex Society: A Marsh and McLennan Public Opinion Survey*. Conducted by L. Harris and Associates for Marsh and McLennan, New York.

Mitchell, R. C. (1980). *Public Opinion on Environmental Issues: Results of a National Opinion Survey*. Council on Environmental Quality, Washington, DC.

Murphy, A. H., and Brown, B. G. (1983). Forecast terminology: Composition and interpretation of public weather forecasts. *Bulletin of the American Meteorological Society*, **64**, 13–22.

Nisbett, R., and Ross, L. (1980). *Human Inference: Strategies and Shortcomings in Social Judgement*. Prentice-Hall, Englewood Cliffs, New Jersey.

Oliver, J. E., *et al.* (1975). Recollection of past weather by the elderly in Terre Haute, Indiana. *Weatherwise* (August), 161–171.

Riebsame, W. E. (1983). News media coverage of seasonal forecasts: The case of winter 1982–83. *Bulletin of the American Meteorological Society*, **64**, 1351–1356.

Ross, L., and Anderson, C. (1982). Shortcomings in the attribution process: On the origins and maintenance of erroneous social assessments. In Kahneman, D., Tversky, A., and Slovic, P. (Eds.) *Judgement Under Uncertainty: Heuristics and Biases*. Cambridge University Press, New York.

Slovic, P., Fischhoff, B., and Lichtenstein, S. (1978). Accident probabilities and seat belt usage: A psychological perspective. *Accident Analysis and Prevention*, **10**, 281–285.

Slovic, P., Fischhoff, B., and Lichtenstein, S. (1979). Rating the risks. *Environment*, **21** (3), 14–20, 36–39.

Slovic, P., Fischhoff, B., and Lichtenstein, S. (1982). Facts versus fears: Understanding perceived risk. In Kahneman, D., Slovic, P., and Tversky, A. (Eds) *Judgement Under Uncertainty: Heuristics and Biases*. Cambridge University Press, New York.

Slovic, P., Kunreuther, H., and White, G. F. (1974). Decision processes, rationality and adjustment to natural hazards. In White, G. F. (Ed.) *Natural Hazards: Local, National, Global*, pp. 187–205. Oxford University Press, New York.

Starr, C. (1972). Benefit-cost studies in sociotechnical systems. In *Perspectives on Benefit-Risk Decision Making*, pp. 17–42. National Academy of Engineering, Washington, DC.

Tversky, A., and Kahneman, D. (1974). Judgement under uncertainty: Heuristics and biases. *Science*, **185**, 1124–1131.

US National Science Board (1981). *Science Indicators 1980*. US National Science Foundation, Washington, DC.

Warrick, R. A., and Bowden, M. J. (1981). The changing impacts of droughts in the Great Plains. In Lawson, M. P., and Baker, M. E. (Eds.) *The Great Plains:*

Perspectives and Prospects. Center for Great Plains Studies, University of Nebraska, Lincoln, Nebraska.

Whyte, A. V. T. (1976). The role of information and communication in the regulation of emissions from a heavy smelter: The case of Avonmouth. *Proceedings of the International Conference on Heavy Metals in the Environment*, pp. 111–127. Toronto, Canada.

Whyte, A. V. T. (1977). *Guidelines for Field Studies in Environmental Perception*. Map Technical Note 5, UNESCO, Paris.

Whyte, A. V. T. (1982). *Perception Studies as a Planning Tool in UNESCO/UNFPA*. ISER Report to the Government of Barbados, January.

Whyte, A. V. T. (1983). Probabilities, consequences and values in the perception of risk. In *Risk: Proceedings of a Symposium on the Assessment and Perception of Risk to Human Health in Canada*, pp. 121–134. The Royal Society of Canada, Ottawa.

Whyte, A. V. T. (1984). Integration of natural and social sciences in environmental research: A case study of the MAB programme. In Di Castri, F., Baker, M., and Hadley, M. (Eds.) *Ecology in Practice*, Part II, pp. 298–323. UNESCO, Paris.

Whyte, A. V. T., and Burton, I. (1982). Perception of risks in Canada. In Burton, I., Fowle, C. D., and McCullough, R. S. (Eds.) *Living with Risk: Environmental Risk Management in Canada*, pp. 39–69. Environmental Monograph No. 3, Institute for Environmental Studies, University of Toronto, Canada.

Whyte, A. V. T., and Harrison, M. R. (1981) *Public Perception of Weather and Climatic Change: Report on a Pilot Study in Ontario, Canada*. Atmospheric Environment Service, Ottawa, Canada.

Climate Impact Assessment
Edited by R. W. Kates, J. H. Ausubel and M. Berberian

CHAPTER 17
Adjustment in Self-provisioning Societies

N. S. JODHA* AND A. C. MASCARENHAS†

* *International Crops Research Institute for the Semi-Arid Tropics*
Andhra Pradesh, India
† *Institute for Resource Assessment*
University of Dar es Salaam
Dar es Salaam, Tanzania

17.1 INTRODUCTION

The impact of climate variation on society and appropriate methods for its assessment will vary according to the society that is being studied. In this chapter we focus on self-provisioning societies, bringing to bear evidence from India and Tanzania. We focus on the variability of climate from one year to the next and over short groups of years, as these variations create pressing problems for self-provisioning societies. We also emphasize drought, which surpasses other climate variations in impact on these societies. For a discussion relevant to the effects of gradual long-term climate change on self-provisioning societies, see Parry (Chapter 14), who offers historical examples from Europe and North America. The question of impacts of seasonality is addressed here in part; for a more extensive treatment, see Chambers (1982). It is also informative to contrast the issues and approaches in this chapter with those in Pilgrim (Chapter 13), who describes social impact assessments undertaken in developed countries.

The two procedural and primary steps in discussing the adjustment mechanisms of self-provisioning societies to climate variability are: to define or identify such societies, and to understand their perception of the phenomenon of climate variability. In place of attempting rigid definitions of the terms, we prefer description of the situations as they obtain in the real-world context.

Literally speaking, a self-provisioning society is one in which its members manage their production and consumption requirements by themselves and the market, or formal exchange transactions, has little place in the system. Such societies, however, are hard to find in the present age except in completely isolated remote habitats. A more meaningful definition of the term would include farming communities where the bulk of production inputs originate from a person's own farm and household and the bulk of output not only is consumed by the household, but also satisfies most of its consumption needs. Market, or formal, exchange plays a very limited role as a link between the farm household's production and consumption activities. Even when dependence on the market is significant (as in the case of small-holder producers of certain cash crops such as cotton), the objective in using the market is largely to support subsistence. Methodologically, using the ratios of: 1. home-supplied inputs to the total inputs used on farms, 2. self-consumed output to the total output of the farm, and 3. the farm's own supplies to the total consumption requirements, one can not only segregate self-provisioning farming communities from highly commercialized farming communities, but also easily rank the communities on the basis of their degree of self-provisioning or subsistence character.

Several field studies, in largely rainfed farming areas of tropical India, have noted the extent of self-provisioning. According to these studies, important own-farm-originated inputs such as human labor, bullock labor, seed, manures, and fodder for draft animals account for 65–90 percent of the total used amount of concerned input (Bharadwaj, 1974). A similar range applies to the share of total consumption items originating from 'own farm'. The extent of self-provisioning is even higher in most parts of Africa (Ruthenberg, 1968, 1976; Collinson, 1972; Lagemann, 1977; Abalu and D'Silva, 1980).

The literature on subsistence or peasant agriculture has discussed at length the features of such communities (see, for example, Krishna, 1969; Mellor, 1969; Wharton, 1969). Two features of such communities which have significant bearing on their adjustment to climate variability are discussed below.

First, to the extent that the household is both a major supplier of production inputs and major final user of the bulk of the output, the production and consumption decisions are quite interlinked. The integration of household (as a family unit) and farm (as a production unit or a firm) helps offer greater internal flexibility for sustaining the impact of climate variability.

Second, lesser dependence of farm households on the market implies their lesser integration with the rest of the economy. This in turn reduces the capacity of farm households to transmit shocks of climatic variability to others, for example, input suppliers and output buyers. (This situation contrasts with that of commercial firms during a crisis period.) Consequently, unless helped by external agencies or public relief, farmers in self-provisioning societies have to bear the weather-induced risk on their own. Further, since their dependence on the market for the purchase of inputs and disposal of products is limited, climate-induced production uncertainties play a more important role than price and technology-related uncertainties in shaping their adjustment strategies (Wharton, 1968).

17.1.1 Climate Variability

The rationale and operational efficacy of farmers' adjustment strategies to climate variability can be appreciated better once one has some idea of farmers' own perceptions of the phenomenon. Rainfall—its amount, timing, and duration—is identified by subsistence farmers as the dominating climate variable. Areas of subsistence agriculture, where rains constitute a principal source of risk, generally are characterized by high interyear and intrayear variability of rains. When rains are normal or higher than normal they seldom get special attention. But rains lower than normal or their unfavorable distribution are considered a cause for concern. Further, the role of rainfall variability is perceived in a rather short-term intrayear or interyear context and defensive measures are adopted accordingly. The varieties of measures adopted in order to meet the short-term situation, however, constitute integral parts of the overall farming systems, which in turn have evolved over generations in response to the long-term behavior of climate variables (especially rainfall) in a given geographical region. Hence, the subsistence farmer's adjustment mechanisms to weather-induced risk can be better understood in terms of the relevant features of his farming system that help accommodate the periodic shocks generated by short-term fluctuations in weather conditions.

To further facilitate the understanding and identification of areas for improving the potential efficiency of adjustment mechanisms, features of farming systems can be grouped in two categories. The first category can be called *adaptations* and includes elements through which farming systems have accommodated to long-term agroclimatic features of the regions. These elements help in harnessing favorable opportunities offered by the environment and also inject preparedness to defend against unfavorable situations created by erratic patterns of rains. The second category includes responses to short-term fluctuations in weather conditions. They are adopted once intraseason weather conditions become unfavorable. We may call them *adjustments*. Adjustments become possible because of the first category of features.

17.1.2 Method of Study

Before we discuss the adaptations and adjustments facilitated by farming systems, a brief digression on methodology to study them in the context of self-provisioning societies may be helpful. In a way, the risk management attributes of a given farming system are largely an outcome of farmers' perceptions of climate-induced risk and efficacy of possible alternatives to handle the risk. Farmers' perceptions, in turn, are largely conditioned by the objective circumstances which generate risk, for example, the pattern of rainfall. Hence, in order to gain understanding of adaptations and adjustments to climate-induced risk, the study-frame should include contrasting situations in terms of rainfall pattern. Climatological data, particularly the extent and distribution of rainfall along with broad information on agricultural activity in the region, can help in the selection of relevant locations for the study (Mallik and Govindaswamy, 1962–63; Sen, 1971; Jodha *et al.*, 1977). Farm surveys of different intensities may be conducted in the selected locations. Data-gathering in self-provisioning or subsistence farming communities requires caution and emphasis on participant observation, as there is likely to be a communication gap between investigators, often urban-trained, and respondents, who are generally illiterate and suspicious. Simple, unstructured questions, supplemented by group discussions, can provide more insight into the rationale behind the components that characterize traditional farming systems (Collinson, 1972; Norman, 1973; Friedrich, 1974; Kearl, 1976; Binswanger and Jodha, 1978).

The information collected should cover farmers' resource bases and their use patterns, types of crop combinations and their time-and-space specific management practices, as well as input–output details, farm production and disposal, and the like. The climate-induced differences between the sets of information relating to areas, years, and seasons with different rainfall patterns can clearly reveal the risk management elements in the farming system. This is illustrated by three studies briefly reported in Tables 17.1, 17.2 and 17.3, contrasting farmer behavior by climate, season and extreme events.

Table 17.1 Diversification strategies to handle climate risk in two areas in semi-arid tropical India

	Akola villages	Sholapur villages
A. *Characteristics of climate risk*		
Annual average rainfall (mm)	820	690
Probability of favorable soil moisture conditions for rainy season cropping	0.66	0.33
B. *Indicators of spatial diversification*		
Number of scattered land fragments per farm	2.8	5.8
Number of split plots per farm	5.0	11.2
Number of fragments per farm by distance from village		
– Zero distance	0.2	0.0
– Up to 0.5 miles (0.8 km)	0.3	1.4
– Up to 1.0 mile (1.6km)	1.1	3.4
– 1 to 2 miles and above (1.6–3.2 km)	0.1	1.0
C. *Indicators of crop-based diversification*		
Number of total sole crops planted	20	34
Number of total combinations of mixed crops planted	43	56
D. *Crop/stock-based mixed farming*		
Crop income/livestock income ratio	94:6	89:11

Source: ICRISAT's village level studies (Jodha *et al.*, 1977); Binswanger *et al.* (1980).

Table adapted from Walker and Jodha (1982).

Table 17.1 contrasts the extent of risk management practices in two areas of India with vastly different amounts of rainfall and probabilities of soil moisture to help the germination of crops. In Sholapur, the more risky area, the farmers resort to more resource-based and crop-based diversification as an insurance mechanism against climate-induced risk.

Table 17.2 contrasts the farming practices followed by a similar group of farmers in the Kilosa area of Tanzania during short (uncertain) and long (certain) rains in the same year. The practices and measures which have greater probability of success with uncertain rainfall, or which can offer partial crop salvage values despite unfavorable rains, are adopted more during the short rains.

Table 17.3 contrasts the measures and farming practices followed by farmers in the arid zone of India during a normal rainfall year and a drought year. The practices having greater potential for protecting the crops, saving the resources and augmenting the supplies (even of inferior products) despite the failure of rains gain significance during the drought year.

Table 17.2 Risk-minimizing farming practices and rain type in four villages of Kilosa, Tanzania during 1980–81

	Short rains	Long rains	Total
A. *Characteristics of climate risk*			
Average rainfall (mm)	260	763	—
No. of rainy days	21	68	—
Chances of crop failure in 10 years	5	1	
B. *Indicators of risk-minimizing strategies*	%	%	
Share of total low lying areas planted in the year	83	17	100
Share of uplands planted	26	74	100
Share of compound plot areas planted	92	8	100
Share of total salvage crops in total crops of season	72	32	—
Share of intercropping in season	95	79	—
Share of staggered planted area in the season	35	69	—

Source: Jodha (1982). Table adapted from Walker and Jodha (1982).

As these illustrations indicate, the main focus of such studies is to capture the contrast among farming practices as dictated by temporal and spatial differences in rainfall patterns. Depending upon requirements, the investigations can be extended to further depths, as will be indicated by subsequent tables.

17.2 RELEVANT FEATURES OF FARMING SYSTEMS

The features of traditional farming systems that have evolved to handle climate-induced risk can be defined as (a) adaptations, and (b) adjustments, classified by their long- or short-term character. These features are interrelated and constitute a complex of crop-based, resource-based and management practice-based measures. Some of them are group-centered, requiring social action, while others are individual-centered, in the control of the farm unit. In some of them traditional technology plays an important role and in others the role of institutional factors is more significant. The actual adoption of a measure or combination of measures is largely a function of farmers' perceptions of a risky situation and the efficacy of a particular measure to meet the situation. Since the ultimate objective of these measures is to cope with a common factor, risk generated by weather or climate, one comes across a broad similarity in adaptation/adjustment measures in different locations such as India and Africa, despite their cultural, infrastructural and demographic differences.

Table 17.3 Loss-minimizing activities during a drought year and a non-drought year in selected villages in the arid zone of India

	Drought year (1963–64)	Normal year (1964–65)
A. *Characteristics of weather risk*		
Rainfall during the year	159 mm	377 mm
Total rainy days	8 days	21 days
B. *Risk/loss-minimizing measures: crop practices*		
Collected weeded material as fodder	53 plots	5 plots
Harvested field borders for fodder	68 plots	6 plots
Harvested premature crops	27 plots	—
Harvested crop byproduct only	49 plots	2 plots
Harvested mature crop	16 plots	144 plots
Interculturing done	7 plots	65 plots
Weeding done more than once	18 plots	—
Thinning done	37 plots	—
Post-sowing operations abandoned	36 plots	—
Hired resource used for post-sowing operations	2 plots	24 plots
Harvested premature *Z. nummulariia* (bush) for fodder	92 plots	—
Lopped trees for fodder/fuel	53 plots	4 plots
C. *Risk/loss minimizing measures: social practices*		
Cases of nonpayment of dues	49	7
Marriages, etc. postponed	9	—
Children withdrawn from school	34 plots	3

After Jodha (1967).

17.2.1 Long-term Adaptations

The evolution of farming systems in climatically unstable areas has bestowed several features which ensure the flexibility and viability of the system in the face of climatic hazards.

17.2.1.1 Diversified Production Strategy

The farming activities are diversified to accommodate the temporal and spatial variability characterizing the natural resource base (land, rainfall, etc.) conditioning the overall production possibilities available to the farmers. The degree of diversification can readily be perceived from the farmers' choice of enterprise combinations (such as mixed farming through cropping and stock farming) with varied capacities to ensure earning in good and bad rain years, and from the choice of crops with varying attributes in terms of maturity period, drought tolerance, input requirements, main product–by–product ratios, end uses of the product, and so forth (Collinson, 1972; Ruthenberg, 1976; Abalu and D'Silva, 1980; Jodha, 1980). Table 17.1 illustrates the relative extent of resource-based and crop-based diversification attempted by farmers in two agroclimatically different areas of semi-arid tropical India.

17.2.1.2 Operational Diversification

Diversification in farming does not end with resource- and crop-based diversification. Traditional agronomic and other management practices also have a significant scope for diversification and flexibility. These practices include lowland-upland (toposequential) planting, staggering of planting and other operations, splitting of plots, splitting of inputs, and skipping certain inputs as warranted by the situation (Jodha, 1967; Ruthenberg, 1968; Collinson, 1972). Table 17.2 illustrates some degrees of diversification attempted by Tanzanian farmers during short and long rains.

Diversification based on resource bases, crops and operations helps generate operations with varying probabilities of success in the face of highly variable weather conditions. The farmers' concentration on specific practices changes according to their comparative advantages in the emerging intraseason weather situation. In the favorable season the options with high payoffs get better attention, whereas in less favorable seasons the options with greater insurance elements are emphasized. This is illustrated in Table 17.2, which shows the priority given to high insurance measures during short (uncertain) rains in the Kilosa area of Tanzania, and Table 17.3 for a drought period in India.

17.2.1.3 Flexible Resource Use Patterns

The degree of diversification and consequent flexibility of the farming system is further strengthened by the diversity and flexibility of resource and consumption patterns. This flexibility is facilitated in turn by the fact that the household is both a production and a consumption unit. Household production and consumption in self-provisioning farming societies are therefore highly interlinked physically, as well as financially. Since the household is a major supplier of production input (human and bullock labor, seed, feed, fodder, manure, etc.) it offers effective control over resource use to contract or expand the farm operations (or their intensity) as required by quick response to emerging weather situations during the season (Jodha, 1967; Collinson, 1972, 1977). A variety of recycling devices, including a limited prior commitment of resources for current production and an accretionary process of asset or capital formation, further help to inject flexibility in resource use (Jodha, 1967).

17.2.1.4 Flexible Consumption Patterns

Similarly, since the household is a major direct consumer of its own farm output (except some cash crops), the fluctuations in production largely get absorbed internally. Highly flexible demand and consumer preference (for example, preparedness to consume damaged grains or even green cobs/pods in place of fully ripe grain, consume normally non-edible stuffs, or drastically cut food intake during poor crop years) helps match the demand situation to the emerging supply situation. The flexibility on the consumption front is further strengthened by on-farm storage and a variety of recycling and food processing devices (which may often convert non-edibles into edibles) (see Jodha, 1967, 1975; Collinson, 1972)

17.2.1.5 Adapting the Environment

The above discussion shows that farmers operating under an unstable environment try several ways of adapting their production and consumption activities to the variability of climate. They also know that greater stability of their farming system could be achieved by some means of adapting the environment to their requirements. Since erratic rainfall is the key variable to determining instability or risk to their farming, any means to manipulate rainfall or other effective moisture to their crops is considered as a permanent or more reliable source of stability. This leads to attempts to place irrigation facilities in at least part of the land. In some drought-prone areas of India, wells or tanks (based on storage of surface runoff) are used as sources of irrigation by a limited number of farmers. In yet other areas, both in India and Tanzania, moisture availability, depending on soil characteristics and topography, is manipulated by means of conservation measures such as contour bunding, field border bunding, ridges and furrows, and the like (Ruthenberg, 1968; Jodha, 1980, 1982).

17.2.1.6 Traditional Forms of Rural Cooperation

Traditional forms of rural cooperation and informal institutional arrangements also have the capability for mutual sharing of risk during bad years and helping to fully harness the potential of bumper crop years (Jodha, 1967; Kirkby, 1974; Wisner and Mbithi, 1974; Hitchcock, 1979). However, under the pressure of modernization and commercialization and institutional interventions by governments, these traditional collective means to facilitate flexibility to the farming system are fast disappearing (Jodha, 1978; Walker and Jodha, 1982).

17.2.2 Short-term Adjustments

The features of farming systems which take the form of responses to short-term climate-induced crises (such as midseason failure of rains) are called adjustments. Adjustment measures, unlike adaptations, are initiated once unfavorable weather performance is known. For example, once midseason failure of rains is certain farmers can initiate two types of action. The first category is directed towards minimizing the losses due to unfavorable weather; we call it specific risk/loss minimizing measures; Berry *et al.* (1972) describe it as measures to modify the loss potential. The second category includes all steps undertaken to manage the losses or adjust to the losses. We designate them specific risk/loss management measures. In Table 17.4 we relate the specific adjustment measures to the adaptive features of farming systems described in Sections 17.2.1.1 through 17.2.1.6. The characteristics of the short-term adjustment measures grouped together under two categories are elaborated in Sections 17.2.2.1 and 17.2.2.2.

17.2.2.1 Risk/Loss Minimizing Measures

Following the intraseason failure of rains, certain measures are adopted for extracting whatever little the adversely affected crops can offer at a minimum of additional input cost. The measures can be further grouped under the following categories.

1. *Salvage operations.* Several recovery efforts, depending on the situation, are made. Examples are: recovery of fodder (byproduct) in the face of the definite impossibility of getting the main product; harvesting green cobs/pods in place of a ripe crop; collection of weeded material (as fodder) rather than allowing it to go to waste; concentration on normally low-value production activities such as harvesting field borders for fodder (for details, see Jodha, 1967 and Table 3: Jodha, 1982).
2. *Midseason corrections/adjustments in operations.* Depending on which crop, plot or operation has higher chances of success in the face of unfavorable rainfall, selectivity and discrimination become important features of the decisions regarding deployment of resources, intensity of operations, etc. for different crops and/or plots. For instance, in the face of a midseason dry spell, plots lying lowest in the toposequence get more attention; intensive weeding and emergency thinning is done in the case of drought-resistant and still promising-looking crops; and depending on the moisture situation, especially after the break of the dry spell, partial resowing and patch cultivation is done (see Jodha, 1967, 1982; Berry *et al.*, 1972; Wisner and Mbithi, 1974; and Tables 17.2 and 17.3 above).

3. *Cutback on resource use.* Cost saving is attempted by reducing dependence on hired resources. Owned resources are used where usually hired resources are employed. Family resources also are spared for alternative earning opportunities outside the family farm. Operations, techniques and priorities are changed for the maximum saving of resources (Jodha, 1967, 1982; Berry *et al.*, 1972; Wisner and Mbithi, 1974).

Table 17.3 summarizes the details of some farm operations which become important only during unfavorable rain years. These operations in their respective ways help the risk/loss minimization measures adopted by the farmer.

17.2.2.2 Risk/Loss Management Measures

Under this category measures are directed towards ensuring the survival and maintenance of the productive capacity of the farm household in the face of a crisis situation caused by the failure of the crop. These measures have been put into five subgroups and are illustrated by detailed data from various drought-affected areas in India. A broadly comparable situation has been observed in the very dry villages of Kilosa, bordering Dodoma in the arid region of Tanzania (Jodha, 1982), but comparable quantitative details could not be collected. Mascarenhas (1973) provides a detailed discussion of relevant issues and problems in the context of Tanzania. Broad similarity in farmers' approaches to meet the consequences of droughts in different countries can be seen from various studies on the subject (Dupree and Roder, 1974; Hankins, 1974; Heijnen and Kates, 1974; Kirkby, 1974; Wisner and Mbithi, 1974).

1. *Reduction in current commitments.* This is attempted through postponement, cancellation, or reduction of expenditures related to current consumption, future production, payment of dues, and so forth. Table 17.5 (adapted from Jodha, 1981) summarizes the situation in drought and post-drought years in three areas of western India. It reveals that consumption expenditures of sample farmers during drought years (compared to non-drought years) declined by 8–13 percent in the affected areas of the states of Gujarat and Rajasthan. The magnitude of decline varied significantly among the different expenditure categories. For instance, decline in the expenditure for total food items was the smallest of all the categories. To prevent further decline in this category, however, expenses on other 'non-essential' consumption items like protective food (including milk, meat, vegetables, sugar, fruits, etc.), education, medicine, clothing and socioreligious ceremonies were curtailed drastically. But despite maintenance of the level of expenses for food in drought years near to those of non-drought years, the per capita food intake (due to high prices) declined by 12–23 percent in different areas. For similar observations in

Table 17.4 Adaptive features of farming systems and farmers' adjustment mechanisms to climate-induced risk

Adaptive features of farming systems	Long-term flexibility or reliability	Short-term adjustments through:									Features observed in	
		(A) risk/loss minimization				(B) risk/loss management						
		Salvage operations	Midseason adjustment	Cut back on hired resources	Change in techniques	Cut in current commitments (consumption etc.)	Resource augmentation (conservation/recycling)	Supplementary earning (migration)	Inventory depletion (assets)	Dependence on others	India	Africa
1. *Diversified production strategy* (through)												
Crop/stock mixed farming	x	x	x				x	x	x	x	x	x
Mixed cropping	x	x	x								x	x
Combining crops of varying maturity, drought tolerance, input needs and end uses	x	x	x								x	x
2. *Operational diversification*												
Toposequential planting		x	x								P	x
Staggered planting			x	x							x	x
Plot scattering/ splitting		x	x								x	x
Varied plant spacing			x	x	x						x	x
Input use skipping/			x	x	x						x	x

3. *Flexible resource use patterns*												
High dependence on own resources	x		x	x	x			x			x	x
Limited ex-ante commitment to a current production			x	x	x						x	x
Accretionery process asset buildup	x			x		x	x		x		x	x
Recycling the resources	x		x		x			x	x		x	x
4. *Flexible consumption patterns*												
Close link between consumption and production	x	x	x			x	x				x	x
Recycling the products	x					x	x				x	
On-farm storage	x					x			x		x	x
Flexible long-term commitment	x					x		x			x	
5. *Adapting the environment*												
Irrigation	x	x	x								x	x
Moisture conservation	x	x	x								x	P
Perennial crops	x		x					x			P	x
6. *Traditional forms of rural cooperation*	x							x		x	x	x

P = partial.

For discussion and evidence see: Jodha (1967, 1978, 1981) for India; Berry *et al.* (1972), Hankins (1974), Heijnen and Kates (1974) and Jodha (1982) for Africa (Tanzania).

Table 17.5 Changes in expenditures and food grain consumption in three drought-prone areas of India

Items	Jodhpur (Rajasthan)			Barmer (Rajasthan)			Banaskantha (Gujarat)		
	63–64†[a] A	64–65* B	(B–A)/B	69–70† A	70–71* B	(B–A)/B	69–70† A	70–71* B	(B–A)/B
	(Rs)	(Rs)	(%)	(Rs)	(Rs)	(%)	(Rs)	(Rs)	(%)
Per-household consumption expenditure on:[b]									
Total food items	1181	1200	−1.6	1183	1153	+2.6	1701	1805	−5.8
Protective foods[c]	291	409	−28.8	235	406	−42.1	501	694	−28.3
Clothing, fuel, etc.	274	327	−16.2	269	316	−14.8	334	483	−30.9
Socioreligious ceremonies[d]	54	148	−63.5	57	110	−48.2	61	88	−30.7
Others[e]	168	259	−35.2	102	175	−41.7	127	98	+29.6
Total	1677	1934	−13.3	1611	1754	−8.2	2223	2474	−10.2
Consumption per adult unit per day(g)									
Total foodgrains	514	594	−13.9	535	606	−11.7	567	740	−23.4
Superior cereals[f]	112	58	+93.1	40	7	+82.5	42	27	+35.7

Source of data: Jodhpur area: Jodha (1975); Barmer and Banaskantha areas: Chaudhuri and Bapat (1975); Sholapur area: data collected under ICRISAT Village Level Studies (Jodha *et al.*, 1977); Aurangabad area: Borkar and Nadkarni (1975). For details of number of sample households see Table 17.7.

[a] †Drought years; *Post-drought (normal) years.

[b] To facilitate comparisons, all rupee values in this and the subsequent tables have been converted into 1972–73 value of the rupee, using index of general prices for agricultural laborers in the respective states.

[c] Includes milk, fats, sugar, jaggery, fruits, etc. These are included in total food items, also.

[d] Socioreligious ceremonies related to deaths, births, marriages and festivals.

[e] Includes education, medicine, recreation, travel, payment to village functionaries for day-to-day services, remittances to children studying outside, etc.

[f] Mainly wheat available through fair-price shops during the drought year.

Adapted from Jodha (1981).

drought-prone areas of Kenya, see Wisner and Mbithi (1974). See Escudero (Chapter 10) for a discussion of the relation of climate variability to nutrition.

2. *Resource augmentation*. This is attempted through the use of hitherto rejected or non-edible produce and the conservation and recycling of food/fodder, using different processing techniques (Jodha, 1967; Berry *et al.*, 1972; Hitchcock, 1979).
3. *Asset/inventory depletion*. During the crisis period it is quite usual to sell or mortgage assets or inventories accumulated over the run of good crop years. The main reason for asset depletion through distress sales is for augmentation of liquid resources to supplement meager income during drought years. Apart from deliberate disposal, asset losses are due also to deaths of animals and to theft, quite common during stress periods. Compared to the respective pre-drought years, assets declined by 15–42 percent in different areas during the drought years, as revealed by Table 17.6. In most cases the productive assets, particularly livestock, had the highest (19–60 percent) decline. Moreover, the recovery of depleted assets in post-drought years was not quick enough. By the time asset losses are fully recouped the next drought may occur. Thus over an irregularly occurring famine cycle the asset depletion–replenishment cycle completes itself without leaving surplus resources for agricultural investment and growth in drought-prone areas (Binswanger, 1978). Besides asset depletion, the drought-affected farm households resort to heavy borrowing through formal and informal land and labor debts during the crisis period. As indicated by Table 17.6, in these areas the incidence of indebtedness increased from 63 to 224 percent within a single drought year. The long-term consequences of such indebtedness include permanent pauperization of the people (for evidence see Jodha, 1981, Table 8).
4. *Other measures for sustenance income*. Other loss management devices during drought years include dependence on public relief works, hiring out of human labor and bullocks, earning during outmigration, remittances from well-off relatives, sale of handicrafts, and various means of mutual risk-sharing.

   Table 17.7 presents the relative contribution of different sources of income towards the sustenance of farmers during a drought year. Public relief works account for the single biggest source of sustenance income in most of the areas. The sale of assets is the next major single source of sustenance income. The data suggest that in the absence of public relief, the farmers' adjustment devices for sustenance are quite weak.
5. *Outmigration*. This is an important measure to adjust to the spatial variability of rainfall. Farmers, with or without animals, travel long distances during stress periods. Jodha (1978) reported that about 37–60 percent of farm households were affected by outmigration during drought years in

Table 17.6 Changes in assets and liabilities in five drought-prone areas of India

	Average per household value of assets and liabilities[a] in (A) pre-drought years[b] (B) drought years and (C) post-drought years in:														
	Jodhpur (Rajasthan)			Barmer (Rajasthan)			Banaskantha (Gujarat)			Sholapur (Maharashtra)			Aurangabad (Maharashtra)	
	(A)	(B)	(C)	(A)	(B)	(C)	(A)	(B)	(C)	(A)	(B)	(C)	(A)	(B)
	62–63[b]	63–64	64–65	68–69	69–70	70–71	68–69	69–70	70–71	71–72	72–73	73–74	71–72	72–73
	(Rs)	(Rs)	(Rs)	(Rs)	(Rs)	(Rs)	(Rs)	(Rs)	(Rs)	(Rs)	(Rs)	(Rs)	(Rs)	(Rs)
Assets														
Livestock[c]	1546	849	1230	1287	786	837	1565	1222	1498	2096	1707	1549	732	464
(% change)[d]		−(45.1)	+(44.9)		−(60.4)	−(6.5)		−(21.2)	+(22.6)		−(18.6)	−(9.3)		−(36.3)
Agrl. implements[e]	409	372	389	202	201	201	645	638	635	496	465	483	751	685
(% change)		−(9.0)	+(4.6)		−(0.5)	—		−(1.1)	−(0.5)		−(6.3)	+(3.9)		−(10.1)
Consumer durables[f]	658	459	486	175	167	164	292	284	284	106	73	85		NA
(% change)		−(30.2)	+(5.9)		−(4.6)	−(1.8)		−(2.7)	—		−(31.1)	+(16.4)		NA
Financial assets[g]	1239	840	726	1226	947	921	1668	1398	1380	310	216	190	296	258
(% change)		−(32.2)	−(13.6)		−(22.7)	−(2.7)		−(16.2)	−(1.3)		−(30.3)	−(12.0)		−(12.8)
Total assets	3852	2520	2831	3590	2101	2123	4170	3542	3797	3008	2461	2307	1779	1407
(% change)		−(34.6)	+(12.3)		−(41.5)	+(1.1)		−(15.0)	+(7.0)		−(18.2)	−(6.3)		−(20.9)
Indebtedness[h]														
Debts outstanding	189	552	637	498	873	949	111	360	302	375	613	651	NA	NA
(% change)		+(192.0)	+(15.4)		+(75.3)	+(8.7)		+(224.3)	−(16.1)		+(63.5)	+(6.2)		

Source of data: See Table 17.5.

[a] Value of assets and liabilities (in Rs) expressed in terms of 1972–73 prices. Assets exclude land and buildings.
[b] Pre-drought year indicates the situation at the beginning of the drought year.
[c] Draft animals, milch stock, sheep, goats, etc.
[d] Percentage change over the preceding period. The change is composed of sales, gifts, losses (of animals due to death, etc.).
[e] Farm equipment, tools, machinery, and handicraft tools.
[f] Consumer durables—only important items like radios, watches, bicycles and modern furniture included.
[g] Includes jewelry, cooperative shares, L.I.C. policies, etc. In Aurangabad, only jewelry is included.
[h] Average per household amount of debts outstanding net of repayments. In Sholapur it excludes old debts imposed on farmers as bunding loans, well loans disbursed under Zaveri Scheme, and Zilla Parishad fodder grants during the past droughts. This amount comes to Rs 723/- per household at 1972–73 prices.

NA: Not available.

Table 17.7 Sources of sustenance income in five drought-prone areas of India

Details	Jodhpur (Rajasthan)	Barmer (Rajasthan)	Banaskantha (Gujarat)	Sholapur (Maharashtra)	Aurangabad (Maharashtra)
Drought year	1963–64	1969–70	1969–70	1972–73	1972–73
Sample households (No.)	144	100	100	80	128
Average amount of sustenance income (Rs/household)[a]	3133	2996	2627	2944	2715
% share of sources in sustenance income					
Cultivation[b]	2.1	—	—	14.4	6.8
Animal husbandry	10.2	7.2	4.8	1.0	NA
Wage income from relief works	24.9	22.4	25.3	46.5	56.2
Institutional help[c]	NA	30.4	6.4	NA	NA
Sale of assets	25.9	12.5	24.9	17.3	13.5
Borrowings (credit)[d]	10.4	12.8	11.7	7.9	6.3
Others[e]	26.5	14.7	26.9	12.9	17.2

[a] Sustenance income is defined as total inflow of cash and kind including borrowing, except term loans unrelated to sustenance during the drought. Value of sustenance income is expressed in terms of 1972–73 prices.
[b] In Aurangabad villages, income is from all household production including cultivation.
[c] This includes free or subsidized supplies of food grain and fodder, including those provided by charitable institutions and the government during the period of migration. In some cases the help also included milk powder, vitamin tablets, medicine, clothing, transport facilities, and water supply, etc.
[d] All borrowings—in cash or kind—taken against mortgage or labor or land-lease contract and others. This does not include the credit in terms of postponement or cancellation of recovery of land revenue and other dues from the farmers. This also excludes term loans not related to loss management during the drought years.
[e] Includes income from other casual or agricultural wage employment (including during the outmigration), handicrafts, transport, remittances and free help from well-off relatives, etc. In the case of Jodhpur villages it includes value of old stocks of food grain and fodder.
NA: Not available.

different areas. The one-way distance covered ranged from 50–243 kilometers. Outmigration involves both real and nominal costs. An important component of the cost is loss of animals through death, desertion or theft. The extent of animals lost by outmigrants in different areas ranged between 28 and 53 percent of the original number of animals. The practice of migration is more common among the pastoralists in Africa, but no details are readily available to quantify the situation.

17.3 THE POTENTIAL OPTIONS FOR RISK STRATEGY

Having learned about the farmers' traditional mechanisms to handle climate-induced risk and their strong and weak points, one can proceed to identify some new measures which can potentially strengthen the farmers' methods (Spitz, 1980). The new measures, of course, are not to substitute for the existing mechanisms. Rather they should help generate more options for the farmer to adapt and adjust to the risky environment. The potential options contain technological and institutional measures that share the insurance elements of traditional measures. Their conceivable superiority, however, lies in providing both insurance and increased capacity for the farmers to more easily withstand periodic stress situations.

The new institutional options indicated are also not very new. The focus of institutional measures (i.e., government policies and programs) is on the need for designing them to adapt to the realities of unstable agricultural situations. Most of the current programs and policies need to be more sensitive to the problems created by climate variability before they can complement the farmers' own measures to handle risk (Wisner and Mbithi, 1974; Jodha, 1981).

17.3.1 Potential Technological Options

Potential technological measures are summarized in Tables 17.8, 17.9, and 17.10. Specific practices and their attributes in terms of potential adaptation and adjustment benefits are indicated, as well as the relevance of the measures to farmers' past experience and resource capacity, helpful in facilitating adoption of the techniques. In keeping with the classification of traditional measures, the new technological measures, which can significantly add to the flexibility and productive capacity of the farming system, can be broadly classified under three groups: Table 17.8, resource measures; Table 17.9, crop measures, and Table 17.10, management practice measures. These measures are at different stages of development and availability to the farmer. Moreover, they are of a general nature and specific changes may be necessary to suit local circumstances in different areas.

17.3.1.1 Resource Measures

These include all the measures in dealing with the improvement, management and manipulation of the resource base—particularly the natural resource base—of farming. Variability of rainfall is the principal source of instability of farming in tropical arid and semi-arid areas. Agricultural scientists maintain that the moisture available in most years, if properly utilized, is sufficient for raising one or (in some areas) two rainfed crops. The main problem is intraseason temporal distribution of rains. It is not uncommon to witness severe flooding and extreme moisture stress for crops in different parts of the same crop season. The distribution of the rains cannot be controlled, but its use pattern can be manipulated to increase its effective availability for crop production. This is attempted through storage of water on the soil surface (in tanks, etc.) and in the profile of the soil. This helps generate the following options to adapt the environment (moisturewise) to the crops planted (see Table 17.8).

Table 17.8 Potential technological options: resource measures

Resource measures: relating to conservation, management of soil and moisture	Attributes									
	Long-term adaptation through:		Short-term adjustment through:		Relevance to:					
	Growth-induced cushion/stability	Flexibility-based options	Salvage operations	Adjustment to intra-season weather	Farmers' experience		Farmers' resource capacity		Specific agroclimatic conditions	Reference for experimental evidence
					India	Africa	India	Africa		
1. *Runoff collection and recycling*	x	x			P				R	1,2
2. *Soil/moisture conservation through*										
– contour bunds	x	x		x	P		P		R,Bm	3
– graded bunds	x	x			P				R,B	3
– broad bed and furrows	x	x		x	P				B	1,2,4,5
– broad-based terraces	x	x							Bd,R	3
– mulching	x	x		x	P	P	P	P	R,B	3
– contour cultivation	x	x				P	P	P	Bd	3
– tie-ridging	x	x		x		P		P	Bm,R	6

References: 1. Ryan *et al.*, 1979; 2. Binswanger *et al.*, 1980; 3. Randhawa and Rao, 1981; 4. Ryan and Sarin, 1981; 5. Virmani *et al.*, 1981; 6. Le Mere, 1972.

Abbreviations: P, partial; R, red soils; Bd, deep black soils; D, dependable rainfall; Bm, medium black soils.

1. *Runoff collection and recycling of water*. By means of proper layout of the landscape on a watershed basis, the facilities for drainage of excess water into small tanks can be arranged. The water thus harnessed during not

infrequent heavy storms can be utilized for supplemental or life-saving irrigation during the midseason drought, or for raising post-rainy season crops. The evidence from experimental work at ICRISAT and the national institutes in India indicates that this measure can make a significant contribution towards the stability and growth of rainfed agriculture in Alfisols (red soil) areas (Ryan *et al.*, 1979; Binswanger *et al.*, 1980). However, the measure may face some problems of an institutional nature as it involves soil and water management on a watershed basis, and even a single small watershed involves a number of small farmers who may or may not agree to a collective decision (Doherty and Jodha, 1979).

2. *Soil/moisture conservation measures*. In addition to traditional field border bunding, experimental work on soil/moisture management has developed further options to suit different soil type and rainfall conditions. A few that have shown promise are graded bunds, broadbeds and furrows, broad-based terraces, land smoothing, contour bunds, tie-ridging, contour cultivation, furrows (in grasslands) and mulching (for details see Le Mere, 1972; Ryan *et al.*, 1979; Binswanger *et al.*, 1980; Randhawa and Rao, 1981; Virmani *et al.*, 1981). Some of these measures, when used with other components of modern technology such as improved seed and fertilizer, can raise production substantially. In areas where water stagnation rather than moisture stress operates as a main constraint, the above-mentioned measures help in better drainage to improve crops.

17.3.1.2 Crop Measures

The new crop technologies offer better and more crop options to the farmer. Certain crops can now be developed, improved or adapted to the environment through scientific research. In some cases the alternative crops available mean the substitution of the traditional crops of one region by traditional crops from other regions. The variety of perennial and annual crops recommended for different agroclimatic zones (Kassam 1976; Spratt and Chowdhury, 1978; Anon., 1979; De Vries and Mvena, 1979; Mukuru, 1980) offer choices for crops to suit different weather conditions, for example, early rain, late rain, inadequate or excess rain, midseason drought, and the like. Depending on their various characteristics, the crops may offer possible stability and higher yields (see Table 17.9).

17.3.1.3 Management Practice Measures

Based on agronomic trials involving knowledge of new crops and their physiology in relation to varying types and levels of inputs, scientists have evolved a range of management practices (De Vries, 1976; Monyo *et al.*, 1976; Keregero *et al.*, 1977; Krishnamoorthy *et al.*, 1977; Virmani *et al.*, 1981). Many

Table 17.9 Potential technological options: crop measures

Crop measures: crop choice/substitution based on crop characteristics*	Attributes							
	Long-term adaptation through:		Short-term adjustment through:		Relevance to:			
	Growth-induced cushion/stability	Flexibility-based options	Salvage operations	Adjustment to intra-season weather	Farmers' experience		Farmers' resource capacity	
					India	Africa	India	Africa
1. Insensitivity to temporal variability of rains (e.g. perennials)	x	x			P	P	P	P
2. Resistance to drought	x	x	x		P	P	P	P
3. Varying maturity periods	x	x	x	x	P	P	P	
4. Responsive to fertilizer (+ moisture)	x				P		P	
5. Moisture use efficiency	x			x	P	P	P	
6. Adapted to new agronomic practices	x	x		x			P	P
7. Resistant to pests/insects	x			x	P	P	P	P

* For experimental evidence see Kassam (1976); Collinson (1977); Krishnamoorthy *et al.* (1977); Spratt and Chowdhury (1978); Anon. (1979); Mukuru (1980), Randhawawa and Rao (1981).

P = partial.

of them involve only changes in husbandry practices rather than substantial input costs. The practices relate to operation at various stages of crop seasons and they are designed to effect efficient use of the environment—soil, moisture, and the like (see Table 17.10).

For instance, the practice of dry seeding eliminates the loss of time involved in traditional systems, where crops are planted after the rains when fields are ready. This period may be as long as 10 days or more in many areas. Dry sowing has several favorable implications for plant stand and growth. The variety of crops with different physiological habits has facilitated the manipulation of

Table 17.10 Potential technological options: management practice measures

Measures relating to management practices*	Attributes: Long-term adaptation through: Growth-induced cushion/stability	Long-term adaptation through: Flexibility-based options	Short-term adjustment through: Salvage operations	Short-term adjustment through: Adjustment to intra-season weather	Relevance to: Farmers' experience: India	Relevance to: Farmers' experience: Africa	Relevance to: Farmers' resource capacity: India	Relevance to: Farmers' resource capacity: Africa
1. Dry seeding	x	x		x	P		P	P
2. Flexible sowing time		x	x	x	P	P	P	P
3. Transplanting some crops		x				P	P	P
4. Plant population and manipulation practices		x		x	P	P	P	P
5. Varying level and selective use of fertilizer	x	x			P	P	P	P
6. Intensive weed management	x		x	x	P	P	P	P
7. Midseason thinning, ratooning, gap filling		x	x	x	P	P	P	P
8. Intercropping with HYVs	x	x				P		
9. Sequential/relay cropping	x	x		x	P			P
10. Post-harvest tillage		x				P		

* For experimental evidence, see footnote to Table 17.9.
P = partial.

sowing dates to suit the timings of rainfall. This has obvious flexibility implications.

Practices relating to plant population, spacing, and midseason changes therein also help better adjustment to emerging weather conditions. Similarly, intensive weed management and the selective use of fertilizer also help to bring about high and stable crop production.

Practices such as intercropping and sequential and relay cropping, involving crops with varying capacity to benefit over time and space from the environment, also add to higher and more stable production.

Post-harvest plowing is one practice which helps weed and moisture control and prepares soil well for effective dry seeding. In some areas this is a traditional practice, but it is often done much after the crop has been harvested. By then soil is completely dry, weeds have already matured and scattered their seeds, and animals are very weak, since it is the dry season.

17.3.2 Potential Institutional Options

Most of the institutional measures discussed below are not new. What is new is renewed emphasis on their reorientation to become more relevant to the situations in areas with drought hazard. The potential institutional measures—public policies and programs conducive to increased effectiveness of farmers' mechanisms to handle weather-induced risk—are summarized in Table 17.11. They are subgrouped under three categories.

1. *Contingency support facilities*. These measures are directed to supplement farmers' own efforts to manage the crisis situation generated by drought-induced scarcities. They are largely short-term measures.
2. *Area-based infrastructure*. This includes long-term and permanent measures to facilitate growth of the regions often hit by droughts.
3. *Schemes supporting adoption of new technology*. These measures include the infrastructure and other support facilities essential for adoption of the new technological options discussed in the preceding section.

The long- and short-term consequences of the measures under the aforementioned three categories are also indicated in Table 17.11. For detailed discussion of the potential role of these measures in helping farmers' traditional adjustment mechanisms see Wisner and Mbithi (1974); Dandekar (1976); Mascarenhas (1979); and Jodha (1981).

17.4 CONCLUSION

Subsistence farmers, through trial and error over a period of generations, have evolved various mechanisms to handle drought-induced risk. The strongest component of these risk-handling mechanisms is the diversification and consequent flexibility of the farming systems. Farmers in low and unstable rainfall areas are faced with very limited production alternatives. They try to multiply the total options by manipulating crop combinations and varying methods of resource use and farm practices. In the process they gain stability in production but do sacrifice the more remunerative opportunities occasionally

Table 17.11 Potential institutional options

Institutional measures: programs/policies*	Attributes in terms of support to: Adaptation	Adjustment		Relevance to current experience of the countries	
	Long-term growth and flexibility	Loss minimization	Loss management	India	Africa
1. *Contingency support facilities:*					
Relief (employment) works		x	x	x	x
Consumption credit			x	P	
Fodder banks		x	x	P	
Seedling nurseries		x		P	
Custom-hire services		x		P	P
2. *Area-based infrastructure:*					
Diversified credit	x		x	P	P
Marketing and transport	x	x		P	P
Price support for crop and stock	x			P	
Crop/stock insurance	x	x			
Non-farm employment	x		x	P	P
3. *Support for new technology:*					
Village/farm centered					
Conservation measures	x	x		x	x
Self/community-managed irrigation	x	x		P	
Local level input/distribution	x	x		P	P

* For detailed discussion and/or evidence on these and related measures see Mascarenhas (1973, 1979); Wisner and Mbithi (1974); Jodha (1981).

P = partial.

presented by the rainfall pattern. In other words, farmers' production strategies are geared largely towards handling the negative aspects of climate, such as droughts, rather than concentrating on positive aspects. This is because climate is recognized more as a source of distress than as a positive resource for production activities.

These mechanisms, which show considerable similarity across geographical, cultural and demographic contexts in the tropical underdeveloped world, have lost part of their effectiveness. Group-based measures to handle risk are fast losing their effectiveness due to increased demographic pressures, commercialization or market orientation of farming, and a number of institutional changes initiated by governments. If, however, climate is considered as a positive factor of production rather than as a mere source of distress and new technological options supported by relevant institutional measures are adopted, the farmers' adjustment mechanisms probably can become stronger than they have ever been.

Although evolved over generations, the traditional structure of options to handle climate-induced risks is fairly static and does not include several new options based on modern scientific advancements in agricultural technology. Traditional technology can at best ensure the balancing of losses and gains at the end of a famine cycle. It does not offer enough scope for generating a surplus for reinvestment and growth to ensure stronger internal cushions for the farmers effectively to sustain the impacts of subsequent droughts. Public relief programs have assumed a significant role in complementing the farmers' own attempts to handle climate-induced risk. Many new options, both technological and institutional, exist in experimental settings or limited practice to assist them also. The best of these options are elaborations of traditional measures and draw inspiration from the underlying principles of diversification and insurance. But they also build up the farmers' long-term capacity to withstand periodic stress and break the cycle of drought pauperization.

REFERENCES

Abalu, G. O. I., and D'Silva, B. (1980). Socio-economic analysis of existing farming systems and practices in northern Nigeria. In *Socio-economic Constraints to Development of Semi-arid Tropical Agriculture*, pp. 3–10. ICRISAT Economics Program, Patancheru, A.P., India.

Anon. (1979). *Improved Agronomic Practices for Dryland Crops in India*. All India Coordinated Project for Dryland Agriculture (ICAR), Hyderabad.

Berry, L., Hankins, T., Kates, R. W., Maki, L., and Porter, P. (1972). *Human Adjustment to Agricultural Drought in Tanzania: Pilot Investigations*. Research Paper No. 13, BRALUP, University of Dar es Salaam, Tanzania.

Bharadwaj, K. (1974). *Production Conditions in Indian Agriculture*. Cambridge University Press, Cambridge.

Binswanger, H. P. (1978). Risk attitudes of rural households in semi-arid tropical India. *Economic and Political Weekly*, **13** (25), A49–A62.

Binswanger, H. P., and Jodha, N. S. (1978). *Manual of Instructions for Economic Investigators in ICRISAT's Village Level Studies*. ICRISAT Economics Program, Patancheru, A.P., India.

Binswanger, H. P., Virmani, S. M., and Kampen, J. (1980). Farming systems components for selected areas in India: Evidence from India. *Research Bulletin No. 2*, ICRISAT, Patancheru, A.P., India.

Borkar, V. V., and Nadkarni, M. V. (1975). *Impact of Drought on Rural Life*. Popular Prakashan, Bombay.

Chambers, R. (1982). Health, agriculture, and rural poverty: Why seasons matter. *Journal of Development Studies*, **18**, 217–238.

Chaudhuri, K. M., and Bapat, M. T. (1975). *A Study of Impact of Famine and Relief Measures in Gujarat and Rajasthan*. Agro-economic Research Centre, Vallabh Vidyanagar, India.

Collinson, M. P. (1972). *Farm Management in Peasant Agriculture: A Handbook for Rural Development Planning in Africa*. Praeger Publishers, New York.

Collinson, M. P. (1977). Demonstration of an interdisciplinary approach to plan adaptive agricultural research programs (drier areas of Mragoro and Kilosa, Tanzania). CIMMYT East African Economics Program, Nairobi.

Dandekar, V. M. (1976). Crop insurance in India. *Economic and Political Weekly*, **11** (6), A61–A80.

De Vries, J. (1976). Has extension failed? A case study of maize growing practices in Iringa, Tanzania. *Rural Economy Research Paper No. 1*, Faculty of Agriculture, Forestry, and Veterinary Science, University of Dar es Salaam, Tanzania.

De Vries, J., and Mvena, Z. S. K. (1979). Sorghum production by small holders in Morogoro district, Tanzania. *Rural Economy Research Paper No. 9*, Faculty of Agriculture, Forestry, and Veterinary Science, University of Dar es Salaam, Tanzania.

Doherty, V. S., and Jodha, N. S. (1979). Conditions for group action among farmers. In Worg, J. (Ed.) *Group Farming in Asia*, pp. 207–224. Singapore University Press, Singapore.

Dupree, H., and Roder, W. (1974). Coping with drought in a preindustrial, preliterate farming society. In White, G. F. (Ed.) *Natural Hazards*, pp. 115–119. Oxford University Press, New York.

Friedrich, K. H. (1974). *The Collecting and Analysis of Micro-economic Data in Developing Countries*. Food and Agriculture Organization, Rome.

Hankins, T. D. (1974). Response to drought in Sukumaland, Tanzania. In White, G. F. (Ed.) *Natural Hazards*, pp. 98–104. Oxford University Press, New York.

Heijnen, J., and Kates, R. W. (1974). Northeast Tanzania: Comparative observations along a moisture gradient. In White, G. F. (Ed.) *Natural Hazards*, pp. 105–114. Oxford University Press, New York.

Hitchcock, R. K. (1979). The traditional response to drought in Botswana. In Hinchey, M. T. (Ed.) *Proceedings of the Symposium on Droughts in Botswana*. Botswana Society and Clark University Press, Worcester, Massachusetts.

Jodha, N. S. (1967). *Capital Formation in Arid Agriculture: A Study of Resource Conservation and Reclamation Measures Applied to Arid Agriculture*. Unpublished thesis, University of Jodhpur, Rajasthan.

Jodha, N. S. (1975). Famine and famine policies: Some empirical evidence. *Economic and Political Weekly*, **10** (41), 1609–1623.

Jodha, N. S. (1978). Effectiveness of farmers' adjustments to risk. *Economic and Political Weekly*, **13** (25), A38–A47.

Jodha, N. S. (1980). Intercropping in traditional farming systems. *Journal of Development Studies*, **16** (4), 227–247.

Jodha, N. S. (1981). Role of credit in farmers' adjustment against risk in arid and semi-arid tropical areas of India. *Economic and Political Weekly*, **16** (42–43), 1696–1709.

Jodha, N. S. (1982). *A Study of Traditional Farming Systems in Selected Villages of Tanzania*. ICRISAT Program Progress Report, Patancheru, A.P., India.

Jodha, N. S., Asokan, M., and Ryan, J. G. (1977). *Village Study Methodology and Resource Endowment of Selected Villages in ICRISAT's Village Level Studies*. ICRISAT Economics Program Occasional Paper No. 16, Patancheru, A.P., India.

Kassam, A. H. (1976). *Crops of the West African Semi-arid Tropics*. ICRISAT, Patancheru, A.P., India.

Kearle, B. (1976). *Field Data Collection in the Social Sciences: Experience in Africa and the Middle East*. Agricultural Development Council, New York.

Keregoro, K. J. B., De Vries, J., and Bartlett, C. D. S. (1977). Farmer 'resistance' to extension advice: Who is to blame. A case study of cotton production in Mara region, Tanzania. *Rural Economy Research Paper No. 5*, Faculty of Agriculture, Forestry, and Veterinary Science, University of Dar es Salaam, Tanzania.

Kirkby, A. V. (1974). Individual and community responses to rainfall variability in Oaxaca, Mexico. In White, G. F. (Ed.) *Natural Hazards*, pp. 119–128. Oxford University Press, New York.

Krishna, R. (1969). Models of family farms. In Wharton, C. R., Jr. (Ed.) *Subsistence Agriculture and Economic Development*, pp. 185–190. Aldine Publishing Co., Chicago, Illinois.

Krishnamoorthy, C., Chowdhury, S. L., Anderson, D. T., and Dryden, R. D. (1977). Crop management in semi-arid farming. *Second FAO/SIDA Seminar on Field Food Crops in Africa and the Near East, Lahore, Pakistan*. All India Coordinated Research Project for Dryland Agriculture, Hyderabad.

Lagemann, J. (1977). *Traditional African Farming Systems in Eastern Nigeria*. Weltforum, Munich.

Le Mere, P. H. (1972). Tie-ridging as a means of soil and water conservation and of yield improvement. *Proceedings of the Second Inter-African Soils Conference*, Ukiruguru, Tanzania.

Mallik, A. K., and Govindaswamy, T. S. (1962–63). The drought problem of India in relation to agriculture. *Annals of Arid Zone*, **1** (1 and 2), 106–113.

Mascarenhas, A. C. (Ed.) (1973). Studies in famines and food shortages. *Journal of the Geographical Association of Tanzania*, No. 8.

Mascarenhas, A. C. (1979). After villagisation what? In Biswark, M., and Cranford, P. (Eds.) *Towards Socialism in Tanzania*. University of Toronto Press, Toronto, Canada.

Mellor, J. W. (1969). The subsistence farmer in the traditional economies. In Wharton, C. R., Jr. (Ed.) *Subsistence Agriculture and Economic Development*, pp. 209–227. Aldine Publishing Co., Chicago, Illinois.

Monyo, J. H., Ker, A. D. R., and Campbell, M. (1976). *Intercropping in Semi-arid Tropics (a symposium report)*. International Development Research Centre, Ottawa, Canada.

Mukuru, S. Z. (1980). *National Sorghum Improvement Program in Tanzania: Annual Report 1979–1980*. ARI, Ilonga, Tanzania.

Norman, D. W. (1973). *Methodology and Problems of Farm Management Investigations: Experiences from Northern Nigeria*. African Rural Employment Study, Zaria, Nigeria.

Randhawa, N. S., and Rama Mohan Rao, M. S. (1981). Management of deep black soils for improving production levels of cereals, oilseeds and pulses in semi-arid regions. *Seminar on Improving Management of Deep Black Soils for Improving Production Levels of Cereals, Oilseeds and Pulses in Semi-arid Regions*, pp. 66–77. ICRISAT, Patancheru, A.P., India.

Ruthenberg, H. (Ed.) (1968). *Smallholder Farming and Development in Tanzania*. Weltforum, Munich.

Ruthenberg, H. (1976). *Farming Systems in the Tropics* (2nd edition). Clarendon Press, Oxford.

Ryan, J. G., and Sarin, R. (1981). Economics of technology options in relatively dependable rainfall regions of the Indian semi-arid tropics. *Seminar on Improving Management of Deep Black Soils for Improving Production Levels of Cereals, Oilseeds and Pulses in Semi-arid Regions*. ICRISAT, Patancheru, A.P., India.

Ryan, J. J., Sarin, R., and Pereira, M. (1979). Assessment of prospective soil-, water-, and crop-management technologies for the semi-arid tropics of peninsular India. *Workshop on Socio-economic Constraints to Development of Semi-arid Tropical Agriculture*, pp. 52–72. ICRISAT, Hyderabad.

Sen, S. R. (1971). Droughts in India—certain dimensional considerations. In Sen, S. R. (Ed.) *Growth and Stability in Indian Agriculture*. Firma K. L. Mukhopadhyay, Calcutta.

Spitz, P. (1980). Drought and self-provisioning. In Ausubel, J., and Biswas, A. K. (Eds.) *Climatic Constraints and Human Activities*, pp. 125–147. Pergamon Press, Oxford.

Spratt, E. D., and Chowdhury, S. L. (1978). *Improved Cropping Systems for Rainfed Agriculture in India*. Field Crops Research, Amsterdam.

Virmani, S. M., Willey, R. W., and Reddy, M. S. (1981). Problems, prospects and technology for increasing cereal and pulse production from deep black soils. *Seminar on Improving Management of Deep Black Soils for Improving Production Levels of Cereals, Oilseeds and Pulses in Semi-arid Regions*. ICRISAT, Patancheru, A. P., India.

Walker, T. S., and Jodha, N. S. (1982). Efficiency of risk management by small farmers. Submitted at *Conference on Agricultural Risks, Insurance, and Credit in Latin America at IICA, San José, Costa Rica*.

Wharton, C. R., Jr. (1968). Risk, uncertainty and subsistence farmer: Technological innovation and resistance to change in the context of survival. Paper presented at the *Joint Session of the American Economic Association and the Association for Comparative Economics, Chicago, Illinois*.

Wharton, C. R., Jr. (1969). Subsistence agriculture: Concept and scope. In Wharton, C. R., Jr. (Ed.) *Subsistence Agriculture and Economic Development*, pp. 6–12. Aldine Publishing Co., Chicago, Illinois.

Wisner, B., and Mbithi, P. M. (1974). Drought in eastern Kenya: Nutritional status and farmer activity. In White, G. F. (Ed.) *Natural Hazards*, pp. 87–97. Oxford University Press, New York.

Part IV

Integrated Assessment

Studies that combine several links in the chain of sensitivity studies, biophysical impact studies, social and economic impact studies, and adjustment responses are integrated assessments. Examples of integrated assessments and problems of linkage between types of studies are found throughout the volume. Part IV explores in depth one major technique for providing linkage in integrated assessment—the use of modeling and simulation. In addition it reviews the experience with both historical and recent integrated assessment.

Integrated assessments involve a scale of activity and a set of complex linkages that encourage the use of modeling and simulation techniques. Such techniques provide an orderly and systematic way to store and analyze large arrays of data, to link data sets together, and to translate different disciplinary approaches into common mathematical language. A special attraction lies in the parallelism with general circulation models (GCMs), the favored tool for exploring dynamics of climate at a global scale. These models, representing the apogee of causal explanation, scale of detail, and massive data handling and computation requiring the most sophisticated of computers, establish a criterion for modeling towards which many biological and social scientists working on integrated assessment seem irresistibly drawn. Thus the opening chapter of Part IV begins with an exploration of global modeling and simulations (Chapter 18).

In Chapter 18 Robinson examines some twenty global models for their potential in climate impact assessment use. She is cautious in her conclusions. Global social, economic and environmental models have not been designed for climate impact analysis, are at best pioneering efforts, and are difficult to use. Nonetheless they can provide insight, data and answers to restricted questions. The pioneering quality of global models is demonstrated by the coevolution biosphere model of the Computing Center of the USSR Academy of Sciences Moisseiev, Svirezhev, Krapivin, and Tarko describe their ambitious, but incomplete, integrative model of climate, ecosystem and society—a model attuned to a time horizon of centuries.

A more modest and limited form of modeling is presented by Lave and Epple (Chapter 20) under the rubric of scenario analysis. They assert three virtues of scenario analysis: stretching the imagination to encompass a wide range of actions and implications: formal modeling of the causes and consequences of climate change and potential adjustments; and interdisciplinary integration to transcend the parochialism of professional method and tradition. Scenarios, and indeed all modeling to date, appear to be exploratory tools, not to be used for reliable prediction but rather to explore the bounds of both the unusual and the possible.

Large-scale modeling is an appealing tool, but still not a broadly realized one. What, then, can be said of other efforts at integrated climate assessment? Wigley, Huckstep, Ogilvie, Farmer, Mortimer and Ingram offer their evaluation of historical climate impact assessment (Chapter 21), considering some 24 examples of historical case studies. To do so they review extensively the methodological underpinnings of historical study, and Chapter 21 should be read jointly with the chapter on historical analysis by de Vries (Chapter 11). Wigley *et al.* note the attractiveness of historical case studies, seemingly free from the complications and confusions of oft polarized current historical explanation. But ironically they find the field suffering from polarities of a different sort, with exaggerated claims and rebuttals for the role of climate in history. Yet within the seeming excess of rhetoric, perhaps half of the studies examined handle both data and assumptions thoughtfully and carefully.

Glantz, Robinson, and Krenz (Chapter 22) examine only five examples of major assessments of climate impacts of recent or future experience, but do so in considerable depth. They focus on comparative issues of study design and length, research staff and linkages between the individual study components, and public presentation of findings. It is clear from these experiences that major climate impact assessments are substantial undertakings, requiring extensive research and time, flexibility in design, and repetition as new data and methods become available. Assessments with strong scientific leadership can advance the state of the art; those organized on a constructive basis can at best attempt only to answer the questions addressed.

The review of integrated assessment concludes with the Chinese proverb: to know the road ahead, ask those coming back. The road ahead is unfolding. In the 4 years that this volume has been in preparation a second generation of integrated studies has been undertaken, and more are planned. Some of these studies simply repeat the past, with little evidence of having sought the advice of those coming back. But most of these half dozen studies evidence a high degree of methodological sophistication, scientific clarity, flexibility in their design, and excellence in their scientific leadership. Another final chapter is being written.

Climate Impact Assessment
Edited by R. W. Kates, J. H. Ausubel and M. Berberian

CHAPTER 18
Global Modeling and Simulations

JENNIFER ROBINSON

Department of Geography
University of California, Santa Barbara,
Santa Barbara, California 93106, USA

18.1 INTRODUCTION

Climate impact assessment often requires foresight and examination of complex patterns of events on a global scale. For example, a global warming manifesting itself over the next 50 years could well occur in the context of a doubling of world population and a transition away from the present petroleum-based industrial system. How might this evolving context relate to the effects of climate change? Would international trade mitigate or amplify climatic influences on agricultural production? What might happen if climatic disturbances were to occur simultaneously with major disruption of the global energy system? Such questions naturally arise, particularly when general circulation models (GCMs) are used to produce climatic scenarios, and drive researchers to look for tools to evaluate the societal consequences of these scenarios.

Can currently existing global social systems models be used for such climate impact assessment? Might the social system models be linked to GCMs? What more limited applications might be feasible? To help answer these questions, this chapter offers a description of available models, how they work, what their purposes and capacities are, and how they do or might represent climate.

Here 'global model' is used to mean a computerized model representing one or more aspects of social systems on a global scale. About 20 models are discussed, all those for which documentation could be acquired at the time of writing. These range from short-term (less than one year) agricultural buffer stock models to very long-term (more than a century) models of interaction among population, resources and environment. Emphasis is on the possibilities, opportunities and difficulties of adapting these existing models to explore globally questions of the effects of climate variability and change. This global analysis should be seen as a companion to the regional and sectoral approach discussed in many chapters of this volume.

18.1.1 Limitations of Global Models

At the outset readers should be cautioned against inflated expectations. Several factors place strict limits on the value of global models for climate impact studies.

The first major limiting factor is that global models have generally not been designed with climate impact analysis in mind. Spatial and temporal resolution and subject matter boundaries render many of them problematic or inappropriate for many climate impact questions. For example, in many models the Soviet Union or all of North America is represented as a single entity, so that calculation of the impacts of changes in finer scale midlatitude precipitation patterns would be difficult or impossible. Similarly, trade models with a time horizon of 1–5 years may provide insight relevant to the examination of a change taking place over 30 years, but they are far from ideal for the purpose. And, obviously, models that do not represent soil fertility or differentiate between forest and desert cannot be used to study desertification. Most models do not represent soil, forest or marine systems, and the few that do provide rather simplistic descriptions.

A second limiting factor is the general scientific basis of global modeling. Although global models are remarkable as pioneering efforts, they are in need of improvement from many sides. Global modeling is scarcely more than a decade old. The disciplines on which it draws, such as economics, demography, ecology and political science, have not achieved great predictive power.

Moreover, the policy and popular audiences to which the results of global models have been addressed have not held modelers to the highest scientific standards of hypothesis formulation, documentation and testing, and the modeling community has yet to agree upon or enforce standards. Global models are often unreliable black boxes; in most cases it is impossible to place confidence limits on their predictions. Realistically, one should not count on global models to accomplish more than

> '... organize existing data in such a way that it provides new insights and facilitates interpretation of the data'.
>
> '... pose new questions and call ... for new facts'.
>
> 'permit us to make important, verifiable predictions'.
>
> 'provide the basis for collecting accurate measurements'.
>
> (Mason, 1976, 3)

A third limiting factor is that global models, like other large, complex models, are difficult to construct, understand, operate, test and maintain. Understanding them, even to gain insight, may be hard work. For example, international trade is extremely complex, and models use various simplifications for attempting to describe it (Neunteufel, 1977, 1979). In assessing how a change in climate or an extreme climatic event may be transmitted through the international trade system, there is little hard evidence to go on as to which trade formulation is most appropriate. Once model results are presented, it may be difficult to sort out whether the findings are artifacts of the model's simplifying assumptions or attributes of the real system.

Liverman (1983) illustrates well the kind of difficulties that arise in applications of global models. Liverman investigated the ability of the International Futures Simulation Model (Hughes, 1982) to replicate the price and trade patterns of 1972–75, given appropriate forcing to represent weather patterns of those years. She concluded that the model did relatively well in estimating slowly changing phenomena such as population, production trends and land in cultivation over 1970–80, but that simulation of crop prices and stocks was poor. This, she concluded, was open to two interpretations: either the model structure inadequately described trade, price and food stock mechanisms, or the observed behavior resulted from distortions of price–trade mechanisms by political decisions and price speculation (p. 266). Such ambiguities are likely to be inescapable in many attempts to employ global models for climate impact analysis.

18.1.2 Ways of Using Global Models

In light of their limitations, how might global models be used in climate impact analysis? The option mainly discussed here is to employ or modify existing models. Two other options deserve mention, however. One is to use no formal global model. The second is to develop a new model specifically for climate impact analysis.

The strongest reasons to refrain from using formal global models arise from critiques of the current state of the art. The refusal to employ and experiment with them, however, may well result in yet weaker development of geographically comprehensive and internally consistent global perspectives of climate impacts.

From a researcher's perspective, the option of developing a new model is highly attractive. Literature on the use of mathematical models has strongly favored *ad hoc* model development with strong user involvement, while repeatedly cautioning against the difficulties of using 'off the shelf' models (see Holcomb Research Institute, 1976; Linstone and Simmonds, 1977). Building a global model is a major effort, however, Assembling data bases, developing theory, and testing and documenting a new model can be expected to take at least 2 years and several hundred thousand dollars. An effort of such scale should be framed around the sharp designation of questions to be explored and the context in which they are to be explored. How it should be done is outside the scope of this chapter.

18.1.2.1 Off-line Use

Global models may be used without new computational work. For example, published model results can be used to establish background scenarios for economic, demographic and resource trends that are to be anticipated concurrently with possible climatic change over the next 2–12 decades. Appropriate models for this purpose could be broadly focused models, such as World 2 (Forrester, 1971), World 3 (Meadows *et al.*, 1972), the World Integrated Model (Mesarovic and Pestel, 1974), the Latin American World Model (Herrera *et al.*, 1976), and the Coevolution Model described by Moisseiev *et al.*, in Chapter 19. Despite disagreements about particulars, global modelers have come to broadly similar findings about agriculture, energy and relationships between rich and poor nations (Meadows *et al.*, 1982; Office of Technology Assessment, 1983).

All global models that have represented limits to land availability and diminishing returns for agricultural inputs (for example, fertilizer) have shown increasing food prices and increasing numbers of people with insufficient food. Regionally disaggregated models tend to show particularly severe agricultural stress in South Asia and non-OPEC Africa. Presuming climatic change has adverse agricultural consequences, one can expect it to amplify incumbent agricultural stress. Also, stressed agricultural systems tend to be associated with intense exploitation of the unmanaged biosphere, shrinking total biomass, and desertification. Reduced biomass also implies transfer of carbon from living organic matter and soil to other pools—predominantly, the atmosphere and the oceans.

Global models generally show the petroleum economy beginning to give way to other energy systems in the next few decades. Model results indicate that most every known non-petroleum energy form is expected to expand, but there are large disparities in the relative rates of growth for coal, nuclear, solar and other renewable energy supplies, and in the growth rates and composition (such as liquid *vs* solid fuel *vs* electricity) of future energy demand.

Lastly, where global models have been used to explore the development prospects of the poorer nations, model results have generally shown that extreme measures will be required for the poor to keep pace with the rich. Many models show the continuation of present trends leading to stagnation or even decline in the poorest nations. As the poorer nations are in general tropical, this finding suggests that climate-induced stress may be harder felt in the tropical than in the temperate or boreal regions.

18.1.2.2 Global Models as Data Sources

Lining up an internally consistent and globally comprehensive body of data is a large chore that may be necessary for some sorts of climate impact analysis. For example, to develop first-order approximations of the susceptibility of various nations to climatic influences in agriculture, fisheries and forestry, one might want to know the fractions of gross national product (GNP) coming from agriculture, fisheries and forestry. Such data are not conveniently assembled on a global basis in common statistical sources (for example, FAO, World Bank, or OECD Annual Yearbooks), but have been assembled in part by several global modeling efforts (for example, Leontief *et al.*, 1977; Bottomley *et al.*, n.d.). Similarly, the analysis of policy options for reducing CO_2 emissions might have use for data on energy capital infrastructure that has accrued in construction of the World Integrated Model or the IIASA energy models (International Institute for Applied Systems Analysis, 1981).

If one is interested in detailed investigation of the agroecological implications of climatic change for developing nations, the UNFPA/FAO/IIASA work is helpful (Food and Agriculture Organization/UN Fund for Population Activities, 1979; Shah *et al.*, 1981). The project, in essence simple models operating on an agroclimatological geographical information system, was established for detailed assessment of the earth's population support capacity, and to date it has absorbed around 500 person-years of labor. It now covers all developing countries except China (which was not in the United Nations when the project began) at a spatial resolution of 1:5,000,000 and a temporal resolution of one month. The data base includes soils and topographical, meteorological, crop-requirement and production data for three levels of

Table 18.1 General attributes of different global models

Model	Authors	Time horizon (years)	Method	Focus
Coevolution Model	Moisseiev *et al.* (Chapter 19, this volume)	Hundreds of years	Dynamic simulation	Society, atmosphere, biogeo-chemical balances
World 2	Forrester (1971)	200	System dynamics	Population, food, soils, industry, pollution
World 3	Meadows *et al.* (1972) Meadows (1974)	200	System dynamics	Population, food, soils, industry, pollution
Latin American World Model	Herrera *et al.* (1976)	100	Dynamic optimization	Allocation of labor and capital to meet basic needs
SARUM	Roberts (1977) SARU (1978)	90	Dynamic simulation, input–output, econometric	Food and mineral resource adequacy
MOIRA 1	Linnemann *et al.* (1979)	45	Algorithmic, optimization, econometric	Hunger, food production, food trade, trade policies
World Integrated Model	Mesarovic and Pestel (1974) Hughes (1980)	25–50	Dynamic simulation, input–output	Population, capital, energy, food, trade, inter-sectoral flows
International Futures Simulation	Hughes (1982)	25+	Dynamic simulation	Population, economic development, energy, agriculture
Grain buffer stock	Eaton *et al.* (1976)	~25	Dynamic stochastic simulation	Rules for managing grain buffer stocks

Table 18.1 (*cont.*)

Model	Authors	Time horizon (years)	Method	Focus
UN World Model	Leontief *et al.* (1977)	25	Input–output (static)	Requirements for pollution generated by UN development targets
Interactive Agricultural Model	Enzer *et al.* (1978)	20	Cross impact, interactive projection	Global food problem, grain trade
Optimal grain reserves	Johnson and Sumner (1976)	~20	Dynamic stochastic optimization	Management of grain reserves
FUGI	Kaya and Onishi (various dates)	~10–15	Econometric, input–output (dynamic)	Macroeconomic detail, energy and resources
USDA Grains, Oils and Livestock	Rojko and Schwartz (1976)	10–20	Econometric (static)	Production, exports, imports, trade of oils, grain and livestock
Input–output	Bottomley *et al.* (n.d.)	~10	Input–output (static)	International interdependence
FAO price equilibrium		~10	Econometric (static)	World agricultural market prices, trade flows
World Food Economy Model	Takayama *et al.* (1976)	1–2	Econometric, quadratic programming	Global agricultural markets and trade
UNFPA/FAO/IIASA	FAO/UNFPA (1979)	1975, 2000	Agroecological analysis, linear programming	Population support, land resources, food production

technological inputs, all computerized to a common format. Adapting it to look at the changes in production potential for a specified set of climatic parameter changes would be straightforward. Alternately, it might be used as a data base from which to develop samples stratified by soil type, topography, and precipitation regime, for example, in monitoring desertification.

The difficulty of gaining access to model data varies. Although some groups (such as Sichra, 1981; Bottomley *et al.*, n.d.) may regard making their data bases available as an important part of their work, not all model data will be accessible. Poor documentation, organizational problems, or modelers' proprietary interest in maintaining control of the rewards from data bases that have taken them many years to put together may make it difficult to extract data from other models. For example, portions of the US Department of Agriculture's Grains, Oils and Livestock Model (Rojko and Schwartz, 1976) were accidentally erased from USDA's computer archives and no backup existed (personal communication, Patrick O'Brien, USDA). Furthermore, the models' data bases may be outdated (for example, energy data may come from 1975 and before).

18.2 MODELS AND MODEL ATTRIBUTES

The following section addresses the question of which models may serve what purpose. It is organized around a two-dimensional matrix, listing models and various attributes of the models (for example, time horizon, method, degree of aggregation). The matrix is presented in Tables 18.1, 18.2 and 18.3. The text describes model attributes in the groups.

18.2.1 Temporal Resolution, Time Horizon, Method, Focus

Table 18.1 lists models in order of increasing time horizon, giving for each model a brief description of method and problem focus. Table 18.2 describes geographical, sectoral and agricultural sector aggregation, while Table 18.3 gives the treatment of climate and food stocks in each of the models. From Table 18.1, one can see that time horizons of existing global models range from 5 to 200 or more years. The character of models, the methods employed, the problems treated, and the model's possible utility for climate work change with time horizon.

Parenthetically, it may be noted that temporal resolution is seldom discussed in the global modeling literature. Most dynamic simulation models are solved at yearly increments and thus ignore seasonal events, an omission that may be significant for agricultural stock, trade and price mechanisms. The system dynamics models, World 2 and World 3, operate with a time increment set sufficiently small as to approximate continuous behavior. The static models are solved for an instantaneous point in time and in some cases, their exogenous

Table 18.3 Treatment of climate and food stocks in different global models

Model	Geographical aggregation	Sectoral aggregation	Aggregation of agriculture sector
Coevolution Model	Land by 500 km^2 grid, coverage of ocean	Agriculture, pollution abatement, mineral, basic capital	Unclear
World 2	Aggregate world	Agriculture, industry resource extraction	1
World 3	Aggregate world	Agriculture, industry resource extraction	1
Latin American World Model	5-region; 20-region may exist	Agriculture, education, housing, capital, other	Livestock, crops
SARUM (1976)	3 regions	10 sectors	4 agriculture products, 1 food product, 3 agriculture inputs
MOIRA 1	106 nations	Agriculture, non-agriculture	1 commodity
World Integrated Model	12 regions (basic), 17 regions (subregional)	7 or more, varies for different regions	5 commodity, 3 land types
Grain buffer stock	Aggregate world	Agriculture only	Aggregate grain
UN World Model	16 regions	40 economic sectors	4 agricultural commodities
Interactive Agricultural Model	10 regions	Agriculture only	Grain as proxy for all foods
FUGI	14–62	15 sectors	4 sector (?)
USDA Grains, Oils and Livestock	28 regions	Agriculture only	Up to 14 commodities
Input–output	90 countries	6 economic	1 agriculture, fisheries and forestry
FAO price equilibrium	28 regions	Agriculture only	18 commodities
World Food Economy Model	20 regions	Agriculture only	8 groups
UNFPA/FAO/IIASA land resources	Much of developing world, 10,000 ha units	Agriculture only, livestock	18 food crops

Table 18.3 Treatment of climate and food stocks in different global models

Model	Treatment of climate	Treatment of food stocks
Coevolution Model	Includes detailed climate model and mechanisms describing anthropogenic climate change	Probably excluded
World 2	Omitted	Excluded
World 3	Omitted	Excluded
Latin American World Model	Omitted	Excluded
SARUM	Generally omitted	Held in regions
MOIRA 1	Production limits = f (photosynthetic potential). Potential estimated from soil maps and climatic maps; past annual harvest variation repeats	Stocks assumed to be held at world market level
World Integrated Model	Omitted	
Grain buffer stock	Random perturbation of yields	?
UN World Model	Omitted	Excluded
Interactive Agricultural Model	Extremes in variation from trend line in past production series ('60–'75) define maximum deviation of random perturbation of yields	
Optimal grain reserves	Estimates made of yield as f (rainfall); model driven with synthetic time-series with mean, variance, autocorrelation structure of past rainfall data (sample years not specified)	Stocks of commodities determine prices in each region
FUGI	Omitted	Unclear
USDA Grains, Oils and Livestock	'Good' or 'bad' weather investigated by raising and lowering yields (for Global 2000 runs)	Regional stocks for each commodity; levels policy controlled
Input–output	Omitted	Excluded
FAO price equilibrium	Exogenous	Unclear
UNFPA/FAO/IIASA land resources	Production functions based on climate inventory and assessment of climate responses of different crops	Excluded

parameters are projected forward and new solutions are derived for future points in time. For example, the UN World Model (Leontief *et al.*, 1977) and the USDA Grains, Oils, and Livestock Model (Rojko and Schwartz, 1976) are solved this way at 5-year intervals. Careful analysis of a model's temporal resolution should be made before applying it to problems involving seasonality.

18.2.1.1 Short- to Medium-term

Models in the 1 to 15-year range are more appropriate for looking at the economic effects of climate variability than for looking at climate change. Such models are typically built by economists and treat economic growth, international trade, balance of payment problems, monetary system behavior and intersectoral flows. They are built to provide detailed and precise descriptions of the global market and monetary systems, and it is logical to look to them for information on the ramifications of climate events through supply, demand, and price and monetary effects for specific commodities and countries or regions.

Such models measure many variables in monetary units, and in many cases use no physical or biological units. Their logic generally combines causality, extrapolation and accounting. They tend to be econometric, are often static, and a large portion of their mathematical operation tends to involve simultaneous equations and linear matrix operations.

These attributes tend to impart the assumption that the biophysical system will remain unchanged, and that most trends observed in the recent past will endure into the future. Linearization inherently assumes either that functional relations are indeed linear, or that systems changes will not be so great as to drive systems relationships far off a line of linear extrapolation. It may, thus, be inappropriate for describing extreme events or for making long-term forecasts.

Caution must be used if short-term models are to be used for looking at either shocks to the system—such as extreme climate events—or deeper underlying change.* Either temporary shocks or underlying change may violate assumptions of temporal and behavioral continuity often implicit in linearization and econometric modeling. To deduce whether a system handles shocks reasonably, it might be well to examine (or ask the modelers to look at) the behavior of climate-sensitive model variables in past years of climatic anomalies to see whether they fit within the model's explantory power, or merely add to the magnitude of its error terms.

* Experience with using large econometric systems to forecast the consequences of the tripling of oil prices of the early 1970s—an event analogous, in some ways, to climatic disturbances—suggests that such models are reasonably good at predicting short-term market effects, but inadequate for representing long-term adjustments. (Personal communication, Bert Hickman, Stanford, California).

18.2.1.2 Longer-term

Models with time horizons of over 50 years are generally built by interdisciplinary groups (engineers, systems analysts, economists, demographers, agronomists, and the like) and focus on biological and physical processes such as population dynamics, resource flows, and creation of physical capital stock. These models generally aspire to describe essential trends in system behavior—not to make point predictions.

Long-term models tend to ignore prices. Hence short-term models are probably preferable for investigating the market-related details of climatic variability and extreme climatic events, while long-term models are preferable for looking at climatic change and at the biological and/or physical consequences of climatic events. For example, while long-term models may be useful for studying the evolution of food scarcity conditions, shorter-term models allow one to study the effect of scarcity on prices and on trade and consumption patterns.

Between the short-term and the long-term models are a variety of trade and/or intersectoral flow models with some feedback between economic development and resource depletion, population growth, and other processes. These typically include both monetary and physical variables and are built by a mixture of economists and experts in physical and natural sciences.

18.2.2 Intersectoral Flows

Intersectoral flows may be important to climate impact analysis, first because they are routes by which indirect implications of climate variations may be felt, and second because intermediate flows (the products that are created in the process of making products rather than meeting end use demand—for example, fuel used in agriculture and industry) account for a large proportion of all economic activity, and it is difficult to keep track of such activity without a device such as an input–output matrix (see also Chapter 12).

Starting from the UN World Model described by Leontief (1977), one branch of global modeling has concentrated on the linkage between intersectoral flows within a country and international trade flows. One might gain information on intersectoral transmission of climate impacts using the UN World Model, the models assembled by Bottomley *et al.*, FUGI (Kaya and Onishi), or the World Integrated Model (Mesarovic and Pestel, 1974). For greater intersectoral detail, the UN model (Leontief *et al.*, 1977) might be preferred. If one simply wants data on intersectoral flows, Bottomley *et al.* have assembled what is probably the largest collection of input–output models in the world. The World Integrated Model (Mesarovic and Pestel, 1974; Hughes, 1980) or its simplified and most accessible version, the International Futures Simulation Model (Hughes, 1982), may be used for looking at intersectoral flows within the context of complex dynamic feedbacks.

18.2.3 Agriculture

The impact of climatic change and variability on agriculture can be observed from many perspectives: management of grain reserves; food and nutrition; trade and balance of payments; or ecological sustainability. Different global models are appropriate for different perspectives.

The section must commence with a caveat. Global agricultural models' market behavior depends heavily on the formulations used for international trade and for reserve management. Systematic comparisons of the various formulations that have been used and their respective strengths and weaknesses have not been published. It is beyond the scope of this paper to address the question of how good a model formulation is—or whether the real system is so complex that no simplification can capture its behavior.

18.2.3.1 Buffer Stocks and Grain Reserves

Grain reserves are an ancient defense against climatic variability. Numerous models have been developed specifically to examine the economics of buffer stock management. Some of these, including the models of Eaton *et al.* (1976) and Cochrane and Danin (1976), are designed to look at global reserves of all grains, others at specific grain commodities. A good review of buffer stock management models is found in Eaton (1980). In addition, as described in more detail below, disaggregated production and trade models have been used to look at stock management questions. An excellent review on this subject is found in Adams and Klein (1978).

To date, buffer stock modelers have tested their models mostly on yield variability such as observed in the recent past. The time-series employed seldom extend back past 1950. Variability has been characterized by such parameters as variance around expected production volume, lagged covariance behavior, and form of random behavior (Eaton, 1980). The problem of bad years occurring in sequence is considered by Eaton, but not by most studies.

The extent to which yield variation is caused by weather, the possibility of climate change, or the possible occurrence of extreme climate events not present in the period from which the model was parameterized are rarely mentioned in the buffer stock model documentation. It would be relatively easy and perhaps rewarding to use existing buffer stock models to look at how actual and/or optimal reserve holding strategies might operate under extreme climate events by altering the yield variability parameters used to drive the model. Under the assumption that climate change will be perceived as variability around a moving average, it may also be possible to use buffer stock models to look at climate change.

18.2.3.2 Production and Trade

Trade patterns are extremely complex (see, for example, descriptions of grain trade in Morgan [1980] and of commodity markets in Labys [1978]) and modeling them necessarily entails simplification. Part of the complexity is institutional and political. Agricultural production and trade policies vary greatly among countries and over time, and they significantly influence national and global agricultural markets.

The physical realities are also complex. Different crops and regions vary greatly in the structure and dynamic behavior of agricultural production. For example, coffee flows from mountainous tropical regions to temperate zones. Coffee production is frost-sensitive and, largely due to the fact that a coffee tree takes 3–4 years to bear fruit, the coffee market is prone to 7- to 8-year price and volume fluctuations. Wheat comes largely from temperate, semi-arid countries, is exported to both developed and developing countries, has much shorter production cycles than coffee, and is affected most by drought. Livestock slaughtering and meat prices are sensitive to grain prices; high grain prices induce increased slaughtering and meat supply in the short term, with meat shortage typically following in a matter of months or years (see, for example, Meadows, 1970).

Further, population and income affect trade. Come 'riches' and the hungry eat more staples (per capita), while those above subsistence are apt to substitute luxuries and animal proteins for staples. Both income fluctuations and secular trends thereby affect the quantity and mixture of food demanded and hence the food that must be imported or may be exported. For many (if not most) regions of the world food demand is changing more rapidly than agricultural production, and the effects of climate change may tend to compound rapidly changing trade balances.

To analyze climate impacts on agricultural trade, a model must first include agricultural trade. This simple criterion selects SARUM (Roberts *et al.*, 1977; SARU, 1978), MOIRA 1 (Linneman *et al.*, 1979), the World Integrated Model (Mesarovic and Pestel, 1974), the UN World Model (Leontief *et al.*, 1977), the Center for Futures Research (CFR) interactive agricultural model (Enzer *et al.*, 1978), the GOL model (Rojko and Schwartz, 1976; US Department of Agriculture, 1978; O'Brien, 1980), and the FAO price equilibrium model (Food and Agriculture Organization, 1971). These all contain mechanisms that account for international agricultural trade.

Most global models represent agricultural demand as a function of income, and can be used to look at income effects. However, substitutions between commodities under income change can be studied only with a multicommodity trade model, such as the GOL or the IIASA/FAP model (Parikh, 1981; Parikh and Rabar, 1981).

Another area in which model structures and behaviors differ is in their handling of reserves. Both the way in which reserve sizes are determined and the way in which reserves affect prices and other behavior seem important. Here it is possible only to mention some of the ways in which representations of reserves differ. At one extreme, there are models, such as the UN World Model, in which reserves are the residual of supply, demand and trade, and where prices are unaffected by reserves. This representation will not show price instability under conditions of short supply. At the other extreme are models, such as the IIASA/FAP model and MOIRA, in which reserves are determined by complex interactions between production and demand responses, trade policies, and (in some cases) reserve policies. Relatively realistic descriptions might be achieved through such representations but, to date, documented validation of the representations is weak. Other models represent reserves as maintained at a policy-specified level through government purchases and sales of grain. In the CFR model (Enzer *et al.*, 1978), reserves are determined off-line in gaming fashion, by the decision of persons playing the role of political decision-makers.

An additional criterion is the model's treatment of the crop or crops of interest. Grain is considered in virtually all global models. In some cases, it is used as a surrogate for all agriculture. On the other hand, none of the models listed is appropriate for exploring the consequences of extreme climatic events on the export earnings of large coffee exporters such as Colombia, Kenya and Brazil.

A few other specialties and climate-relevant aspects of various models are worth a mention. The GOL Model (Rojko and Schwartz, 1976) was constructed to study the medium-term (~10 years) interaction among the global grain, oils and livestock markets from the perspective of the United States as a large grain exporter. This model has been used to study the effects of changes in mean values of climatic parameters as transformed into changes in grain production (National Defense University, 1983), although published documentation of the experiment is not very illuminating on the subject of how secondary effects of climate change were transmitted through the livestock and oilseed markets. There are, in published output, no signs that the GOL model has been run stochastically using assumptions of varying weather patterns, and without performing the experiment it is difficult to say how realistically the model would behave if it were and portions of the model's source code (as described above) were accidentally destroyed.

MOIRA (Linneman *et al.*, 1979) was constructed to study the effect of a doubling of population on the world food system. As described under demographic behavior, the model is particularly rich in its description of demand; it includes six separate income groups and calculates the dietary adequacy of each group as a function of income and price. Hence MOIRA is particularly good for studying income aspects of agricultural variations. MOIRA also attempts to model national agricultural trade policy in a more refined fashion than most other global models. However, the behavior of its trade mechanisms is essentially unvalidated and difficult to follow because more than a hundred nations are presented.

The IIASA Food and Agriculture Program (FAP) model, when complete, will be a set of mutually compatible national models, interlinked by a trade mechanism (Parikh, 1981; Parikh and Rabar, 1981). It will feature both multiple commodities (major grains, livestock, and non-food commodities) and detailed descriptions of agricultural policy levers available to different nations. The modelers intend to study the linked system's response to various shifts and disruptions, including climate shocks. The difficulty with the FAP model is that it is very large, extremely complex, and to date inadequately tested or documented. Development of the model has been slower than anticipated. Modeling work was scheduled for termination in 1984.

18.2.4 Energy

The impact of energy systems on climate, particularly on fossil fuel through CO_2 generation, has been analyzed using many models (Jäger, 1983; see also Chapter 9) and will not be described here. The tendency of firewood and dung fuel systems to contribute to local climate alteration, soil impoverishment, and desertification is outside the boundaries of existing global models because critical variables, such as firewood usage and forest growth, are not included in global models. Verbal attempts to describe and quantify such effects, however, were made in both the group IIASA energy model (International Institute for Applied Systems Analysis, 1981) and the Global 2000 study (Barney, 1980, 1981).

The main effects of climate on energy systems appear to be through the effect of weather on energy use for heating and air conditioning and on the supply of renewable energy sources, such as hydropower, firewood, and so forth (Quirk, 1981). The effect on heating fuel demand is relatively important in developed, temperate regions. The effect on renewable energy supply is most important in tropical, less developed regions.

By assuming a relationship between climate parameters and energy demand and/or supply, one could translate climate scenarios into scenarios of energy supply and demand, and use these to drive global energy models. However, for long-term climate change the effects of a change of a few degrees in average temperature are likely to be small in comparison to changes such as increased building insulation and increased efficiency of air conditioning.* It would be difficult to study the effects of climate on the supply of renewable energy sources other than hydropower (for example, wood fuel and solar energy) because treatment of the unmanaged biosphere is sparse in most global models, and representation of solar energy development remains highly tenuous. One might conceivably use the World Integrated Model (Mesarovic and Pestel,

* Personal communication, William Quirk, Lawrence Livermore National Laboratory, Livermore, California.

1974; Hughes, 1980), the IIASA energy models (IIASA, 1981), the International Energy Evaluation System (IEES) (Barney, 1980, 1981), or other models containing international energy trade and fuel infrastructure development to study the impact of climate on energy demand, as transmitted through international fuel markets, intersectoral flows, and long-term effects on capital formation and resource depletion. Of the models listed, only IEES contains sufficient detail on energy demand to study the effects of short-term climate variation. IEES has not been updated since the late 1970s, and thus may not reflect adjustments made in the energy system since then.

18.2.5 Demography

Population growth, being an important and relatively predictable part of social systems development (at least as compared to energy futures or economic growth) is accounted for in virtually all global models. How various global models treat demographic variables is shown in Table 18.4.

18.2.5.1 Migration

It appears that no global model has yet dealt with migration between nations. Both MOIRA (Linneman *et al.*, 1979) and the Latin American World Model (Herrera *et al.*, 1976) describe rural–urban migration. A peculiarity in the optimization routine in the Latin American World Model causes all regions to move towards 100 percent urban at an incredibly rapid speed,* and its formulation cannot be taken very seriously. MOIRA shows rural–urban migration as a function of relative per capita income in and outside of agriculture. In simulation, rates of rural to urban migration follow food price; when prices are high, farmers are better off and migration is less; when prices are low, farmers are poorer and there is more urban migration. It would be an interesting test of the model to see if it would, given severe weather shock in rural areas, replicate observed patterns of urban migration in times of famine.

Other than MOIRA and the Latin American World Model, global models can look at migration only by inference, as none explicitly includes migration.

One can, however, infer heavy migration as a plausible outcome of food deficits. It is possible to use most global models to look at food availability.

18.2.5.2 Other Demographic Parameters

Several global models relate population growth to economic development and food supply (World 2 and 3, some versions of WIM, the Coevolution Model

* The model maximizes life expectancy at birth. Its functional relationships were developed by statistically relating several variables, including education, fraction of population urban, and others to life expectancy. Because there is a strong correlation between life expectancy and urbanization, the model favors rapid urbanization.

Table 18.4 Demographic aspects of different global models

Model	Aggregation structure	Demography	Migration
Coevolution Model	In development	Probably none	
World 2	No disaggregation	Fertility = *f* (income-cap) mortality = *f* (food, income)	None
World 3	5–15 age cohorts	Fertility = *f* (income, services)	None
Latin American World Model	By region rural–urban (*n*) age cohorts	Fertility and mortality are *f* (basic needs fulfillment); model maximizes life expectancy	Rural–urban at rate to maximize life expectancy
SARUM	Exogenous	None	
MOIRA 1	By nation (106), rural–urban income group (6)	Growth exogenous	Rural–urban *f* (income in agriculture, income outside agriculture)
World Integrated Model	By region (12+), age cohort (85)	Exogenous	None
Grain buffer stock	No disaggregation	Exogenous	None
UN World Model	No disaggregation	Exogenous	None
Interactive Agricultural Model	Probably none	Exogenous	None
Optimal grain reserves	Probably none; exogenous	None	None
FUGI	Undisaggregated	Exogenous	None
USDA Grain, Oils and Livestock	No disaggregation	Exogenous	None
Input–output		Demography excluded, implicit in demand projections	
FAO price equilibrium	Undisaggregated	Exogenous	None
World Food Economy Model	No disaggregation	Exogenous	None
UNFPA/FAO/ IIASA land resources		Exogenous	None

described in Chapter 19), environmental conditions (World 2 and 3, the Coevolution Model), and social welfare (the Latin American World Model and World 3). Models such as these may be pertinent to climatic analyses in three ways. First, where they indicate tight food supplies, one can presume that the potential disturbance caused by the effects of climatic change and variability on yields will be amplified. Second, where they indicate population pressure, one can infer shrinkage of the unmanaged biosphere and deterioration of soil organic matter (through fire, overgrazing, and the like), and thus creation of harsher microclimates and increased CO_2 release. Third, if one had reason to believe climatic change would directly affect mortality or fertility, one could rerun the models with the changed parameter to see the consequences of the larger system.

The choice of models for looking at population–environment interaction depends on purpose. For moderately fine regional resolution, at the cost of rather difficult to interpret results, the World Integrated Model or the International Futures Simulation may be most appropriate. The WIM contains many (12 or more) regions and 1-year age cohorts; the IFS, 10 regions and 5-year age cohorts. Either could be adapted to look at the effects of food supply, income and population control on population dynamics, as well as to approximate the way in which climate change in one region might be transmitted to affect other regions through international trade. For example, suppose that a global warming causes lowered agricultural productivity in the Great Plains, but slight increases in rice yields in most tropical regions. The WIM or IFS could be used to investigate whether this will result in more or less hunger and starvation in India, Indonesia, and/or Africa. One could also use MOIRA for this purpose; this would give results that take uneven income distribution into account, but would not permit looking at the feedback to population growth, as population growth in MOIRA is exogenous.

For less detail, but more inclusive structure, World 2 and World 3 might be recommended. Both see longevity and fertility as affected by food supply, economic resources per capita, and pollution. Neither disaggregates the world into regional populations. World 3 normally uses five age cohorts, and has, in structural testing, been disaggregated to 15 age cohorts, leading to the finding that model results are not sensitive to the degree of age disaggregation (Meadows, 1974).

The pollution term in World 2 or World 3 could be adapted to describe the effects of anthropogenic climate change on morbidity and mortality. The part of model structure describing soil deterioration might be expanded or adapted to show shrinking of the biosphere, thus CO_2 generation and desertification. The Coevolution Model described in Chapter 19 employs demographic formulations similar to that of World 2 and World 3, and may eventually be expanded to account for population–resource–environment interaction on a 500 km^2 grid.

Theoretically, the Latin American World Model, which finds the allocation of labor and capital among various sectors (housing, education, agriculture, capital formation and other) that maximizes life expectancy at birth, might be used to study strategies for meeting basic needs in the face of climate variability or change. However, introducing climate parameters into the Latin American World Model might not produce meaningful results. Its representations of trade and agricultural production are probably unequal to the problem, and it has some wildly unrealistic tendencies (Office of Technology Assessment, 1983).

18.2.6 Political Ramifications

Climate change is generally expected to benefit some and cost others. The costs of anthropogenic climate change are apt to be borne by groups other than those causing the change. Control of many of the economic and social forces contributing to climate change (for example, deforestation, CO_2 emission) will in many cases require cooperation. Because these considerations have strong political implications, one might want a formalized model to examine them.

The importance and difficulty of representing political decisions is almost routinely discussed at global modeling conferences (see Meadows *et al.*, 1982). The Wissenschaftzentrum group in West Berlin, under the directorship of K. Deutsch, has been working on a politically oriented global model that deals with both domestic stability and international economics and politics. According to recent reports (Bremer, 1981; Ward and Cusack, 1981), it will employ a five-sector aggregation within which it would be very difficult to specify climate impacts (the sectors represented are household, government, capital production and foreign trade). However, extension and adaptation of the model will eventually be possible.

The Center for Futures Research (CFR) (Enzer *et al.*, 1978) has constructed an interactive food model, in which persons playing policy-makers adjust production targets, reserve targets, import, export, aid, and other features in each year of simulated time. The model divides the world into ten regions, each of which is represented by a decision-maker. It has introduced stochastic weather effects on yields, and political responses to weather variation on the order of that observed over 1950–75 have been studied. The model, which is used in conjunction with the 'player's' decisions about political reactions to changing circumstances, might be described as a simple simulation model (it can be simple, as modeling social and political decision-making is one of the more difficult aspects of social system model building, and the CFR model's interactive format absolves modelers of the need to be sophisticated in their human behavioral equations). It is parameterized using Delphi survey techniques.

The Climate Task of the Resources and Environment Group at the International Institute for Applied Systems Analysis tried a gaming approach to look at strategic and political aspects of CO_2-induced climate change. Two games were constructed, a computer game and a board game. They consider the strategic and economic aspects of coal extraction and trade and policy measures available for containing and/or adapting to CO_2 in the context of the evolution of highly uncertain and unevenly distributed costs and benefits arising from climate change (Stahl and Ausubel, 1981; Robinson and Ausubel, 1983). The games might be regarded as initial attempts to build interactive 'climate-centered' global models.

18.3 CONCLUSION

A crude model used with good scientific practice is more enlightening than poor scientific practice and a good model. Existing models are pioneering efforts, not perfected tools. Both imagination and scientific discipline are needed to obtain meaningful results for them. Essential activities include:

1. Defining the problem one wishes to explore and translating it into terms that are consistent with an existing model or models.
2. Critically examining model method, structure, and parameters, perhaps extending the criticism to include structural testing and model validation not produced by the modelers themselves, and comparison of the model with other models.
3. Analyzing model output, studying it both mathematically and in terms of real world significance, and comparing it to what is known from other sources.
4. Documenting one's findings in a fashion that makes them accessible to critical review and examination by others.

If one prefers to contract research with an existing global modeling group, much of the work mentioned above can be conducted by the modelers themselves. It must be realized, however, that global modeling has not generally adhered to (or been rewarded for adhering to) rigorous scientific standards. Unless one is willing to insist on—and pay for—upgrading the standards of practice, one is likely to end up with results that will not withstand critical review.

REFERENCES

Adams, F. G., and Klein, S. A. (Eds.) (1978). *Stabilizing World Commodity Markets*. Lexington Books, Lexington, Massachusetts.

Alexandrov, V. V., Lotov, A. V., Moisseiev, N. N., and Svirezhev, Yu. M. (1981). Global models. The biospheric approach: Theory of the noosphere. Paper presented at the *Global Modeling Forum, Global Modeling at the Service of the Decision Maker*. International Institute for Applied Systems Analysis, Laxenburg, Austria, 14–18 September 1981.

Barney, G. O. (Study Director and Ed.) (1980). *The Global 2000 Report to the President of the U.S.: Entering the 21st Century. Vol. 2: The Technical Report*. Pergamon Press, New York.

Barney, G. O. (Study Director and Ed.) (1981). *The Global 2000 Report to the President of the U.S.: Entering the 21st Century. Vol. 3: Documentation on the Government's Global Sectoral Models: The Government's 'Global Model'*. Pergamon Press, New York.

Bottomley, A., Ergatoudes, M., Carpenter, L. C., and Lloyd, M. (n.d.). A six sector input–output world model. Undated manuscript. This and other data and material are available through Dr. A. Bottomley, University of Bradford, Bradford, W. Yorkshire BD 7IP, UK.

Bremer, S. A. (1981). *The GLOBUS Project: Overview and Update*. IIVG/dp 81-109. International Institute for Comparative Social Research, Wissenschaftszentrum, Berlin.

Cochrane, W. W., and Danin, Y. (1976). *Reserve Stock Grain Models for the World, 1975–1985*. In Eaton, D. J., *et al.* (Eds.) *Analysis of Grain Reserves, A Proceedings*. USDA Economic Research Service Report No. 634. US Dept. of Agriculture, Washington, DC.

Eaton, D. J. (1980). *A Systems Analysis of Grain Reserves*. US Dept. of Agriculture, Economics, Statistics and Cooperatives Service, Technical Bulletin No. 1611, Washington, DC.

Eaton, D., Steele, W. S., Cohon, J. L., and ReVelle, C. S. (1976). A method to size world grain reserves: Initial results. In Eaton, D. J., and Steele, W. S. (Eds.) *Analysis of Grain Reserves, A Proceedings*. USDA Economic Research Service Report No. 634. US Dept. of Agriculture, Washington, DC.

Enzer, S., Drobnick, R., and Alter, S. (1978). *Neither Feast Nor Famine*. Lexington Books, Lexington, Massachusetts.

Food and Agriculture Organization (FAO) (1971). *A World Price Equilibrium Model*. General Commodity Analysis Group, Commodities and Trade Division and Research Division, UNCTAD. Rome.

Food and Agriculture Organization/UN Fund for Population Activities (FAO/UNFPA) (1979). *Report on the Second FAO/UNFPA Expert Consultation on Land Resources for Populations of the Future*. Food and Agriculture Organization of the United Nations, Rome.

Forrester, J. W. (1971). *World Dynamics*. Wright and Allen Press, Cambridge, Massachusetts.

Herrera, A. O., Scholnik, H. D., *et al.* (1976). *Catastrophe or New Society? A Latin American World Model*. International Development Research Center, Ottawa.

Holcomb Research Institute (1976). *Environmental Modeling and Decision Making*. Praeger, New York.

Hughes, B. B. (1980). *World Modeling*. Lexington Books, Lexington, Massachusetts.

Hughes, B. B. (1982). *International Futures Simulation: User's Manual*. University of Denver, Denver, Colorado.

International Institute for Applied Systems Analysis (IIASA) (1981). *Energy in a Finite World: A Global Systems Analysis*. Ballinger, Cambridge, Massachusetts.

Jäger, J. (1983). *Climate and Energy Systems*. Energy Systems Program, International Institute for Applied Systems Analysis, Laxenburg, Austria.

Johnson, D. G., and Sumner, D. (1976). An optimization approach to grain reserves for developing countries. In Eaton, D. J., and Steele, W. S. (Eds.) *Analysis of Grain Reserves: A Proceedings*. USDA Economic Research Service Report No. 634. US Dept. of Agriculture, Washington, DC.

Kaya, Y., and Onishi, A. FUGI is documented in short articles in many places (for *listing see Meadows et al.*, 1982), but no single document stands out as a key reference. For complete documentation, it may be advisable to contact Kaya and/or Onishi directly. Kaya is at the Dept. of Electrical Engineering, University of Tokyo, Hong 7-3-1, Bunkyolku, Tokyo 113; Onishi at the Dept. of Economics, Soka University, 1-236 Tangi-machi, Hachioji-shi, Tokyo 192.

Krapivin, V. F., Moisseiev, N. N., Svirezshev, J. M., and Tarko, A. M. (n.d.). *On a Systems Approach to Risk Assessment of Human Impact on the Environment* (manuscript, translation from Russian). Laboratory of Mathematical Ecology of the Computing Center of the USSR Academy of Sciences, Moscow.

Labys, W. C. (1978). Commodity markets and models: The range of experience. In Adams, F. G., and Klein, S. A. (Eds.) *Stabilizing World Commodity Markets*. Lexington Books, Lexington, Massachusetts.

Leontief, W. (1977). Structure of the world economy: Outline of a simple input-output formulation. *American Economic Review* (December 1977), 823–834.

Leontief, W., Carter, A., and Petri, P. (1977). *Future of the World Economy: A United Nations Study*. Oxford University Press, New York.

Linnemann, H., De Hoogh, J., Keyzer, M. A., and Van Heemst, H. D. J. (1979). *MOIRA: Model of International Relations in Agriculture*. North-Holland, Amsterdam.

Linstone, H. A., and Simmonds, W. C. (1977). *Futures Research: New Directions*. Addison-Wesley, Reading, Massachusetts.

Liverman, D. M. (1983). *The Use of a Simulation Model in Assessing the Impacts of Climate on the World Food System*. University of California, Los Angeles-National Center for Atmospheric Research Cooperative Thesis No. 77. NCAR, Boulder, Colorado.

Mason, R. O. (1976). The search for a world model. In Churchman, C. W., and Mason, R. O. (Eds.) *World Modeling: A Dialogue*, Volume 2. North-Holland, Amsterdam (Elsevier, New York).

Meadows, D. H., *et al.* (1972). *The Limits to Growth*. Universe Books, New York.

Meadows, D. H. (1974). The population sector. In Meadows, D. L., *et al.*, *Dynamics of Growth in a Finite World*. Wright Allen Press, Cambridge, Massachusetts.

Meadows, D. H., Richardson, J., and Bruckmann, G. (1982). *Groping in the Dark: The First Decade of Global Modeling*. John Wiley & Sons, New York.

Meadows, D. L. (1970). *The Dynamics of Commodity Production Cycles*. Wright Allen Press, Cambridge, Massachusetts.

Mesarovic, M., and Pestel, E. (1974). *Mankind at the Turning Point*. E.P. Dutton, New York.

Morgan, D. (1980). *Merchants of Grain*. Penguin Books Ltd., Middlesex, England.

National Defense University (NDU) (1983). *World Grain Economy and Climate Change to the Year 2000: Implications for Policy*. National Defense University, Fort Lesley J. McNair, Washington, DC.

Neunteufel, M. (1977). *The State of the Art in Modelling of Food and Agriculture Systems*, Research Memorandum 77–27. International Institute for Applied Systems Analysis, Laxenburg, Austria.

Neunteufel, M. (1979). Modelling food and agricultural systems, a state-of-the-art study. *Food Policy*, **4** (2), 87–94.

O'Brien, P. (1980). Food projection. In Barney, G. O. (Study Director and Ed.) *The Global 2000 Report to the President of the U.S.: Entering the 21st Century. Vol. 2: The Technical Report*. Pergamon Press, New York.

Office of Technology Assessment (OTA) (1983). *World Futures and Public Policy*. Technology Assessment Board, OTA, Washington, DC.

Oltmans, W. L. (1974). *On Growth*. Putnam, New York.

Parikh, K. (1981). *Exploring National Food Policies in an International Setting: The Food and Agriculture Program of IIASA*. WP-81-12. International Institute for Applied Systems Analysis, Laxenburg, Austria.

Parikh, K., and Rabar, R. (Eds.) (1981). *Food for All in a Sustainable World: Summary of Presentations at IIASA Food and Agriculture Program Status Report Conference, February 1981*. SR-81-2. International Institute for Applied Systems Analysis, Laxenburg, Austria.

Quirk, W. J. (1981). *Energy Supply Interruptions and Climate*. UCRL-86254, Rev. 1. Lawrence Livermore National Laboratory, Livermore, California.

Roberts, P. C. (1977). *SARUM 76—Global Modelling Project*. Research Report No. 19. UK Departments of Environment and Transport, London.

Roberts, P. C., Norse, D., Phillips, W. B. G., Jones, J. B., Parker, K. T., and Goodfellow, M. R. (1977). The SARU model 1976. In Bruckmann, G. (Ed.) *SARUM and MRI: Description and Comparison of a World Model and a National Model: Proceedings of the Fourth IIASA Symposium on Global Modelling, 20–23 September 1976*. Pergamon Press, Oxford.

Robinson, J., and Ausubel, J. H. (1983). A game framework for scenario generation for the CO_2 issue. *Simulation and Games*, **14** (3), 317–344.

Rojko, A. S., and Schwartz, M. W. (1976). Modeling the world grains, oilseeds and livestock economy to assess the world food prospects. *Agricultural Economics Research*, **28**, 89–98.

SARU (Systems Analysis Research Unit) (1978). *SARUM Handbook*. UK Departments of Environment and Transport, London.

Shah, M. M., Fischer, G., Higgins, G. M., and Kassam, A. H. (1981). *Food Production Potential and Assessment of Population Supporting Capacity: Methodology and Results for Africa*. CP-81-6. International Institute for Applied Systems Analysis, Laxenburg, Austria.

Sichra, U. (1981). The FAP data bank. In Parikh, K., and Rabar, F. (Eds.) *Food for All in a Sustainable World: Summary of Presentations at IIASA Food and Agriculture Program Status Report Conference, February 1981*. SR-81-2. International Institute for Applied Systems Analysis, Laxenburg, Austria.

Stahl, I., and Ausubel, J. H. (1981). Estimating the future input of fossil fuel CO_2 into the atmosphere by simulation gaming. In Fazzolare, R. A., and Smith, C. B. (Eds.) *Beyond the Energy Crisis: Opportunity and Challenge*. Pergamon Press, Oxford.

Takayama, T., Hashimoto, H., and Schmidt, S. (1976). *Projection and Evaluation of Trends and Policies in Agricultural Commodity, Supply, Demand, International Trade, and Food Reserves*. University of Illinois Agricultural Extension Service, 4405 (Part 1), 302 pages. University of Illinois, Urbana-Champaign.

US Department of Agriculture (1978). *Alternative Futures for World Food in 1985*, Volume 1 of *World GOL Model Analytical Report*. Foreign Agricultural Economics Report No. 146, US Government Printing Office, Washington, DC.

Ward, M. D., and Cusack, T. R. (1981). *Report on the State of GLOBUS: Modeling Domestic Economic and Political Interface in Contemporary Nations*. International Institute for Comparative Social Research, Wissenschaftszentrum, Berlin.

Climate Impact Assessment
Edited by R. W. Kates, J. H. Ausubel and M. Berberian

CHAPTER 19
Biosphere Models

N. N. MOISSEIEV, YU. M. SVIREZHEV, V. F. KRAPIVIN AND A. M. TARKO

Computing Center of the USSR Academy of Sciences
40 Vavilova Str.
Moscow 117333, USSR

19.1 INTRODUCTION

Questions about conditions of human life in the biosphere belong to an untraditional science, global ecology. The problems of global ecology are unprecedented from a scientific point of view. At global scales, investigation of the state of the biosphere must be the basic object of the investigation of ecological processes. The homeostasis of humankind as a species is a key question for analysis.

A model described below demonstrates the possibility in principle for the joint numerical analysis of biospheric processes. It offers the possibility of coordinating information from different physical domains. It also offers a common language for contacts between researchers with different fields of interest. The ocean, land and atmosphere are selected as objects of analysis. The land is subdivided into regions, while the ocean is subdivided into aquatories, and the atmosphere is described by a point model. Climatic conditions influence various processes on land and in the oceans, and are themselves partly formed by them.

19.2 COEVOLUTION OF MAN AND THE BIOSPHERE

There is a deep interdependence of all processes taking place on Earth—geological, chemical and biological. Vernadski (1926, 1944) was among the first to show that the entire Earth, its landscapes as well as its hydrosphere and atmosphere, are indebted to living processes, to the living components of the biosphere. As the development of our planet proceeds, the role of life becomes increasingly influential in its further destiny.

In arriving at a logical completion of the evolutionary system of the biosphere, Vernadski developed the concept of the noosphere, that is, the sphere of the human mind. Gradually, man's mind creates a civilization that is able to influence the natural course of the Earth's evolution in a purposeful way. Gradually, that part of our planet which is becoming accessible to the active will of individuals is transforming itself into an organism, a system possessing its own goals together with the possibilities for achieving them. The outcome of Vernadski's teachings proved to be a conception of the integral unity of man and the biosphere and of the integral unity of man within the biosphere, according to which a natural stage in the development of the biosphere is its gradual transformation into a single element that may be described as a system possessing common developmental goals.

Uncovering mechanisms governing development, as scientists have done over the past centuries, begins to make it possible to glimpse into the future and to provide corresponding quantitative as well as qualitative estimates. It is hardly possible to overestimate the importance of this fact, for if man is able to forecast the outcome of his achievements, then he also acquires the possibility of guiding the course of events in a purposeful, goal-oriented way. And it is natural that he begin to make use of this power. But guidance has meaning only when the objectives for which particular actions are taken are clearly perceived.

It follows from Vernadski that the general objective of the development of civilization is to provide for a coevolution of man and the biosphere. The power of human civilization and its ability to influence the course of the planet's evolution is becoming so significant that, in principle, it can disrupt the condition that has already been established that we may call its equilibrium. Today, of course, man is still not able to destroy the entire biosphere. But under the influence of man the biosphere can shift to a new equilibrium mode and to conditions within which there may be no room for mankind. This is why it is important to have evaluations of alternative forms of human activity that will not disrupt the homeostasis of mankind as a species and that, instead of destroying man's joint development with the biosphere, will enhance it. These propositions serve as a point of departure for developing a scientific research program within which the conditions of man's coevolution with the biosphere can be studied from a wide diversity of points of view.

19.2.1 Style and Objectives

One of the principal difficulties with such studies is their broad systemic character. It is difficult to localize any particular object of the biosphere and study it independently. The biosphere is an integral whole, a single system possessing a high degree of mutual interdependence. For example, industrial discharges of CO_2 into the atmosphere may, after several decades, produce shifts in the behavior of the atmosphere over all the regions of the planet and changes in photosynthesis. Similarly, it is not possible to describe many processes taking place on land without considering ocean waters thousands of kilometers away. Thus, within the design of a research program, purely methodological problems will assume great importance. In what way should research on a system whose elements represent objects that are so varied in their physical content be organized? How should one cope with the fantastic dimensions of the problems that emerge? How should one create a simplified version of the models? Today we do not have sufficient experience, and its acquisition will turn out to be an important step in the realization of such a global research program.

The study of the properties of the biosphere as a single integral unit is still primitive. It is certainly insufficient for expressing any definitive opinion about the choice of strategies for the development of society. For this reason, providing for coevolution studies will require an extensive scientific program of interdisciplinary research. It will have to include the participation of specialists of the most varied professions—biologists and climatologists, physicists and economists, mathematicians and agriculturists, and more. It will also require the participation of specialists in the humanities. The manifold activity of these efforts should be united, the information should be integrated, and international programs should be established to solve particular problems. On what basis can such a unification of the efforts of researchers be achieved?

First, a common language should be created, which will enable specialists in various fields of knowledge to know about the research of their colleagues working in other areas of the same program, as well as to understand concrete objectives (and in a number of cases contents). Such a language can only be a language based on a formalized description, that is, the language of mathematical models.

Thus, we consider our first objective to be the creation of a system of models which would serve as a framework for a general scheme of concrete research, a basis for the planning and management of international research programs directed towards the elaboration of strategies providing for the coevolution of mankind and the environment. This system of models should be created in a form for use in a conversational 'man-machine' mode. We consider one of the results of our activities to be the fact that the still primitive system of models that has been developed until now has already helped us in our contacts with specialists in the fields of biology, soil science and other areas. The creation of an informational base has, in essence, turned into a discussion of plans for joint work.

To summarize, the present stage of the history of the civilization of *Homo sapiens*, in our view, requires the intervention of the human intellect in the formulation of principles for its further development. But this intervention may be justified only if a new scientific discipline emerges, that is, the dynamics of the noosphere, the study of the conditions of coevolution. It is the elaboration of the initial positions for the creation of the instruments of this discipline that is the principal objective of the efforts that we have undertaken.

19.3 GENERAL DESCRIPTION OF THE MODEL

Studies connected with the construction of a system of mathematical models simulating the functioning of the biosphere are being developed in three directions: simulation of processes of climate, of the biota, and of human activity. These studies are, to some extent, carried out independently. The organization of such studies is conditioned by the difficulties of complex investigation and by the necessity to refine some principles and to understand special features and methods of description. Simultaneously, it generates certain duplication. For example, a description of geochemical cycles is impossible without taking into account their impact on the climate, while a description of human activity must be connected to the conditions of environmental evolution. Therefore, one has to introduce deliberately some simplifications, parametrization of large blocks by hypothetical relationships that should be specified later when developing other studies. Each of the three main directions has the possibility of being pursued independently. We call our current descriptions, or models of the blocks, study training versions, as they are used today mainly for training the investigators themselves.

The principal obstacle that has been encountered is the poor state of knowledge concerning the principles that govern ecological and climatic processes, and the insufficiency of information concerning corresponding relationships. It is therefore important to consider not only the variability of model parameters, but also, in some cases, the possibility of representing the dynamic processes themselves in alternative ways.

Biogeocenosi are the biosphere's elementary units. Because the biosphere is made up of interacting blocks, each of which may be represented in varying degrees of detail and may embody varying degrees of autonomy, it is possible to increase the level of detail within individual blocks without interfering with others and without altering the overall model structure.

Time horizon extends to decades and centuries. These are the characteristic time spans of man–environment interactions. We do not consider processes lasting millions of years (the duration of geological ages), for it is unlikely that the formation of mountains or the accumulation of mineral resources will produce substantial changes during the time intervals that are being considered. Similarly, processes lasting less than one year are omitted. By accepting average yearly values of changes as our measure of precision, we also select one year as the time-step to employ in simulation experiments.

The following spatial blocks are distinguished within the biosphere: the atmosphere, the ocean, and regions within the land mass. The land mass is subdivided into regions that may be related to natural, economic or political boundaries. The atmosphere and the ocean are common to all. Because mixing processes in the atmosphere and in the surface layers of the ocean are rapid, no problems result from such an assumption. (The characteristic time of mixing processes in the atmosphere is of the order of several months.)

The state of each block (see Figure 19.1) is defined by a set of variables that jointly constitute a vector of major state variables. The selected variables have been well studied in aggregated form; we know their qualitative and sometimes also their quantitative relations. However, we are in an early stage of synthesis of available information that reflects the structure of both internal and external biogeochemical, ecological, social and economic relations. We have sought to make full use of information available in the literature on biospheric processes and their qualitative characteristics.

The model itself is formalized as a Cauchy problem for a system of ordinary nonlinear differential equations designed to reflect all major relations between components of the biospheric regions under consideration. It is programmed in FORTRAN-IV and is employed as a learning program at the Computing Center of the USSR Academy of Sciences and at Moscow University. The model contains more than 400 coefficients requiring quantitative data and approximately 200 relations requiring mathematical descriptions. Table 19.1 contains a list of model parameters that are common to all blocks. Quantitative values used in the model are not shown, but have been established from numerous sources (Kovda, 1975; Singer, 1975; Krapivin *et al.*, 1982). A more detailed description of the model is contained in Moisseiev *et al.* (1978).

19.3.1 The Atmosphere Block

The basis for the atmosphere block is given by a description of the flux of solar radiation. The atmosphere regulates the flux, secures the stability of the radiation balance, and creates necessary conditions for life on our planet.

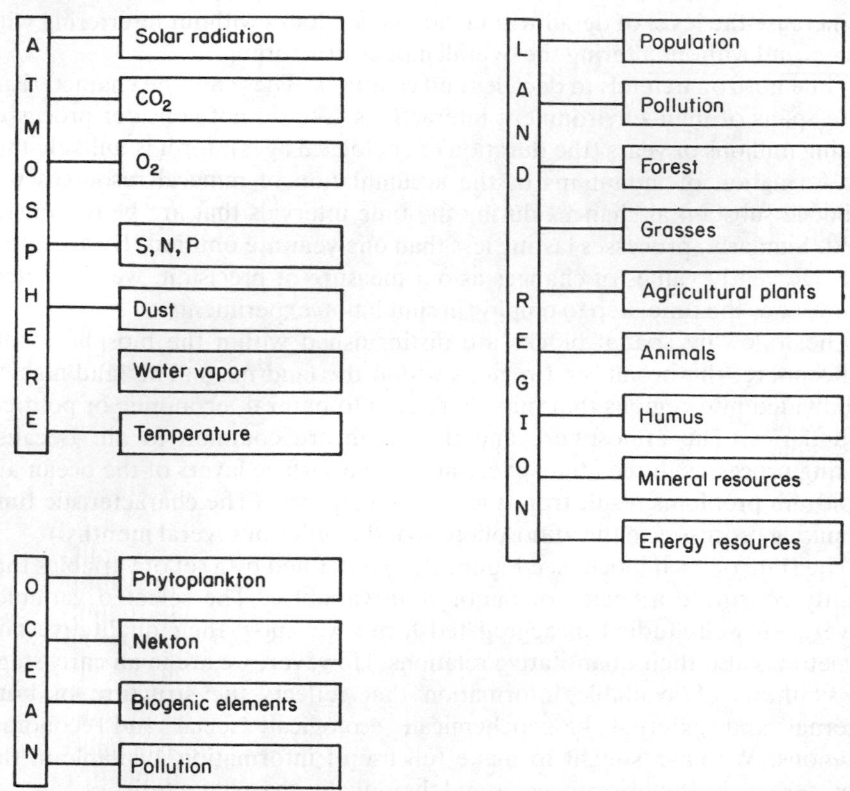

Figure 19.1 Spatial blocks within the biosphere and the variables

The upper layers of the atmosphere receive radiation from the sun, evaluated on the average by $E_o(t) = 1.94$ cal cm^{-2} min^{-1}. The atmosphere reduces this flux, partially absorbing and reflecting it. Analysis of experimental data allows the following approximation for description of this process: $E(t) = E_o(t)\exp(-\alpha B(t) - \beta)$, where α is the coefficient of the solar energy absorption by the atmosphere due to dust and cloudiness, β is the transparency index of the clear atmosphere, and B is the measure of atmospheric dust. According to available climatological data we take $\alpha = 0.144$, $\beta = 0.477$.

Propagation of solar radiation energy through the ocean depends on water transparency and can be described by the following relation: $E(z,t) = E(t)\exp(-\alpha_1 z)$, where z is the depth (in meters) and α_1 is the vertical radiation reduction, α_1 being an integral characteristic of water quality.

A great influence on the general radiation balance of the Earth is exerted by its surface albedo. The albedo depends on the state of the land and ocean—area of vegetative cover and urbanized surface, and roughness of

Table 19.1 Selected model parameters

1. Area of biosphere: land, ocean, forests, arable land, land under cultivation, land under human structures
2. Level of solar radiance at Earth's surface at which glaciation may occur
3. Efficiency of solar energy utilization: in the biosphere, on land, in oceans
4. Albedo of the Earth–atmosphere system
5. Total mass of atmosphere
6. Average temperature of atmosphere at Earth's surface
7. Volume of dust in atmosphere
8. Volume of dust entering the atmosphere
9. Biomass of the biosphere: land, forests, nekton
10. Humus in soil
11. Net primary production of biosphere: on land, in the ocean
12. Consumption of primary products: by nekton, by animals, by human population
13. Volume of food products extracted by man: on land, in the ocean
14. Volume of nekton catch
15. Reserves of animal food
16. Global reserves of coal
17. Energy consumption per capita in Europe
18. Rate of annual increase in average per capita energy consumption
19. Rate of increase of consumption of nonrenewable resources
20. Volume of industrial wastes released into the atmosphere
21. Volume of pollutants entering the ocean
22. Population growth in 1970: birth rate, death rate

Based on published accounts (Kovda, 1975; Singer, 1975; Krapivin *et al.*, 1982)

ocean surface, for example. In a strict formulation, albedo should be a function of state variables of a model, but we initially consider it as a constant, assuming that its changes during the time periods under consideration are negligible.

Changes in the amount and distribution of biota alter the cycling of carbon and can generate climatic changes that, in their turn, can influence biotic processes. The current state of climate research does not allow accurate prediction of regional climatic changes due to anthropogenic influences. Therefore, we consider the part of the model that simulates the climate as a point model, that is, the state of the atmosphere is described by values of solar radiation energy, by the general amount of CO_2 and other constituents of the atmosphere, and by the mean global atmosphere temperature near the Earth's surface. Such a choice of variables allows us to take into account, more or less exactly, the influence of human activity on the climate of the planet and of climatic factors on biological processes of the land and ocean.

The sensitivity of atmospheric temperature to changing CO_2 content has been calculated (Krapivin *et al.*, 1982); the results are generally consistent. For this model the relation is taken in accord with the calculations of

Rakipova and Vishnyakova (1973). The influence of changing temperature on the moisture and precipitation regime has not yet been included.

Among anthropogenic factors, there are several that might impact on climate. These include the burning of fossil fuels and changing land use, which release CO_2; other industrial and agricultural activities that may release 'greenhouse' gases; industrial activity that results in increasing aerosol release into the atmosphere and disturbs the thermal regime of the planet; and power plants that excrete a great amount of heat (10^{20} Jy^{-1}).

The presence of oxygen in the atmosphere defines many metabolic processes. Although according to current evaluations the amount of oxygen in the biosphere is stable, the model takes into account the actual regimes of oxygen exchange between different biosphere components and regions, as well as the possibility of equilibrium disturbances (see Table 19.2).

Table 19.2 Selected parameters of the oxygen cycle

Content of O_2 in the atmosphere by volume, weight
Volume of O_2 released through photosynthesis: land plants (including forests), aquatic plants
Volume of O_2 consumed each year on burning fuel
Yearly loss of O_2
O_2 required to oxidize 1 mg of petroleum
Reduction in coefficient of exchange of O_2 between water and the atmosphere in the presence of a 40-μm layer of petroleum

One of the most intensive and important biospheric processes is the circulation of carbon. The presence of carbon in the atmosphere, mainly in the form of CO_2, defines to a considerable extent the climate of the planet. The climatic conditions and amount of CO_2 in the atmosphere at any given time are, in their turn, factors on which depend the intensity of both the atmospheric CO_2 assimilation by plants and its release into the atmosphere, resulting from decomposition of dead organic matter in the soil. The CO_2 content in the atmosphere is defined by the balance between these two processes, as well as processes in the oceans. It is also necessary to take into acount CO_2 release due to volcanic activity and the influence of anthropogenic factors.

When coal, oil, gas or wood burn, CO_2 is released into the atmosphere (during one year an amount at present equal to about 0.7 percent of the amount of atmospheric CO_2, or about 10 percent of the CO_2 assimilated for construction of plant biomass). Destruction of forests and reduction of other areas covered by vegetation increase the scale of anthropogenic influence on the cycle of CO_2. The gas exchange between the atmosphere and the ocean is described by Machta's (1971) model, and the land-atmospheric exchange by Tarko's (1977) model. Some characteristics of global carbon processes are given in Table 19.3. In addition to carbon, the representations are derived from available literature (Krapivin *et al.*, 1982) about the nitrogen, sulphur and phosphorus cycles in the biosphere using rather simplified schemes of their circulation.

Table 19.3 Selected parameters of the carbon dioxide cycle

Concentration of CO_2 in the atmosphere in 1970
Mass of CO_2 in the atmosphere
Assimilation of CO_2: by land plants, by phytoplankton
Average yearly addition of CO_2 (1962–65, 1970–71) through burning of fossil fuels
Rate of production of CO_2 by human population
Rate of addition of CO_2 from the Earth's core (volcanic CO_2)
Rate of addition of CO_2 from humus decomposition and respiration
Mass of CO_2 dissolved in the hydrosphere
Volume of CO_2 contributed by a person or animal through respiration
Optimal concentration of CO_2 for respiration and photosynthesis

The atmosphere also contains water in a gaseous state, and the character of the water cycle is determined by many biospheric processes. The relative content of water in the atmosphere is not large, but it is of great importance as one of the basic factors of atmospheric turbidity (cloudiness), defining climate in many respects. Moreover, the productivity of plants depends substantially on rates and distribution of precipitation and humidity.

In the model under consideration we have accepted a simplified scheme of the water cycle, taking into account water vapor fluxes between land regions and between land and ocean; atmospheric precipitation; evaporation from the ocean and land surfaces; and the transpiration of plants.

19.3.2 The Ocean Block

Like the atmosphere block, the ocean block is represented by a point model, that is, a model with ideal mixing. It contains the following variables: contents of phytoplankton, zooplankton, nekton, and biogenic nutrient elements; the extent of ocean pollution; and CO_2 content in the upper mixed layer and in the deep ocean.

The importance of the ocean in determining the composition of the atmosphere is also taken into account in the model, mainly in the exchange of CO_2. The exchange of CO_2 between the atmosphere and the ocean depends to a large degree on the temperature of the atmosphere and on the CO_2 content in the air, which has been increasing during recent years. In other words, the CO_2 exchange depends directly on human activities.

While the ocean is of great importance as a food source for man, it contributes only about 1 percent (energy equivalent) of the total gross production of food (Vinogradov and Monin, 1976). Biospheric processes in the ocean are described by equations obtained by Vinogradov *et al.* (1975). Fishing is defined by the catch strategy chosen by a land region. Effects on reproduction of fish resources are not included. Photosynthesis is the main source of organic matter in the ocean. Photosynthetic rate is partly a function of light intensity and decreases with deviation of light intensity from an optimum value. Photosynthesis also depends on temperature, concentration of nutrient and other elements, phytoplankton biomass, and other parameters. Equations for phytoplankton behavior can be derived by using the various phenomenological relations.

We assume that Liebig's law of limiting factors is correct (Taylor, 1934). According to Liebig's law, the photosynthetic rate at each moment depends only on one factor, although the factors themselves can change in the process of system development.

The model includes vertical cycling of biogenic nutrient elements. It is assumed that biogenic element reserves are not limited in deep ocean layers (deeper than 200 meters), so that they can limit photosynthesis only in upper layers. Biogenic elements are taken out of the upper layer partly through the sinking of feces and dead organic matter. The assimilation rate of biogenic elements by phytoplankton is proportional to the photosynthetic intensity.

The rate of change of the nekton biomass is regulated by the temperature of the environment, by the character and intensity of trophic relations (ratios), by natural mortality, by mortality due to pollution, and by mortality due to population exploitation (fishing).

The introduction of a generalized element like nekton into the model is only a rough approximation for the real structure of the system. However, the approximation seems to be satisfactory for the initial version of the model, since nekton is the highest level in the trophic pyramid in the ocean ecosystem and its energy content is rather small in comparison with the other levels. Accuracy of the nekton description influences the global element and energy cycles only to a small extent. On the other hand, nekton is a human food resource, and further detail in its description will be necessary for a more refined representation of the structure of food relations in the dynamics of human populations.

19.3.3 Land Regions

Land vegetation is characterized by a large variety of species and by wide ranges of productivity and rates of exchange with other media. The gross primary production of land varies from 28,140 kcal m^{-2} y^{-1} in tropical woods to 489 kcal m^{-2} y^{-1} in a desert. The concept of a land region is introduced in the model to take into account all the variety of vegetation forms and to keep the point character of the model blocks. It allows an approximate account of the zonal character of the Earth's vegetation without introducing space coordinates. The model involves three types of vegetation: woods, agricultural plants and natural grass vegetation.

Change in biomass for each type of vegetation is described by a first-order differential equation in which the photosynthetic rate is a complex function of light intensity, humidity, humus state, the amount of fertilizers put into the soil, the composition of the atmosphere, and so forth. The rate of decay of plants is also taken into account, as well as consumption of plant biomass by animals and people.

Photosynthetic production is consumed by the animals of the Earth. They are treated in the model as a single-state variable of the biosphere. The variable is characterized by average growth and death rates. Animal bioproduction is in turn consumed by man.

Soil-forming processes are of great importance in the global element and energy cycles. The numerous stages of soil humus formation, which is the last link in a chain of biogeochemical organic matter transformations, are described by one component—'humus'.

There are several estimates of the amount of humus in the biosphere (Kovda, 1975), but the estimates do not differ much. An average estimate equal to $2.3 \cdot 10^{12}$ tons has been taken in the model. The model is not very sensitive to variation in this parameter.

It must be noticed that the humus is important for the composition of the atmosphere. The model assumes that the rate of change of humus depends on the intensity of accumulation of the organic waste of plants. It is assumed that the decomposition rate is directly proportional to the mass of decomposing substance, and that it increases exponentially with temperature. It is taken into account that a considerable rise or fall in temperature depresses the activity of soil microorganisms and, hence, decreases the decomposition rate. It has also been taken into account that the humus decomposition rate decreases with deviations of humidity from some optimum value.

The description of the demographical block is the most difficult part of the simulation. While relations of cause and effect, including a mortality coefficient (more exactly, an index of mean-life duration), have been specified, there is no unique understanding of the processes describing the birth-rate dynamics in the *Homo sapiens* population. Extrapolations based upon demographical statistics are not appropriate because the evolution of the global system is computed for many decades ahead, and previous demographic experience shows how rapidly and inexplicably basic demographic parameters can change. Thus, the demographic dependences accepted in the model are hypothetical to some extent. The model representation is, however, more sophisticated than a simple model of exponential growth. Parameters are verified by demographic data and, in contrast with some other models, they allow for inverse cause-effect relationships.

19.3.4 Human Activities

It is only in recent years that man has begun to play one of the leading roles in the biosphere. And while until recently it was possible to speak of evolution of man as an *element* of the biosphere, his increasing independence from the biotic environment creates a 'technological' environment for his existence. Also the increasing load on the biosphere, as well as the comparable roles of anthropogenic and biogenic cycles of matter and energy, make it now appropriate to refer to the *joint* evolution (coevolution) of the biosphere and of *Homo sapiens*.

Let us consider some figures. During the last 30 years man has consumed as much mineral raw material as during his entire preceding history. Between 1950 and the early 1970s, the production of energy per capita was growing exponentially, with an average yearly increment of 4–5 percent. Man retrieves from ocean and land ecosystems $17 \cdot 10^6$ tons per year and $1.3 \cdot 10^9$ tons per year, respectively, of food products. And while the biosphere still possesses a certain reserve for increasing the productivity of its ecosystems, the current level of their exploitation is already quite intense. It is in order to reflect these processes that other blocks representing human activities besides the 'population' block are included, namely pollution generation and energy production.

Anthropogenic influences on the environment are quite diverse. It was already noted in describing the atmosphere block that corresponding releases of CO_2 and aerosols into the atmosphere influence the planet's thermal regime.

But the influence of human activities is not limited to the atmosphere. Changes in the overall carbon dioxide cycle also influence humus formation. Another major anthropogenic influence relates to the depletion of easily accessible mineral resources. The influence of changes in the environment on human activities is equally diverse.

All of these factors are parametrized within the model by defining the following hypothetical time relations:

1. the rate of generation of pollution per capita (this is a characteristic of both the standard of life and the technology of social production);
2. an indicator of the rate of dissipation of pollution;
3. the intensity of mineral source utilization;
4. indicators of the rate at which available arable land is brought into the agricultural sphere and of the growing productivity of agricultural ecosystems;
5. the share of capital investment in agricultural development, the replenishment of resources, the development of new types of resources, and antipollution measures.

The human activities block embodies possibilities for changing over time the coefficients of mutual exchanges in resources and food among regions of the land mass, as well as for specifying similar changes in other parameters.

The large volumes of matter and energy that circulate in the biosphere are accompanied by releases of large quantities of components that can be harmful to the environment. Before 1800, humans had access only to those forms of solar energy that were released in the course of various biological processes. But during the last century and a half this situation has changed fundamentally and still newer prospects have emerged during the past 20 years, following the utilization of nuclear energy. For each person living in the United States of America, 10^4 W d^{-1} are expended, and this number could continue to increase. The average per capita consumption of all types of energy in Europe is 70,000 kcal d^{-1}. The total volume of energy that is produced has doubled several times in this century in a period of 10 years or so.

The world's growing population and associated growth in energy production causes a pollution of the atmosphere, of soil, and of water. In particular it tends to increase the environment's temperature. Available estimates indicate that a total of 2.7 billion tons of coal and 1.6 billion tons of petroleum are burned on our planet each year, while 1.25 million tons of pesticides are added to the soil. Each year $1.5 \cdot 10^{10}$ tons of CO_2 are added to the atmosphere and $9 \cdot 10^9$ tons of O_2 are expended. The biosphere's declining absorption capacity causes a yearly growth in the concentration of CO_2 and a reduction in the volume of O_2.

A generalized component called 'pollution' is considered in the model for purposes of simplification. It expresses only averaged characteristics of numerous types of pollutants. While this approach helps to keep the model simple, it is justified more by the absence of reliable knowledge concerning the influence of various types of pollutants on biogeocenotic processes.

Forrester's equation (1971) is employed in determining the rate of change in the concentration of pollutants in a region. It reflects increases that are proportional to population growth, as well as a coefficient dependent on time. It is assumed to increase as the material level of the population increases.

Pollutants operate as inputs in processes of natural decomposition and pollution neutralization activities. The latter proceed at a rate that is linearly dependent on the share of capital that is assigned to these purposes and is inversely proportional to the expenditures required to neutralize a unit of pollutant.

The ocean's pollution results from the total volume of all pollutants entering it from all land regions. They reduce its transparency, attenuate the growth in the biomass of its live components, and increase the coefficient of mortality.

A generalized Forrester type of equation is employed in the model to describe the depletion of all non-renewable resources in the biosphere. Possible changes in the initial conditions of the corresponding differential equation may be entered either discretely or gradually. Shifts to new mineral resources are represented as changes from one set of initial conditions to a series of others at specified time intervals.

19.4 EXPERIMENTS WITH THE MODEL

Preliminary computer calculations point to the logical consistency of model behavior but also to a need for greater precision within the existing structure blocks and a need for additional blocks. Experiments with the model require the specification of various scenarios. The manner in which they must reflect alternatives in the future development of society, alternative control policies, and estimates of rates of progress in science and technology call for a separate article. At the present time we shall simply describe the findings of the main experiment that has been carried out with the model so far. It addressed the following question: how will the biosphere evolve if existing trends in anthropogenic influences and scientific and technical progress are maintained?

The results are shown in Figure 19.2. The curves of the first of these figures (19.2A) describe the dynamics of CO_2 in the atmosphere, as well as those of average global temperature. The trajectories are carried to the year 2470, that is, over a relatively prolonged period of time. While these forecasts are extremely hypothetical, they do appear to provide a certain quantitative image of possible trends in the system's evolution. In particular we see that while the average temperature fluctuates in a period of approximately 200 years, it remains within an interval between 13 and 17 °C. Let us recall that some climatologists believe that should these boundaries be exceeded, a 'climatic catastrophe' could result. It is interesting that a situation arises in the

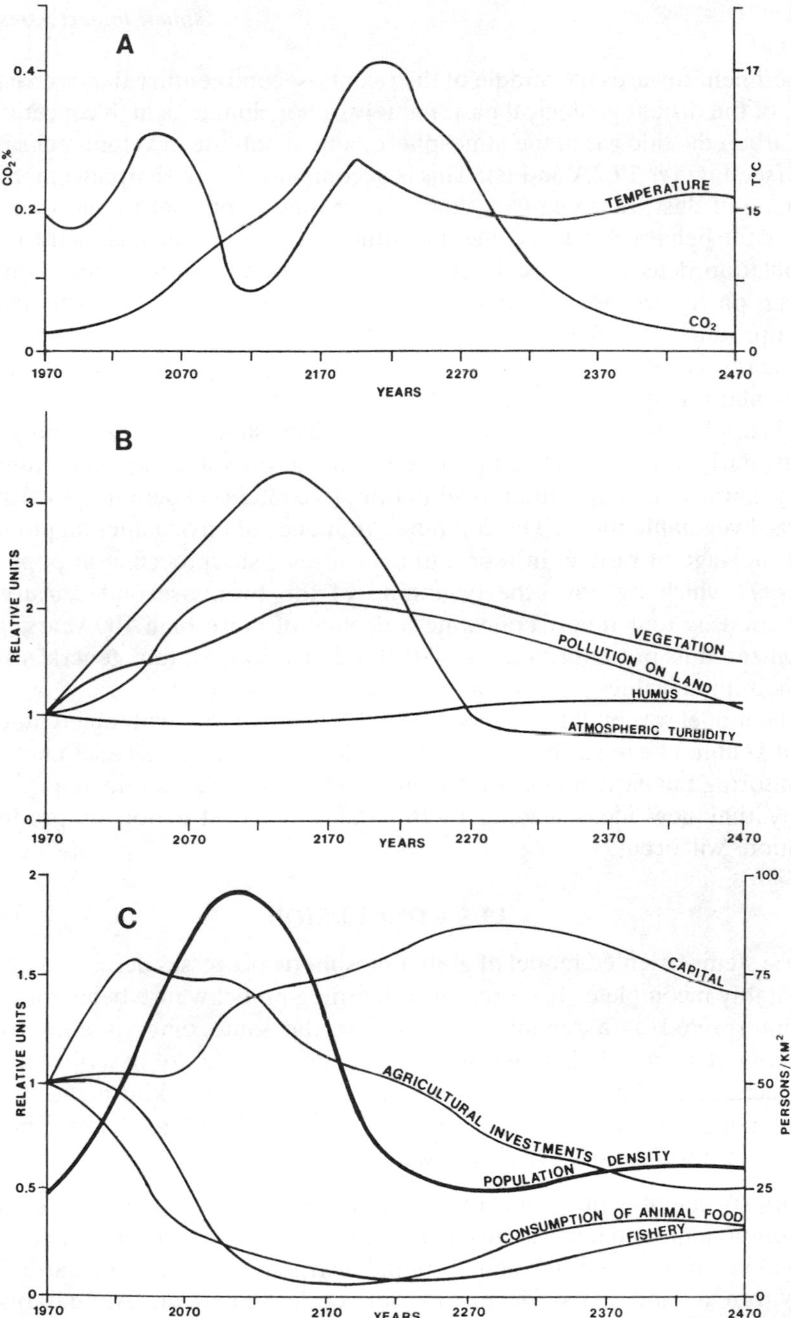

Figure 19.2 A: Projected evolution of the atmosphere under continuation of present trends; B: Projected state of the biosphere: hot climate, high concentration of oxygen, and intense development of plant life; C: Population density, certain economic factors, and evolution of agriculture and diet under assumption of initial maintenance of current trends

experiment towards the middle of the twenty-second century that is characteristic of the distant geological past, namely, a hot climate, a high concentration of carbon dioxide gas in the atmosphere, and an intense development of plant life (see Figures 19.2A and B). This is accompanied by a sharp increase in the volume of dust, reaching four times larger than at present by the year 2125. While the behavior of dust content in atmospheric pollution is similar to that of population density (Figure 19.2C), that of land pollution is more complex. This is partly because such pollutants are more 'durable', and partly because their production is closely associated with the economy. The curves in Figure 19.2C represent the behavior of population and certain economic factors. In particular it may be seen that rapid population growth causes the adequacy of food supplies to fall by nearly 50 percent. The shortage of proteins is then particularly serious. And while in terms of its caloric value the food ration does not generally meet the required minimum, its contents experience a sharp shift toward vegetable foods. The combined influence of environmental pollutants and shortage of protein in food rations causes a sharp decline in population density, which by now, the beginning of the twenty-second century, has increased by four times. Following a decline of more than 100 years it then stabilizes at its present level. By 2470 all other variables, too, generally return to their initial values (except for nonrenewable resources).

The model was built for the sake of experiment, and the obtained modeling results cannot be regarded as predictions. Moreover, if for decades mankind is monitoring the early phases of changes similar to those described above, it is likely that new ideas leading to the radical reconstruction of production relations will occur.

19.5 CONCLUSION

The systems-oriented model of global biospheric processes described above is inevitably incomplete. It represents a learning model whose behavior cannot be interpreted as a reliable forecast. At the same time, it does express fundamental causal linkages in nature and is able to describe relevant interactions in relatively flexible ways. The authors are working in the direction of further sophistication and detailing of the model and its blocks. The main guidelines for this work are as follows.

1. Improvement of functional forms for different relationships between model components and more accurate determination of their coefficients.
2. Taking into account the actual spatial distribution of plant formation types within a land mass for the carbon cycle submodel. Annual primary production and the rate of dead organic matter decomposition within a given cell are proposed to be a function of annual temperature and precipitation for each cell. Tentative results are described in Bazilevitch *et al*. (1982).

3. Development of a spatial climatic model that could provide annual temperature and precipitation for defined cells.
4. Addition of blocks reflecting economic and political factors.
5. Recognition of the multiplicity of objectives and corresponding differences among various groups, countries, and regions. The biosphere encompasses regions that differ with regard to their strategies and tactics in achieving social and economic development and in exploiting resources.

Finally, it is appropriate to conclude by mentioning the problem of critical parameters. In the evolution of any complex nonlinear system, it is possible to distinguish the relatively 'calm' course of the process when its parameters are not in the vicinity of their bifurcation values. In this situation a chance perturbation may disturb the state of equilibrium (to be more exact, the quasi-equilibrium) of the system only temporarily. The inner forces of damping that are inherent to nature return the system to its initial state (or else close to the initial one). It is another matter with situations that are close to the values of the 'carrying capacity of the biosphere', where new states of equilibrium may emerge.

There may be many states of equilibrium, and it may be impossible to predict to what state of equilibrium a system will come—it may depend on unforeseen or chance factors. In the case of the biosphere, we do not know of any other states of equilibrium except those that are observed. For this reason the study of carrying capacities is an important problem standing before science—we do not have any guarantee that the present-day load on the biosphere is sufficiently removed from a critical one. Indeed, there is an indirect basis to suppose that a number of characteristics of the biosphere are close to their critical values. For example, an increase in the average temperature by 3 or 4 °C could lead to the beginning of an irreversible melting of glaciers. What will the properties of the new state of equilibrium of the biosphere be like; will they permit the existence of man? We do not know.

ACKNOWLEDGMENT

We have the pleasure of thanking M. I. Bazilevitch and P. I. Medow for helpful comments.

REFERENCES

Bazilevitch, N. I., Vilkova, L. P., and Tarko, A. M. (1982). The model of biospheric processes to spatial distribution of land ecosystems. In Izrael, Yu A. (Ed.) *The Problems of Ecological Monitoring and Ecosystems Simulation*, Vol. 5, pp. 181–199. Gidrometeoizdat, Leningrad (in Russian).

Forrester, J. W. (1971). *World Dynamics*. Wright-Allen Press, Cambridge, Massachusetts: 189 pages.

Kovda, V. A. (1975). *Biochemical Cycles in Nature and Their Disturbance by Man*. Nauka, Moscow (in Russian).

Krapivin, V. F., Zvirezhev, Yu M., and Tarko, A. M. (1982). *The Mathematical Modeling of Global Biospheric Processes*. Nauka, Moscow: 272 pages (in Russian).

Machta, L. (1971). The role of the oceans and the biosphere in the carbon dioxide cycle. In Dryssen, D., and Jagner, D. (Eds.) *The Changing Chemistry of the Oceans*, pp. 121–145. John Wiley, New York.

Moisseiev, N. N., *et al.* (1978). Realizatsia na evm globalnoi modeli biosfery (Implementing a global model of the biosphere on a computer). In *Voprosy Matematicheskogo Modelirovania* (*Problems of Mathematical Modeling*). Institut Radiotekhniki i Elektroniki AN SSSR, Moscow.

Rakipova, L. R., and Vishnyakova, O. N. (1973). Influence of carbonic gas variations on the thermal regime of the atmosphere. *Meteorology and Hydrology*, No. 5, 23–31 (in Russian).

Singer, S. F. (1975). *The Changing Global Environment*. Reidel, Dordrecht.

Tarko, A. M. (1977). Global importance of the 'atmosphere-plants-soil' system in compensation of impacts on the biosphere. *Docl. of the U.S.S.R. Academy of Sciences*, **237**, 234–237 (in Russian).

Taylor, W. P. (1934). Significance of extreme or intermittent conditions in distribution of species and management of natural resources, with a restatement of Liebig's law of the minimum. *Ecology*, **15**, 274–379.

Vernadski, V. I. (1926). *The Biosphere*. Nautchtechizdat, Moscow (in Russian).

Vernadski, V. I. (1944). *Problems of Biogeochemistry: The Fundamental Matter-Energy Difference Between the Living and the Inert Natural Bodies of the Atmosphere* (translated by George Vernadsky, edited and condensed by G. E. Hutchinson). Connecticut Academy of Arts and Sciences, New Haven, Connecticut.

Vinogradov, M. E., Krapivin, V. F., Fleishman, B. S., and Shushkina, E. A. (1975). Use of a mathematical model for analysis of behavior of a pelagic ecosystem in ocean. *Oceanology*, **10** (2), 313–320 (in Russian).

Vinogradov, M. E., and Monin, A. S. (1976). On the way to the oceanic aquaculture. In *Science and Humanity*, pp. 112–125. Znanie, Moscow (in Russian).

Climate Impact Asessment
Edited by R. W. Kates, J. H. Ausubel and M. Berberian

CHAPTER 20
Scenario Analysis

LESTER B. LAVE AND DENNIS EPPLE

Graduate School of Industrial Administration
Carnegie-Mellon University
Pittsburgh, Pennsylvania 15213 USA

20.1 INTRODUCTION

Scenario analysis is a tool for addressing the magnitude and consequences of climate change and the steps that can be taken to prevent the change or mitigate the effect (Bell, 1964; Jantsch, 1967; Kahn and Wiener, 1967; Polak, 1971; Durand, 1972; Bunge, 1976). It is a style of analysis that has three principal uses, to be elaborated below. The first is jarring people out of a mindset that climate is fixed, that no actions can alter it or mitigate its effects; the tool can be used to 'stretch' people's minds to encompass a wider range of actions and their implications. The second is formal modeling of the causes and consequences of climate change and of actions that can alter them. The third is a method of integrating the contributions of such various disciplines as physics and law so that diverse experts can work together without having to learn all the intricacies of other areas. While the method can be applied to many areas, climate will be the focus here, with the effects of increased carbon dioxide in the atmosphere receiving special attention.

20.2 STOKING THE IMAGINATION

Anyone who thinks he or she can predict, or even characterize, the future is invited to perform the following exercise. Imagine what someone in early 1929 would have said about the following 50, or even the following 10 years. Try the same exercise for 1942 or even 1962. In 1929, economic conditions were wonderful, in the United States at least, and the recent past had seen marvelous innovations including the telephone, automobile and airplane. These hopes were dashed by the Depression and Second World War. In the early 1940s, the technologies that have since shaped our world were on scientists' benches—television, jet aircraft, computers, satellites, modern highways and a modern telephone system— but the devastations of war preoccupied our thinking. In 1962, one could have identified these technologies as important, but one could hardly have foreseen their consequences. On scientists' benches today are a host of inventions that we know little about. We can predict that microprocessors and recombinant DNA will recast the future, but it is impossible to know exactly how the world will be shaped by these technologies.

The three most unlikely predictions of the future are:

1. that it will be like the present,
2. that it will get better and better in every way, or
3. that it will be a steady descent to a state of misery and starvation.

We know that climate change itself will prevent the status quo from persisting. A glance at the past is convincing evidence that change comes with large parts of both good and ill; change brings problems as well as opportunities. Finally, one of the most prominent forecasts of doom, that of Thomas Malthus in 1778, has proven to be a totally incorrect description of the past two centuries.

Rather than thinking about what has happened recently, we need a tool to focus on what has happened in the past century, or longer, and what could easily happen in the future. A number of areas—from scientific innovation to demography, economic institutions, and sociology—are at issue. Only by looking at the confluence of these changes and considering the likely ramifications in other areas, and especially in behavioral changes, can one begin to get a reasonable picture of what the future might be.

20.3 FORMALIZING IMAGINATION INTO SCENARIOS

The first step in formalizing imagination is to bring people of different disciplines together to perform the exercise. To be useful, such an imagination must be disciplined, formalized into models. Perhaps the loosest form of such modeling is to create a scenario, or sketch of a future pathway, using some initial assumptions about the nature of a change. One could have as many scenarios as there are major differences in assumptions about the future. Experts from various disciplines would provide a balanced capability and could account for interactions among areas.

The exercise would begin by gathering experts from half a dozen disciplines, from technologists to demographers and economists. A particular issue, such as future drought in the Sahel, would be needed to provide a useful focus (Picardi and Seifert, 1976). A time period, such as the end of this century, must be specified. The participants would then attempt to sketch out the economic, demographic and social conditions of the region at that time, as well as the immediate consequences of the drought and related conditions in other countries. The drought might take various forms, depending on factors such as precipitation, temperature, persistence and variability.

Having humans in the system complicates the modeling, since they will react to the drought and its consequences. Food could be shipped to the region, grazing animals could be removed and other positive steps taken; alternatively, crime could increase and even war could result. A first step might be to estimate the effect on the region if a few actions were taken to adapt to the climate change or relieve the resulting misery. Then, one might attempt to specify the range of adaptive behavior that could help to mitigate the problem, beginning first with local residents, then gradually expanding the focus to include international institutions. The participants might attempt to find actions that would curtail the resulting ecological damage and human misery, as well as to define social policy for each type of drought. Presumably, the scenario exercise would identify some steps that should be taken today, even though we have no idea when the next major drought will occur.

Such an exercise is extremely stimulating to the participants, since it forces them to broaden their thinking and exposes each to the expected ramifications in other areas. The value of the exercise is generally to those taking part in it; perhaps there is also value to some who find the resulting scenarios interesting. Such exercises are not more generally useful because no attempt is made to set out the scenarios rigorously or to evaluate them.

Scenarios can also serve as a training device for policy-makers. They provide a low-cost way of trying out a variety of responses to problems—much like the aircraft simulators used to train pilots. One value of the model can be that it demonstrates time lags so long that there will be an enormous endurance cost to society until adaptation can occur. If so, an investigation is prompted into ways of modeling the process so that earlier, more tentative data are sufficient to recognize the problem and initiate action, thereby shortening decision time and inducing remedial action more quickly.

20.4 IMPROVING ON IMAGINATION

Informed judgment formally elicited can be used to improve upon, supplement or constrain imagination. For example, in the case of a Sahelian drought, Glantz (1977) sought to elicit all possible contingency responses by way of a questionnaire that asked what might be done if there were a perfect warning (6 months in advance) of a future drought. Such an exercise seeks to elicit the full range of contingency actions available to Sahelian inhabitants after the initiation of a drought.

The method of eliciting possible responses has been much improved upon. Haas *et al.* (1981) developed a conventional method of using a quasi-realistic simulation of an earthquake warning to elicit the behavior of key actors in each stage of a causal response chain. Gaming and simulation have a long history of use for training or for eliciting responses to a scenario. Ausubel and his colleagues at the International Institute for Applied Systems Analysis (IIASA) have developed a simple game to simulate possible responses to CO_2 increase (Robinson and Ausubel, 1983). Experience with it to date suggests that the game is a useful device for teaching players about the physical aspects of CO_2 and the sociopolitical complications of the problem.

Another common use of informed judgment to strengthen scenarios is the choice of an analogue. This tool has been used by White and Haas (1975), who base their description of a hypothetical hurricane Betsy sweeping up the Miami coast on the effects of previous hurricanes. Climatologists have been seeking analogues to a carbon-increase-warmed earth because causal models (general circulation models) of the atmosphere cannot yet produce reliable regional distributions of changed climate. Flohn (1981) has suggested the use of historical and polar climate analogues of 1000 years ago for a 1.0 °C average increase, 6000 years ago for a 1.5 °C average increase, and 2.5–12 million years ago for a 4.0 °C average increase. Similarly Kellogg (1977) mapped the Altithermal period (4000–8000 years ago) as an analogue. Williams (1979) and Wigley *et al.* (1979) use as their analogue recent years that were unusually warm, and Kellogg and Schware (1981) have blended all approaches (including climate models) in a map reproduced as Figure 20.1.

Informed judgment can be used as the scientific basis of an entire assessment. As described in Chapter 22, the National Defense University, with a panel of 24 climatologists, used a formal elicitation procedure to derive aggregate subjective probabilities for five possible climate scenarios for the year 2000: large and moderate global cooling or warming, and no change. In the second phase of the study the impacts of these scenarios were calculated for 15 key crops, using yield-effect estimates made by an agricultural panel of 35 experts. In the final phase the yield changes were examined for policy implications, using an existing world agricultural model.

Ericksen (1975) used informed judgment to construct a flood scenario for Boulder, Colorado. A hypothetical, but realistic, 1 percent probability storm

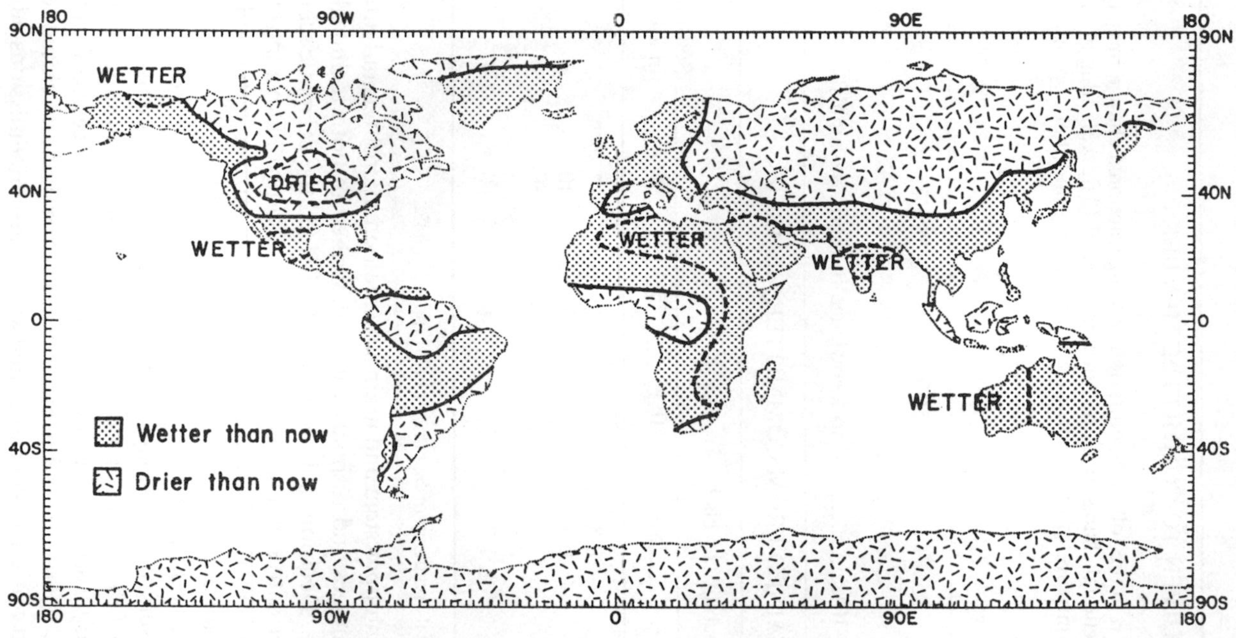

Figure 20.1 Example of a scenario of possible soil moisture patterns on a warmer Earth. It is based on paleoclimatic reconstructions of the Altithermal Period (4500–8000 years ago), comparisons of recent warm and cold years in the northern hemisphere, and a climate model experiment. Where two or more sources agree on the direction of the change the area of agreement is indicated with a dashed line and a label. (Reproduced by permission of Westview Press from Kellogg and Schware, 1981)

Table 20.1 Quantifying expert judgment on link between increase in ultraviolet radiation and skin cancer incidence

THE QUESTION GIVEN TO COMMITTEE MEMBERS:

Assess the constant α in the following equation relating fractional change in weighted UV radiation to fractional change in skin cancer incidence. (The weighting of the UV radiation is a 50:50 mixture of that appropriate for sunburn and that appropriate for DNA damage.)

$$\Delta Ca/Ca = \alpha \cdot \Delta UV/UV$$

________	________	________	________	________
1%	25%	50%	75%	99%

Linearity?________
Self-rating index (0–10)________[0 = no knowledge beyond a layman's]

ESTIMATES OBTAINED FROM COMMITTEE MEMBERS:

Respondent	Probability levels 1%	25%	50%	75%	99%	Self-rating index
A	0.5	0.8	1	2	3	5
C	0	0.4	1	2	10	8
D	0	1	2	4	10	4
E	0.5	0.8	1	1.2	2	2
F	0.6	1.7	2.2	6	60	9

Adapted from: National Academy of Sciences (1975), pp. 335, 342.

was traced through upper stem and main stream flood hydrology and into the city. Probable damages and responses were described in both technical and dramatic scenarios. Decision-makers in Boulder judged the dramatic scenario as helpful and informative, but not necessarily representative. Downing (1977) furthered Ericksen's work by writing scenarios that compared different levels of emergency preparedness.

Subjective, informed judgment can be elicited, analyzed and displayed in ways that enable users of such scenarios to make their own judgments about the knowledgeability of the judges and the degree of consensus in judgments. In one such exemplary exercise, a National Academy of Sciences panel (1975) sought to measure the degree of uncertainty in key parameters related to the chain of events and consequences associated with jet aircraft emissions into the stratosphere. These emissions were alleged to cause ozone depletion and thus increase the incidence of skin cancer. Table 20.1, which links increased ultraviolet radiation to increased skin cancer, shows how the experts assessed the probability levels. This display method provided subjective probabilities of different estimates (a considerable improvement over single numbers), a self-assessment as to expertise, and exhibited the variance among five judges. These judgments have stood up well; a later report based on considerable scientific effort on the question in the ensuing 7 years placed the multiplier between 2 and 5 (National Academy of Sciences, 1982).

Informed judgment also can be used in a recursive process utilizing a Delphi technique (Pill, 1971). An entire field of analysis using informed judgment and based on subjective probability assessments has been consolidated under the rubric of decision analysis (Raiffa, 1968; Keeney and Raiffa, 1976).

20.5 FORMAL MODELING

Scenarios can be made quite formal (Epple and Lave, 1980; Nordhaus and Yohe, 1983). Rather than begin with some events that seem interesting, one can attempt to characterize a range of events deemed plausible or worthy of exploration. Rather than ask people to guess consequences, causal models can be used to spell out implications in each area. Repeated runs can ensure consistency among the various aspects of each scenario so that each becomes an 'if–then' statement based on carefully stated assumptions and a system of cause and effect models. Rather than use a visceral reaction to the desirability of each scenario, one can define a measure of scenario outcome and seek actions that would improve this outcome for a given set of initial conditions.

Rigor is introduced into these formal scenarios to help them do more than stimulate the imagination of participants. Since they are replicable by others, they become scientific explorations of the implications of each set of assumptions. The formalization, however, does not necessarily make them better predictions of the future. At this point the formal scenario models merge with other forms of modeling efforts. A distinctive feature of scenario modeling, however, is gaining an integrated picture of the events and their consequences.

The elements of a formal scenario consist of:

1. *a model*—a set of functional relationships, often hierarchical, with component submodels and a macromodel. These models can be causal (embodying what are believed to be cause-effect relationships) or descriptive (embodying empirical associations between variables from recent history);
2. *an objective function*—a function which puts a quantitative, undimensional value on each outcome detailing its social desirability;
3. *exogenous variables*—factors determined outside the model which influence events of interest (some of which are boundary conditions);
4. *policy variables*—a subset of exogenous variables over which policy-makers have control (or partial control);
5. *endogenous variables*—those whose values are determined within the model (the set of outcome);

6. *parameters*—the functional forms used in the model and the values of parameters determine the quantitative relationship between inputs and outputs, or between exogenous variables and endogenous variables; both functional forms and parameters can be varied to produce different scenarios, although parameters are typically much easier to shift.

The degree of aggregation of a model can vary along several dimensions. It is possible to affect the detail of a model by choice of time unit (for example, day, year, decade), by the degree of subclassification of variables (for example, total agricultural production, production of individual commodities), and by geographic unit (for example, regions, nations). For these reasons, macroeconomic models vary in detail from a dozen equations to thousands of equations. Although there are exceptions, a good rule of thumb is that increasing the degree of detail of a model does not result in more accurate prediction of aggregate variables. If annual values are of interest, then modeling the variables on a monthly or quarterly basis will not improve predictions of variables at the annual level, and detailed modeling of subclasses will not improve the prediction of the class aggregates. Thus, the degree of detail should be indicated by the extent to which the detailed results are of interest in their own right.

Formal or algorithmic models produce an outcome when values of the exogenous variables and of the parameters have been specified; informal models require some sort of informal process, such as the thinking of an expert. Some scenarios are used to explore the effects of policy choices in the near future, for example, the effect of a cut in federal taxes on consumers. Although there is uncertainty concerning shifts in consumer behavior that might result under these conditions, a scenario would be expected to produce answers that were qualitatively correct, and quantitatively accurate to within a few percentage points. In contrast, a scenario attempting to explore the implications of current or near-term policy actions designed to alleviate a possible problem in the mid-twenty-first century faces the considerable difficulty that the world will be quite different than the present; thus, the model will be an inadequate description of the world then.

While informal scenario analysis has been widely used for problems of defense and for foreign affairs, formal scenario analysis emerged with the energy crisis. Energy production and demand relationships are sufficiently complicated that informal modeling was deemed to be of little help. Energy scenarios are useful to illustrate the design problem of formal climate impact modeling, not only because the art is well advanced but also because scenarios of energy use are a major input into scenarios of CO_2-induced climate change. A series of elaborate energy-economic models were constructed and these became the core of a number of energy scenario analyses (see the reviews by Just and Lave, 1979a,b). These models attempted to relate the availability of various energy sources (at assumed dates and costs) to the demand for energy and the level of economic activity. Outputs of the model consisted of energy use over time (and thus of the date when some types of energy supplies would run out), levels of economic activity, the mix of fuels, and possible environmental and risk consequences. Models such as PIES (1974) were highly detailed as to the region where each type of fuel was produced and used, as well as the characteristics of fuel in each region. Other models, such as ETA (Manne, 1976), utilized only two sources of energy and produced highly simplified, stylized results.

The output of the model will typically be values of a series of many endogenous variables such as production, consumption and price of each fuel, perhaps including the technologies used for production, transport and utilization and their geographical locations. To judge whether one set of outcomes is preferred to another, some valuation or objective function is needed. What relative weights should be given to endogenous variables viewed as desirable? How should endogenous variables viewed as undesirable be treated?

To formulate an objective function, we first need to know whose preferences count. Is it all humans who will ever live on the Earth, all people currently alive, or a few policy-makers? What outcomes can be treated quantitatively (for example, a one-third chance of unchanged precipitation patterns in the Corn Belt in 2010 and a two-thirds chance of reduced precipitation)? What outcomes must be treated qualitatively (such as, the possibility of the extinction of humans on Earth)?

Several considerations affect whether a variable is treated as exogenous or endogenous. Some variables are uncontrollable and thus easily classified as exogenous. These may be either perfectly predictable (for example, time) or partially or completely random (for example, sun spots). Other variables may interact with and be partially determined by the endogenous or policy variables, but still be treated as exogenous. This may occur for two reasons. It may be that satisfactory endogenous treatment of the variable would require a much more elaborate model, and such elaboration may be deemed not worth the added cost. To take another example, the future price path of electricity might be taken as exogenous in a study of the viability of electric car technology. Alternatively, it may be that no satisfactory theory is available for modeling the feedback to the variable in question from other variables in the system. For example, the rate of advance of fusion reactor technology may affect the future demand for fossil fuels, but no satisfactory model for predicting the advance of fusion technology may be available. Hence the rate of advance in fusion technology may be treated as exogenous despite the fact that it is at least partially controllable.

The longer the time horizon of the model, the more important it becomes to limit the use of exogenous variables. For short-term forecasts, many variables can be treated as exogenous because they adjust slowly. Thus, in forecasts of energy consumption with a time horizon of 1–5 years, the energy-using capital stock (the stock of automobiles, appliances, industrial boilers, etc.) may be taken as exogenous. Use of historical depreciation and replacement rates and conversion efficiencies will give a sufficiently accurate characterization of the exogenous capital stock variables over such a time interval. Over a period of 20 or more years, however, the bulk of the existing stock will wear out and the character of the capital stock may change dramatically. The stock variables may still be taken as exogenous, but the range of changes in depreciation rates, replacement rates and conversion efficiencies that would feasibly occur over such a time period would be quite large. Endogenous modeling of these variables will often be the best way to reduce the range of uncertainty about future values of variables in long-term forecasts.

20.5.1 Sources and Treatment of Uncertainty

Uncertainty in formal modeling arises from several sources:

1. errors in specifying the model,
2. misestimated parameters,
3. incorrect projections of exogenous variables, and
4. stochastic elements in the model.

We discuss each in turn.

The foundation of the scenario is the cause–effect relationship in each of the modules. Insofar as they are incorrect or are approximations over the current range, the entire effort is subject to fundamental errors. For example, clouds have not been accounted for yet in the global climate models. Insofar as they are of first-, not second-order importance, the results might change even qualitatively. Although there are fundamental laws of physics that apply throughout the universe, virtually all of the models actually used are gross simplifications that assume that all but a few effects are of second-order importance. Occasionally, there are nasty surprises when a second-order effect turns out to be of first-order importance, or some heretofore fixed variable shifts. In a case like carbon dioxide, where a large change is being considered, it is prudent to check carefully the nature of the approximation concerning variables assumed to be fixed and effects assumed to be second-order.

Even assuming that the underlying model is correct, it is difficult to estimate the parameters. For example, climate data have been measured and recorded in detail only comparatively recently and only in a few parts of the world. Within this narrow band of experience we do not observe vast heating or cooling, and aspects of climate have tended to vary together. If, for example, temperature and precipitation have varied together for the recorded historical period, analysis of these historical data cannot provide estimates of the effects of either factor by itself. As long as climate changed little, the inability to separate these two factors would make little difference. However, when a change of the magnitude contemplated for carbon dioxide takes place, the two are unlikely to continue to vary together.

Even assuming the relationships within the model can be estimated satisfactorily (historical data or data from other regions may provide sufficient variation to improve parameter estimates), and assuming that development of theory may improve the underlying models, scenarios still require projections of exogenous (or driving) variables.

What will be the pattern of fossil fuel use? This depends on the growth rate of population, the growth of economic activity, the distribution of people by climate zone, the growth of other energy sources, from nuclear to solar, and on the extent of energy conservation. Individual analyses of each of these driving variables is possible; while each can be guessed at in various ways, the further one projects into the future, the greater the uncertainty becomes. Further, the uncertainty is compounded when all variables are entered into the model simultaneously. Often uncertainties about two variables serve to expand the range of possible outcomes rather than to offset each other.

The fourth source of uncertainty, stochastic elements in the model, are usually considered to be of minor importance. But if interest resides in the climate patterns of small regions, or such factors as the date of the last and first freezes of the season or amount of summer rainfall, the stochastic elements may dominate. The global climate models are not currently intended to produce day-to-day information about major regions. To do so would require not only a major increase in computation, but also a different level of modeling, taking into account minor features in terrain and the like. For practical purposes, one cannot reduce the size of the stochastic element indefinitely.

In some cases, where it is virtually impossible to forecast exogenous variables and other sources of uncertainty seem to dominate, one can simplify the structure by assuming that one of several states of the world will occur. For example, it is unlikely that energy use would increase markedly while economic activity was falling. Thus, two relevant states would be one high economic- and energy-growth path and one low-growth path. When the future seems to fit into one of a few future states, the state preference approach may be used. One needs to characterize the physical outcomes in each future probability occurring at each state, and a value or utility measure for outcomes. Where objective probability estimates are lacking, subjective judgments may be used to weight the outcomes in various states. Finally, one must search for the crucial actions that would lead to the highest expected value of outcomes.

Indeed, the whole point of the exercise is to determine how our actions can lead to a better future. If we could not affect the future, scenarios might help us to reconcile ourselves to the invevitable, but they would not lead to any change. Thus, it is important to determine the actions and policies that will have the greatest influence on the future and to focus on how these factors might affect it.

A moment's thought is sufficient to indicate that uncertainties grow with time, probably exponentially. Thus, we are likely to be able to predict fossil fuel use in 1988 much better than fossil fuel use in 2000, and that much better than in 2020. The further we peer into the future, the more likely are such catastrophic events as nuclear war and the more likely are such beneficial events as a cheap alternative energy source. Fossil fuel use could decline dramatically after either event, or continue to increase exponentially as population increases and the Third World struggles to develop.

20.5.2 Formal Modeling of Carbon Dioxide-induced Climate Change

Increasing concentrations of carbon dioxide in the atmosphere act to prevent radiation of heat into space, thus warming the atmosphere and surface. A doubling of carbon dioxide concentrations from preindustrial levels is predicted to lead to between a 1.5 and 4.5 °C (mean of 3 °C) increase in mean temperature around the Earth, with the greatest effect at high latitudes: perhaps no change at the equator and a 10 °C increase at the poles. However, rather than some uniform warming, some regions would get much warmer and some colder, as suggested in Figure 20.1. Precipitation would generally increase, although some regions would be expected to become deserts and some deserts would get more precipitation.

While the melting of all polar ice would take perhaps a thousand years, shorter-term effects of vast importance could occur. Perhaps the greatest short-term effect could be a disintegration, over perhaps two centuries, of the West Antarctic ice sheet, which is grounded below sea level. Without the protection of surrounding sea ice, wave action could break up this ice sheet, raising the sea level by perhaps 6 meters. The result would be the flooding of many of the world's largest cities, since many are ports, and much of the fertile farmland. An even shorter-term effect would be the melting of sea ice in the Arctic, which would change the Earth's albedo and perhaps accelerate melting. Lessening the temperature gradient between the equator and the poles might lessen average wind velocity and ocean currents. Such changes would have profound effects on microclimates, possibly making some of the most productive farmland barren.

None of the above effects constitutes a scenario. Rather, each presents an opportunity to develop what might be enlightening scenarios. Each physical effect must be developed in greater detail to get the time profile of effects. Then the implications of the effects and behavioral adaptations must be modeled.

20.6 CONSTRUCTING A SCENARIO

One might begin to sketch a scenario for climate change due to carbon dioxide by postulating some path of world economic activity, with its implied level of total energy use by type of fuel, for the next century. The resulting fossil fuel use would result in a carbon dioxide emission rate that could be calculated. The carbon cycle experts would take this number, along with estimates of forest clearing, to estimate atmospheric concentrations of carbon dioxide during the century. The atmospheric physicists would then run their climate models to calculate, roughly, temperature and precipitation for large regions. Given these, the oceanographers and glaciologists would attempt to infer changes in circulation patterns, melting and breakup. Then marine and terrestrial biologists would attempt to infer ecological effects, focusing on species of particular concern to humans, such as pests and species providing food. Agronomists would examine the relatively small set of plants and animals cultivated by people. Finally, economists would attempt to examine the implications of these changes for economic activity generally and fuel use in particular; sociologists would attempt to examine the extent of social tension and any resulting disruptions. These forecasts of the implied level of economic activity complete the cycle, providing feedback to the initial assumptions about economic activity and fuel use. One would iterate this cycle until it converged on an internally consistent scenario.

Presumably, there would be more than one internally consistent scenario. For example, a high-economic-growth, high-energy-use scenario would lead to more climate change and more feedback on economic and social institutions; lower economic growth, or at least lower fuel use, would lead to less climate change. Even within the two levels of economic activity, fuel mix could change radically. One extreme would be a fuel mix centered on coal, with large quantities of coal being burned to produce electricity or to be converted into synthetic gases and liquids. The other extreme would be a fuel mix using nuclear and solar energy, with little or no use of coal. Thus, six scenarios could be specified—three levels of economic activity, with either high or low coal use.

This scenario structure provides a framework within which each expert can get the inputs needed for his/her calculation and receive requests for the inputs needed by others. Since there is no feedback, no area is 'superior' to others; all experts must get inputs from someone and supply outputs to others. All can think of additional interactions, such as changes in agronomy changing the Earth's reflectivity and thus altering climate. No expert need understand the details of others' models in order to interact, but all must learn a bit about the nature of other models, the inputs and outputs needed, and the way uncertainty is described and its source of origin. The linking of uncertainty from models based on different principles (cause, correlation, informed judgment, etc.) is difficult.

One should not be under the illusion that causal models exist that would give confident answers to any of the areas set out above. For example, climate modelers must provide not only mean temperature for each latitude, but also details about precipitation for each region, when the precipitation will occur, what will be the length of the growing season, and so on. Agronomists will be asked how various plants will function under climates and soils quite different than those currently experienced—and how pests will react. Economists will be asked to forecast economic growth, fossil fuel use, and the effect on economic growth of various policies to curtail fossil fuel burning. Sociologists will be asked about social disruption stemming from climate change. We cannot pretend that any of these disciplines have causal models that can answer all these questions. However, all have models that provide partial answers and give clues for the additional inputs.

One objective of the scenario analysis would be to sketch out the implications of current policies. A second objective would be to find policies that involve lower social cost and to estimate the improvement in social welfare. A third objective is to isolate the crucial areas of uncertainty as an aid to structuring research programs. If these objectives can be attained, even with great uncertainty, vast progress will have been made.

20.7 SCENARIO EVALUATION

For informal scenarios whose principal purpose is to enlighten the participants, the evaluation should be in terms of how much they feel they learned from the exercise. A scenario intended to integrate the efforts of many experts should be evaluated in terms of the extent to which it allowed people to work independently, but still accomplished the objective. Formal scenarios should be evaluated in terms of the extent to which they have captured known cause-effect relationships, turned out elegant models, and provided enlightening results.

For all three types of scenarios, there is an element of looking for enlightening results. What hypothetical occurrences, aspects of structure and reactions did the scenario analysis illuminate?

It is particularly difficult to know how to evaluate scenarios of complicated problems set in the distant future. We cannot wait to see if the future actually follows one of the scenarios. We cannot see if the actions that seem best within the framework turn out to perform well in the world. Rather, the criteria must be theoretical. Do people feel they understand the problem better after the analysis? Does it more clearly identify the conditions under which each policy would be useful? Does it indicate critical data needs and uncertainties to be resolved? Where formal models are used, one can attempt to assess the extent to which those models would have predicted outcomes observed in the past. However, there are no simple means for evaluation.

It is important to realize that scenarios are not truth. Insofar as the scenarios generate confidence that a particular solution or policy is optimal, they have probably done a disservice. Part of the evaluation must include not having people put unwarranted faith in the scenarios and their analysis.

20.8 INTERDISCIPLINARY RESEARCH

The progress of science has led to its fragmentation. In assessing a global problem, such as increasing carbon dioxide, there is a need for a coordinating mechanism to draw these fragments together to form a comprehensive picture. The major difficulty in research involving people from different disciplines is facilitating communication. Experts must share a common language and must know something about each others' disciplines. The object is not to make each scientist a universal genius, knowing everything about every field. Rather, it is to teach each enough about the other areas that effective communication is possible.

Formal scenario analysis provides just such a tool. It focuses on preconditions, thus getting participants to talk about what each sees as possible future occurrences in his or her area. It requires each participant to discuss what data and parameters he or she requires as inputs and what he/she can give as outputs. No one need know a great deal about the causal models used in each area. Instead, the interaction is focused on inputs and outputs from each area, on integration of results, and on aspects of the problems that are not captured within each formal model. The exercise has an outcome and evaluations and thus provides a common goal. Since each participant has a clearly defined task, namely, to cover a particular area, there is no need for people to learn the details of each others' models.

This practical strength of formal scenarios as a synthesizing tool for integrating disciplinary contributions is also a weakness in a larger epistemological sense. Even if each participant could understand the details of each other's model, assessing the likelihood of a particular scenario is probably beyond the present state of the art. This is the problem, referred to in Chapter 1, of linking models that are inferential, rather than causal, with those based on clearcut physical principles and those based on informed judgment. These linkages, necessary to constructing many scenarios, have never been closely examined and reviewed and are a task for future research.

20.9 CONCLUSION

Scenario analysis has much to add to climate modeling. It can help make researchers more imaginative and educate policy-makers. It can provide a framework within which to structure interdisciplinary research on climate and to isolate the critical research issues. Finally, formal scenarios can define internally consistent scenarios that show the consequences of current policy and help isolate policies that are superior, as well as the remaining problems. The scenarios are not predictions of the future or direct guides to policy. Instead, they represent a systematic process that uses available theory, facts and judgments to explore the implications of hypothesized conditions. A major advantage is highlighting crucial uncertainties and being able to incorporate new information to improve the models and results.

The paper began with three general purposes of scenarios: stretching people's minds, formal modeling, and integrating people of different disciplines. Each purpose suggests a way in which climate scenarios might be especially useful.

The nature of scenario development means that the tool should not be chosen for relatively well understood problems. Scenario analysis is not likely to be a good tool for 8-hour weather predictions, or for anticipating the consequences of such predictions. More powerful tools are already in use. The comparative advantage of the scenario approach is in exploring issues beyond the ones normally dealt with. Thus, exploring the twenty-first century effects of climate change due to carbon dioxide is a natural topic for scenarios.

Formal modeling is useful when there are cause-effect relationships known for virtually all aspects of the issues. This parametric modeling is then used to explore consequences and go beyond intuition or back-of-the-envelope calculations. This modeling is most useful for problems that are generally well understood, but are too complicated for straightforward exploration.

The integration exercise is most useful when a problem requires experts from several disciplines. The larger and more diverse the problem, the more helpful is this integrative framework likely to be. Exploring carbon dioxide effects in the twenty-first century is probably an ideal application. Another example might be exploring the effects of a future Sahel drought.

ACKNOWLEDGMENT

We thank the editors and unknown reviewers for comments. Dr. Lave thanks the National Science Foundation for support.

REFERENCES

Bell, D. (1964). Twelve modes of prediction—a preliminary sorting of approaches in the social sciences. *Daedalus*, **93** (3), 845–880.

Bunge, M. (1967). *Scientific Research II: The Search for Truth*. Springer-Verlag, New York.

Downing, T. E. (1977). *Warning for Flash Floods in Boulder, Colorado*. Natural Hazards Research Working Paper 31, Institute of Behavioral Science, University of Colorado, Boulder, Colorado.

Durand, J. (1972). A new method for constructing scenarios. *Futures* (December), 325–330.

Epple, D., and Lave, L. B. (1980). Helium: Investments in the future. *The Bell Journal of Economics*, **11** (2), 617–630.

Ericksen, N. J. (1975). *Scenario Methodology in Natural Hazard Research*. Institute of Behavioral Science, University of Colorado, Boulder, Colorado.

Flohn, H. (1981). *Life of a Warmer Earth, Possible Climatic Consequences of a Man-Made Global Warming*. International Institute for Applied Systems Analysis, Laxenburg, Austria.

Glantz, M. H. (1977). The value of a long-range weather forecast for the West African Sahel. *Bulletin of the American Meteorological Society,* **58**, 150–158.

Haas, J. E., Hutton, J. R., Mileti, D. S., and Sorenson, J. H. (1981). *Earthquake Prediction Response and Options for Public Policy*. Team Monograph No. 1, Institute of Behavioral Science, University of Colorado, Boulder, Colorado.

Hudson, E. A., and Jorgenson, D. W. (1974). U.S. energy policy and economic growth, 1975–2000. *The Bell Journal of Economics,* **5** (2), 461–514.

Jantsch, E. (1967). *Technological Forecasting in Perspective*. Organization for Economic Cooperation and Development, Organization for European Economic Cooperation, Paris.

Just, J., and Lave, L. (1979a). Review of government energy scenarios. *Energy Systems and Policy,* **3** (3), 271–307.

Just, J., and Lave, L. (1979b). Review of scenarios of future U.S. energy use. *Annual Review of Energy,* **4**, 501–536.

Kahn, H., and Wiener, A. J. (1967). *The Year 2000: A Framework for Speculation on the Next Thirty-Three Years*. Macmillan, New York.

Keeney, R. L., and Raiffa, H. (1976). *Decisions with Multiple Objectives: Preferences and Value Tradeoffs*. John Wiley & Sons, New York.

Kellogg, W. W. (1977). *Effects of Human Activities on Global Climate*. Technical Note No. 156, World Meteorological Organization, Geneva.

Kellogg, W. W., and Schware, R. (1981). *Climate Change and Society: Consequences of Increasing Atmospheric Carbon Dioxide*. Westview Press, Boulder, Colorado.

Malthus, R. R. (1778). *An Essay on Population*.

Manne, A. (1976). ETA. A Model for energy technology assessment. *The Bell Journal of Economics,* **7** (2, Autumn) 379–406.

National Academy of Sciences (1975). *Environmental Impact of Stratospheric Flight*. NAS, Washington, DC.

National Academy of Sciences (1982). *Biological Effects of the Increased Solar UV Radiation*. NAS, Washington, DC.

Nordhaus, W. D., and Yohe, G. W. (1983). Future paths of energy and carbon dioxide emissions. In National Research Council, *Changing Climate (Report of the Carbon Dioxide Assessment Committee)*. National Academy Press, Washington, DC.

Picardi, A. C., and Seifert, W. W. (1976). A tragedy of the commons in the Sahel. *Technology Review,* **78** (6), 42–51.

PIES (1974). *Project Independence: A Summary*. US Federal Energy Administration, Washington, DC. Available from US Government Printing Office.

Pill, J. (1971). The Delphi method: Substance, context, a critique and an annotated bibliography. *Socio-Economic Planning and Sciences*, **5**, 57–71.

Polak, F. L. (1971). *Prognostics: A Science in the Making Surveys for the Future*. Elsevier, New York.

Raiffa, H. (1968). *Decision Analysis*. Addison-Wesley, Reading, Massachusetts.

Robinson, J., and Ausubel, J. (1983). A game framework for scenario generation for the CO_2 issue. *Simulation and Games,* **14** (3), 317–344.

White, G. F., and Haas, J. E. (1975). *Assessment of Research on Natural Hazards*. MIT Press, Cambridge, Massachusetts.

Wigley, T. M. L., Jones, P. D., and Kelly, P. M. (1979). Scenario for a warm, high-CO_2 world. *Nature,* **283**, 17–20.

Williams, J. (1979). Anomalies in temperature and rainfall during warm arctic seasons as a guide to the formulation of climate scenarios. *Climate Change,* **2**, 249–266.

Climate Impact Assessment
Edited by R. W. Kates, J. H. Ausubel and M. Berberian
© 1985 SCOPE. Published by John Wiley & Sons Ltd

CHAPTER 21
Historical Climate Impact Assessments

T. M. L. Wigley, N. J. Huckstep,
A. E. J. Ogilvie, G. Farmer,
R. Mortimer and M. J. Ingram*

Climatic Research Unit
University of East Anglia
Norwich NR4 7TJ England

* *Now at Department of Modern History*
The Queen's University of Belfast
Belfast NT7 1NN
Northern Ireland

21.1 INTRODUCTION

The dictum that the past is the key to the future has frequently been used to justify scientific research (see de Vries, Chapter 11). It is particularly appropriate to the present study. Climate is one of the many complex and subtly changing elements of the environment upon which human societies depend for their survival. Precisely because of this complexity and subtlety, it is difficult to determine the degree to which climatic fluctuations affect society. To give a recent and well known example, there is still considerable controversy concerning the Sahelian famine of the late 1960s and early 1970s. Some hold that this was the direct result of climatic variation, while other see the causes in the underdeveloped economic infrastructure of the countries affected, external interference in those countries' affairs, or simply in inflexible techniques of animal and crop husbandry.

One can argue that, by looking at the ways in which climate has affected societies in the historical past, it should be possible to identify more precisely the potential impacts (and successful adaptive strategies) that present and future climatic fluctuations can have on human societies. Such studies are free from contemporary political controversies and other obscuring factors.

> Global and regional climate changes will affect different societies and different segments within societies in a wide variety of ways. One means to determine the range of impacts is to undertake case studies on ways in which families, political institutions, and social sectors such as agriculture have been affected by changing or varying climates .... Historical case studies of climatically vulnerable areas such as Iceland and the Great Plains of the United States may be particularly useful in understanding societal adaptations.
>
> (American Association for the Advancement of Science, 1980, 14.)

It is the purpose of this paper to suggest ways in which such case studies might be carried out and to discuss some of the existing work in the field.

Although most historians, historical geographers and archaeologists have long recognized the possible importance of the impact of climate on human affairs, the literature reveals considerable disagreement about how seriously the possibility should be taken. It is generally accepted that short-term (intra-annual, annual and interannual) variations in climate and weather, having an immediate effect on harvests and other economic activities, are relevant to short-term economic fluctuations. But long-term climatic influences (on time-scales of decades or more) have been commonly regarded as of little or no historical interest, on the basis of one or more of the following assumptions: first, that climate has been essentially stable in historical times; second, that the magnitudes of past long-term climatic shifts, though striking from a scientific point of view, have been too small to warrant their consideration as significant variables in the processes of economic and social change; third, that the resilience and adaptability of past societies has been sufficient to absorb the effects of periodic short-term climate-induced stresses and so to minimize the cumulative effect on longer time-scales; fourth, that lack of detailed information on past weather and climates and imperfect understanding of the complex processes of climate–history interactions precludes any serious study of the subject.

Nevertheless, for a long time, a minority of historians and larger numbers of archaeologists and natural and environmental scientists have been convinced of the importance of climate as a major independent variable affecting the development of human societies. The most extreme exponents may be labeled climatic determinists, of whom Ellsworth Huntington (1907, 1915) is the best known. Although the term 'climatic determinism' does not necessarily imply a belief that the whole course of history is explicable in terms of climate (Pearson, 1978), it certainly implies that climatic factors have been among the most important influences on the development of civilizations (see Riebsame, Chapter 3). Such views are still held today, for example by Chappell (1970), and in a less extreme form by Bryson and Murray (1977), Lamb, and others. For example, Lamb has asserted that 'climatic history must be central to our understanding of human history' (1969, 1209).

Others have been more cautious, eschewing grand generalizations about the significance of climate in world history, yet strongly urging recognition of the importance of climatic factors in particular areas and periods. Among historians they include Utterström (1955), Braudel (1949, 1973), and more recently Parker and Smith (1978) and Parker (1979); but these contributions, though interesting, are marred by methodological weaknesses (see, for example, the extensive criticisms of Utterström by Le Roy Ladurie, 1972, 8–11).

In recent years a new breed of historians interested in climate–history interactions has emerged. These scholars, of whom Pfister (1975, 1978, 1981, 1984), de Vries (1977, 1980 and Chapter 11), Post (1977, 1980) and Parry (1978, 1981a,b and Chapter 14) are some of the most outstanding, admit the possible importance of climate variations on human affairs both in the long and short term, make serious attempts to investigate the possibility with the aid of all the available methodological resources of history, economics and climatology, but demand the highest standards of proof. In particular, these historians have been able to impart to climatologists the importance of a rigorous analysis in evaluating climate data found in historical documents, together with an appreciation of the complexity of climate–society links. Their work has transferred the study of climate–history impact onto a higher plane of methodological rigor.

A review and synthesis of the methodologies for historical climate impact study is therefore appropriate. We begin this chapter with a discussion of the underlying philosophy related to *explanation* in the historical climate impact context. We then discuss briefly the problems of data availability, on both the societal and climate side, before introducing a model-oriented framework for climate impact studies. The major statistical tools which can be used for testing impact models, together with some of the pitfalls which one should strive to avoid, are then described. Alternative strategies, less axiomatic and more empirical in their approach, are seen to play a valuable supporting role, but these must be used cautiously. Having described the methods available, we then consider more specific examples of historical climate impact divided into short, medium and long time-scale effects. Some of the most convincing studies are those of marginal areas, and those which consider the issue of climate as only one of many factors in the milieu of potential determinants of societal impact. We therefore devote considerable space to discussing human adaptation to climate stress. Finally we present a brief, selective summary of a number of published climate impact studies, with the important features of these studies categorized in terms of our recommended methodological framework.

21.2 METHODOLOGY

21.2.1 Philosophical Assumptions

A philosophical issue basic to the study of the impact of climatic variations on human societies is on whom the onus of proof should lie. It is plain that climatic variations must have had *some* influence on economies and societies. Hence, some writers feel justified in accepting and extrapolating poorly substantiated 'demonstrations' of the importance of climatic effects in particular cases, and casting the onus of disproof on the sceptics. The importance of climatic influences cannot be assumed, however, and to demand, either explicitly or implicitly, that the burden of proof should lie with those who doubt the importance of climatic effects is inadmissible.

Notwithstanding this, because we are dealing with human behavior, which eludes quantification, the concepts of 'proof' and 'explanation' in the historical context should not be assumed to be the same as in the context of the physical sciences. It would be unfair to demand precisely the same standards of generality, definition of model, or even statistical rigor in studying the linkages between climate and society as one might require in other disciplines.

21.2.2 The Problem of Data

One difficult question which anyone attempting to investigate possible climate–history links in particular cases must ask is whether the available data are sufficient to justify the attempt. The amount and variety of available data on past climate and weather is extensive (reviewed in Wigley *et al.*, 1981). However, the kinds and quality of data which may satisfy the purposes of scientists seeking only to extend the climate record back in time may be inadequate for the purpose of demonstrating climate–history interactions. The latter purpose often demands detailed information about individual seasons, months, even weeks, and such data have been, unfortunately, absent or sparse for large regions of the world until the very recent past.

Data on the complexities of human economic and social activities are also highly variable, both in quantity and quality, in different locations and at different periods; and, in general, the further back into the past one penetrates, the fewer are the available data. It is impossible to summarize the extent of the available information on the varieties of human activity most likely to have been affected by climate. A few examples taken from Western Europe, a region whose history over the past thousand years or so is relatively well documented, will illustrate the problems.

Grain production is and has been an activity of major economic importance. Over large areas and for long periods, a considerable proportion of cereal production and distribution was controlled by large-scale secular and religious landholders whose activities are comparatively well recorded. Much information on grain production accordingly survives. Yet, for the purposes of the economic historian, it suffers from many imperfections. While abundant grain *price* data exist (e.g. Beveridge, 1921, 1922, 1929; Bowden, 1967), information which bears more directly on agricultural productivity is much less common. In fact, its very scarcity has led Hoskins (1964, 1968; see also Harrison, 1971) to use price data to reconstruct, year-by-year, the quality of the harvest. As Appleby (1980), de Vries (1980, especially pp. 620–621) and others have pointed out, this may be valid in a simple 'closed' agricultural economy, but it is highly debatable for developed, open economies. Flohn (1981), using data from Titow (1960, 1970) has demonstrated a yield–price link for medieval England (1211–1448); but examples from fifteenth-century Netherlands (Tits-Dieuaide, 1975) and eighteenth-century Switzerland (Pfister, 1975) show more complex relationships, and there are many instances in England from the sixteenth century onwards of harvest failures which were not reflected in price increases (Appleby, 1980). Of course, price data, although imperfect as a proxy for harvest data, may be used directly as an economic indicator which might in turn be statistically related to climate. Anderson (1981) exposes the pitfalls which simplistic analyses along these lines might encounter, but some excellent examples of a *rigorous* approach are given by de Vries (1980) and Lee (1981).

As another example, demographic data, in many ways fundamental to an understanding of social change, are distressingly sparse and imprecise. In many parts of Europe before the sixteenth or seventeenth centuries, the available information suffices only to indicate, with considerable margins of uncertainty, gross trends in population. Nevertheless, some excellent demographic studies have been undertaken. The scholarly and detailed study by Imhof (1976) on population in the Nordic countries during 1720–50 is a notable example. (In this study, Imhof claims a close connection between climatic variability and mortality through disease.) On the whole, however, demographic data become abundant and satisfactory only in the nineteenth century.

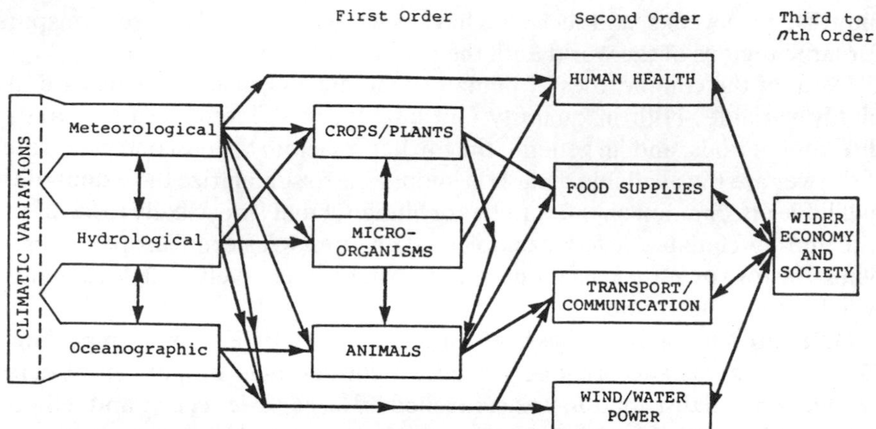

Figure 21.1 First- and higher-order effects of climate on aspects of society. As the particular impact becomes more removed from the climatic 'cause', more and more interactions intervene to disguise and modify the link. The two-way arrows in the last links symbolize possible adaptive feedback processes. (Reproduced by permission of Cambrige University Press from Ingram *et al.*, 1981, Figure 1.2)

For localities outside Europe the problems are, in many cases, even greater. This is not to say that *no* data adequate to investigate climate–history links are available, but it is as essential to be aware of the gaps and problems in the record of human activities as it is to recognize the imperfections in the climatic evidence. It must be accepted that, for many parts of the globe over long periods, the data currently or potentially available on human activities which are likely to show a discernible climatic impact are so sparse or poor as to vitiate any attempt to demonstrate it.

21.2.3 Preliminary Models

In discussing specific examples of assessments of climatic impact, we must distinguish between effects on different time-scales. However, all attempts to identify and measure climate–history interactions must rest on models of the processes involved in order to identify, in general terms, how climate may be expected to affect human history. A simple model of a type which, explicitly or implicitly, underlies the majority of studies of climatic impact in the past is shown in Figure 21.1 (from Ingram *et al.*, 1981). Variations in the atmospheric circulation manifest themselves as meteorological, hydrological and oceanographical phenomena. These may have a direct (first-order) effect on biophysical processes important to man, including crop and animal survival and growth, marine and other aquatic life, and the activity of microorganisms capable of causing disease in plants, animals and man. In addition, they may have effects on aspects of the purely physical environment of importance to human societies; for example, causing rivers to freeze or flood. These direct biological and physical effects may have economic or social (second-order) significance, affecting food and raw material supplies derived from agriculture or animal husbandry; human health; the performance of machines (such as industrial and food processing mills) driven by wind and water; transport and communications; and military and naval operations. Depending on their magnitude, these effects may ramify into the wider economy and society (third- to *n*th-order). For example, food shortages may (in association with other factors) lead to rebellions, which in turn may help to undermine political systems.

Such a conceptual model serves only to focus attention on the indirect and complex nature of the links between climate and the majority of economic and social phenomena, and to aid in the identification of relevant data. It does not include the many other variables which, apart from climate, might affect human activity; nor does it specify in detail which climate variables are critical for each activity.

These complexities, which are considerable, may be illustrated by reference to the effects of weather on agriculture, perhaps the most obvious link between climate and human activities. Different crops are sensitive to different climatic factors. In any particular study of the effects of climate on agriculture, it is necessary to establish precisely in what respects the relevant crops are sensitive to climate. This is not an easy task, since crops may respond to a number of climate variables and the response may be markedly nonlinear (McQuigg, 1975; Thompson, 1975; Starr and Kostrow, 1978; Wigley and Tu, 1983). Furthermore, the important variables and critical times are not always obvious, and the direction of the relationships between yield and climate may vary from place to place for the same crop. In southern England, for example, mild winters are an important determinant of good winter wheat yields today, but higher yields are also favored by cooler springs and summers, contrary to common conceptions. In England, too, cold summers have been correlated with higher hay yields (Hooker, 1922), whereas in Iceland precisely the opposite correlation holds (Bergthórsson, 1966, 1976). In low-lying areas of western Europe, rainfall, especially in winter, may be the most important climatic factor in determining crop yields (de Vries, 1980; de Vries also describes the excellent work by Baars, 1973). Further information on crop-climate links in the historical past is given by Slicher van Bath (1977). Our understanding of these relationships can be put into perspective by noting that, today, even the best crop-climate models rarely account for much more than 50 percent of the variance when long-term technology trends have been factored out, although there are notable exceptions (for example, Bergthórsson's work [1966, 1976] on hay yields in Iceland). In evaluating climatic impact in the past, the links between yield and climate are generally even more diffuse. One should, therefore, take heed not only of any relevant crop-climate linkages established using modern data, but also of the possibility that past relationships may have differed from those of today.

It is obvious that climate is not the only variable affecting agricultural production. Account must also be taken of such factors as variations in the extent of land under cultivation and of the level of investment in seed, fertilizer, technology and labor; the possible effects of disease (which may or may not itself be influenced by climate); and even the intrusion of non-agricultural human activities such as devastating warfare. In any given case, the effects of climate will be masked by the operation of these other variables.

The complexities of the relationships between climate and all other human activities are at least equal to those affecting crop yields, and in many cases greater. In particular, it is evident that the more remote the relationship between climate and a given activity, the greater the number of complicating variables. A rebellion or grain riot, for example, is separated from a variation in the atmosphere by a complex mesh of causality.

21.2.4 Modeling a Climate–Society Link

In climate impact studies, the real issue is not whether the climate has had an influence, but to establish what and how strong the influence was, and whether it was of any real significance. Even strong identifiable primary impacts (see Figure 21.1) may be of little consequence if diffused by other non-climatic factors as they cascade through the socioeconomic system to second-, third- and *n*th-order effects. The basis of any analysis, therefore, must be a reasonable model of the processes, including those which govern climate impact and those which might act to ameliorate such impact. Figure 21.2 gives a model framework which will be discussed further below. Development of a model helps to identify the variables which might be examined to define cause and effect, and to establish the realism of any cause-effect relationship, although the complexity of the model itself may be determined by the availability of data. As de Vries (1980, 608) points out, although an econometric model of the variables (including climate) affecting rye prices at Utrecht in the seventeenth and eighteenth centuries could be devised, this would be of little use because data on several of the key variables are lacking.

In many cases the problem is not only that of inadequate data, but also an imperfect understanding of the economic and social processes involved. The feasibility of constructing an adequate model depends on several factors.

1. The complexity of the task varies according to the number of economic and social variables included in the investigation. The extreme case would be to attempt to gauge the effects of climatic variations on all aspects of life in a given society. Less ambitious investigations, focusing on one or a small number of economic or social variables, are more realistic.

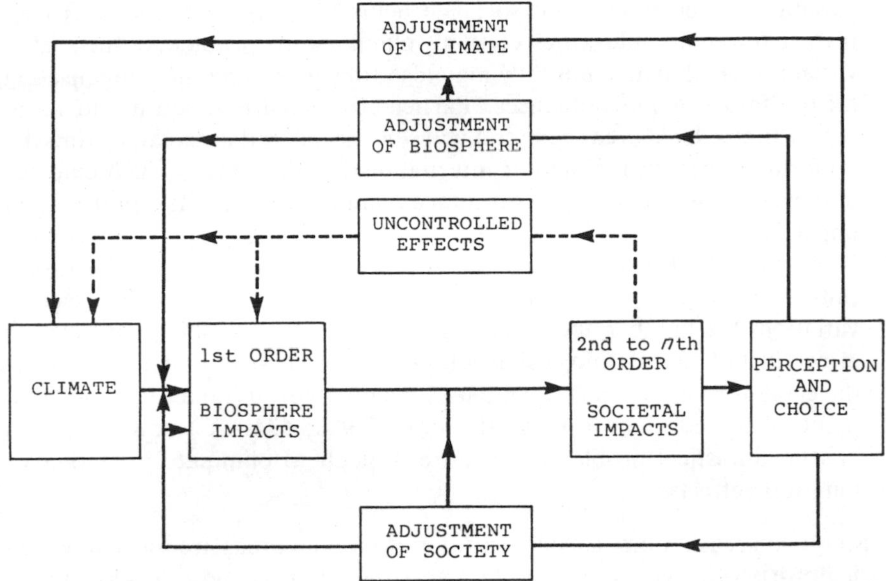

Figure 21.2 Conceptual model of the impact of climate on man and society showing some of the possible feedbacks via adaptive strategies. The box 'uncontrolled effects' refers to all inadvertent, unplanned or uncontrolled modifications of the biosphere and/or climate (carbon dioxide pollution and acid rain are examples). 'Adjustment of the biosphere' refers to changes in land use and/or deliberate exploitation of living resources. 'Adjustment of climate' might include cloud seeding, although man's success in this direction has been equivocal. On a smaller scale, however, man has, for years, successfully adjusted the microclimate by building shelter belts, wind breaks and greenhouses, by combating frost with smudge pots, etc. 'Adjustment of society', which covers a host of possible adaptive measures to perceived stresses, may lead directly back to influences on second- and higher-order impact variables, or, through the market economy and adjustments in demand, may lead to modifications of the way climate affects the biosphere or to direct changes of the biosphere. (Reproduced by permission of Cambridge University Press from Ingram *et al.*, 1981, Figure 1.3)

2. First- and second-order climatic effects (see Figure 21.1) can be modeled more easily than the links between climate and more remotely connected human activities. The more remote the activity, the greater the number of complicating variables, and the less easy it is to demonstrate a unique causal link.
3. The problems vary according to the geographical scope of the investigation. In general, it would appear easier to model the effects of climate for a small area than for a larger one. In studying a large area, such as an entire country, weather and climate patterns may vary markedly over the area, their influence will vary according to the varieties of regional landform and even local topography, and it may be necessary to take account of regional

variations in economic and social structures. Smaller-area studies may involve fewer variables (they certainly eliminate the problem of the spatial variability of climate), but they may also transform *internal* compensating factors into *external* influences. Particularly in early-modern and more recent times, local areas cannot always be considered as isolated from the buffering effects of regional or international trade; indeed, such external influences may, in some circumstances, amplify rather than buffer local impacts.

4. It is plain that it is easier to investigate the effects of short-term (intra-annual, annual and interannual) climatic fluctuations. For secular variations the number and complexity of the factors masking any climatic effect, including technological development and changes in the structure of the economy, are normally so great as to defy rigorous quantification. This point is argued vigorously by de Vries (1980, 624–625) and Anderson (1981). Statistical problems are more difficult to eliminate in studies of long-term effects.

Marginal areas require specific mention. Marginal areas are those in which agriculture or other economic activities are conducted in conditions close to the climatic limits beyond which such activities are physically unviable. By definition, such marginal areas should be particularly vulnerable to the effects of climatic variation. In such circumstances, the number of complicating variables may be smaller, and the impact of climate itself correspondingly easier to gauge. This applies regardless of the time-scale of the study. In studying the possible influences of secular climatic variation, marginal areas may prove valuable as 'laboratory' test cases. If climatic impact is absent (or small) in such marginal situations, one might conclude that secular climatic variations in more environmentally sheltered societies may be safely ignored. If, on the other hand, the effects of such climate variations on marginal societies appear important, this will provide at least some basis for arguing the more general significance of climatic changes in human history and may encourage further consideration of the admittedly more complex cases of societies relatively well sheltered from climatic stress. These ideas are strongly urged by Parry (1978, 1981a,b) and Ogilvie (1981).

21.2.5 The Preferred Strategy

The primary aim is to determine the magnitude and understand the mechanisms of climatic impact. To determine the magnitude of climatic impact, only a quantitative and rigorous statistical analysis will suffice. Our emphasis in this paper is on this type of approach. This should not be taken to be a condemnation of qualitative investigations of particular events, which can undeniably provide both useful insight into the mechanisms of climatic impact and valuable supporting documentation for quantitative analyses. There is no single methodological framework which can be used as a basis for all historical climate impact studies. Since the interactions between climate and society are so complex, the application of simplified models and quantitative techniques may give a false air of rigor and generality to such analyses, of which one should be wary.

Nevertheless, we consider the quantitative, model-oriented approach to be the basic skeleton on which historical climate impact studies should be fleshed. The method we advocate is first to construct a realistic (but not necessarily complex) model of climate–society interaction, and then to test such a model, using time-series data of chosen dependent (i.e. impact) and independent (i.e. climate) variables. The model itself may be constructed *a priori* (for example, based on modern agronomic data or sociological information), or may be derived from a preliminary analysis of the historical data (such as de Vries' analysis of butter prices at Leiden; de Vries, 1980, 609–610).

For both model testing and preliminary analysis, three simple statistical techniques offer considerable scope: regression or correlation analysis, comparison of means, and the use of contingency tables. (For further details on the uses and abuses of statistics in climate impact analysis see Wigley, 1983.)

21.2.5.1 Regression Analysis

The technique is to relate a dependent or predictand variable (Y) to a single or to a set of predictor variables (X_1, X_2, etc.) via an equation of the form

$$Y = a + bX_1 + cX_2 + \dots$$

Generally, Y is the impact estimator (crop yield, prices, etc.) while X_i are the climate variables. Although this is a linear equation, the predictors, X_i, may themselves be nonlinear functions (as are often used in crop-climate regression models). The predictors may, of course, be lagged variables: for example, mortality in a particular year may be related to climate in an earlier year or years. The regression coefficients $a,b,c\dots$ are usually evaluated using least-squares techniques. The magnitude and significance of the relationship is determined by the multiple correlation coefficient, R. R^2, which is called the coefficient of determination, is a measure of the amount of variability of the predictand variable which is explained by the predictors.

The method is a commonly used statistical tool which is described in most standard textbooks but there are some points which need to be kept in mind in applying it to climate impact studies.

1. Results which appear to be significant can occur more easily by chance when the data analyzed are strongly autocorrelated (that is, when the value of a variable in year n is statistically correlated with its value in year $n + 1$).

Price data, for example, tend to be strongly autocorrelated (even after long-term trends have been removed), so that spurious price-climate links might occur if the climate data were also autocorrelated. Such autocorrelation, even if not in the raw data, can be induced by considering moving-averaged data, or filtered data, so care must be taken when using such data.

2. Intercorrelated regressors—one of the major problems in regression modeling involving climate variables is that these variables are often strongly correlated—for instance, seasonal rainfall and temperature values invariably show strong statistical links. If predictors are intercorrelated, it is often impossible to interpret individual regression coefficients. Thus, the sign of the statistically determined regression coefficients may not be in accord with preconceived or independently determined ideas. Furthermore, regression analyses using intercorrelated variables tend to be unstable (for example, changing a single data point, or adding or subtracting a predictor, may change the regression coefficients dramatically). This latter problem may sometimes be circumvented using more elaborate techniques like principal components regression (see, for example, Gunst and Mason, 1980), but the difficulty in the interpretation of results is more difficult to avoid. The best advice is to be aware of potential problems and to make a judicious choice of climate predictor variables.
3. Autocorrelated residuals—the residuals are the deviations of the value of Y predicted from the regression equation and its observed value. Autocorrelation of the residuals is generally a sign that the original data were autocorrelated (point 1 above) or that the choice of model (that is, the form of the regression equation, or the variables used) was inappropriate in some way. A simple statistic, the Durbin-Watson statistic, can be calculated for any regression analysis in order to test for autocorrelated residuals. Caution is advised if the value of Durbin-Watson statistic indicates a statistically significant degree of autocorrelation in the residuals.
4. Multiplicity—if many regression models are derived, then the probability of obtaining a statistically significant result by chance increases. To select the single most significant relationship from a large number of trial relationships would be a procedure of dubious statistical validity.
5. Cause and effect—while regression analysis can demonstrate the existence of, and quantify, a relationship between variables, this does not necessarily prove a cause-effect link. Two correlated variables may have a common cause; this circumstance may arise when the climate data used in a relationship are derived from indirect evidence (such as yield data or the geographical distribution of crops, vines, etc.). When information comes from diverse sources the possibility of circular argument must be considered, particularly in medieval and earlier analyses. The value of

regression or correlation results depends intimately on the realism of the model they are used to support.

21.2.5.2 Comparison of Means

For examination of possible climate impact relationships on longer time-scales, it may be sufficient to look for significant changes in the average value of an impact variable which parallel those expected on the basis of an impact climate model. An early example of this approach is Hoskins' test of Utterström's contention (1955) that 'there may have been a fundamental climatic change in the mid-sixteenth century over most of northwest Europe which adversely affected the quality of harvests in the second half of the sixteenth century' (Hoskins, 1964, 30). Let us accept the reality of a significant climate difference before and after 1550 (a point open to independent testing), and also accept Hoskins' (1964) price-derived harvest quality data (open to some doubt, as pointed out earlier). Hoskins compared the ratios of above-average to average to below-average yields for the periods 1480–1549 and 1550–1619, expecting to find a difference if there had been any climatic effect. This essentially amounts to comparing the average yields in these two periods. Hoskins found no differences and concluded that there was no influence of climate. His argument is, however, specious, and his comparison was a futile one because long-term trends had been removed from the price data before estimates of yield were made.

A slightly different application has been given by de Vries (1980) who compared average butter prices at Leiden for years with frosty Marches to those for other years over the period 1658–1757. He found a statistically significant difference, with prices in the 'cold' years about 12 percent above those in other years. This technique, in which a sample is stratified on the basis of a climate character chosen according to an expected relationship (or model), may be useful in studying both short- and long-term impact.

The appropriate statistical test here is a t-test where the magnitude of the test statistic

$$t = \frac{|\bar{X}_1 - \bar{X}_2|}{\sigma\sqrt{1/N_1 - 1/N_2}} \quad \text{where } \sigma = \sqrt{\frac{N_1 s_1^2 + N_2 s_2^2}{N_1 + N_2 - 2}}$$

is compared with the value expected by chance, on the assumption of no difference in means, using Student's t-distribution. Here $\bar{X}_1$, $\bar{X}_2$ are the means, s_1^2, s_2^2 are the variances, and N_1, N_2 are the sample sizes. An approximate method based on the Normal distribution may be used if the sample sizes are large (this is the method used by de Vries).

This particular statistical method can also give spurious results if either sample has any significant autocorrelation, although there are simple methods available to account for autocorrelation effects (see, for example, Mitchell *et al.*, 1966).

Autocorrelation problems are less likely to arise in the stratified sample application.

21.2.5.3 Use of Contingency Tables

In some cases, data may be available only in qualitative form and/or as incomplete time-series. These data deficiencies do not, however, preclude the use of rigorous statistical techniques for testing the *reality* of climatic impact, although quantifying the magnitude of the impact is still difficult. The appropriate technique is best illustrated with an example from Ogilvie (1981; see also Ogilvie, 1984). As one aspect of her study, she examines the impact of climate on grass growth and hay yield in seventeenth- and eighteenth-century Iceland. The reality of a link in the twentieth century has been established using agronomic studies and regression analysis (see, for example, Bergthórsson, 1966, 1976). Contemporary historical documents suggest that grass growth and hay yields are adversely affected by temperature and rainfall during the growing season. Although only qualitative descriptive accounts are available, these can be tested by grouping them into cold, average and mild for thermal conditions (or wet, average and dry for precipitation) and poor, average and good for the grass growth or hay yield. An example comparing spring temperature and grass growth in the north of Iceland is given in Table 21.1 below. The table shows the observed number of occurrences in each joint category and the expected number of occurrences (in parentheses), obtained by multiplying the observed column and row frequencies together. The expected numbers so calculated are those which would occur if there were *no* relationship.

A significant difference between observed and expected frequencies, based on the value of the test statistic

Table 21.1 Spring temperature and grass growth in northern Iceland, 1601–1780. Numbers denote number of occurrences, numbers in parentheses are estimates based on the assumption of no crop-climate relationship. $\chi^2 = 43.3$; significance level better than 0.001.

Grass growth \ Spring temperature	Cold	Average	Mild	Totals
Poor	46(31.0)	9(18.1)	9(14.9)	64(39.3%)
Average	26(32.5)	30(18.9)	11(15.6)	67(41.1%)
Good	7(15.5)	7(9.0)	18(7.5)	32(19.6%)
Totals	79(48.5%)	46(28.2%)	38(23.3%)	163(100%)

$$\chi^2 = \sum_{i=1}^{9} \frac{(O_i - E_i)^2}{E_i}$$

(where the sum, $i = 1$ to 9, is over the number of boxes in the contingency table), indicates a relationship between variables. The value of χ^2 is tested using a χ^2 distribution. In this example, the χ^2 value is 43.3, highly significant and indicating a strong relationship. From Table 21.1 it can be seen that the observed incidence of poor grass growth following cold springs is much larger than would be expected if cold weather had no influence.

21.2.6 Alternative Strategies

21.2.6.1 Semidescriptive Case Studies

The difficulties of constructing detailed causal models to identify and measure climatic impact, especially for larger areas, a wide range of economic and social variables, and extended periods, prompts the use of less precise methods which may nevertheless be capable of yielding worthwhile results. One approach is the use of detailed semidescriptive case studies, of which the most notable so far completed are by Post (1977) and Pfister (1975, 1981). The method is to concentrate on particular periods of climate 'crisis' in which atmospheric variations appear to have been associated with marked changes in the economy and society of a given area. As far as possible, the links between climate variations and the society in question are specified rigorously. Variations in crop yields are examined in relation to a detailed analysis of such meteorological variables as temperature and precipitation. However, at the point where rigorous modeling ceases to be feasible (on account of the large number of variables involved, the complexity of the links with climate, imperfections of the data, etc.), the case studies take on a more impressionistic character, hence the term 'semidescriptive'.

Such studies can be immensely stimulating and perceptive. However, their value is inevitably limited to the extent that rigorous analysis is abandoned. In some ways, too, they may be misleading. Concentration on particularly acute periods of crisis may give a false impression of the importance of climate in human affairs. Although admittedly overstating the case, de Vries draws the parallel that 'short-term climatic crises stand in relation to economic history as bank robberies to the history of banking' (1980, 603). Moreover, the descriptive element in the method can easily degenerate into the cataloging of detail after detail of disruption and misery. The abundance of such detail, and the use in the description of such essentially rhetorical epithets as 'crippling', 'threatening', 'disastrous' (all of these examples are taken from Post, 1977) may, in the mind of the unwary reader, obscure the lack of a precise framework for gauging the importance of climatic impact. Semidescriptive studies, therefore, need to be carried out (and evaluated) with care. When properly qualified, however, they can shed useful light on climate–society links, suggest new hypotheses, and add considerably to any parallel quantitative study. They can, further, frequently lead to the discovery of important new climatological and socioeconomic data. Pfister's meticulous work (1981) on Switzerland in the Little Ice Age is a perfect exmple of the positive attributes of studies of this kind.

21.2.6.2 Occam's Razor

Anderson's work (1981) suggests an approach based on Occam's razor (the principle that, for the purposes of explanation, things not known to exist should not, unless it is absolutely necessary, be postulated as existing) in order to gauge the possible long-term importance of climatic variations in circumstances in which rigorous testing and assessment are not possible. The method is to scrutinize long-term social and economic changes and to examine how far they are explicable in non-climatic terms. The aim is to isolate possible explanatory lacunae which appear to require the invocation of climatic change as a relevant variable. If none is found, it may be concluded that climatic change was of negligible importance. Clearly, the method is highly subjective. Nevertheless, it may be valuable as a means of eliminating cases in which climatic change was almost certainly of negligible importance and focusing attention on cases where the issue of significant climate impact is more open to contention.

21.3 TIME-SCALES OF IMPACT

Climate impact studies in history may be classified according to the time-scale on which the impact occurs—short-term (annual or intra-annual); medium term (interannual); and longer-term (extending over periods of decades or centuries).

21.3.1 Short-term Influences

Short-term influences may be divided into two categories: the impact of isolated climatic events, including such phenomena as single storms or periods of storminess, and individual months or seasons of aberrant weather; and time-series analysis of the impact of seasonal or annual climatic fluctuations.

21.3.1.1 Isolated Climatic Events

Clearly, the impact of such events in normal circumstances is likely to be slight in relation to the totality of economic and other historical processes. However, there may conceivably be grounds for ascribing greater importance to a few isolated climatic events which happened to impinge on key historical situations. For example, it might be argued that the North Sea storms which helped to shatter the Spanish Armada in the summer and autumn of 1588 exerted an important influence on the course of English history.

The number of events which can plausibly be regarded as significant in this way appears to be small, and their level of significance is a matter of debate. For example, many factors contributed to the failure of the Armada in 1588, and the fleet had already been decisively defeated before bearing the main brunt of the storms. Even if the Armada had landed it is by no means certain that the subsequent history of England, far less that of Europe, would have been significantly altered (Parker, 1976). Thus, the argument that the storms had an important impact on human history is tenuous at best. In general, it would seem reasonable to regard the impact of isolated weather events on history as of fleeting and random importance only.

21.3.1.2 Seasonal and Annual Fluctuations

A much better case can be made for the importance of the short-term impact of month-to-month, season-to-season and year-to-year variations in the weather. Even historians sceptical of the significant impact of long-term climate changes commonly admit the importance of short-term variations (for example, Le Roy Ladurie, 1972, 119). Two widely accepted notions are that the economic history of *ancien régime* Europe is, through harvest success or failure, dependent on year-to-year climate fluctuations, but that this predominant direct effect of climate on society has diminished as society has become more industrialized in recent centuries (Braudel, 1973; de Vries, 1980, 601). Nevertheless, comparatively little research has been devoted to investigating in precise detail just how important such fluctuations were.

An attempt to fill this gap has been made by de Vries (1980) who correlated time-series of weather data with grain and butter prices, burial statistics, and transport data for Holland in the seventeenth to nineteenth centuries. His results throw some doubt on the traditional assumption that weather fluctuations were self-evidently of major economic importance. Many of his analyses show no statistically significant results, and those that do, show climate to account for relatively little of the 'impact' variance. He suggests that, although the idea of weather-dominated economies may hold true for closed, technologically primitive subsistence societies, in most areas of early-modern Europe the level of economic integration—including trade, markets, inventory formation and even futures trading—was sufficient to loosen greatly the asserted links between weather and harvests and between harvests and economic life more generally (de Vries, 1980, 602).

Two objections to this are immediately obvious. First, de Vries may have chosen the wrong variables to relate; in other words his original model may be faulty. He is, of course, well aware of this (the constraint of data availability has, in part, determined the choice of model). He has, for instance, in the case of butter prices, analyzed another model which gives more significant results (de Vries, 1980). Second, Post (1980, 721) has pointed out that the case of the Netherlands considered by de Vries may be atypical because of the advanced and sophisticated nature of the Dutch economy relative to many of her contemporaries. This objection needs to be tested by rigorous statistical studies for less economically sophisticated areas of preindustrial Europe. At present, such studies are lamentably lacking.

Something of an antidote to de Vries' scepticism is offered by Lee (1981). He has considered the impact of meteorological variables on short-run fluctuations in births, deaths and marriages in seventeenth- to nineteenth-century England, a country which in the seventeenth century rivaled Holland's economic development and subsequently exceeded it. Lee found no significant relationship between vital rates and rainfall data, unfortunately available to him only in the form of an inhomogenous annual series from 1727 (Nicholas and Glasspoole, 1931; Wigley *et al.*, 1983). However, striking effects of temperature variation were apparent. Mortality was increased by cold temperatures in the months from December to May ('winter'), and by hot temperatures in the June to November period ('summer'). The main effect of winter temperatures was contemporaneous, but for summer temperatures the effect was delayed by one or two months. (Lee speculates that low winter temperatures killed older elements in the population by means of rapidly lethal diseases, whereas hot summer weather tended to kill infants and small children through debilitating digestive tract diseases which took longer to cause death.) A 1 °C warming of winter temperatures would reduce annual mortality by about 2 percent; a 1 °C cooling of summer temperatures would reduce annual mortality by about 4 percent. Over the period 1665–1744 temperature explained a smaller proportion of the variance in annual mortality than did prices, but temperature and prices were equally important from 1745–1834. With regard to fertility the effects of temperature variation were more muted, about a quarter to a third the size of those for mortality.

Lee concluded that, overall, the effects of temperature variations on vital rates were quite striking (especially given the fact that the climate of England is relatively moderate), to the extent that he felt justified in speculating on the long-term implications of his results (see below). Lee's work exposes three factors which might have bearing on longer time-scale climatic impact. His analysis of grain price-mortality links shows that mortality increases occur mainly in the second and third year after a price maximum, and that contemporaneous mortality occurs only during price 'events' above a certain threshold (an interesting nonlinear effect). He also shows that relationships are not solely the result of extreme events. Even when the most extreme cases are removed from the analysis, the links between prices and mortality still hold. This can be compared with similar results obtained by Palutikof (1983) for the impact of climate and climatic extremes on the level of industrial production in England over the past few decades: the links hold even after the most extreme events are removed. If links between climate and socioeconomic factors hold in general (that is, not just for extreme events) then the cumulative influence of such linkages might provide a mechanism for the long-term effects of climate on society. However, such a hypothesis would require stability of the climate impact linkage and Lee, in analyzing different time periods, finds that some relationships are manifestly unstable—for instance, the price-mortality link reverses in sign after 1745!

21.3.2 Medium-term Influences

In analyzing medium-term and long-term effects, a new modeling difficulty arises. Possible causal links mentioned and discussed above are invariably of a transitory nature, and a mechanism must be proposed whereby such impacts are translated to longer time-scales. For instance, this might occur if short-term events are so closely spaced or if their impact is sufficiently long-lasting that recovery from one event cannot occur before the next impact event. Alternatively, a relationship must be established where the impact is not solely the result of extreme events (such as Lee's relationship between prices and mortality).

In this light, a number of authors have considered the effects of clusters of extreme seasons or years, which would appear likely to have a greater impact than continuous weak effects, even though these might operate over a range of values of a particular causal parameter. Post (1980), for example, argues that, whereas preindustrial societies might to a large extent adjust to cope with annual fluctuations, after a *succession* of years of severe weather the systems of adaptation and adjustment would be overchallenged, leading to higher death rates and a decline in economic activity.

A number of recent studies have focused in detail on the economic effects of anomalous weather conditions on interannual time-scales. Pfister (1975, 1981) has analyzed conditions in Switzerland during the Little Ice Age period, especially in the years 1570–1600, 1768–71, and 1812–17. Bowden (1967, 630–633) has examined clusters of anomalous years in England in 1546–71 and 1618–25, periods similar to each other in that in each case a run of exceptionally good harvests was followed by a succession of crop failures. Unfortunately Bowden lacked complete data on climatic fluctuations, but his brief analysis is nonetheless suggestive: he concluded that a climatically induced sequence of bounty and dearth had the effect of impoverishing first one section of the community and then other sections, and thus seems to have provided the optimum conditions for the onset of trade depression.

The possibility that the impact of severe climatic events might be magnified when these events occur after a long sequence of good years has been suggested by de Vries (1980), but such a possibility would be difficult to demonstrate in a statistically rigorous manner. In support of de Vries' hypothesis, the severe winters of 1709, 1740 and 1795 in Europe, which all followed long periods of favorable climate, were times of considerable economic stress; yet 1729, 1766 and 1780, while climatically comparable, had minimal impact. Interestingly, in the former three years there was also a contemporaneous impact on prices—prices rose during the harvest year, seemingly in anticipation of a climate-induced harvest failure, which then led to a further increase in prices. The influence of perception and adaptive measures needs to be examined further, but the absence of any simple general relationship weakens the case for any possible medium- or long-term impact through the cumulative effect of shorter-term crises. A relevant example is given by Sutherland (1981) in considering the effects of adverse (though admittedly not spectacularly severe) weather in the 1780s on a community in Upper Brittany. In the face of droughts in 1782 and 1785–86 and of a poor harvest in 1788, the social and economic structure of the area around the town of Vitré proved remarkably elastic. The worst-hit victims of economic stress were infants and small children, who experienced high rates of mortality; but because general fertility was high these children could easily be replaced, and their loss from a social structural point of view was relatively insignificant. Other symptoms of a crisis in rural society were absent. Sutherland's conclusion is that 'many peasant communities were not as vulnerable to the weather as we generally think' (1981, 434).

Another viewpoint is presented by Post (1983a,b) who strongly emphasizes the important issue of the relationship between climatic stress and disease, in particular the spread of epidemic diseases. Post calls for further research to 'specify the pathways along which the common infectious diseases spread in premodern and preindustrial Europe' (1983a, 159), as a way of discovering more about the role of climatic change in both economic and demographic spheres. In his own work, Post finds a link between climatic variability and the high incidence of infectious disease in 1740s Europe, but suggests that this was social rather than physiological; epidemics arose primarily through the social disorder that was largely a second-order effect of climatic stress (1983b). The effects of climate on human and animal disease incidence have also been examined by Ogilvie (1981), Pfister (1984) and others. There is considerable scope for further work on climate–disease interactions, both in terms of quantification of cause and effect and in terms of interpretation in a wider social context (see Escudero, Chapter 10 of this volume). There are, further, important instances of disease-related crop failures (such as the European potato blights in the 1840s [Lamb, 1977; Grigg, 1980] and the effects of *Fusarium nivale* on spelt and rye in Switzerland [Pfister, 1978]) which have been associated with both medium- and shorter-term climate fluctuations.

Opinion on the medium-term influence of climatic stress is thus divided. Any attempt to come to an overall conclusion would be premature because not enough detailed studies have been carried out. As in the case of short-term climatic influences, the need for further research is obvious.

21.2.3 Longer-term Influences

The possibility that climatic change is an important independent variable affecting the course of human history over extended periods of decades or centuries is a seductive one, and there have been a number of attempts to prove the connection. Utterström (1955), for example, argued that climatic changes may well have been of 'decisive' importance in influencing population movements in medieval and early modern Europe. Braudel (1949) hinted that the economic and social difficulties which were experienced in the Mediterranean area at the end of the sixteenth century, and which heralded the slow decline in the economic importance of this region in the seventeenth century, were partly caused by climatic change. Subsequently he stressed the importance of climatic change for the whole of Europe: 'the "early" sixteenth century was everywhere favored by the climate; the latter part everywhere suffered atmospheric disturbance' (Braudel, 1973). More recently, Parker (1979) has firmly asserted a relationship between the 'crisis' allegedly observable in the economy and society of seventeenth-century Europe and climatic 'deterioration'—the ultimate blame for which he rashly seems to attach to sunspots!

Suggestions of long-term climate effects have provoked scepticism and even sharp criticism. Le Roy Ladurie (1972, 292–293), noted the relatively small magnitude of long-term climate variations in Europe since about AD 1000 and questioned whether 'a difference in secular mean temperature ... [of about] 1 °C [could] have any influence on agriculture and other activities of human society, especially given the fact of human adaptation'. In response, Post has argued that it is misleading to concentrate on mean temperature values as such: 'interdecennary or even *annual* thermometric values and precipitation levels are insufficient data for understanding the dynamics of ecological effects. The micro-aspects are essential: annual temperature means often conceal critically large seasonal variations, which frequently mask destructive monthly deviations' (Post, 1973, 728). This observation is supported by the evidence presented by Lamb (1977, 465; 1981) that in Europe, in the early onset stages of the Little Ice Age between about 1300 and 1450, and in the climax phase in the sixteenth and seventeenth centuries, the *variability* of the weather was particularly marked (see also Pfister, 1981). However, the evidence for these changes in variability is still inconclusive. It is self-evident, for example, that a depression of the mean temperature will produce a greater frequency of severe cold events if the variability, as measured by the variance, remained unchanged (Sawyer, 1980; Ingram *et al.*, 1981), so an increased frequency of extreme events need not indicate a change in variability, but may, nonetheless, amplify the effect of a change in the mean.

Secular changes in climate may well involve changing levels of variability and such changes may be more important than changes in the mean. Changes in variability may or may not be statistically related to changes in the mean; we need only assume that they can occur. Parry (1981a) explains how an increased frequency of extreme events (which might result from either a change in variability or a change in the mean) may have a magnified impact by dramatically increasing the probability of *successive* (or closely spaced) extremes. If the probability of an extreme month (or season) is, for example, 1/10 and increases to 1/4 as a result of a secular change of climate, then the probability of two successive extremes will increase from 1/100 to 1/16, a six-fold reduction in return period. The question of assessing the impact of longer-term climatic shifts may resolve itself, then, into the problem of measuring the importance of individual clusters of extreme weather events of the type discussed in the previous subsection, and devising some means of assessing the cumulative effects of a succession of such clusters. Given that the economic and social impact of even a single cluster is difficult to establish, and that on secular time-scales the effects of climatic stress will be obscured by a multitude of other variables, the practical problems involved in these operations are clearly enormous.

21.4 HUMAN ADAPTATION TO CLIMATIC STRESS

The relationship between climatic stress and economic and social life should not be conceived as a one-way process. Man is a highly adaptive animal, capable of devising and deploying a wide range of technologies and social strategies to cope with a wide variety of environmental conditions. In view of this fact, and given the comparatively small range of climatic variations in historic times, it may be assumed that past human societies have, to a considerable extent, had the potential to successfully adapt to changes in climate. Thus the interrelationships between climate and human society may be conceived in terms of a two-way model involving elaborate feedback mechanisms, a simple version of which is presented in Figure 21.2 (see also Kates, Chapter 1 of this volume).

Approaches to climate–history studies that seek to assess the impact of climate variations on human economic and social systems in the manner discussed in the preceding section frequently obscure the processes of human adaptation to climatic stress. Depending on whether the particular researcher's aim is to advocate or to reject the importance of climate in human affairs, the processes of adaptation may be either neglected as an embarrassing complication, or emphasized in order to argue that worthwhile assessments of climatic impact are inordinately difficult.

An alternative approach is to focus attention on the processes of adaptation (or *failure* to adapt) explicitly. This antideterministic approach has been advocated by de Vries (1980, 625–630), and is well represented among papers in Wigley *et al.* (1981). Such studies make it clear that societies subjected to climatic stress must not be regarded as passive victims of external forces, but rather that such stress is in the nature of a challenge to which a variety of responses are possible. This allows a more subtle appreciation of the role of climate in human affairs.

From this point of view, cases in which societies appear to have been seriously damaged by, or even totally succumbed to, climatic stress should not be taken to demonstrate the determining influence of climate. It is essential to consider ways in which these societies might have coped better, and to focus on the political, cultural and socioeconomic factors which inhibited them from doing so. Such studies are potentially of great relevance to the problems of the modern world. Identification of the factors which prevent successful adaptation could well aid planners in their attempts to avert future disasters.

This question has been explored by Mooley and Pant (1981) in their discussion of the socioeconomic impact of droughts in India over the past 200 years. They argue that the extent of human suffering and mortality occasioned by Indian drought was largely conditioned by inflexible and exploitative systems of social and economic organization, in part the result of foreign rule in the period before independence (1947). They suggest that changes in agricultural practices and in other aspects of the rural economy, complemented by planned programs to accumulate food reserves and cash funds to provide, when necessary, swift and effective relief in the drought-prone areas, could in the future significantly reduce the suffering occasioned by monsoon failure (see also Jodha and Mascarenhas, Chapter 17 of this volume).

Mooley and Pant's discussion is not specifically related to long-term climatic changes (on the contrary, they show statistically that the incidence of drought in India in the period 1771–1977 is random); and the climatic stresses which they studied, albeit enough to inflict immense suffering on substantial sections of society, were not of a severity sufficient to destroy the entire society. By contrast, McGovern's work on Norse Greenland (1981) deals both with long-term climatic shifts and the extinction of a substantial part of the Norse society.

McGovern argues strongly that, even in this manifestly marginal case, it would be wrong to explain the collapse of the more northerly of the two Norse settlements—the so-called 'Western Settlement'—solely in terms of a deteriorating climate, or even of climate in association with other external pressures. The Norse could have survived by shifting the economic balance of their society away from stockraising towards greater exploitation of seals and other marine resources, and by adopting the use of elements of Inuit eskimo culture and technology (skin boats, clothing, etc.) in order to facilitate the shift. But instead of pioneering these adaptive strategies, the evidence suggests that the political and religious élite in Greenland persisted in maintaining existing and increasingly inappropriate economic and cultural patterns. Their failure as flexible managers of the community's scanty resources was the ultimate cause of the extinction of the Norse colonies.

The question of adaptation can be approached from other directions. In cases where marked climatic stress appears (in terms of gross economic or other indicators) to have had little impact, it may be valuable to investigate the processes of adaptation whereby the climatic stress was minimized or absorbed. Again, the insights which might be derived from such studies could prove valuable in the context of present-day planning.

A significant contribution to the study of such adaptation has been offered by de Vries (1980). He focuses primarily on economic responses and specifically on changes in agricultural practices and technology. He suggests (pp. 625–626) that certain changes in agricultural practices, such as variations in crop mixes, may sometimes be interpreted as adaptive strategies developed in response to climatic stress. It is possible, for example, that the tendency towards crop diversification in midland England in the period 1550–1650 (hitherto explained in terms of shifts in demand towards cheaper grains, induced by a reduction in the purchasing power of much of the population; Skipp, 1978, 44–49) could represent farmers' attempts to reduce their vulnerability to a climate which had become more frequently threatening towards winter crops.

The approach by Sutherland (1981) is more general. He focuses not on changes in agricultural practices, but on the complex of social mechanisms which enabled the Vitré area of Brittany in the 1780s (and by implication other peasant societies) to cope with climatic stress. In Vitré the poor were cushioned by secondary sources of income which were only partly dependent on the grain harvest, while employment for farmhands continued to be available throughout the period of meteorological stress. The richer peasantry, with substantial grain surpluses even in bad years, actually benefited from the high prices associated with poor harvests. In addition, relief systems for the poor and other charitable mechanisms, such as the willingness of landlords to postpone or waive demands for rent, helped to distribute the costs of climatic stress. Overall, a range of social and economic factors was responsible for absorbing the shocks administered by the climate and helped to maintain the social fabric more or less intact.

These approaches leave a number of questions unanswered. It is not clear how climatic stress-induced adaptations can be distinguished from other elements of technological and economic change, except in certain specific cases. In most cases the problems of detecting adaptive strategies would appear to be analogous to, but possibly even more complex than, those involved in identifying and measuring climatic impact. In those studies where adaptive or cost-distributing factors have been shown to have been effective (for example, Sutherland, 1981) there remains the vital question which such an analysis inevitably provokes: what were the limits to the society's capacity to absorb stress? In other words, what degree of meteorological adversity would have been required to precipitate serious social and economic dislocation?

To generalize from these specific comments, it is plain that in order to advance the study of human adaptation to climatic variations to the point where the findings may be of real use to modern planners, it will be necessary to try to specify more clearly what degrees and types of climatic stress impose the greatest problems of successful response, and to seek to identify more rigorously the key features of more- and less-adaptive societies. To do this it will be necessary to attempt extensive cross-cultural comparisons. A preliminary step in this direction has been taken by Bowden *et al.* (1981) in a comparative study of agriculture in the US Great Plains in the period AD 1880–1979, the droughts in the Sahel region of Africa between 1910–15 and 1968–74, and the societies of the Tigris-Euphrates valley over six millennia. They discuss the complementary hypotheses that, over time, societies adapt to cope with 'minor' climatic stresses (defined as events with a return period of the order of less than 100 years), but that thereby they do little to decrease, and may actually increase, their vulnerability to 'major' stresses of rarer frequency.

Because of inadequate data, the results from the Tigris-Euphrates study proved inconclusive and the part of the hypothesis relating to 'major' stresses could be handled only on a speculative level. However, it is argued strongly that both the Sahel and the US Great Plains have become less vulnerable to 'minor' climatic stress. In the case of the Great Plains, wheat yield evidence was analyzed to discover whether the lessening of vulnerability could be explained by reference to developments in agricultural technology; the analysis failed to substantiate this possibility. Both in the Sahel and in the Great Plains the key factor in reducing vulnerability appeared to be the integration of each area into a wider economic system. The sufferings of the Sahel in 1968–74, though massive and perhaps partly precipitated by political rather than climatic factors (García, 1981), were, nonetheless, alleviated by external aid. On the Great Plains, distress was reduced after about 1930 when the Union as a whole accepted the responsibility of providing relief for drought-afflicted areas, thus in effect sharing the costs of climatic stress. It proved possible to absorb even a 'major' climatic stress, the drought of the 1930s, in this way. Massive Federal relief ensured that the potential catastrophe of these years took the form of a large ripple through the national economy, rather than a tidal wave of disaster located in the Great Plains region itself. However, Bowden *et al.* caution that in the future there may prove to be a limit to the effectiveness of such integrative mechanisms.

Ingram *et al.* (1981) have noted an important point which emerges from the study by Bowden *et al.* This is that newly settled regions (such as the Great Plains in the late nineteenth century) and areas where the economic system is subjected to rapid modification (as in the Sahel in the twentieth century) face particular problems of vulnerability to climatic stress. In such unfamiliar circumstances the nature of the climatic regime, and its likely hazards for the type of economic system which is being implemented, may be poorly understood and mismanagement (leading sooner or later to disaster) is particularly likely to occur. The same point has been emphasized in a paper by Smith *et al.* (1981) on climatic stress and Maine agriculture in the nineteenth century. Here, a brief warming period in the 1820s raised unrealistic expectations that the region could support large-scale commercial agriculture, and the process of adjustment to the normal cooler conditions which soon reasserted themselves was a painful one which took several decades to accomplish.

The complex question of human adaptation to climate is one whose systematic study has only just begun. The value of this field of endeavor is obvious: apart from its intrinsic interest, it is of major relevance to the problems of modern world planning. Further research is urgently required. The studies so far completed, though limited in their scope and achievement, are of considerable interest and have added a major new dimension to the climate impact debate.

21.5 REVIEW OF HISTORICAL CLIMATE IMPACT STUDIES

One of the main concerns of this chapter is to describe a methodology or framework for historical climate impact studies. To do this we have synthesized what we consider to be the best elements of other physical and social scientists' work, and our own views. To complement this framework we now consider briefly and selectively some of the published historical climate impact analyses available in the literature. The majority of these are specific case studies, but we have also included McGhee's (1981) review of archaeological work. In trying to summarize the important features (both good and bad) of past studies, we have, admittedly, introduced an element of unfairness by categorizing these works using a set of rules with which the workers themselves may not agree. Nevertheless, our summary provides an integrated review which is a guide, not only to the types of study carried out in the past, but also to the quality of these studies, and to the pitfalls which future workers might strive to avoid.

Our summary is presented in Table 21.2. It is far from comprehensive, not least because there is a vast pool of studies (using this word loosely) which we consider to be little more than poorly documented and generally unsupported speculations, and which, perforce, warrants neglect. In addition, works which have been subsequently superseded have also been omitted.

The studies considered below fall into three broad categories:

1. Climatic factors used casually to support an argument.
2. Climatic factors seriously considered, but in insufficient depth to permit an assessment of their influence.
3. Publications concerned solely with the impact (or lack of impact) of climate on historical societies.

We have categorized these studies under the following headings.

Reference number and author
The complete bibliographic reference is given in the general reference list.

Time period studied

Time-scale of impact
Short, medium or long—defined above. Some studies cover more than one time-scale.

Region

Main variables examined
These are subdivided into socioeconomic impact variables and climate variables according to the following number scheme:

Society	1	Agriculture
	2	Demography
	3	Economy
	4	Industry–technology–transport
	5	Politics
	6	Culture
	7	Adaptability
Climate	1	Temperature
	2	Precipitation
	3	Other

Data quality
Also subdivided into socioeconomic and climatic data; 'n.a.' stands for 'not applicable'.

Conceptual model used
Our identifiers here are measures of the realism and complexity of the model used (or implied). A *causal* model is one which considers only cause and effect, with little or no consideration of other possible causes or complicating factors. In some cases such a model may be appropriate, but generally this category implies an oversimplified analysis of the problem. A *realistic* model is one where a reasonable attempt has been made to place the problem in a firmly based conceptual framework, often with some thought given to adaptive measures.

Table 21.2 Summary of selected works on historical climate impact assessment. See text for explanation of numerical identifiers of variables and statistical methods

No.	Author(s)	Period	Time scale	Region	Variables used	
					Society	Climate
1	Abel (1966)	13–19th c.	short	Europe	1–5	1,2
2	Bell (1971)	~2000 B.C.	long	Egypt	1–7	3
3	Beveridge (1921)	16–19th c.	short	Europe	1	3
4a	Bowden et al. (1981)	0–6000 B.P.	long	Mesopotamia	1,2,4,5	1–3
4b		1890 on	short/med.	USA	1–3,7	1–3
4c		20th c.	short/med.	Sahel	1–3,5,7	1–3
5	Brandon (1971)	1340–1444	short	SE England	1	1–3
6	Braudel (1949, 1973)	4th c. B.C. to 20th c.	short	Mediterranean	1–7	1–3
7	Bryson and Murray (1977)	various	short to long	numerous	1,2,6	1–3
8	de Vries (1980)	14th c. on	short/med.	W Europe	1–3,7	1–3
9	Hoskins (1964, 1968)	1480–1759	short	England	1	2
10	Kershaw (1973)	1315–1322	short	England	1	2
11	Lee (1981)	1540–1840	short	England	2	1,2
12	McGhee (1981)	0–5000 B.P.	long	numerous	1,6,7	1–3
13	McGovern (1981)	10–15th c.	med./long	Greenland	1–3,5–7	1–3
14	Mokyr (1977)	18/19th c.	short/med.	W Europe	1,3,7	unspecified
15	Ogilvie (1981)	9–18th c.	short/med.	Iceland	1–7	1–3
16	Parry (1978)	14–18th c.	med./long	Scotland, Scandinavia	mainly 1	1–3
17	Pfister (1975, 1981)	16–19th c.	short/med.	Switzerland	mainly 1	1–3
18	Post (1977)	1815–1819	short	W Europe	1–7	1–3
19	Puiz (1974)	16–18th c.	short/med.	Geneva	1	1,2
20	Shaw (1981)	Roman period	long	N Africa	1–3,7	1–3
21	Utterström (1955)	16–19th c.	long	W Europe	mainly 1,2	1–3
22	Walton (1952)	18th c.	short/med.	Scotland	1,2	1–3
23	Whyte (1981)	17th c.	short/long	Scotland	mainly 1	1–3
24	Wright (1976)	14–16th c.	long	a Derbyshire village	1,2,4	1–3

Data quality		Model	Stat. methods	Inferred strength of link	Comments
Society	Climate				
good	suspect	none	2	weak	a classic work on agricultural history
good	good	none	2	strong	a thorough study by a non-historian (see also Bell, 1970)
good	good	none	*	strong	*relationship based on coincidence of cycles
suspect	variable (often suspect)	causal	2	moderate	
good	good	realistic	2	moderate	
good	good	causal	2	moderate	
good	good	realistic	2	moderate	
good	good	none	1	weak but important	one of the first historical works to take climate/society links seriously; mainly restricted to 16th c.
variable (often suspect)	variable (often suspect)	none	1,2	strong	far reaching, but generally unconvincing
good	good	realistic	3	weak or none	one of the most important papers on the subject
poor (yields deduced from prices)	poor	none	2	strong	see also Harrison (1971)
variable	good	none	1	strong	
good	good	none	3	weak but important	an important paper on demography and climate
good	good	n.a.	n.a.	weak or none	review of archaeological work
good	good	n.a.	n.a.	indirect	an important work on adaptability
n.a.	n.a.	*	n.a.	weak	*an economic model with climate as a fundamental forcing factor
good	good	realistic	3	moderate	one of the most detailed and rigorous climate impact studies available
good	variable (mainly good)	realistic	2	moderate	a major work on marginal societies (see also Parry, 1981)
good	good	realistic	2	moderate	an important and detailed study (see also Pfister, 1978)
good	good	none	1,2	strong	an important and detailed study (see also Post, 1980)
good	suspect to good	none	1,2	strong	
good	good	n.a.	n.a.	none	demolishes some myths
fair to good	suspect to good	none	1,2	strong	an important, but dated, early work on climate impact
suspect	good but local	realistic	1,2	moderate	
good	good	realistic	none	weak	
suspect	suspect	none	2	weak	

Statistical methods employed

Very few studies have employed rigorous statistical methods. In some cases, where a strong correspondence between impact and climate supports a well thought-out conceptual model, a detailed statistical analysis may not be necessary; but such cases are rare. We have classified statistical methods as:

0 none
1 a single coincidence
2 qualitative parallels drawn
3 rigorous statistical methods used
n.a. not applicable

Inferred strength of link(s)

This category gives the strength of the relationship(s) as claimed in the original reference. The descriptors used are self-explanatory. A *claimed* link does not necessarily mean that the link has been convincingly demonstrated; in fact, according to our criteria many such claims are subject to considerable doubt.

Comments

21.6 CONCLUSION

Viewpoints on the extent to which past variations in climate have affected societies are polarized. A number of physical scientists (and a few historians) are unshakeable in their belief that climate *must* have affected society in the past, and would be quick to quote standard examples such as the decline of the Roman empire, the failing of the Norse settlement in Greenland, the influence of the Little Ice Age in Europe, and so forth. Few rigorous studies have been made of these textbook examples, however, and closer examination shows the 'demonstration' of climate's importance to be based often on deficient or inadequate data, oversimplified arguments, and/or a tendency to ignore non-climatic factors. Historians, on the other hand, tend to the opposite viewpoint. With notable exceptions they have chosen to ignore the possible importance of climate on the development of society.

Of the more detailed studies, the most convincing have been those dealing with short time-scales (such as the work of de Vries and Lee, and some of Pfister's work) or with marginal societies (in particular the studies by Parry and Ogilvie). All of these analyses have involved clearly defined impact variables and have used reliable, quantitative data. Each study has produced results which are specific to the time and place considered, and it is difficult to draw any broad conclusions from them. This is, in fact, the essence of the difference in viewpoints noted above. Historians tend to eschew broad generalizations, partly because it is the detail, the differences from one case to another, which is central to historical research. Physical scientists, however, tend actively to seek broad generalizations.

What, then, can be learned from historical impact studies? Perhaps the most important message is to maintain an open and moderate stance, to accept and try to account for the complexities of society, and, especially, to give due consideration to the flexibility of the social system and to the many possible modes of adaptation to imposed stress which are available from the level of the individual up to the level of the whole society. Equally important are the lessons to be learned from examples of the failure to adapt.

Human nature has changed little over the centuries; crops are still affected by the vicissitudes of the weather, and many parts of the world today have parallels with seventeenth- to nineteenth-century Europe. Each case study tells us something which may be of value today in building up our body of information of societies' experiences and perceptions of, and reactions to, external stresses, and a continued research effort into all three aspects of impact is demanded.

Modern climate impact studies are manifestly interdisciplinary in nature, and historical impact studies, by virtue of their historical perspective and requirements of the historian's expertise, are even more so. There is a temptation to classify a study by the ground that it attempts to cover rather than by the disciplines deployed. In this sense there have been many interdisciplinary historical climate impact studies, spawned mostly by the intellectual wanderings of scientists or historians perhaps bored with the myopia of specialization. For such wanderers

> there are great risks [in] invading an unfamiliar turf [since] ... the complications of the new subject may not conform readily to the insights the newcomer brings from some other discipline. (McNeill 1981, 642)

We end this review, knowing full well that we, as individual authors, are guilty of such trespasses, with an appeal for more rigorous and truly interdisciplinary studies, studies in which individual expertise from different fields is brought together by closely collaborative efforts. In no field is the need for this so great as in historical studies of the impact of climate on societies.

ACKNOWLEDGMENTS

This chapter owes much to the papers presented in the Climate and History conference held in Norwich in 1979 which was the brainchild of Professor H. H. Lamb, and to historical climate research carried out in the Climatic Research Unit using funds from the Rockefeller Foundation and the European Economic Community's Climate Programme.

REFERENCES

American Association for the Advancement of Science (AAAS) (1980). *Workshop on Environmental and Societal Consequences of a Possible CO_2-Induced Climate Change, Annapolis, Maryland, April 2–6, 1979*. US Dept. of Energy Carbon Dioxide Effects Research and Assessment Program 009, CONF-7904143, UC-11, Washington, DC.

Abel, W. (1966). *Agrarkrisen und Agrarkonjunktur* (second edition). Parey, Hamburg/Berlin.

Anderson, J. L. (1981). History and climate: Some economic models. In Wigley, T. M. L., Ingram, M. J., and Farmer, G. (Eds.) *Climate and History*, pp. 337–355. Cambridge University Press, Cambridge.

Appleby, A. B. (1980). Epidemics and famine in the Little Ice Age. *Journal of Interdisciplinary History*, **10**, 643–663.

Baars, C. (1973). De geschiedenis van de landbouw in de Beijerlanden. Centrum voor Landbouwpublikaties en Landbouwdocumentatie. *Verslagen van landbouwkundige onderzoekingen 801* (105.2 V6 IV), Wageningen, Netherlands.

Bell, B. (1971). The dark ages in ancient history, 1: The first dark age in Egypt. *American Journal of Archaeology*, **75**, 1–26.

Bergthórsson, P. (1966). Hitafar og búsæld á Íslandi. *Veðrið*, **11** (1), 15–20.

Bergthórsson, P. (1976). Töðufall á Íslandi frá aldamótum. *Freyr*, **LXXII** (13–14), 250–263.

Beveridge, W. H. (1921). Weather and harvest cycles. *Economic Journal*, **31**, 421–453.

Beveridge, W. H. (1922). Wheat prices and rainfall in western Europe. *Journal of the Royal Statistical Society*, **85**, 411–478.

Beveridge, W. H. (1929). A statistical crime of the seventeenth century. *Journal of Economic and Business History*, **1**, 503–533.

Bowden, M. J., Kates, R. W., Kay, P. A., Riebsame, W. E., Warrick, R. A., Johnson, D. L., Gould, H. A., and Weiner, D. (1981). The effect of climate fluctuations on human populations: Two hypotheses. In Wigley, T. M. L., Ingram, M. J., and Farmer, G. (Eds.) *Climate and History*, pp. 479–513. Cambridge University Press, Cambridge.

Bowden, P. (1967). Agricultural prices, farm profits, and rents. In Thirsk, J. (Ed.) *The Agrarian History of England and Wales*, Vol. 4: *1500–1640*, pp. 593–695. Cambridge University Press, Cambridge.

Brandon, P. F. (1971). Late-medieval weather in Sussex and its agricultural significance. *Transactions of the Institute of British Geographers*, No. 54, 1–17.

Braudel, F. (1949). *La Mediterranée et le Monde Mediterranéen á l'Époque de Philippe II*. Librairie Armand Colin, Paris.

Braudel, F. (1973). *The Mediterranean and the Mediterranean World in the Age of Philippe II*. 2 vols. Fontana/Collins, London.

Bryson, R. A., and Murray, T. J. (1977). *Climates of Hunger*. University of Wisconsin Press, Madison, Wisconsin.

Chappell, J. E. (1970). Climatic change reconsidered: Another look at 'The Pulse of Asia'. *Geographical Review*, **60**, 347–373.

de Vries, J. (1977). Histoire du climat et économie: des faits nouveaux, une interprétation différente. *Annales: Économies, Sociétés, Civilisations*, **32**, 198–227.

de Vries, J. (1980). Measuring the impact of climate on history: The search for appropriate methodologies. *Journal of Interdisciplinary History*, **10**, 599–630.

Fischer, D. H. (1980). Climate and history: Priorities for research. *Journal of Interdisciplinary History*, **10**, 821–830.

Flohn, H. (1981). Short-term climatic fluctuations and their economic role. In Wigley, T. M. L., Ingram, M. J., and Farmer, G. (Eds.) *Climate and History*, pp. 310–318. Cambridge University Press, Cambridge.

García, R. V. (1981). *Nature Pleads Not Guilty. Vol. 1 of Drought and Man: The 1972 Case History*. Pergamon Press, Oxford.

Grigg, D. B. (1980). *Population Growth and Agrarian Change: An Historical Perspective*. Cambridge University Press, Cambridge.
Gunst, R. F., and Mason, R. L. (1980). *Regression Analysis and Its Application, a Data-Oriented Approach*. Marcel Dekker, Inc., New York.
Harrison, C. J. (1971). Grain price analysis and harvest qualities, 1465–1634. *Agricultural History Review,* **19**, 135–155.
Hooker, R. H. (1922). The weather and crops in Eastern England, 1885–1921. *Quarterly Journal of the Royal Meteorological Society,* **48**, 115–138.
Hoskins, W. G. (1964). Harvest fluctuations and English economic history, 1480–1619. *Agricultural History Review,* **12**, 28–46.
Hoskins, W. G. (1968). Harvest fluctuations and English economic history, 1620–1750. *Agricultural History Review*, **16**, 15–31.
Huntington, E. (1907). *The Pulse of Asia*. Houghton Mifflin, Boston, Massachusetts.
Huntington, E. (1915). *Civilisation and Climate*. Yale University Press, New Haven, Connecticut, and Oxford University Press, London.
Imhof, A. E. (1976). *Aspekte der Bevölkerungsentwicklung in den nordischen Ländern 1720–1750*, 2 vols. Francke-Verlag, Berlin.
Ingram, M. J., Farmer, G., and Wigley, T. M. L. (1981). Past climates and their impact on Man: A review. In Wigley, T. M. L., Ingram, M. J., and Farmer, G. (Eds.) *Climate and History*, pp. 3–50. Cambridge University Press, Cambridge.
Kershaw, I. (1973). The great famine and agrarian crisis in England 1315–1322. *Past and Present*, No. 59, 3–50.
Lamb, H. H. (1969). The new look of climatology. *Nature,* **223**, 1209–1215.
Lamb, H. H. (1977). *Climate: Present, Past and Future. Vol. 2, Climatic History and the Future*. Methuen, London.
Lamb, H. H. (1981). An approach to the study of the development of climate and its impact in human affairs. In Wigley, T. M. L., Ingram, M. J., and Farmer, G. (Eds.) *Climate and History*, pp. 291–309. Cambridge University Press, Cambridge.
Lee, R. (1981). Short-term variation: Vital rates, prices and weather. In Wrigley, E. A., and Schofield, R. S. (Eds.) *The Population History of England 1541–1871, a Reconstruction*, pp. 356–401, Edward Arnold, London.
Le Roy Ladurie, E. (1972). *Times of Feast, Times of Famine: A History of Climate Since the Year 1000*. George Allen and Unwin, London.
McGhee, R. (1981). Archaeological evidence for climatic change during the last 5000 years. In Wigley, T. M. L., Ingram, M. J., and Farmer, G. (Eds.) *Climate and History*, pp. 162–179. Cambridge University Press, Cambridge.
McGovern, T. H. (1981). The economics of extinction in Norse Greenland. In Wrigley, T. M. L., Ingram, M. J., and Farmer, G. (Eds.) *Climate and History*, pp. 404–433. Cambridge University Press, Cambridge.
McNeill, W. H. (1981). Trap for the unwary (review of book by P. Colinvaux). *Nature*, **290**, 642–643.
McQuigg, J. D. (1975). *Economic Impacts of Weather Variability*. Atmospheric Sciences Department, University of Missouri, Columbia, Missouri.
Mitchell, J. M., Jr., Dzerdzeevskii, B., Flohn, H., Hofmeyr, W. L., Lamb, H. H., Rao, K. N., and Wallen, C. C. (1966). *Climatic Change*. World Meteorological Organization Technical Note No. 79 (WMO-No.195. T.P.100). Geneva, Switzerland.
Mokyr, J. (1977). Demand vs. supply in the Industrial Revolution. *Journal of Economic History*, **37**, 981–1008.
Mooley, D. A., and Pant, G. B. (1981). Droughts in India over the last 200 years, their socio-economic impacts and remedial measures for them. In Wigley, T. M. L.,

Ingram, M. J., and Farmer, G. (Eds.) *Climate and History*, pp. 465–478. Cambridge University Press, Cambridge.

Nicholas, F. J., and Glasspoole, J. (1931). General monthly rainfall over England and Wales, 1727 to 1931. *British Rainfall,* **1931**, 299–306.

Ogilvie, A. E. J. (1981). *Climate and Society in Iceland from the Medieval Period to the Late Eighteenth Century*. Unpublished PhD dissertation. University of East Anglia, Norwich, UK.

Ogilvie, A. E. J. (1984). The impact of climate on grass growth and hay yield in Iceland: A.D. 1601 to 1780. In Mörner N.-A., and Karlén, W. (Eds.) *Climatic Changes on a Yearly to Millennial Basis*, pp. 343–352. D. Reidel Publishing Co., Dordrecht.

Palutikof, J. P. (1983). The impact of weather and climate on industrial production in Great Britain. *Journal of Climatology,* **3**, 65–79.

Parker, G. (1976). If the Armada had landed. *History,* **61**, 358–368.

Parker, G. (1979). *Europe in Crisis, 1598–1648*. (Fontana History of Europe.) Fontana Books, London.

Parker, G., and Smith, L. M. (Eds.) (1978). *The General Crisis of the Seventeenth Century*. Routledge and Kegan Paul, London.

Parry, M. L. (1978). *Climatic Change, Agriculture and Settlement*. William Dawson & Sons, Folkestone, UK

Parry, M. L. (1981a). Climatic change and the agricultural frontier: A research strategy. In Wigley, T. M. L., Ingram, M. J., and Farmer, G. (Eds.) *Climate and History*, pp. 319–336. Cambridge University Press, Cambridge.

Parry, M. L. (1981b). Evaluating the impact of climatic change. In Delano Smith, C., and Parry, M. L. (Eds.) *Consequences of Climatic Change*, pp. 3–16. Department of Geography, University of Nottingham, UK.

Pearson, R. (1978). *Climate and Evolution*. Academic Press, London.

Pfister, C. (1975). Agrarkonjunktur und Witterungsverlauf im Westlichen Schweizer Mittelland. *Geographica Bernensia*, **G2**. Bern.

Pfister, C. (1978). Climate and economy in eighteenth century Switzerland. *Journal of Interdisciplinary History,* **9**, 223–243.

Pfister, C. (1981). An analysis of the Little Ice Age climate in Switzerland and its consequences for agricultural production. In Wigley, T. M. L., Ingram, M. J., and Farmer, G. (Eds.) *Climate and History*, pp. 214–248. Cambridge University Press, Cambridge.

Pfister, C. (1984). *Das Klima der Schweiz und Seine Bedeutung in der Geschichte von Bevölkerung und Landwirtschaft*. 2 vols. Haupt, Bern.

Post, J. D. (1973). Meteorological historiography. *Journal of Interdisciplinary History,* **3**, 721–732.

Post, J. D. (1977). *The Last Great Subsistence Crisis in the Western World*. John Hopkins University Press, Baltimore, Maryland.

Post, J. D. (1980). The impact of climate on political, social and economic change: A comment. *Journal of Interdisciplinary History,* **10**, 719–723.

Post, J. D. (1983a). Climatic change and historical discontinuity. *Journal of Interdisciplinary History,* **14**, 153–160.

Post, J. D. (1983b). Climatic variability and epidemic disease in preindustrial Europe. *Journal of Interdisciplinary History* (submitted).

Puiz, A. M. (1974) Climat, récoltes et vie des hommes à Genève, XVIe-XVIIIe siècle. *Annales: Économies, Sociétés, Civilisations,* **29**, 599–618.

Sawyer, J. S. (1980). Climatic change and temperature extremes. *Weather,* **35**, 353–357.

Shaw, B. D. (1981). Climate, environment, and history: The case of Roman North Africa. In Wigley, T. M. L., Ingram, M. J., and Farmer, G. (Eds.) *Climate and History*, pp. 379–403. Cambridge University Press, Cambridge.

Skipp, V. (1978). *Crisis and Development. An Ecological Case-study of the Forest of Arden, 1500–1674*. Cambridge University Press, Cambridge.

Slicher van Bath, B. H. (1977). Agriculture in the vital revolution. In Rich, E. E., and Wilson, C. H. (Eds.) *The Cambridge Economic History of Europe*, Vol. 5, pp. 42–132. Cambridge University Press, Cambridge.

Smith, D. C., Borns, H. W., Baron, W. R., and Bridges, A. E. (1981). Climatic stress and Maine agriculture, 1785–1885. In Wigley, T. M. L., Ingram, M. J., and Farmer, G. (Eds.) *Climate and History*, pp. 450–464. Cambridge University Press, Cambridge.

Starr, T. B., and Kostrow, P. I. (1978). The response of spring wheat yield to anomalous climate sequences in the United States. *Journal of Applied Meteorology*, **17**, 1101–1115.

Sutherland, D. M. G. (1981). Weather and the peasantry of Upper Brittany, 1780–1789. In Wigley, T. M. L., Ingram, M. J., and Farmer, G. (Eds.) *Climate and History*, pp. 434–449. Cambridge University Press, Cambridge.

Thompson, L. M. (1975). Weather variability, climatic change and grain production. *Science*, **188**, 535–541.

Titow, J. Z. (1960). Evidence of weather in the account rolls of the Bishopric of Winchester, 1209–1350. *Economic History Review*, 2nd series, **12**, 360–407.

Titow, J. Z. (1970). Le climat à travers les rôles de comptabilité de l'Evêché de Winchester, 1350–1450. *Annales: Économies, Sociétés, Civilisations*, **25**, 312–350.

Tits-Dieuaide, M. J. (1975). *La Formation des Prix Céréaliers en Brabant et en Flandre au XVe Siècle*. Éditions de l'Université de Bruxelles, Brussels.

Utterström, G. (1955). Climatic fluctuations and population problems in early modern history. *Scandinavian Economic History Review*, **3**, 3–47.

Walton, K. (1952). Climate and famine in North East Scotland. *Scottish Geographical Magazine*, **68**, 13–21.

Whyte, I. (1981). Human response to short- and long-term climatic fluctuations: The example of early Scotland. In Delano Smith, C., and Parry, M. L. (Eds.) *Consequences of Climatic Change*, pp. 17–29. Department of Geography, University of Nottingham, UK.

Wigley, T. M. L. (1983). The role of statistics in climate impact analysis. In *Proceedings of the Second International Meeting on Statistical Climatology, Lisbon, 8.1.1–8.1.10*. Instituto Nacional de Meteorologia e Geofísica, Portugal.

Wigley, T. M. L., Ingram, M. J., and Farmer, G. (Eds.) (1981). *Climate and History*. Cambridge University Press, Cambridge.

Wigley, T. M. L., Lough, J. M., and Jones, P. D. (1983). Spatial patterns of precipitation in England and Wales and a revised homogeneous England and Wales precipitation series. *Journal of Climatology*, **4**, 1–25.

Wigley, T. M. L., and Tu Qipu (1983). Crop-climate modelling using spatial patterns of yield and climate: Part 1, Background and an example from Australia. *Journal of Climate and Applied Meteorology*, **22**, 1831–1841.

Wright, S. M. (1976). Barton Blount: Climatic or economic change? *Medieval Archaeology*, **20**, 148–152.

Climate Impact Assessment
Edited by R. W. Kates, J. H. Ausubel and M. Berberian

CHAPTER 22
Recent Assessments

MICHAEL H. GLANTZ,* JENNIFER ROBINSON† AND MARIA E. KRENZ*

* *Environmental and Societal Impacts Group*
National Center for Atmospheric Research‡
Boulder, Colorado 80307 USA

†*Department of Geography*
University of California, Santa Barbara
Santa Barbara, California 93106, USA

‡ The National Center for Atmospheric Research is sponsored by the National Science Foundation.

22.1 INTRODUCTION

Integrated assessments, those that involve one or more connected links in the chain of climate–society impacts, were undertaken with increasing frequency in the decade of the 1970s. The authors have selected five studies for the purpose of identifying their successes and their problem areas in the hope of assisting those who undertake similar assessments in the future. The readers should keep in mind that the review of these five assessments is meant to be suggestive and in no way is meant to provide definitive guidelines. The review is designed to:

1. identify the origins of, or reasons for, each study,
2. evaluate the internal organizational factors, and
3. highlight the study's impacts (that is, how it was received).

The studies chosen met totally or in part the following criteria:

1. each was an attempt at an integrated climate-related impact assessment,
2. each represented a multidisciplinary effort,
3. each had been requested by policy-makers,

4. the research activity was completed in the 1970s, and
5. each study was done at a high cost and/or received considerable coverage by the media and the attention of policy-makers.

Although we tried to deal with the studies systematically, we were limited by the varied amount and type of information available about each study. Our discussions are, for the most part, based on longer working papers on each of the impact assessments as well as on the discussions held with several of the principal investigators at the Workshop on Improving the Science of Climate-Related Impact Studies at Oak Ridge Associated Universities, Institute for Energy Analysis, in late June 1981. Given the limitations of time, space and resources, however, emphasis was placed on the important aspects and highlights of each study.

22.2 MASSACHUSETTS INSTITUTE OF TECHNOLOGY SUDAN-SAHEL STUDY

The US Agency for International Development (AID) gave $1 million to the Massachusetts Institute of Technology (MIT) Center for Policy Alternatives between September 1973 and September 1974 to prepare a framework for evaluating long-term strategies for the development of the Sahel-Sudan region of West Africa. The Center's final report consisted of 12 volumes, two of which were treated as the final report and distributed by AID (Matlock and Cockrum, 1974; Seifert and Kamrany, 1974), while the others, labeled 'Annexes', were not widely distributed.

The report and its annexes discussed such topics as agricultural development; economic considerations; health, nutrition and population; industrial and urban development; sociopolitical factors; systems analysis of pastoralism; technology, education and institutional development; transportation; water resources; and energy and mineral resources.

22.2.1 Background

Most scholars as well as foreign assistance donors familiar with the region believed that this particular study was the direct result of concern about the plight of inhabitants in the West African Sahel caused by the cumulative effects of prolonged regional drought between 1968 and 1973. Internal AID documents, however, indicate that there was a nascent interest within AID as early as 1971 to address regional problems related to long-term, low-grade, but cumulative environmental degradation (AID, 1972, 1976a). The national and international news coverage of drought in West Africa, especially after March 1973, heightened public and congressional interest within the United States (Morentz, 1980) and presented AID with an opportunity to take a leading role in the long-term development of a region formerly within the French sphere of interest (Glantz, 1976).

At a United Nations meeting at Geneva held in June 1973, AID proposed the appointment of a major American university to develop a framework for evaluating mid- and long-term development programs for the Sahel–Sudan region of West Africa (Brown, 1973). The proposal was accepted, but not without skepticism voiced by European representatives about its chances for success (AID, 1976a).

Massachusetts Institute of Technology was the university selected by AID under 'noncompetitive' procurement procedures because (according to M. J. Harvey, AID assistant administrator for legislative affairs)

> ... it had experience in systems approaches, especially with regard to water systems. It alone, of the institutions investigated, offered to devote the attention of senior people over long periods to the task, and it had the stature needed to attract French, African, and other U.S. academic cooperation in the task as well as to substantiate the validity of such an approach. (Harvey, 1974)

However, at the beginning of the project, none of the members of the MIT team had had experience with, or first-hand knowledge of, the Sahel-Sudan region.

The 1-year period allowed for completion of the project, agreed to by MIT, was a direct result of AID's desire to meet a UN timetable for a series of meetings, a timetable to which UN eventually did not adhere (AID, 1976a).

22.2.2 Internal Factors

22.2.2.1 Design Approach

Nine disciplinary teams were established. A parallel design approach was adopted, with each team simultaneously researching a problem area (such as agriculture; hydrology; and the social, political and economic context for development) and each preparing a report. The researchers selected were associated primarily with MIT's Center for Policy Alternatives, but there were some participants from other universities as well. Many researchers assessed the problem using formal techniques such as linear programming, statistical analysis, and systems dynamics modeling, all of which, in retrospect, were too theoretical to be of direct value to policy-makers in their attempts to resolve the region's development problems (NAS, 1975a, 13; AID, 1976a).

Because several months were spent defining the problem and formulating a plan before additional members were recruited to initiate the component studies, some groups had only 6 months or less to complete their work. For example, although water resource projects had been given the highest priority by AID representatives, the hydrology group was among the last to be organized by project managers; thus, some of the potentially most relevant development alternatives were examined belatedly and in a cursory fashion (AID, 1976a).

22.2.2.2 *Time Factor*

Although MIT and AID officials settled on a 1-year time period for the project, both groups later agreed that it was inadequate for the task. Three of the major time-related problems were:

1. No work plan was prepared until long after the research effort was under way. According to AID, AID 'should have insisted on a work plan earlier ... and a written explanation of methodology ... and managed the contract strictly against these measures of progress' (AID, 1976a, 27).
2. MIT researchers, unfamiliar with the region, were constantly in need of more information before they could conceptualize the problem (to the point of requiring an extension of the contract by 4 months).
3. Few senior scholars from MIT or elsewhere could be attracted to the project on such short notice.

22.2.2.3 *Size of Research Staff*

Of the 72 people listed as 'personnel associated with the project', approximately one-third were students, a quarter were listed as consultants, and the remainder were professors, research assistants and advisors (the advisory committee met only once and was viewed by MIT as hostile to the project). While disciplinary subgroups were considered necessary because of the size and the scope of the study, they apparently contributed to the difficulty of integrating the components and, hence, to the isolation of disciplinary perspectives.

22.2.2.4 *Integration of Component Studies*

The lack of successful project conceptualization, in addition to time constraints, appears to have been a major obstacle to integrating the components of the MIT study, in spite of the fact that the study was supposed to have relied on systems analysis methodology. The study groups were allowed considerable freedom to define their own research approaches, and the project managers did not provide sufficiently strong leadership to bring them together effectively. The summary report, written by the project managers, was, in essence, a compilation of some of the primary findings from each component study, but was not a well-integrated document.

22.2.2.5 *General Observations*

The experience of the MIT Sahel-Sudan study suggests that an interdisciplinary systems analysis approach to climate impact assessment requires a well-defined focus, careful attention to project integration, and sufficient time to achieve the desired research objectives. The fact that the researchers, including the project managers, had little, if any, experience with the region was a major obstacle to successful project conceptualization and execution, even if it did, as the project director noted, permit MIT to approach the study 'with no biases' (IEA, 1981).

22.2.3 Impact of the Study

When the study was completed in February 1975, AID requested only a few hundred copies of the final report. The report was generally ignored by the media. In its retrospective assessment, AID noted that

> ... the study did not live up to expectations (as well as to MIT's verbal and written commitments). Its reception has been reflective of that fact. Little reaction to it or demand for further effort by MIT have resulted once the report was distributed to AID, international organizations, and African governments in February 1975 (AID, 1976a, 1).

AID representatives believed that MIT study managers were unable to conceptualize the Sahelian problem, despite repeated explanations by AID representatives (see Adams, 1973). For example, they continually treated the Sahel as six individual West African states, each with its own problems. AID, on the other hand, sought to develop new approaches to long-term development in West Africa that included coastal and Anglophone states as well as the six Francophone Sahelian states.

AID representatives expressed concern early that the MIT managers might not deliver what they expected: MIT did not convene a proposed senior think-tank retreat; the vital water systems study was not mentioned in MIT's interim report; and the two major alternative development strategies MIT focused on were viewed by AID as 'unexciting' (AID, 1976a). Thus, none of the reasons suggested by AID for its selection of MIT's Center for Policy Alternatives proved to be valid. In late 1973 AID asked the National Academy of Sciences to appoint a panel to 'advise AID with respect to the critical medium- and long-term natural resource management problems of the drought-stricken region of West Africa', and to critique the draft reports of the MIT group (NAS, 1978, 156; see also NAS, 1975b).

Midway through the project, AID stated its concerns in an internal memo (AID, 1974). Some changes were so critical to the project's success, the memo noted, that continuation of the contract to completion should be made conditional to implementing those changes. 'The result is that MIT is not inspiring confidence and thus may do damage to future cooperation of Africans and other academics, and ultimately to the acceptability of the project', it concluded (p. 3). However, few substantive changes either in personnel or in project conception were forthcoming.

For their part, MIT study managers continue to believe that AID did not properly advise them during the project (IEA, 1981) and viewed AID's attempts to keep the research effort on track as interference that hindered the research process (Holloman, 1974, 3).

AID representatives constantly sought to salvage some parts of the MIT effort to present to the international development assistance community, but they felt that only the first two volumes (Summary Report and Agricultural Development Report) would serve to some extent. MIT has questioned this view, however, noting that many excerpts from the entire 1974 MIT report were used in an AID proposal to the US Congress for a long-term comprehensive development program for the Sahel (AID, 1976b, *passim*). In private correspondence, W. W. Seifert, project manager, stated that the MIT study team felt that 'AID panned us but then presented our results to the Congress without reference [to the MIT study]'.

22.3 NATIONAL DEFENSE UNIVERSITY: CLIMATE CHANGE TO THE YEAR 2000

The National Defense University (NDU) study on *Climate Change to the Year 2000* (NDU, 1978a), the first of a three-part assessment of the impacts of climate change on agriculture, began in the fall of 1976. The final report was released to the public in February 1978. The second part of this study, *Crop Yields and Climate Change to the Year 2000* (NDU, 1980), was published in late 1980 and distributed in mid-1981. A progress report of the crop yields study had been issued in August 1978 (NDU, 1978b). The third part, *The World Grain Economy and Climate Change to the Year 2000: Implications for Policy*, was released for limited circulation in June 1981 in manuscript form; the final version of the third part was distributed in the spring of 1983, bringing to completion this project (Johnson, 1983).

The project represented the first attempt to quantify in a comprehensive way perceptions about climatic change. The task of *Climate Change to the Year 2000* was 'to define and estimate the likelihood of changes in climate during the next twenty-five years, and to construct climate scenarios for the year 2000' (NDU, 1978a, vii). *Crop Yields and Climate Change* indicated a broadened view of the task: 'A secondary goal of this interdisciplinary effort was to advance the art of making climate impact assessments' (NDU, 1980, iii).

The entire study was supported by the US Department of Defense through the Defense Advanced Research Projects Agency. An estimated $100,000 was given, primarily to the Institute for the Future, to assist with the study design. This excluded approximately nine person-years of contributed research by a small, multidisciplinary staff detailed from the Department of Defense (DOD), the US Department of Agriculture (USDA), and the National

Oceanic and Atmospheric Administration (NOAA), as well as the contributions of numerous expert panelists and advisers who received nominal honoraria.

22.3.1 Background

The pressure by US policy-makers for information about climate change began in the early 1970s with the debate among climatologists about whether the earth's atmosphere would become warmer or cooler during the next few decades. The debate was fueled by weather anomalies in 1972 that adversely affected food production and availability in some regions of the world, as well as by the publication of such popular articles as 'Ominous changes in the world's weather' (Alexander, 1974), 'What's happening to our climate?' (Matthews, 1976), and the Central Intelligence Agency reports on climate and agricultural production (CIA, 1974a,b, 1976). These publications have been explicitly acknowledged as having stimulated government agencies to consider the climate factor in planning (Gasser, 1981).

In light of this debate, USDA was criticized for not having considered possible future climate scenarios and their effects on US and global grain production and trade (Shapley, 1976). This led a research fellow from USDA to recommend to the NDU research directorate that a climate change study be undertaken, noting that it would relate to strategic planning and to the management of resources for national security (Gasser, 1981; Johnson, 1983, v).

22.3.2 Internal Factors

22.3.2.1 Design Approach

NDU study managers undertook three separate substudies, referred to as 'tasks', on climate change, climate-crop yields, and policy implications. They used a sequential approach, with the output of one substudy consistent with the input needs of the next one. The objective of Task I was to 'seek from those who were thoroughly familiar with the state of research and knowledge subjective probability judgments about the likelihood of occurrence of certain well-defined climatic events in the future', using a questionnaire survey. The responses of 24 climate experts from seven countries were used by the NDU team to develop five scenarios of climate change to the year 2000. The scenarios were designated as large cooling, moderate cooling, same as the last 30 years, moderate warming, and large warming.

In the second substudy, 35 agriculturalists were surveyed to quantify their expectations about how various combinations of changes in annual temperature and precipitation might affect crop yields in 15 country–crop combinations. (For example, how might a 2 °C warming combined with a 10 percent increase in precipitation affect wheat yields in Australia? How might a 1 °C cooling and a 20 percent decrease in precipitation affect US corn yields?) Responses in this survey were used with information from the previously developed climate scenarios to create climate and yield scenarios. The climate-response model devised for the study predicted not only changes in average yield but also changes in the interannual variability of yields. In the third substudy, the climate and yield scenarios were used to drive a USDA econometric model of international agricultural demand, production, and trade in order to generate information on the potential implications of climate change for international grain trade and agricultural policies.

22.3.2.2 Time Factor

Because the NDU study managers reported to NDU officials and not to an outside agency, the project had only self-imposed deadlines (IEA, 1981). However, personnel attrition at NDU and the return of all but one resident study manager to their parent organizations by 1979 apparently affected the degree of internal support for the project, contributing to the long delay between the publication of the first report (February 1978) and the third one (May 1983).

22.3.2.3 Size of Research Staff

The size of the core study group was relatively small, with apparently well-established lines of communication among the staff. The second of the study's reports commented on those involved in the NDU research effort:

> The focal point of the endeavor was a small, interdisciplinary staff drawn from the several branches of the Government. Assisted by the Institute for the Future, the resident staff conducted a brokerage operation, planning the study around futuristic techniques for the solicitation and analysis of nonexistent information [sic], and orchestrating advice, 'data' and insights from a host of volunteers. (NDU, 1980, xv)

22.3.2.4 Integration of Component Studies

The sequential approach facilitated the integration of the component studies. The first study produced climate scenarios; the second integrated those scenarios with information on crop yield responses to produce climate-crop scenarios; and the third study used the latter scenarios to produce projections on the global dynamics of the agricultural sector.

22.3.2.5 General Observations

The NDU study was well defined in its scope and survey methodology. It was limited in general to the agricultural effects of climate change to the year 2000 or, more specifically, it assessed those effects only on 15 country–crop combinations.* Adaptive measures for countering adverse impacts or for capitalizing on favorable ones were excluded, as were societal or ecological effects. With respect to their survey methodology, T. Stewart, in an internal NCAR memo (1981), expressed concern about the lack of safeguards to protect against possible judgmental bias and inconsistency and about the pooling procedure which ignored individual differences among experts.

22.3.3 Impact of the Study

Climate Change to the Year 2000 was released at the 1978 annual meeting of the American Association for the Advancement of Science (AAAS) in Washington, DC. Later, insight about the intended audience was gained from the preface to the second part: 'In drafting this report we envisioned an inhomogeneous audience of meteorologists, climatologists, agronomists, economists, futurists, model builders, and policy-makers, to name a few' (NDU, 1980, xvii). The findings of the report were presented to the press at the AAAS meeting and were subsequently reported by the media in a descriptive way, with little or no analysis of the report, its methodology or its conclusions (see Kraemer, 1978; *Science News*, 1978; UNFAO, 1979; Sellers, 1979). During the World Climate Conference convened by the World Meteorological Organization in Geneva in February 1979, the NDU report was one of the very few reports unofficially available in large quantity to the conference participants, who represented developing as well as developed countries.

The impact of the first part of the study was evident in the Council on Environmental Quality's (CEQ) *The Global 2000 Report to the President: Entering the Twenty-First Century* (1980), in which the two climate-related chapters were heavily dependent on the NDU report. *The Global 2000 Report* has been republished in five languages and its sales have totalled well over half a million, hence it has greatly extended the circulation of the NDU study's findings. The foreword to Part III of this assessment offers additional insight into the impact that the NDU expected it to make on policy-makers:

> The team was pleased to note that the Carter administration's responses to the Soviet invasion of Afghanistan took cognizance of the conclusions of draft copies of the reports sent to high-level policy-makers in key agencies. NDU is equally pleased that a draft of this concluding report is being integrated into the report of the Reagan administration's Global Issues Working Group. (Johnson, 1983, v)

Soviet scientists and policy-makers apparently took interest in the first report of the NDU study and included it as part of a survey article on the influence of climate on man's economic activity (Gruza, 1979).

* The agriculture panelists were also asked to forecast the influence of technological changes on yield trends, assuming no change in climate. The responses to this inquiry enabled the staff to assess the potential importance of technological change for crop yields relative to the estimated effects of a range of climatic change.

In our opinion, four aspects of the NDU study may have affected its usefulness and, therefore, its long-term impact.

1. The report concludes that 'the most likely event will be a climate which resembles the average of the past thirty years ...' (NDU, 1978a, xix). Statements such as these, among other factors, seemingly diminished interest in the final reports of Parts II and III. This view was reiterated in Part III: 'The significance of this study is that the United States can consider its proper role in the world food situation without great concern that climatic changes during the rest of this century will upset its calculations' (Johnson, 1983, 4). Yet, the negative conclusion about the agricultural implications of midrange climate change may also be a useful finding for policy-makers. One of the study's principal conclusions is that 'technology, rather than climate, is likely to be the chief determinant of most crop yields in the last quarter of the twentieth century' (NDU, 1980, 2).
2. The NDU study has been referred to as 'science by consensus' (NDU, 1978a). Study managers feel that this accusation is unjustified, although this perception continues to exist. In their report they noted

   > The experts' aggregated subjective probabilities do not reflect a consensus on any narrowly defined climatic issue, but a large majority of the climate panelists were in broad agreement, for example, that the average global temperature is not likely to change more than half a degree Celsius by the year 2000. (NDU, 1978a, xix)

3. The relatively long time lag between publication of the first and the successive two parts of the study suggests a possible change in interest at the NDU or Department of Defense for this research activity, thereby detracting from the value of the entire three-part study. Changes in personnel involved with the study (as people retired or were reassigned) as well as apparent changes in concern by policy-makers about climate (that is, consideration of climatic factors, such as interannual variability, as being of more immediate concern than climatic change) have perhaps made the reports of the last two parts of the study less important to policy-makers (except in the use of the methodology) than might have been the case had they been published closer together.
4. Discussions of the NDU research effort inevitably include comments on the 'packaging' of this scientific study. Some people have noted (IEA, 1981) that the cover graphics of the first volume were designed to attract attention to the report but were not suitable for a serious scholarly report. The 'packaging' of the report has had a positive aspect, however. Most people recall the study, many have retained it for their library (partly because of its content and partly due to the way it is packaged), and those who have saved it know where it is on their library shelf. This may not be the case for other climate-related impact studies of similar length, no matter how important their research findings.

22.4 INTERNATIONAL FEDERATION OF INSTITUTES FOR ADVANCED STUDY: DROUGHT AND MAN: THE 1972 CASE HISTORY

The 'Drought and Man' study was sponsored by the International Federation of Institutes for Advance Study (IFIAS) in Stockholm, Sweden, and was carried out under the auspices of the director of the Food and Climate Program of the Aspen Institute for Humanistic Studies (United States). It was funded by UNEP (Kenya) and three private foundations at a cost of approximately $380,000, with an estimated $750,000 in contributed research.

The study resulted in a three-volume report. The manuscripts for all three volumes were completed in 1979. Volume 1, *Nature Pleads Not Guilty*, was published in August 1981 (García, 1981); Volume 2, *The Constant Catastrophe: Malnutrition, Famines, and Drought* (García and Escudero, 1982) was published the following year; and Volume 3, *Case Studies*, is scheduled for future publication.

The study examined in depth the impact of the climate anomalies of 1972—specifically drought in many parts of the world—and their impacts on the production and availability of food, encompassing both social and physical aspects of the situation. The year 1972 was selected for assessment because in that year droughts adversely affected food production in the Soviet Union, China, Eastern Europe, Latin America, and sub-Saharan Africa.

22.4.1 Background

The underlying reasons for this study were similar to those for the National Defense University study, and to some extent for the MIT study: with the occurrence of the climate anomalies of 1972 which resulted in the first major decline in total food production since 1945, climatic factors in food and energy issues took on a greater importance than had previously been the case.

In 1972 a new interdisciplinary, international, nongovernmental organization, IFIAS, was created for the purpose of ensuring that science be used to improve the quality of human life. IFIAS established a 'climate and the quality of life' project under the leadership of W. O. Roberts, the representative of one of its member institutes, the Aspen Institute for Humanistic Studies (AIHS). Its mandate was clear:

> The project is not to be viewed as a contribution to physical climate research. The project instead will focus on its main task—the social, ethical, and humanistic implications of changes in global or regional climate. (IFIAS, 1974, 2)

The 'Drought and Man' study was one of the three projects undertaken. With the selection of Rolando García as the senior study author, the study officially began in the spring of 1976.

22.4.2 Internal Factors

22.4.2.1 Design Approach

García was given total independence for conducting the study, which included reconceptualizing the research problem several months after the project began. This was necessary, according to García (IEA, 1981), because it was not clear to him that the 'official version' of the direct cause-and-effect relationship on which his investigations were expected to have been based (between drought and famine, and drought and malnutrition) was, in fact, valid. García not only redefined the research problem but, after establishing research groups in Latin America, Africa and South Asia, he also coordinated the research activities, and finally integrated all of the project's research results. Separate case studies of specific aspects of the 1972 droughts were also undertaken by individuals in the United States and Europe.

The study groups were encouraged to follow what García termed 'a structural approach', which entailed research at three levels of analysis (García, 1981, Chapter 6). Basic to this approach were the interdisciplinary assessments of the events of 1972, particularly in terms of atmospheric anomalies, soil systems, agricultural systems, and social and economic systems. At the second level, researchers identified processes perceived to be responsible for the observed events, including such trends as global increase of meat and cash crop production, urbanization, and industrialization. The third level of analysis involved identifying the causes of the processes observed at the second level.

22.4.2.2 Time Factor

García has stated that time was not a serious constraint in carrying out the study (IEA, 1981). However, there were two extensions to the study: the first from August to December 1978 to allow for the completion of the regional studies, and the second to March 1979. The time element of importance with regard to the IFIAS study was the delay in the publication of its results (see Verstraete, 1984).

22.4.2.3 Size of Research Staff

The project involved a small core group under the leadership of García and several regional study groups headed by local study authors who were to submit their reports to García. The regional study groups were spread over several continents, making communication and the exchange of information between them difficult and often delayed.

22.4.2.4 Integration of Component Studies

Intellectually, the study was laid out so that component studies formed pieces of an integrated whole. Logistically, however, the project was difficult to integrate, because some regional groups were late in producing their reports or produced reports inconsistent with the study's lines of inquiry, and the geographical separation of contributors handicapped the effective integration of component studies.

22.4.3 Impact of the Study

The first volume of the study, *Nature Pleads Not Guilty*, was published in August 1981. To date, there have been reviews of the project based on the earliest working manuscripts (Cusack, 1981; Ruttenberg, 1981) as well as formal reviews of the final reports (Palutikof, 1983; Verstraete, 1984; Talbot and Olorunsola, 1984).

García has prepared reports on various topics stemming from the 'Drought and Man' study (García 1977, 1978). In spite of these presentations, it appears that his strategy was to publish all of the findings at the completion of the research instead of piecemeal throughout the project as the conclusions arose. As a result, other authors, undertaking independent research on topics related to those in the IFIAS study, have in a sense preempted (but also reinforced) some of the conclusions drawn earlier by the IFIAS study, conclusions that had not been published until 1981.

The IFIAS study had at least two important impacts:

1. it considered and discarded the hypothesis that we are now witnessing a period of profound climatic changes;
2. it raised doubts about the validity of what García referred to as the 'official' or generally accepted version of the sequence of events to explain the crisis of 1972 and thereafter.

That version maintained that, as a result of the decline in their agricultural production, the Soviet Union purchased large quantities of grain from the United States, which led to the depletion of US grain reserves, to shortages in the international marketplace, and to the exceptional price increases of grains and other foodstuffs. García asserted that in addition to an assessment of the impact on society of 'natural forces' such as climate fluctuations, 'human structures' (societal, political, economical) must be examined, and their interactions studied to reveal the actual forces at play. It is this set of forces, perhaps triggered by a physical disaster, that determines in the end what will be the effects on man and his structures (García, 1981, 4). In fact, one reviewer noted that 'the influence of climate in general is largely discounted' (Palutikof, 1983, 635).

Such challenges to the 'official' version often become labeled as radical or iconoclastic assessments and become part of the polarized debate over whether people (as policy-makers) or nature (as climate anomalies) represent the primary cause for worldwide food crises or shortages that are usually perceived as climate-related.

22.5 DOT CLIMATE IMPACT ASSESSMENT PROGRAM

The Department of Transportation Climatic Impact Assessment Program (CIAP) was authorized by the US Congress in 1971 and was funded at a total cost of about $20 million, with an unofficially estimated $40 million of contributed research (Dotto and Schiff, 1978). The study was a 3-year, multidisciplinary effort 'to determine the regulatory constraints necessary to safeguard that future flights in the stratosphere do not result in adverse environmental effects' (Grobecker, 1974, 179). It was to assess the potential impact of fuel emissions from a large fleet of high-flying supersonic transports (SST) on stratospheric ozone concentrations and of the hypothesized effects of the resulting ozone depletion on the incidence of skin cancer as well as on climate. The US Department of Transportation had responsibility for the entire program and provided the principal investigator and project manager.

CIAP publications consist of six monographs (Bauer, 1974; Caldwell, 1974; Daly, 1974;, Hidalgo, 1974a,b; Oliver, 1974) and a *Report of Findings*, (Grobecker *et al.*, 1974) which summarizes the monographs. An executive summary of the *Report of Findings* and a press release (DOT, 1975) were issued at a press conference in mid-January 1975, the *Report of Findings* in March 1975, and the monographs from September to December 1975. Additional supporting CIAP material was issued after the final CIAP conference in February 1975 (Hard and Broderick, 1976).

22.5.1 Background

In the early 1960s, public controversy developed concerning the potential value to America of supersonic aircraft. The debate on whether to develop an American SST centered primarily around the economic and political costs and benefits (Primack and von Hippel, 1972) and around two environmental issues—noise pollution in the vicinity of airports and sonic booms (Shurcliff, 1970). At the end of the 1960s, environmental concern expanded to include the possibility of stratospheric ozone depletion (SCEP, 1970), caused not only by American SSTs but also by the European Concorde and the Soviet Tupolev. Scientists raised the possibility that trace gases (at first water vapor and later oxides of nitrogen) from fuel emissions during SST flights in the stratosphere could reduce the concentration of stratospheric ozone (Crutzen, 1971;

Johnston, 1971), thereby reducing its effectiveness in shielding the ground from biologically damaging ultraviolet radiation (UV-B). It was also suggested that there was a link between increased UV-B and the incidence of skin cancer (a 1 percent reduction in stratospheric ozone was predicted to cause a 2 percent increase in skin cancer).*

In March 1971, Congress refused to fund the Boeing SST prototype program, despite European and Soviet intentions to develop theirs. The ozone depletion aspect of the SST debate appears to have been an 'eleventh hour' consideration, used by opponents of the SST to block its development. In 1971 Congress established CIAP to investigate more closely the potential impact of SSTs on the stratosphere. Although the American SST debate had essentially been concluded before the CIAP program began, US decision-makers were still in need of more information on SST emissions and stratospheric ozone because they had to contend eventually with a decision about landing rights in the United States for the British–French Concorde (DOT, 1976).

From the beginning of the study, CIAP managers considered this an international research activity and invited hundreds of scientists from more than ten countries (but predominantly from the United States) to participate in CIAP conferences to discuss research activities and findings.

22.5.2 Internal Factors

22.5.2.1 Design Approach

CIAP project managers embarked on a 'crash' program to combine basic research done in the past with the stimulation of new research and with a climate impact assessment. A parallel approach was taken for the basic research activities in the areas of atmospheric science, aircraft propulsion and biological science, and in modeling activities. A sequential approach was used later in analyzing the monograph data to assess the impact of climatic changes resulting from propulsion effluents of vehicles in the stratosphere, as projected to 1990. Concurrent analyses, for which all data collected by CIAP were made available, were undertaken by the National Academy of Sciences (NAS, 1975c), and by groups in France and Belgium (COVOS, 1976), England (COMESA, 1975), USSR (Budyko and Karol, 1976), and Canada (*Atmosphere*, 1976).

* At the conclusion of the project, CIAP's project manager cited three instrumental factors that in his view led to the development of CIAP. 'During the discussion of the U.S. supersonic transport project in 1970, the question was raised, notably by (a) James McDonald (who linked ozone depletion to skin cancer), (b) the SCEP (Study of Critical Environmental Problems), and (c) Harold Johnston (who linked a trace gas in SST fuel emissions to ozone depletion), whether impurities resulting from aircraft flight high in the stratosphere could alter the proportions of atmospheric trace constituents, with harmful results to the earth's environment' (Grobecker, 1976, 1).

22.5.2.2 Time Factor

CIAP managers were given 3 years to produce their final report. During this period, much basic research was required, particularly on various aspects of stratospheric chemistry and on biological responses to increased UV-B.

Allocation of time among the six research topics encompassed by each of the monographs favored atmospheric research. The first three monographs were given several months more time than the last three. Two years for designing, conducting the research for, and compiling the results of the biospheric and economic components (monographs 5 and 6, respectively) proved too short for the task.

Time became an important factor also at the end of the study; pressures of a congressional deadline caused the project managers to release the final version of the executive summary without peer review, 2 months before issuing the full *Report of Findings*. The ensuing controversy about the timing and the wording of the executive summary overshadowed the impact of the report itself, and, in fact, led to congressional hearings on the matter (Carter, 1975; US Congress, 1976).

22.5.2.3 Size of Research Staff

Because large amounts of information had to be gathered from a great number of sources, the CIAP study involved hundreds of scientists, organized into disciplinary subgroups and sub-subgroups, which tended to complicate the task of integrating component studies, thereby contributing to isolation of disciplinary perspectives.

22.5.2.4 Integration of Component Studies

CIAP considered two types of activity: data gathering and integrative analysis of the data. The data gathering was accomplished by representatives of six groups working in parallel to assemble all data that seemed relevant to their particular task. Preparation of each monograph began with a conference attended by 25–100 scientists. Existing information was assembled and suggestions for additional research were made. Coherence of the contributions within each major group was achieved under the editorial direction of the group leader, who was responsible for preparing the monograph. The monographs were further refined and expanded through an iterative process that involved the circulation of monograph drafts and the convening of annual workshops.

The integrative analysis of the monograph data, accomplished in the final year of the study, consisted of several steps, some under the direction of the responsible monograph chairmen, and others by the CIAP staff. Summaries of each monograph and a complete description of the analysis were compiled into a single volume, *Report of Findings*.

22.5.2.5 General Observations

Although initially viewed with great skepticism, CIAP's atmospheric research has since received general praise. The community of atmospheric scientists needed relatively little incentive to bridge the gaps between various atmospheric science-related disciplines, as tropospheric scientists crossed disciplinary lines to work on stratospheric research and scientists who researched the unperturbed stratosphere also undertook research on the perturbed atmosphere. In addition, interdisciplinary communication was aided by the fact that study structure cut across the deepest disciplinary line within the atmospheric sciences—that between dynamicists and chemists.

The biospheric and economic component studies were less successful in achieving interdisciplinary coordination or in developing a viable set of research results, in part because both of these studies were under relatively severe time constraints—having only 2 years to design the studies, conduct the required research and compile the research results. Since both studies took place simultaneously, the economists were unable to receive the biologists' research results until late in their research schedule. Perhaps more important was that the biologists could not provide the information that the economists most needed—the magnitude of the effect of UV radiation on crop plants.

In summary, CIAP's use of different study designs apparently worked well within the atmospheric sciences, where there was a higher degree of initial momentum, enthusiasm among scientists, spontaneous communication among those working in some component studies, and a relatively generous allocation of research time and funds. The mix of design approaches was less effective, however, in the biological and social science studies because:

1. studies were undertaken in parallel under circumstances where one study required the other's results,
2. little communication developed between biologists and social scientists, and
3. those scientists involved in both studies were required (expected) to conduct basic research but had been allocated insufficient time to complete their research task.

22.5.3 Impact of the Study

In January 1975, a press conference was held to announce the completion of the Climatic Impact Assessment Program and to issue an executive summary of the project. The contents of the summary were immediately challenged by some of the scientists who had worked on various aspects of the CIAP reports (for example, Donahue, 1975; Johnston, 1975a). They felt that the tone* of the

* On the question of tone, tenor, or mood of scientific reports in general, and their potential impact on the effectiveness of a report, see Martin (1979).

summary disagreed with the tone of earlier drafts of the executive summary and with the tone of the complete *Report of Findings*† in that it appeared to indicate that the report found that the effects of existing SSTs (that is, the effects of a small number of lower-flying Concordes and Tupolevs) were negligable and those of a large, hypothetical fleet were technologically preventable. Also, the 'scientific conclusions' listed in the summary were described in words different from those of the *Report of Findings* and were listed in an order which seemed, again, to bring out an up-beat (favorable) opinion regarding the SST. In addition, some topics that were identified as important in the *Report of Findings* (such as skin cancer effects) received little or no mention in the summary. Many of the scientists who participated in CIAP activities felt that their credibility had been permanently damaged; they cited news headlines and editorials that appeared after the release of the executive summary, such as 'Scientists Clear the SST', (*Christian Science Monitor*, February 5, 1975) and 'World SST Fleets Said Not To Damage the Ozone Blanket' (*New York Times*, January 21, 1975). One scientist reacted by issuing his own executive summary and principal scientific conclusions (Johnston, 1975b).

CIAP study managers declared that they had been misquoted in local editorials based on an Associated Press dispatch issued as a result of the January 1975 press conference. They argued that their summary simply emphasized the positive aspects of technological improvement, such as engine design changes to lower trace gas emissions. They contended that their conclusions were consistent with those in the National Academy of Sciences report on the environmental impact of the SST, issued in late March (NAS, 1975c), and with those of a number of other studies abroad (COMESA, 1975; COVOS, 1976). After the congressional hearing, a review of the issue in *Science* (Carter, 1975) concluded that the summary of conclusions was what may have been expected from a study by engineers who anticipate the best (rather than the worst) that technology might offer for the future.

According to CIAP's project manager (IEA, 1981), the controversy over the executive summary could have been avoided if time had been taken to distribute the final version to scientists for comment (as had been done with other parts of the study and earlier drafts of the summary).

The CIAP scientific findings, as an assessment of knowledge at that time and as presented in the *Report of Findings*, have been generally praised (except monograph 6) as a good summary of the state of the art, even by most scientists who vehemently criticized the executive summary. In fact, an outgrowth of CIAP, the Lawrence Livermore Laboratory's High Altitude Pollution Program (HAPP), has been supported since 1975 by the Federal Aviation

† For example, the August 20, 1974, version of the summary ('Executive Conspectus', US Department of Transportation, Climatic Assessment Program, Review Draft) was the last one sent to Harold Johnston for comment (private communication). Johnston was one of those scientists who strongly criticized the final version of the summary.

Administration (Luther, 1980; Wuebbles, 1981; Knox *et al.*, 1983). CIAP also appears to have stimulated research on the effects that trace gases from sources other than SSTs (for example, chlorofluorocarbons from aerosol cans, refrigerants, etc., oxides of nitrogen from agricultural fertilizers) might have on the concentration of stratospheric ozone or on climate (Mormino *et al.*, 1975).

As a corollary to the CIAP studies, DOT requested the US National Academy of Sciences to issue an independent report, *Environmental Impact of Stratospheric Flight* (NAS, 1975c). Thus, in spite of the controversy surrounding the executive summary, CIAP mobilized researchers in an area of atmospheric sciences that had been relatively neglected—the stratosphere (Hoffert and Stewart, 1975; MacDonald, 1976; Bastian, 1982).

22.6 NAS STUDIES ON THE EFFECTS OF CHLOROFLUOROCARBON RELEASES ON STRATOSPHERIC OZONE

The US National Academy of Sciences (NAS) Panel on Atmospheric Chemistry was appointed in March 1975 to assess 'the extent to which man-made halocarbons, particularly chlorofluoromethanes (CFMs), and potential emissions from the space shuttle might inadvertently modify the stratosphere' (NAS, 1976a, vii). The panel issued its report, *Halocarbons: Effects on Stratospheric Ozone*, in 1976 (after a delay of some months caused by the panel's need to evaluate newly developed scientific information). The panel's parent Committee on Impacts of Stratospheric Change (CISC) undertook a study to address 'the question of biological and climatic effects of ozone reduction and the appropriate policy consequences of both our present knowledge and the knowledge we are likely to have in the future' (NAS, 1976b, viii). CISC issued its report, *Halocarbons: Environmental Effects of Chlorofluoromethane Release*, at the same time as the Atmospheric Chemistry panel issued its report. The two reports were produced at a total cost of more than $300,000 and were funded by the National Aeronautics and Space Administration (NASA), National Science Foundation (NSF), Environmental Protection Agency (EPA) and National Oceanic and Atmospheric Administration (NOAA). Computer studies were supported by the Federal Aviation Administration (FAA).

22.6.1 Background

Fluorocarbons, Freon 11 ($CFCl_3$) and Freon 12 (CF_2Cl_2), are manufactured worldwide for use in aerosol spray cans, refrigerators, air conditioners, certain types of plastic foams, solvents and cleaning agents, among other uses. As late as the mid-1970s, about 50 percent of the fluorocarbons in the United States had been used as aerosol propellants and up to 90 percent of those were used in such products as hairsprays, deodorants and antiperspirants.

In the early 1970s, scientific research activities that would eventually tie the CFCs closely to stratospheric ozone depletion were under way in different places and in different disciplines, such as the activities of James Lovelock, a scientist in England interested in developing highly sensitive instrumentation to measure fluorocarbons in the atmosphere. A few atmospheric chemists (for example, Molina and Rowland, 1974; Stolarski and Cicerone, 1974), working independently focused attention on chlorine in the stratosphere. Molina and Rowland suggested that manmade chlorofluorocarbons, while inert in the lower atmosphere, were diffusing upward into the stratosphere where they would photodissociate, releasing free chlorine to react catalytically with ozone, thereby significantly depleting the ozone layer. They suggested that, given the long time lag between the emission of CFCs in the atmosphere and their diffusion to the stratosphere, there had already been a serious atmospheric build-up of CFCs. Estimates from different sources of eventual ozone depletion resulting from CFC releases ranged from 5 to 30 percent, based on projected rates of CFC production.

Interest in the effects of trace gases on stratospheric ozone was heightened by the CIAP research efforts and by concern about fuel emissions from the American space shuttle. In the summer of 1974, while involved in the preparation of the final report for CIAP, the Climatic Impacts Committee determined that the impact of photodissociated chlorine atoms released from CFCs was potentially a more serious threat to the ozone layer than space shuttle exhaust. This concern led to the establishment of an ad hoc panel to ascertain whether the chlorine issue was important and worthy of a more serious investigation. In 1975 NAS charged CISC with the task of investigating the impact of CFCs on stratospheric ozone depletion and the effects of ozone depletion on the lower atmosphere and at ground level.

22.6.2 Internal Factors

22.6.2.1 Design Approach

The two studies (NAS, 1976a,b) assessed the following scientific contentions:

1. CFC releases to the atmosphere can lead to ozone depletion; and
2. ozone depletion might have serious, deleterious effects on life on earth.

The studies used both a sequential and a parallel approach. During the first few months of the studies, a 13–member panel on atmospheric chemistry sorted through evidence on the possible links between manmade CFC releases and stratospheric ozone depletion. Once this panel was working, a multidisciplinary committee, also with 13 members, representing various physical (but no social) science disciplines was established to investigate the broader problems related to ozone depletion.

The procedures used in these studies were informal and ad hoc. Panel and committee members were chosen because they had a neutral viewpoint, a high standing in the scientific community, and access to the latest information on questions of concern to the study. Monthly and bimonthly meetings were held to update and exchange information and to work toward a consensus. The committee chairman wrote the main body of the report, which was then reviewed by all committee members as well as by an outside body of scholars.

In 1977, the US Congress called on the Academy to conduct an additional study, covering not only the physical sciences, including biology, but also the 'health and welfare effects', as well as 'methods for control ... including alternatives, costs, feasibility, and timing' (US Congress, 1977). For these CFC studies (NAS, 1979), the research for the physical and biological sciences section was conducted by CISC, the same committee that had conducted the previous Academy study of CFCs. A second committee, the Committee on Alternatives for the Reduction of Chlorofluorocarbon Emissions (CARCE), was formed to look at the regulatory aspects of the CFC issue. CISC in its physical science studies used an approach similar to the one used in its 1976 study. Stratospheric chemistry was first assessed, followed by an interdisciplinary assessment of climatic, biological and human health impacts. Small panels (fewer than five) conducted disciplinary substudies, reported them back to the full committee, and each contributed one chapter to the committee's report (NAS, 1979).

The societal impact component (undertaken by CARCE) concerned possible options for CFC emission control. This component proved to be unavoidably political. Rather than attempt to establish a committee of 'unbiased' scientific experts (as it had done with the 1976 report), the Academy sought to include a balance of viewpoints. Committee membership included representatives from corporations, conservation, labor and consumer groups—in addition to university professors with backgrounds in engineering, economics and law. CARCE divided its inquiry into topics related to industrial technology and socioeconomic impacts, and panels were selected to investigate each topic.

Both CARCE and CISC worked closely together through informal communication links and a number of joint meetings. Their respective chairmen maintained regular communication, and their executive secretaries shared an office at the Academy. The committee issued a joint report, *Protection Against Depletion of Stratospheric Ozone by Chlorofluorocarbons* (NAS, 1979) that represented a consensus of both committees and was reviewed by an outside panel in accordance with the Academy's review procedure.

The Academy's approach to these studies apparently balanced the treatment of the complex physical aspects of the CFC issue with an assessment of possible options for CFC control, as well as with an assessment of the societal impacts that included discussion of the potential effectiveness of different control options. While covering both policy and physical science considerations, the approach left large gaps between these two areas of research. For example, the paucity of information on the effects of UV radiation on plant growth meant that CISC would be unable to assess meaningfully the non-human biological consequences of ozone depletion. As for CARCE, no attempt was made to assess the economic and social costs of ozone depletion. While the US Environmental Protection Agency wanted the Academy to undertake a full cost-benefit analysis, the Academy chose instead to review critically three studies on costs and benefits of CFC emissions regulation.

22.6.2.2 *Time Factor*

The NAS studies are different from the other studies considered in that they are part of a series of assessments. Time constraints do not appear to have affected the effectiveness of these studies. While it does have deadlines to meet, the Academy's prestige and the potential authority of its reports gives it an additional degree of flexibility. As a further step in this ongoing assessment (biennial review), two other Academy studies related to the CFC-ozone depletion issue were published in early 1982 (NAS 1982a,b).

22.6.2.3 *Size of Staff and Integration of Component Studies*

The NAS deliberately keeps its committees small (15 members or fewer) to increase the flexibility of its studies and the ease of achieving integrated findings. Where specialized disciplinary study is needed, it forms panels that report back to the parent committee. This format seems to aid multidisciplinary exchange. Integration was achieved through cooperation of those responsible for the study, both committee and panel members. The general report of each NAS study on the CFC issue was produced by a multidisciplinary group as a consensus position.

22.6.2.4 *General Observations*

In early 1976, a few months before the report was published, the resiliency of the NAS study process as well as the potential importance of its reports was tested; new scientific questions arose, which needed to be accounted for in the study's findings. The time needed to evaluate these new inputs delayed the release of the NAS report by several months (Handler, 1976). The panel's ability to assess new scientific information and to secure agreement on delaying the release of the final report shows a high degree of flexibility afforded to NAS studies that might be difficult to find in most other climate-related impact assessments.

The NAS recruitment policies for the panel gave an image of integrity and impartiality to the panel and the final reports. As noted earlier, the report represented a consensus and was carefully worded to avoid a strong position favoring any side of the CFC dispute. The report asserted that CFCs might pose a serious hazard, but also suggested that regulations on some uses and releases of CFCs be delayed for up to 2 years, based on the view that 'the impact on the world of waiting a couple of years before deciding whether or not to regulate the uses and releases of Freon 11 and Freon 12 is small although we are uncertain just how small' (NAS, 1976b, 9).

22.6.3 Impact of the Studies

The Academy reports appear to have had a major impact in policy-making circles. Interest in them was evident even before the reports were issued. For example, Bastian noted that the recommendations of the interagency task force on the Inadvertent Modification of the Stratosphere (IMOS, 1975) were tied to the NAS study: '[IMOS] set forth a timetable for decision-making by the Executive Branch ... [calling for] initiation of rulemaking for some type of restrictions on fluorocarbon use following issuance of the NAS report' (Bastian, 1976, 1–2). The study was expected to have significant influence on regulations of CFC emissions. Bastian noted that agencies such as the Consumer Product Safety Commission and the Food and Drug Administration 'have been petitioned by the Natural Resources Defense Council and several states to take some action, and have not done so, waiting for the results of the NAS study to be available' and that 'the various industries potentially affected by any regulation are aware of the importance of the NAS study' (Bastian, 1976, 2).*

The NAS study confirmed the concern of atmospheric scientists, who since 1975 had been looking into the possible effects on climate of fluorocarbons and other infra-red-absorbing gases in the troposphere through the greenhouse effect (Ramanathan, 1975, 1980; Dickinson and Chervin, 1979; Ramanathan and Dickinson, 1979).

Although scientists recognize that there is a large degree of scientific uncertainty surrounding the interactions of CFCs with stratospheric ozone and that there is still no general agreement on the scientific issue (UK Department of Environment, 1976, 1979; Harris and Kosovich, 1981; UNEP, 1981), worldwide research and regulatory activity has been steadily developing since the mid-1970s. The Environment Committee of the Organization for Economic Cooperation and Development (OECD) agrees that 'CFCs constitute a potentially serious problem which should be reviewed periodically' (Harris and Kosovich, 1981, 4–5) and is preparing reports on the scientific, economic and legal aspects of this issue.

* As a later example of the importance of NAS reports on this issue, Don Clay, Director of the Office of Toxic Substances of the EPA, testifying to Congress, noted with trespect to the 1982 NAS reports that 'EPA is awaiting the NAS assessment, the NASA/WMO report and other pertinent scientific information, and is reviewing and analyzing the issues raised .... I can assure you that further EPA action would be based upon reasonable scientific evidence and the most sound economic analysis available.' (Hearings of the House Subcommittee on Health and Environment of the Committee on Energy and Commerce, November 5, 1981)

All major CFC-producing nations and many users (including the ten member nations of thc European Economic Community, Canada, Japan and several other nations) have taken actions to control CFC emissions, by reducing or banning aerosol propellant uses of CFC-11 and -12 and/or limiting production capacity for all uses of those CFCs to present levels. Many of these nations are considering further emission controls for various CFC uses and are working with industry on the technology required to do so. In 1977, the United Nations Environment Programme (UNEP) established a Coordinating Committee on the Ozone Layer (CCOL) to assess the significance of research and other information relating to stratospheric ozone protection and to recommend further efforts. The UNEP Governing Council has also initiated work on a global framework convention for the protection of stratospheric ozone. Studies have been completed within the World Meteorological Organization Ozone Project on the possible impact of ozone variability on climate (UNEP, 1981; WMO, 1982).

In the United States, the first phase of CFC regulatory activity ended in 1978 when the EPA ordered the termination of the manufacture and use of most CFC-propelled aerosols. A second phase in CFC regulations is presently under consideration as the US Congress and Federal agencies debate the need for additional regulatory action on nonaerosol uses of CFCs (Nautilus, 1981).

22.7 CONCLUSIONS

Discussions on ways to improve the science of climate-related impact studies took place at a workshop convened by the Institute for Energy Analysis, Oak Ridge Associated Universities, 29–30 June 1981, in which the principal investigators of four of the five studies participated (no representative of the NAS chlorofluorocarbon studies was able to attend). In addition to the successes and problem areas that become evident from the studies reviewed, it was suggested that these studies could serve to highlight important aspects of conducting climate-related impact assessments, from project conceptualization to dissemination of the research findings. The reader should keep in mind that the following are suggestions and are not meant to be definitive guidelines.

22.7.1 MIT Sudan-Sahel Study

1. All five studies faced delays either in project completion or in the dissemination of results, but for the MIT study the time factor proved to be an insurmountable problem. A year after the project had been completed, AID acknowledged that '... in one year, no group could have been expected to produce a once-and-for-all conceptual framework for the entire region. The task is much more a gradual one, a learning process systematically pursued

over a much longer period' (AID, 1976a, 9). Time constraints were even more critical in this specific case because MIT personnel were unfamiliar with the region they were researching, no matter how capable they may have been in their own disciplines. These and other time constraint problems, precipitated by the eagerness of AID to acquire research results and the eagerness of MIT to undertake the research project and to comply with its own 1–year time frame, could have been minimized had MIT convened the retreat of senior scholars and policy-makers requested by AID. Another facet of this problem was that those negotiating the contract with AID were not directly involved in the project and had no precise knowledge of what research efforts and how much time might be required to complete the project successfully.

2. Under certain circumstances researchers not directly familiar with the topic to be investigated will be selected for involvement in a project. While this may be a valuable tactic (i.e., no preconceived biases, a chance for 'new thinking'), measures need to be taken to assure that additional researchers familiar with the topic are associated with the project from the outset and in a significant way; for example, authoritative (as opposed to nominal) advisers on an advisory council that is to be convened early in the project, one that meets often and to whose advice serious consideration is given.

3. At the outset of the research effort, study objectives, problem conceptualization and design approach must be made explicit and agreed to by the contractor and the group undertaking the research activity. As the project develops, the need may arise to modify the original research agenda including, possibly, the methodological approach or the underlying assumptions. Such re-evaluation some months into the project can be healthy and there must be enough flexibility (of time, of funds, of perceptions of principal investigators) to accommodate such changes in research design if they are deemed necessary.

4. The selection of the principal investigator is of major importance. The principal investigator will be, directly or indirectly, the driving force for the study. For example, he/she will be instrumental in the selection of other project managers and researchers, as well as in the conceptualization of the research problem and in determining the research design.

22.7.2 NDU: Climate Change to the Year 2000

1. The NDU study's methodological approach and the packaging of its final report have attracted much attention. Its methodological approach has been viewed as useful because policy-makers often have to make decisions relying on expert judgment, even though uncertainties surrounding a particular issue may be very large. With the methodology chosen, the NDU staff sought to identify and quantify the subjective views of selected climate experts on climate-related issues. In their third and final volume they suggest that the methodology developed in this study could be applied to CO_2 impact

assessments and to other similar problems (Willett, 1981, vi). Yet the methodological approach did also generate criticism that the study represented 'science by consensus' rather than evaluative judgment of climate experts concerning future climate scenarios.

As for the packaging of the NDU report, few can argue that the report stands out visually from the others; for groups without well-established reputations there may be some value in packaging their results so as to attract the attention of their target audiences.

2. The strengths and weaknesses of the chosen methodological approaches need to be explicitly identified so that the research findings can be judged appropriately. Once a study's methodological underpinnings have been questioned, the credibility of its conclusions, no matter how valid they may be, will be subject to challenge.

3. While good packaging cannot replace 'good science', it can assist in the distribution of 'good science'.

22.7.3 IFIAS: Drought and Man

1. The IFIAS project highlights at least two potential problems that project managers might encounter during these types of studies: a need to redefine the study after it has begun and a delay in publication of the study's findings. Delays of only a few months in the publication of the 1975 NAS chlorofluorocarbon reports caused concern among those awaiting the panel's findings. Policy-makers awaited the NAS reports and had to accept the delay. Such may not be the case for policy-makers awaiting the results of other climate-related impact assessments.

2. It is important to identify and make explicit the underlying assumptions on which a research project is developed. If some of those assumptions are proven not to be valid, they can be corrected early in the research process. This, too, reinforces the need for flexibility of the research timetable, as well as of the research team, to readjust activities part way through the project.

3. Reports need to be published in a timely fashion to avoid long delays between submission of the manuscript and publication and distribution of the final project. Likewise, the lag time between publication of multiple volumes from the same research project should be minimized if they are to be useful for policy-makers.

22.7.4 DOT Climatic Assessment Program

CIAP's problem with its executive summary highlights in general the importance of executive summaries. It was with respect to this final stage of the CIAP process (that is, the preparation and dissemination of the executive summary) that charges of political interference in the scientific aspects of CIAP arose.

Other reports, too, had problems with their executive summaries. For example, the MIT Summary Report (Volume 1) provides less information on the West African Sahel than already existed (for example, UNFAO, 1962). The executive summary of the NAS report on CFCs and stratospheric ozone depletion was challenged by industry representatives on the ground that it omitted necessary caveats, thereby giving the appearance of a higher degree of scientific certainty about the chlorofluorocarbon issue than was warranted. In contrast, the British reports on stratospheric ozone depletion provided conclusions but not executive summaries. As for the IFIAS study, García noted (IEA, 1981) that he did not produce an executive summary in order to avoid what he termed misinterpretation of his report's findings.

Executive summaries are often more important to policy-makers and the public than the reports that they summarize. They are often designed for the busy decision-makers who have little time (and little expertise) to delve into the lengthier technical report from which they would otherwise have to draw their own, probably less informed, conclusions. These summaries must accurately reflect the tone and the content of the longer report as well as the degree of uncertainty that surrounds its conclusions.

22.7.5 NAS Chlorofluorocarbon Studies

The NAS studies on stratospheric ozone raised the issue of 'term' (that is, one-time) studies as opposed to ongoing assessments. It is important to qualify what might be viewed as an ongoing assessment. In the case of the NAS ozone depletion studies, although several studies have been done since 1970 in which the general topic of ozone depletion has been addressed, the NAS staff members as well as the scientists involved in the preparation of the individual reports changed from one assessment to the next. The NAS assessment process, however, does seem to have more flexibility than the one-time impact assessment. One might argue that an ongoing climate-related impact assessment program can follow up on its earlier findings by taking advantage of new developments in instrumentation, data analysis techniques and enhanced scientific understanding.

'Term' or one-time projects that are expected to present definitive results within a specific period of time may not have this degree of research flexibility. On the other hand, it could be argued that once important issues and areas for concern have been identified in a one-time impact assessment, it is possible to focus on those few issues and to establish smaller but more specific follow-up assessments, such as was done by establishing the High Altitude Pollution Program as a result of the CIAP effort.

The objectives for undertaking an assessment must match the way the assessment will be conducted. Certain research objectives (for example, to determine the state of the art with respect to a climate-related problem) might require a one-time study, while others might call for an ongoing assessment (for example, if the range of uncertainties surrounding our knowledge were so great that periodic assessments could serve to monitor the state of knowledge as it changed over time). While either approach is valid, both types should be considered and matched to a study's objectives.

22.7.6 Final Observations

There is much to learn from past experience of previous climate impact assessments. In fact, a few reviews of such programs now exist (Mormino *et al.*, 1975; AID, 1976a; Gasser, 1981), in addition to reviews of sets of environmental studies (Mar, 1977; McHale, 1981). It would be valuable for managers and funding agencies of future climate impact assessment studies if retrospective reviews of each project were to become a required component in a formal climate-related impact study process. A Chinese proverb is appropriate: To know the road ahead, ask those coming back.

ACKNOWLEDGMENTS

The authors of this paper wish to acknowledge the support, financial and moral, of the scientists and staff at the Oak Ridge Associated Universities Institute for Energy Analysis for the NCAR/IEA workshop on 'Improving the Science of Climate-related Impact Studies'. They would also like to thank the principal investigators of the impact studies reviewed as well as the scientists and staff at the National Center for Atmospheric Research for their support and encouragement. Special appreciation is due to the many scientists who took the time to critique the several drafts of this paper, without whose comments this paper could not have been written.

An earlier version of this paper, which was commissioned for the SCOPE project, appeared in W. C. Clark (Ed.) (1982) *Carbon Dioxide Review: 1982*, Oxford University Press.

REFERENCES

Adams, S. C. Jr. (1973). Letter to Director, Center for Policy Alternatives, MIT, from Adams, AID's Assistant Administrator for Africa (November 9).

AID (Agency for International Development) (1972). *Development and Management of the Steppe and Brush-Grass Savannah Zone Immediately South of the Sahara.* AID In-House Report, Washington, DC: 161 pages.

AID (Agency for International Development) (1974). *Internal AID Review of MIT Interim Report*: Memo (March). Washington, DC.

AID (Agency for International Development) (1976a). *Retrospective Assessment of the MIT Study*. AID internal memo (P. Lyman). Washington, DC.

AID (Agency for International Development) (1976b). *A Report to Congress: Proposal for a Long-Term Comprehensive Development Program for the Sahel.* Part II, Technical Background Papers. Washington, DC.

Alexander, T. (1974). Ominous changes in the world's weather. *Fortune* (February), 90–95, 142, 146, 150–152.

Atmosphere (Journal of the Canadian Meteorological Society) (1976). Entire Vol. **14**, No. 3 deals with findings of the Canadian program.

Bastian, C. (1976). National Science Foundation memo to files (August 2, 1976).

Bastian, C. (1982). The formulation of federal policy. In Ward, R. (Ed.) *Stratospheric Ozone and Man*. CRC Press, Boca Raton, Florida.

Bauer, E. (1974). CIAP Monograph 1: *The Natural Stratosphere of 1974*. US Dept. of Transportation, Climatic Impact Assessment Program, Washington, DC.

Brown, D. S. (1973). U.S.A. statement. In *Final Report on the Meeting of the Sudano-Sahelian Mid- and Long-Term Programme, 28–29 June 1973, Geneva*. UN Special Sahelian Office, New York.

Budyko, M. I., and Karol, I. L. (1976). A study of CIAP. *Meteorologiya: Gidrologiya* (Soviet Meteorology and Hydrology), No. 9, 103–111 (in Russian). Translated by Allerton Press Journal Program, New York, 1977, pp. 82–91.

Caldwell, M. M. (1974). CIAP Monograph 5: *Impacts of Climatic Change on the Biosphere*. US Dept. of Transportation, Climatic Impact Assessment Program, Washington, DC.

Carter, L. J. (1975). Deception charged in presentation of SST study. *Science*, **190** (November 28), 861.

CEQ (Council on Environmental Quality) (1980). *The Global 2000 Report to the President: Entering the Twenty-First Century*. US Government Printing Office, Washington, DC.

CIA (Central Intelligence Agency) (1974a). *A Study of Climatological Research as it Pertains to Intelligence Problems* (August). Washington, DC.

CIA (Central Intellingence Agency) (1974b). *Potential Implications of Trends in World Population, Food Production, and Climate* (August, OPR-401). Washington, DC.

CIA (Central Intellingence Agency) (1976). *USSR: The Impact of Recent Climate Change on Grain Production* (October, ER 76-10577 U). Washington, DC.

COMESA (1975). *The Report of the Committee on Meteorological Effects of Stratospheric Aircraft, 1972–75*. Meteorological Office, United Kingdom.

COVOS (1976). *Comité d'Études sur les Conséquences des Vols Stratosphériques*. Meteorological Society of France, Boulogne, France.

Crutzen, P. J. (1971). Ozone Production rates in an oxygen-hydrogen-nitrogen atmosphere. *Journal of Geophysical Research,* **76** (30), 7311.

Cusack, D. (1981). Variabilidad climática y el hambre mundial: Solución técnica o solución política? *Interciencia*, **6** (July–August), 288–291.

Daly, G. (1974). CIAP Monograph 6: *Economic and Social Measures of Biologic and Climatic Change*. US Department of Transportation, Climatic Impact Assessment Program, Washington, DC.

Dickinson, R., and Chervin, R. M. (1979). Sensitivity of a general circulation model to changes in infrared cooling due to chlorofluoromethanes with and without prescribed zonal ocean surface temperature change. *Journal of the Atmospheric Sciences,* **36** (December), 2304–2319.

Donahue, T. M. (1975). The SST and ozone depletion: Letter. *Science,* **187** (March 28), 1142–1143.

DOT (Dept. of Transportation) (1975). *Press release, Office of the Secretary* (January 21). Washington, DC.

DOT (Dept. of Transportation) (1976). *The Secretary's Decision on Concorde Supersonic Transport* (February 4). Washington, DC.

Dotto., L., and Schiff, H. (1978). *The Ozone War*. Doubleday & Co., New York.

García, R. (1977). *Drought, Desertification and the Structural Stability of Ecosystems—The Case of Latin America*. Paper prepared for UNEP's Desertification Secretariat. Geneva: 40 pages.

García, R. (1978). Climate impacts and socioeconomic conditions. In National Academy of Sciences, *International Perspectives on the Study of Climate and Society*, pp. 43–47. NAS, Washington, DC.

García, R. (1981). *Drought and Man: The 1972 Case History. Vol. 1: Nature Pleads Not Guilty*. Pergamon, New York.

García, R., and Escudero, J. C. (1982). *Drought and Man: The 1972 Case History. Vol. 2: The Constant Catastrophe: Malnutrition, Famines, and Drought*. Pergamon, New York.

Gasser, W. R. (1981). Climate change to the year 2000 and possible impacts on world agriculture: A review of the National Defense University Study. Paper presented to the *Institute for Energy Analysis/NCAR Workshop on Improving the Science of Climate-Related Impact Studies, June 30–July 2, 1981, Oak Ridge, Tennessee*.

Glantz, M. H. (Ed.) (1976). *Politics of a Natural Disaster: The Case of the Sahel Drought*. Praeger, New York.

Grobecker, A. J. (1974). Research program for assessment of stratospheric pollution. *Acta Astronautica*, **1**, 179–224.

Grobecker, A. J. (1976). The CIAP report of findings: The effects of stratospheric pollution by aircraft. In Hard, T. M., and Broderick, A. J. (Eds.) *Proceedings of the Fourth Conference on the Climatic Impact Assessment Program*. NTIS, US Dept. of Commerce, Springfield, Virginia.

Grobecker, A. J., Coroniti, S. C., and Cannon, R. H., Jr. (1974). *Report of Findings: The Effects of Stratospheric Pollution by Aircraft*. US Dept. of Transportation, Climatic Impact Assessment Program, Washington, DC.

Gruza, G. V. (1979). Fluctuations of climate and man's economic activity. *Hydrometeorology*, Vol. 3 (translation from the Russian by the National Science Foundation).

Handler, P. (1976). Letter to H. G. Stever, National Academy of Sciences (April 16).

Hard, T. M., and Broderick, A. J. (Eds.) (1976). *Proceedings of the Fourth Conference on the Climatic Impact Assessment Program*. NTIS, US Dept. of Commerce, Springfield, Virginia.

Harris, F. A., and Kosovich, J. (1981). International action to protect the ozone layer. Paper presented to the *Air Pollution Control Association Annual Meeting, June 21–26, Philadelphia, Pennsylvania*.

Harvey, M. J. (1974). Letter to US Congressman Diggs from Harvey, AID's Assistant Administrator for Legislative Affairs (April 23).

Hidalgo, H. (1974a). CIAP Monograph 3: *The Stratosphere Perturbed by Propulsion Effluents*. US Dept. of Transportation, Climatic Impact Assessment Program, Washington, DC.

Hidalgo, H. (1974b). CIAP Monograph 4: *The Natural and Radiatively Perturbed Troposhere*. US Dept. of Transportation, Climatic Impact Assessment Program, Washington, DC.

Hoffert, M. I., and Stewart, R. W. (1975). Stratospheric ozone—fragile shield? *Astronautics and Aeronautics* (October), 42–55.

Holloman, J. H. (1974). Letter to Adams, AID's Assistant Administrator for Africa (February 14).

IEA (Institute for Energy Analysis) (1981). Notes from discussions, *Workshop on Improving the Science of Climate-Related Impact Studies*. IEA, Oak Ridge, Tennessee.

IFIAS (1974). *The Impact on Man of Climate Change: Report of an IFIAS Project Workshop, Meteorological Institute, University of Bonn, May 6–10, 1974*. International Federation of Institutes for Advanced Study, Stockholm, Sweden.

IMOS (1975). *Fluorocarbons and the Environment: Report of Federal Task Force on Inadvertent Modification of the Stratosphere (IMOS) by the Council on Environmen-*

tal Quality (CEQ) and Federal Council for Science and Technology (FCST). US Government Printing Office, Washington, DC.

Johnson, D. G. (1983). *The World Grain Economy and Climate Change to the Year 2000: Implications for Policy*. National Defense University Press, Washington, DC.

Johnston, H. S. (1971). Reduction of stratospheric ozone by nitrogen oxide catalysts from supersonic transport exhaust. *Science,* **173** (August 6), 517–522.

Johnston, H. S. (1975a). Supersonic transports: Letter. *Chemical and Engineering News* (April 25), 5.

Johnston, H. S. (1975b). Panel discussion. In Hard, T. M., and Broderick, A. J. (Eds.) *Proceedings of the Fourth Conference on the Climatic Impact Assessment Program*, p. 35. NTIS, US Dept. of Commerce, Springfield, Virginia.

Knox, J. B., MacCracken, M. C., Dickerson, M. H., Gresho, P. M., Luther, F. M., and Orphan, R. C. (1983). *Program Report for FY 1982 Atmospheric and Geophysical Sciences Division of the Physics Department*. Lawrence Livermore National Laboratory, University of California, Livermore, California.

Kraemer, R. S. (1978). Meeting reviews: Session on climatic futures at the annual meeting of the AAAS, 17 February 1978. *Bulletin of the American Meteorological Society,* **59** (7), 822–823.

Luther, F. M. (1980). *Annual Report of Lawrence Livermore National Laboratory to the FAA on the High Altitude Pollution Program—1980*. University of California, Livermore, California.

MacDonald, G. J. (1976). Panel discussion of CIAP report of findings. In Hard, T. M., and Broderick, A. J. (Eds.) *Proceedings of the Fourth Conference on CIAP*. NTIS, US Dept. of Commerce, Springfield, Virginia.

Mar, B. (Principal Investigator) (1977). *Regional Environmental Systems: Assessment of RANN Projects*. University of Washington, Dept. of Civil Engineering, Seattle, Washington.

Martin, B. (1979). *The Bias of Science*. Society for Social Responsibility in Science, O'Connor, Australia.

Matlock, W. G., and Cockrum, E. L. (1974). *A Framework for Evaluating Long-Term Strategies for the Development of the Sahel-Sudan Region; Vol. 2, A Framework for Agricultural Development Planning*. MIT Center for Policy Alternatives, Cambridge, Massachusetts.

Matthews, S. W. (1976). What's happening to our climate? *National Geographic,* **50** (5), 576–615.

McHale, M. C. (1981). *Ominous Trends and Valid Hopes: A Comparison of Five World Reports*. Hubert H. Humphrey Institute of Public Affairs, University of Minnesota, Minneapolis, Minnesota.

Molina, M. J., and Rowland, F. S. (1974). Stratospheric sink for chlorofluoromethanes: Chlorine atomic-catalyzed destruction of ozone. *Nature,* **249**, 810–812.

Morentz, J. W. (1980). Communication in the Sahel drought: Comparing the mass media with other channels of international communication. In National Academy of Sciences, *Disasters and the Mass Media*, pp. 158–186. National Academy of Sciences, Washington, DC.

Mormino, J., Sola, D., and Patten, C. (1975). *Climatic Impact Assessment Program: Development and Accomplishments, 1971–75*. US Dept. of Transportation, Washington, DC: 206 pages.

NAS (National Academy of Sciences) (1975a). *Arid Lands of Sub-Saharan Africa: Staff Final Report, July 1974–December 1974*. NAS, Washington, DC.

NAS (National Academy of Sciences) (1975b). *Arid Lands of Sub-Saharan Africa: Staff Progress Report, September 1973–June 1974*. NAS, Washington, DC.

NAS (National Academy of Sciences) (1975c). *Environmental Impact of Stratospheric Flight*. NAS, Washington, DC.
NAS (National Academy of Sciences) (1976a). Panel on Atmospheric Chemistry. *Halocarbons: Effects on Stratospheric Ozone*. NAS, Washington, DC.
NAS (National Academy of Sciences) (1976b). Committee on Impacts of Stratospheric Change. *Halocarbons: Environmental Effects of Chlorofluoromethane Release*. NAS, Washington, DC.
NAS (National Academy of Sciences) (1978). *Programs of the Board on Science and Technology for International Development (BOSTID): Summary of Activities, 1970–78*. NAS, Washington, DC.
NAS (National Academy of Sciences) (1979). *Protection Against Depletion of Stratospheric Ozone by Chlorofluorocarbons*. NAS, Washington, DC.
NAS (National Academy of Sciences) (1982a). *Biological Effects of the Increased Solar UV Radiation*. NAS, Washington, DC.
NAS (National Academy of Sciences) (1982b). *Chemistry and Physics of Stratospheric Ozone Depletion*. NAS, Washington, DC.
Nautilus (1981). A decision to further regulate CFCs hinges on a value judgment. *Weather and Climate Report,* **4** (8), 3.
NCAR (National Center for Atmospheric Research) (1981). Internal memo, T. Stewart.
NDU (National Defense University) (1978a). *Climate Change to the Year 2000: A Survey of Expert Opinion*. NDU, Washington, DC.
NDU (National Defense University) (1978b). *Crop Yields and Climate Change: The Year 2000—Progress Report*. NDU, Washington, DC.
NDU (National Defense University) (1980). *Crop Yields and Climate Change to the Year 2000*, Vol. 1. NDU, Washington, DC.
Oliver, R. C. (1974). CIAP Monograph 2: *Propulsion Effluents in the Stratosphere*. US Dept. of Transportation, Climatic Impact Assessment Program, Washington, DC.
Palutikof, J. (1983). Drought without water. *Nature*, **203**, 635.
Primack, J., and von Hippel, F. (1972). Scientists, politics and SST: A critical review. *Bulletin of the Atomic Scientists* (April), 24–30.
Ramanathan, V. (1975). Greenhouse effect due to chlorofluorocarbons: Climatic implications. *Science,* **190** (October 3), 50–52.
Ramanathan, V. (1980). Climatic effects of anthropogenic trace gases. In Bach, W., *et al.* (Eds.) *Interactions of Energy and Climate*, pp. 269–280. D. Reidel Publishing Co., Netherlands.
Ramanathan, V., and Dickinson, R. (1979). The role of stratospheric ozone in the zonal and seasonal radiative energy balance of the earth-troposhere system. *Journal of the Atmospheric Sciences,* **36** (6), 1084–1104.
Ruttenberg, S. (1981). Climate, food and society. In Slater, L. E., and Levin, S. K. (Eds.) *Climate's Impact on Food Supplies*, pp. 23–37. Westview Press, Boulder, Colorado.
SCEP (Study of Critical Environmental Problems) (1970). *Man's Impact on the Global Environment*. MIT Press, Cambridge, Massachusetts.
Science News (1978). The 25-year forecast: Group prediction. **113** (8), 116.
Seifert, W. W., and Kamrany, N. M. (1974). *A Framework for Evaluating Long-Term Strategies for the Development of the Sahel-Sudan Region; Vol. 1, Summary Report: Project Objectives, Methodologies, and Major Findings*. MIT Center for Policy Alternatives, Cambridge, Massachusetts.
Sellers, W. D. (1979). Climate change to the year 2000: A book review. *Journal of Applied Meteorology,* **60** (6), 686.
Shapley, D. (1976). Crops and climatic change: USDA's forecasts criticized. *Science,* **193** (September 24), 1222–1224.
Shurcliff, W. A. (1970). *SST and Sonic Boom Handbook*. Ballantine Books, New York.

Stolarski, R. S., and Cicerone, R. J. (1974). Stratospheric chlorine: A possible sink for ozone. *Canadian Journal of Chemistry,* **52** (8) (Part 2), 1610–1615.

Talbot, R. B., and Olorunsola, V. A. (1984). An essay prompted by Rolando García's edited work, *Nature Pleads Not Guilty. Climatic Change,* **6**, forthcoming.

UK Dept. of Environment (1976). *Chlorofluorocarbons and Their Effect on Stratospheric Ozone* (Pollution Paper No. 5). Her Majesty's Stationery Office, London.

UK Dept. of Environment (1979). *Chlorofluorocarbons and Their Effect on Stratospheric Ozone* (Pollution Paper No. 15). Her Majesty's Stationery Office, London.

UNEP (United Nations Environment Programme) (1981). *Environmental Assessment of Ozone Layer Depletion and Its Impact.* Bulletin No. 6. Nairobi, Kenya.

UNFAO (United Nations Food and Agriculture Organization (1962). *Africa Survey.* FAO, Rome.

UNFAO (United Nations Food and Agriculture Organization) (1979). Scanning the future for climatic change. *CERES* (January-February), 7.

US Congress (1976). *FAA Certification of the SST Concorde.* Hearings before the Committee on Government Operations (HR: 94th Congress, 1st and 2nd Sessions; November 13, 1975). US Government Printing Office, Washington, DC.

US Congress (1977). *Clean Air Act Amendment of 1977.* PL 95-95, Section 153 (d). US Government Printing Office, Washington, DC.

Verstraete, M. M. (1984). Review of *Drought and Man: The 1972 Case History; Vol. 1, Nature Pleads Not Guilty*, by Rolando García. *Climatic Change,* **6** (forthcoming).

Willett, J. W. (1981). Preface to *Climate Change and the World Grain Economy to the Year 2000: Some Implications for Domestic and International Agricultural Policy—Report on the Third Phase of the [NDU] Climate Impact Assessment* (D. G. Johnson). Draft manuscript dated April 1979, cleared for release June 1981. National Defense University, Washington, DC.

WMO (World Meteorological Organization) (1982). *Report of the Meeting of Experts on Potential Climatic Effects of Ozone and Other Minor Trace Gases, Boulder, Colorado, September 13–17, 1982.* WMO Report 14. Geneva.

Wuebbles, D. J. (1981). *Chlorocarbon Emission Scenarios: Potential Impact on Stratospheric Ozone.* Lawrence Livermore National Laboratory, University of California, Livermore, California.

Author Index

Subject Index

262155

LINKED